Springer Series on Environmental Management

David E. Alexander
Series Editor

Springer
New York
Berlin
Heidelberg
Barcelona
Hong Kong
London
Milan
Paris
Singapore
Tokyo

Springer Series on Environmental Management
David E. Alexander, Series Editor

Gradient Modeling: Resources and Fire Management (1979) S.R. Kessell

Disaster Planning: The Preservation of Life and Property (1980) H.D. Foster

Air Pollution and Forests: Interactions between Air Contaminants and Forest Ecosystems (1981) W.H. Smith

Natural Hazard Risk Assessment and Public Policy: Anticipating the Unexpected (1982) W.J. Petak and A.A. Atkisson

Environmental Effects of Off-Road Vehicles: Impacts and Management in Arid Regions (1983) R.H. Webb and H.G. Wilshire (eds.)

Global Fisheries: Perspectives for the '80s (1983) B.J. Rosthschild (ed.)

Heavy Metals in Natural Waters: Applied Monitoring and Impact Assessment (1984) J.W. Moore and S. Ramamoorthy

Landscape Ecology: Theory and Applications (1984) Z. Naveh and A.S. Lieberman

Organic Chemicals in Natural Waters: Applied Monitoring and Impact Assessment (1984) J.W. Moore and S. Ramamoorthy

The Hudson River Ecosystem (1986) K.E. Limburg, M.A. Moran, and W.H. McDowell

Human System Responses to Disaster: An Inventory of Sociological Findings (1986) T.E. Drabek

The Changing Environment (1986) J.W. Moore

Balancing the Needs of Water Use (1988) J.W. Moore

The Professional Practice of Environmental Management (1989) R.S. Dorney and L. Dorney (eds.)

Chemicals in the Aquatic Environment: Advanced Hazard Assessment (1989) L. Landner (ed.)

Inorganic Contaminants of Surface Water: Research and Monitoring Priorities (1991) J.W. Moore

Chernobyl: A Policy Response Study (1991) B. Segerståhl (ed.)

Long-Term Consequences of Disasters: The Reconstruction of Friuli, Italy, in its International Context, 1976-1988 (1991) R. Geipel

Food Web Management: A Case Study of Lake Mendota (1992) J.F. Kitchell (ed.)

Restoration and Recovery of an Industrial Region: Progress in Restoring the Smelter-Damaged Landscape near Sudbury, Canada (1995) J.M. Gunn (ed.)

Limnological and Engineering Analysis of a Polluted Urban Lake: Prelude to Environmental Management of Onondaga Lake, New York (1996) S.W. Effler (ed.)

Assessment and Management of Plant Invasions (1997) J.O. Luken and J.W. Thieret (eds.)

Marine Debris: Sources, Impacts, and Solutions (1997) J.M. Coe and D.B. Rogers (eds.)

Environmental Problem Solving: Psychosocial Barriers to Adaptive Change (1999) A. Miller

Rural Planning from an Environmental Systems Perspective (1999) F.B. Golley and J. Bellot (eds.)

Frank B. Golley, Editor
Institute of Ecology
The University of Georgia
Athens, Georgia

Juan Bellot, Editor
Department of Ecology
University of Alicante
Alicante, Spain

Rural Planning from an Environmental Systems Perspective

With 135 Illustrations

CIHEAM

The International Centre for Advanced Mediterranean Agronomic Studies (CIHEAM) has both stimulated and co-ordinated the preparation of this book, whose origin lies in the Advanced Course on Rural Planning in Relation to the Environment developed by its Mediterranean Agronomic Insititute of Zaragoza.

Springer

Frank B. Golley
Institute of Ecology
The University of Georgia
Athens, GA 30602
USA

Juan Bellot
Universidad de Alicante
Facultad de Ciencias
Departamento de Ecología
Apartado 99
03080 Alicante
Spain

Series Editor:
David E. Alexander
Department of Geology and Geography
University of Massachusetts
Amherst, MA 01003
USA

Cover photo: Development of new land uses in the semiarid Central Ebro Valley, Spain. Red colours correspond to irrigated crops: small plots are the old irrigated lands close to the river, and big plots are the new irrigated lands (center pivots and other sprinkling engines) using water pumped from the river. White and yellow areas correspond to bare soils on gyprock. Pale green colours are mostly rain-fed winter cereals. Green dark tones correspond to xerophytic vegetation. [False colour composition (RGB 453) of a Landsat TM image, from 18 July 1996. Original image © Eurimage. Image processing by M.A. Casterad and T. Martín-Ordóñez in the Agricultural Research Service; Government of Aragón, Zaragoza (Spain).]

Library of Congress Cataloging-in-Publication Data
Golley, Frank B.
Rural planning from an environmental systems perspective / Frank B. Golley, Juan Bellot.
p. cm. — (Springer series on environmental management)
Includes bibliographical references and index.
ISBN 0-387-98519-0 (alk. paper)
1. Regional planning—Environmental aspects. 2. Rural development—Environmental aspects. 3. Land use, Rural—Planning. I. Bellot, Juan. II. Title. III. Series.
HT391.G5485 1998
333.76'17—dc21 98-15160

Printed on acid-free paper.

Production coordinated by Impressions Book and Journal Services, Inc., and managed by Steven Pisano; manufacturing supervised by Nancy Wu.
Typeset by Impressions Book and Journal Services, Inc., Madison, WI.
Printed and bound by Edwards Brothers, Inc., Ann Arbor, MI.
Printed in the United States of America.

9 8 7 6 5 4 3 2 1

ISBN 0-387-98519-0 Springer-Verlag New York Berlin Heidelberg SPIN 10676447

Foreword

The International Centre for Advanced Mediterranean Agronomic Studies (CIHEAM), established in 1962, is an intergovernmental organization of thirteen countries: Albania, Algeria, Egypt, France, Greece, Italy, Lebanon, Malta, Morocco, Portugal, Spain, Tunisia and Turkey.

Four institutes (Bari, Italy; Chania, Greece; Montpellier, France; and Zaragoza, Spain) provide postgraduate education at the Master of Science level. CIHEAM promotes research networks on Mediterranean agricultural priorities, supports the organization of specialized education in member countries, holds seminars and workshops bringing together technologists and scientists involved in Mediterranean agriculture, and regularly produces diverse publications including the series "Options Méditerranéennes." Through these activities, CIHEAM promotes North/South dialogue and international cooperation for agricultural development in the Mediterranean region.

This book is based on almost twenty years of experience in teaching a course on rural planning and the environment at the Mediterranean Agronomic Institute of Zaragoza (IAMZ) (Spain) of the CIHEAM. The course has been taught by specialists from universities, research institutions, and companies in Europe and North America. The course was one of the first to integrate land planning and ecosystem analysis for the advanced training of professional scientists and planners. It has incorporated the dramatic changes in the fields of planning, systems analysis, and systems ecology over the past twenty years so that the approach is current and up-to-date. The text is more general than the course but retains this dynamic approach to the subject. While the rural areas treated focus mainly on the countries and landscapes of the Mediterranean region, the concepts, the methods and the approach are relevant to any region of the biosphere.

The book is directed toward postgraduate students and professionals, providing a source of information and concepts for those concerned with land and water policies and practice. It presents an integrated approach using practical and applicable models and methods, and takes a middle position between an elementary conceptual approach to land and water management and a highly mathematical advanced treatise based exclusively on system modelling.

We express our gratitude to those who, directly or indirectly, have contributed toward this endeavour, and particularly to the contributors of the chapters and the lecturers of the course. Professor Golley, from the beginning of the programme at the IAMZ, has been its main scientific inspirator and, together with Dr. Bellot, has made this book possible through careful planning and editing. We are most indebted to both of them for their enthusiasm and their perseverance.

The collaboration with Springer-Verlag has been easy and effective and we thank those entrusted with this edition for their interest and attention.

At the CIHEAM, we hope that this book, produced in a context of international cooperation, will serve, besides its specific content, as a good example to encourage a cooperative attitude amongst its readers.

Miguel Valls
Director
Mediterranean Agronomic Institute of Zaragoza

Series Preface

This series is concerned with humanity's stewardship of the environment, our use of natural resources, and the ways in which we can mitigate environmental hazards and reduce risks. Thus it is concerned with applied ecology in the widest sense of the term, in theory and in practice, and above all in the marriage of sound principles with pragmatic innovation. It focuses on the definition and monitoring of environmental problems and the search for solutions to them at scales that vary from the global to the local according to the scope of analysis. No particular academic discipline dominates the series, for environmental problems are interdisciplinary almost be definition. Hence a wide variety of specialties are represented, from oceanography to economics, sociology to silviculture, toxicology to policy studies.

In the modern world, increasing rates of of resource use, population growth, and armed conflict have tended to magnify and complicate environmental problems that were already difficult to solve a century ago. Moreover, attempts to modify nature for the benefit of humankind have often had unintended consequences, especially in the disruption of natural equilibria. Yet, at the same time, human ingenuity has been brought to bear in developing a new range of sophisticated and powerful techniques for solving environmental problems, for example, pollution monitoring, restoration ecology, landscape planning, risk management, and impact assessment. Books in this series will shed light on the problems of the modern environment and contribute to the further development of the solutions. They will contribute to the immense effort by ecologists of all persuasions to nurture an environment that is both stable and productive.

David E. Alexander
Amherst, Massachusetts

Contents

Contributors

M. Teresa Aguinaco, International Centre for Advanced Mediterranean Agronomic Studies (CIHEAM), Mediterranean Agronomic Institute of Zaragoza (IAMZ), Carretera de Montañana 177, 50059 Zaragoza, Spain

Juan Bellot, Universidad de Alicante, Facultad de Ciencias, Depto. de Ecología, Ctra. San Vicente del Raspeig s/n, 03080 Alicante, Spain

Alan Belward, European Commission, Space Applications Institute, Joint Research Centre, 21020 Ispra, Varese, Italy

François Bonnieux, Institut National de la Recherche Agronomique (INRA), Centre de Rennes, Unité d'Economie et Sociologie Rurales, 65 rue de Saint Brieuc, 35042 Rennes Cedex, France

Robert G.H. Bunce, Institute of Terrestrial Ecology, Merlewood Research Station, Windermere Road, Grange-over-Sands, Cumbria LA11 6JU, England

Felisa Ceña, Universidad de Cordoba, Escuela Técnica Superior de Ingenieros Agrónomos y de Montes, Depto. de Economía, Sociología y Política Agrarias, Alameda del Obispo s/n, 14080 Córdoba, Spain

M. Angeles Ceñal, DENDROS S.L., Avenida Pablo Iglesias 50, 1º, 1ª A, 28039 Madrid, Spain

Luisa M. Frutos, Universidad de Zaragoza, Facultad de Filosofía y Letras, Depto. de Geografía y Ordenación del Territorio, Ciudad Universitaria, Pedro Cerbuna 12, 50009 Zaragoza, Spain

Germán Glaría, Escuela Universitaria de Ingenieros Técnicos Forestales (EUITF), Cátedra de Proyectos, Ciudad Universitaria s/n, 28040 Madrid, Spain

Frank B. Golley, Institute of Ecology, The University of Georgia, Athens, GA 30602-2202, USA

Carlos A. Gracia, Universidad de Barcelona, Facultad de Biología, Departamento de Ecologa, Avda. Diagonal 645, 08028 Barcelona, Spain

Wolf Dieter Grossmann, Environmental Research Centre Leipzig/Halle (UFZ), Department of Applied Landscape Ecology, Permoserstrasse 15, D-04318 Leipzig, Germany

Hervé Guyomard, Institut National de la Recherche Agronomique (INRA), Centre de Rennes, Unité d'Economie et Sociologie Rurales, 65 rue de Saint Brieuc, 35042 Rennes Cedex, France

Arwyn Jones, European Commission, Environment Institute, Joint Research Centre, 21020 Ispra, Varese, Italy

Vernon Meentemeyer, Department of Geography, Climatology Research Laboratory, The University of Georgia, Athens, GA 30602-2502, USA

Francisco Pellicer Corellano, Universidad de Zaragoza, Facultad de Filosofía y Letras, Depto. de Geografía y Ordenación del Territorio, Ciudad Universitaria, Pedro Cerbuna 12, 50009 Zaragoza, Spain

José Luis Salinas, International Centre for Advanced Mediterranean Agronomic Studies (CIHEAM), Mediterranean Agronomic Institute of Montpellier (IAMM), 3191 Route de Mende, 34093 Montpellier Cedex 5, France

Caroline Saunders, Department of Agricultural Economics and Food Marketing, Centre for Rural Economy, University of Newcastle-upon-Tyne, Newcastle-upon-Tyne NE1 7RU, England

Malcolm E. Sumner, Department of Crop and Soil Sciences, College of Agricultural and Environmental Sciences, The University of Georgia, Athens, GA 30602-7272, USA

John B. Thornes, Department of Geography, King's College of London, University of London, Strand, London WC2R 2LS, England

Ramón Vallejo, Universidad de Barcelona, Facultad de Biología, Departamento de Biología Vegetal, Unidad de Fisiología Vegetal, Avda. Diagonal 645, 08028 Barcelona, Spain

Marc Van Liedekerke, European Commission, Environment Institute, Joint Research Centre, 21020 Ispra, Varese, Italy

Martin Whitby, Department of Agricultural Economics and Food Marketing, Centre for Rural Economy,University of Newcastle-upon-Tyne, Newcastle-upon-Tyne NE1 7RU, England

Part I
Introduction

1

Planning As a Way of Achieving Sustainable Development

Frank B. Golley and Juan Bellot

1.1 Introduction

This book will focus on planning at a relatively large scale of space and time. It will focus on rural areas, which are changing so rapidly globally. It is based on the principle that if we have an adequate context to organize information and we bring the appropriate information to the table, the success of planning will be enhanced and the degree of controversy reduced. The context will be that of systems analysis in which the earth is treated as a mechanical model, which acts in predictable ways under the laws and principles of the natural and social sciences. This approach is a way to begin the planning process; it identifies gaps in knowledge and areas of conflict.

1.1.1 The Context: A Systems Approach

A system is defined as a collection of parts which are organized to function as a whole. Each part is a subsystem with its own characteristic behaviours. However, when it is combined with other subsystems into a system, a new behaviour emerges. This new behaviour modifies the subsystem behaviours, so that their combined actions form a whole. These processes can be expressed mathematically or symbolically, as well as verbally and pictorially. This means that systems may be analyzed and predictions of future states can be expressed quantitatively.

The system is a powerful organizing idea and has become so widespread in modern thinking that it has replaced many of the older forms of organization. Because the system concept can utilize quantitative information it is widely used in scientific investigation. Systems analysis has its own special language and techniques. It is a technical discipline. In this book we do not intend to teach systems analysis; rather we will use the systems analytical approach to help us assemble the important factors in rural regions in order to plan their future.

The system approach is widely useful; however, it is not without limits and biases. The structures of the models we create to organize information constrain the outputs. Frequently, we are obliged to focus on the central tendency of data and not consider the range of variation in the model. Many factors cannot be converted into quantitative data easily and yet they must be represented within the model. Thus, the output from a systems analysis of a rural area is not a fixed, determined statement but is conditional. It is expressed in the form: "given the information and the modelling approach chosen to study these data, the output from the model is...."

It is important that we recognize that the

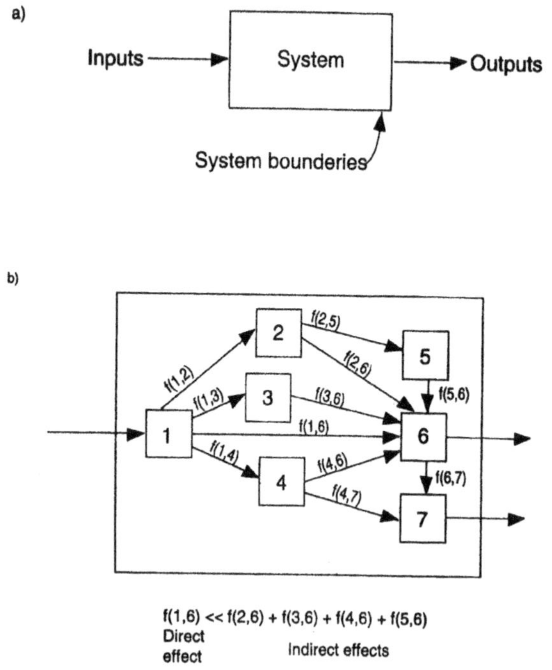

FIGURE 1.1. A diagram of a simple system, showing inputs and outputs to and from the system (a), and compartments within the system that are linked together through exchange of energy, matter and information (b). The direct flow from compartments one to six is contrasted to indirect flows from two, three, four and five to six.

system output is a function of the input data and the methods used to integrate the data. There are many choices. The result of the systems analysis is still human decision making, but with a fuller, more complete, and objective database organized in a form needed for decisions. It is at this point that the planner must depart from the mechanics of models and deal with reality, social need, tradition, and culture using appropriate political methods. These topics are crucial because they frequently dictate if a plan will be a success or not.

Systems analysis is useful because it rises above the enormous detail in the system and focuses on how the individual parts interact with one another to create higher-level structures. Within a rural system there are thousands, even millions, of species or organisms, and an array of chemical species all organized into a great number of physical-chemical and social subsystems. There is no way we can take into account this teeming mass of interactions. Rather, it is necessary to determine the objectives of the analysis and then aggregate the parts into structures that have a common behaviour within the context of the exercise.

We will find that as we consider direct effects of one object or process on another that we are unable to understand the consequences of the connection. There is more than is revealed by the direct interactions. The reason is that the elements in the system interact through one another so that there are secondary, tertiary, and multiple interactions (Figure 1.1). In many models the numbers of potential interactions continue endlessly. These models are complex! Yet, only some of the interactions are significant to the normal behaviour of the system. We will concentrate on these interactions, remembering that unique, special cases may occur and that chance is always operating. We can never rule out the surprise conjunction that might occur but could not be predicted. Surprise is a form of natural creativity within systems.

The consequences of systems analysis are that simple, short-term analyses will almost always mislead us. Yet detailed analytical studies made by system engineers are also out of the realm where the ordinary planner and specialist can understand the process. Somewhere between the too simple and the too complex is where we want to focus. The exact position of the focus depends upon the training and technical skills of the planners. Some planners are not highly quantitative individuals. Mathematically inclined people are seldom attracted to planning. This means that we must form teams of specialists to carry out the different analyses of the planning activity. Through teamwork we can incorporate a much deeper skill in quantitative analysis.

1.1.2 The Environment

The rural systems treated in this book are open systems. This means that they receive inputs from outside the system and produce outputs that leave the system. They are connected to an environment. The environment consists of the rest of the world, which is not within the system boundaries. It will include physical-chemical processes such as climate and water flow; it will involve receipt and discharge of solar energy; and it will incorporate social, political, cultural, and economic factors. How the system will function and how it will accomplish its purposes depends to a large extent upon its environment, to which it is coupled in input-output relationships.

Recognition of the environment of the system raises questions about the system boundaries. Where do we place boundaries in space-time? This question is difficult to answer. There are several points to consider in developing an answer. First, choice of the boundary will influence the analysis by including or excluding and by creating arbitrary divisions where none occur. Second, since most natural processes form gradients of change over space-time, setting a boundary is necessarily arbitrary. One seeks out natural thresholds or breaks in a gradient and uses these to locate bounds. The environment built by humans is highly bounded and so it is less difficult to approach in this way. Human-created boundaries often have a historical origin and do not coincide with natural systems. Yet they control what can be done. Finally, the boundary area is a special habitat. Ecologists call it an ecotone. It often has its own fauna and flora and fills special roles in natural ecosystems. A boundary is not a line; it is a maze, a sieve or filter, and a barrier.

1.1.3 Ecological Hierarchies

In the above discussion we suggested, first, that systems can be constructed by aggregating subsystems together to make the system. Then second, we placed the system, made up of subsystems interacting together as a whole, within an environment and we pointed out how the system is linked to the environment through inputs and outputs (Figure 1.1). It should be clear that we could change our focus and consider one subsystem by itself. This subsystem could then be treated as a system and we can identify its parts and examine how they interact to create the behaviour of the subsystem. This process of reduction can go on deeper until we consider individual organisms or their chemistry and physics.

Similarly, we could aggregate systems together to create supersystems, until we eventually ended our aggregation when we reached the planet Earth. The strings of aggregated and disaggregated systems that we create for each analysis are hierarchies of nested systems (Figure 1.2). The process of creating hierarchies improves the efficiency of analysis and reduces likelihood of error. These strings do not imply control passing up or down the hierarchy, although we could add control processes if that was necessary. Additionally, there are as many hierarchies as we require to

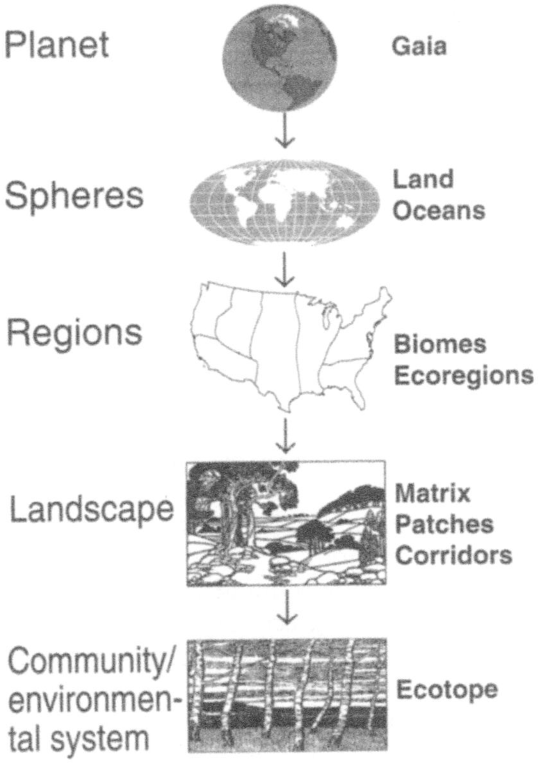

FIGURE 1.2. A rural spatial hierarchy showing an arbitrary set of nested elements including the Planet Earth, the spheres of land and oceans, regions and biomes, landscapes, and at the finest scale, fields of crops, pastures and patches of forest. These fine scale systems are called ecotopes.

understand the system. Hierarchies may anastomose and separate; they do not need to be linear. Individuals might be members of many hierarchies.

Systems appear to be organized hierarchically in nature and in human social systems. The hierarchical concept is helpful because it is a way to connect systems that are separated in space-time or are of disparate size and character. Different systems, such as those of hydrology and economics, are linked through the landscape system, in which both function as subsystems. For this reason the book is organized on hierarchical principles.

1.2 The Rural Concept

We focus on rural planning because of the rapid change occurring in rural areas worldwide. Rural areas are declining in population, causing land to be abandoned. Individuals become isolated without adequate social services. Residents often face poverty and social conflict, as well as environmental degradation. The rural poor may cause deforestation, soil erosion, and water degradation. While rural areas frequently do not contain resources sufficient to manage change, planning can be useful in showing how to use resources efficiently and how to direct and control processes. Planning is not a cure-all but is an essential part of the overall process of creating sustainability in a finite word of resources (Lassey, 1977).

What do we mean by the term "rural"? Rural derives from the Latin word for the country. It usually is applied as having to do with farming as an economic activity. Rural contrasts with urban. These dictionary definitions, however, are only a partial help in developing a practical working definition of the concept.

Traditionally, humans in the past have lived in a world of three parts. They lived in a familiar area, in which they could find the resources for life as hunters and gatherers. Next, they began to rearrange the resources. Gradually the familiar area was reconstructed into a humanized landscape. Outside of or beyond the humanized landscape was nature. Nature might provide resources, opportunity for expansion of humans, grazing, escape from authorities, adventure, and so on. Nature might be feared or be seen as hospitable, but it was outside the humanized area. Man and nature became separated conceptually.

Within the humanized landscape, humans constructed concentrations of buildings and structures, which eventually grew into a distinct place-a town. Towns provided storage of resources or the residence of religious and authority figures. They might be set off by walls, which could be protected and defended. Thus the urban became distinct from the rural which was distinct from the natural. Obviously these boundaries are and always were fuzzy.

In many countries political power became concentrated in urban areas. Around this political power was then concentrated governmental functions, services, and intellectual sources. In this way the rural/urban division was translated into dualisms of traditional/modern, backward/progressive, ignorant/intellectual. The urban person held rural labour in contempt. Or, alternatively, a romantic and unrealistic view of rural life was developed by urban intellectuals who were reacting against overcrowding, pollution, and crime in the city.

These mistaken definitions of rural landscapes and rural people are always unfortunate but have been especially apparent in the current decades when urban people have sponsored projects to aid rural people in other nations. In many of these projects, the need, the approach, and the activity, created, identified, and carried out by urban professionals from developed countries, impacted the rural society negatively. Rural people could not resist development and were often injured, so that after the project their condition was less sustainable than before. Planners frequently had no sensitivity to rural social forms, traditional knowledge, or historical land use patterns.

In this book we will introduce social, environmental, economic, and physical–biological approaches to counteract these biases. Obviously, no one individual has all the skills necessary to understand and advise residents of a rural region. But teams can bring a variety of points of view to a project.

Our task is to integrate knowledge from team members so that information can be focused on the task. This is the reason why we have emphasized a systems approach to planning and why we have introduced the hierarchical approach at the outset. Nature, the rural landscape, and the city are all systems, arranged hierarchically in space-time. Our task will be to relate the hierarchical systems approach to rural planning.

1.3 Rural Resources and Environment

1.3.1 Natural Capital

In the rural landscape we encounter a variety of resources that can be used by the inhabitants for economic and social purposes. Land, air, and water are the main elements of the natural capital of the rural landscape. Land provides the source of chemical elements that are required to build biological structures through the biological production process, the support for plant roots so that plants can compete for solar energy, the place where soil organisms breakdown dead organic material and renew the chemical capital which is recycled in production, a source of building material, and materials that can be mined and used for energy or mineral industries. Land is usually divided into property. The western concept of land is that human labour mixed with the natural capital of land results in economic value. In our concept, land is natural capital with or without the addition of human labour because it is necessary to support human life in the natural state.

Water is a second component of natural capital. It moves from the oceans to the land through the hydrologic cycle. Water in precipitation and snow-fog-ice runs off sloping surfaces into streams and eventually returns to the ocean. Water also is evaporated and becomes a gas that mixes with other constituents of the atmosphere. Yet another part passes through the surface layer of the land and enters the groundwater. Groundwater moves downslope, accumulates underground, and forms artesian springs in discharge regions of watersheds.

Water is the universal solvent and as it moves across and through the land it dissolves and moves chemical atoms. The process leads to erosion and leaching of the products of the land to the oceans and their accumulation in the ocean depths. Where the protective surface of green plants has been removed from the land, rapid erosion may occur, with concomitant loss of productivity and loss of fertility.

Air is another element of natural capital. Air contains essential oxygen and nitrogen sources for plant and animal life. Air can be polluted with industrial waste and with dust. Air sheds fit above the landscape and shape the rate at which water and other atmospheric materials are delivered to the land.

Unlike land, air and water are usually not owned by individuals, although water is so important a resource that complex laws have been developed in most rural areas to determine who has or has not access to water. Planners must understand the laws relating to the distribution and inheritance of land and the allocation of water.

The combination of the elements of natural capital with human labour, knowledge and material life (seeds, technique, and machinery) result in rural production. The base of rural production is the process of natural biological production. Humans manipulate this process to increase the yield to humans and reduce the yield to competitors, such as insects. Most of the advances in agricultural technology are designed to provide water at the proper time, adequate nutrients for plant and animal growth, reduction of competition by plants and animals, and preservation, storage and conversion of organic product into food and industrial raw materials. Agricultural technology is complex because it has adapted to the characteristics of land, water, and air in each landscape and to the plant and animal species and varieties of the region.

There has been a tendency to ignore the agricultural diversity that has evolved in human history and replace it with a few varieties that provide the industrial user exactly the properties the machines require. Not only does this policy result in great loss of genetic information but also it makes rural producers vulnerable to diseases and pests due to the homogeneity of the crops and environment.

1.3.2 The Cultural Capital

The rural landscape has other attributes in addition to the production of food and industrial raw materials to support a rural population. Rural regions provide scenery, recreation, and connection with traditional cultural activities and customs. They are attractive places for vacations, and rural land is commonly sold for second homes and vacation resorts and facilities. Indeed, in some environments a human presence is necessary to maintain physical structures of the environment, such as terraces, hedgerows, and ponds. These capital resources must be managed properly to retain their value so that rural residents receive benefits from them.

1.4 The Rural Planning Process

1.4.1 Evolution of the Different Approaches in the Planning Process

Rural planning in the context of this book is an attempt to change land use to produce an improvement in the quality of life of the rural people. Historically, rural planners have used a variety of approaches to obtain and treat information related to the natural environment and socioeconomic aspects of people. Whilst in the last few decades rural planning has become especially relevant, its antecedents are situated in the nineteenth century. Two principle approaches have developed: (i) integrated studies of environmental units as if they were homogeneous with respect to their contents; and (ii) those which analyse the environment by looking at each element or independent sector (geology, water, economics) and later integrate the information.

The integrated studies of an environmental unit are focused basically on land elements which are studied as a whole, in the same way that environmental analysis is carried out where all the elements are considered together. This unitary conception of the territory was used by the Commonwealth Scientific and Industrial Research Organization (CSIRO) in Australia (Christian, 1958) and consisted of a physiognomic description of the territory (using land units and land systems) that recognized repetitive models of landscapes arranged hierarchically. Whilst the method is simple in concept, it is also well adapted in order to identify the scientific features and human developments of a region on a small scale.

An advance in this methodology occurred in North America when, in 1961, Hills proposed studying territorial areas divided into zones with similar macroclimates (Hills, 1961). These areas can be further divided into subzones based on geological, morphologic, and hydrologic criteria. They are divided into physiographic classes when the microclimate is considered and into physiographic types when we incorporate microrelief. Finally, when we incorporate vegetation we obtain land units, which are considered to be a spatial projection of ecosystems.

In Europe the integrated approach has utilized the idea of the landscape as a natural unit, following the German concept of landschaft which associates elements such as relief, climate, vegetation, soil, and so on. The French incorporated an evolutionary dimension into the landscape (Bertrand, 1968), suggesting that a spatial unit may evolve as a whole. They elaborated the first maps of ecological risk and used the concept of planning as an intention to modify the dynamics of systems. In Spain, the integrated examination of environmental units was made by COPLACO (1975) in the Special Plan to Protect the Physical Environment of the Province of Madrid. This project was a pioneer in taking the ecology into account in rural planning and use allocation. In this case the environmental units were a synthesis of many characteristics organized into a relatively homogeneous system as a whole. This concept is like a representation of an ecosystem, defined by easily identified indicators (vegetation, geomorphology, crops, and so on).

The second major approach developed has been the independent analysis of environmental components by sectors followed by the subsequent integration of information. In this case, the integration is realized at a more advanced step of the planning process. Steinitz and his colleagues at the Harvard University Graduate School of Design presented a relevant methodology (GRID and IMGRID programs) to the development of this approach during the 1960s and 1970s (Steinitz and Sinton, 1967, 1975; Steinitz and Rogers, 1968; Steinitz, 1976). This was possible after the development of concepts like the "environmental corridor" (Lewis, 1964) and "design with nature" (McHarg, 1969).

A further advance was that of Johns (1973), in Manitoba, who developed the concepts of the intrinsic capacity of a territory and also used impact studies to produce planning alternatives. But it was in Israel where the interest in the evaluation of environmental impacts to establish the sensitivity of a territory for concrete projects was tested and confirmed (Amir, 1976).

In Europe, planners in both the Netherlands and in England developed ecological bases of physical planning, demographic growth, and economy of the territory in the 1960s (van der Maarel and Vellena, 1977). These operate through the generation and evaluation of alternatives, obtained through a consideration of the relative capacity of a territory to support activities for and uses of the soil.

In Spain, precedence in applying this type of approach is given to the Comisión de Planeamiento y Coordinación del Area Metropolitana de Madrid (COPLACO) which, in 1973, incorporated the concept of support capacity. This had substantial importance in later studies, through the evaluation of landscape resources (González Bernáldez et al., 1974; Gómez Orea, 1975; Ramos, 1979).

Later, a second generation of methods based on planning with an ecological foundation appeared, which utilized the concept of support capacity, although not that of impact, under the direction of the National Institute for the Conservation of Nature (ICONA).

Years later some works utilized the concept of aptitude and impact (Villarino, 1978). Within this approach an optimal distribution of uses and activities of the territory was proposed, which maximizes the aggregated aptitudes of the resources and, at the same time, minimizes the global impacts.

From the technical point of view there was also an advance in the utilization of the Geographical Information Systems (GIS). In turn the Modelo de Asignación de Usos del Suelo en Areas Rurales (MAUSAR model) (Gómez Orea, 1992) permitted the integration of information and the criteria for generating alternatives with distinct support capacities. Conceptually, it facilitates linking the compatibility of uses with the Evaluation of Environmental Impacts Assessments (EIA), setting a precedent in using the EIA as an instrument in planning.

From this background, we consider that the final result derived from study of the physical environment supporting planning, independent of being previously or later integrated, physically or economically, is *the optimal distribution of the support capacity of the territory.* To arrive at this final point we follow a series of steps, each producing outputs and distinct data that can enlarge and give continuity to the planning process.

The working method used in the course on rural planning held at the Mediterranean Agronomic Institute of Zaragoza (IAMZ), which has served as a basis for the development of this book, is an approximation to the independent analysis of the environment of Steinitz and Sinton (1976), with a later integration using techniques such as systems analysis, the GIS method, and the evaluation of multicriteria (Roy, 1973).

As a conceptual approach it follows the concepts of environmentalists and ecologists such as H.T. Odum, F. González Bernáldez, R. Margalef, and W. Haber. The final objective is the optimal distribution of the land use allocation, incorporating values of the support capacity, aptitudes, impacts, and the sustainability of the proposed alternatives.

1.4.2 Steps in the Applied Methodologies in the IAMZ Course

The planning process is usually conceived of as a linear series of steps (Figure 1.3). The following steps are essential: (i) clear statement of goals and objectives; (ii) inventory of relevant information; (iii) diagnosis of the system problems; (iv) development of alternative ways to solve problems; (v) evaluation of proposed alternatives; (vi) implementation of selected alternatives in a plan; and (vii) monitor the implemented plan.

1.4.2.1 Step 1: Goals and Objectives

The first step of developing a clear statement of goals and objectives is exceptionally important because frequently the motivation for planning comes from outside the territory. For this reason, the problem-setting stages require field visits, meetings with both authorities and local populations, and review of the key literature and documents.

There is also a conceptual element in goal setting that is crucial. The goals will be expressed in language that represents needs, desires, objectives, and hopes for the future. Often local people have difficulty expressing their view of the situation. Partly this difficulty is due to their not usually being part of a discussion substructure in society and partly because they are "inside" the problem. Yet the outsider-the planner or administrator-may, with the best intentions, use language in a very special technical sense that misinforms the public and introduces concepts which are technically unproved or problematical. Meeting, discussing, and observing with an open mind is a crucial part of the goal-setting step.

For example, in the United States, the Corps of Engineers, which is responsible for construction of the water plans for regions of the country, adopted a "fishbowl" planning process to revise the objectives of the northwest water plan. In fishbowl planning, they asked every citizen organization, and interested individual to describe how water was

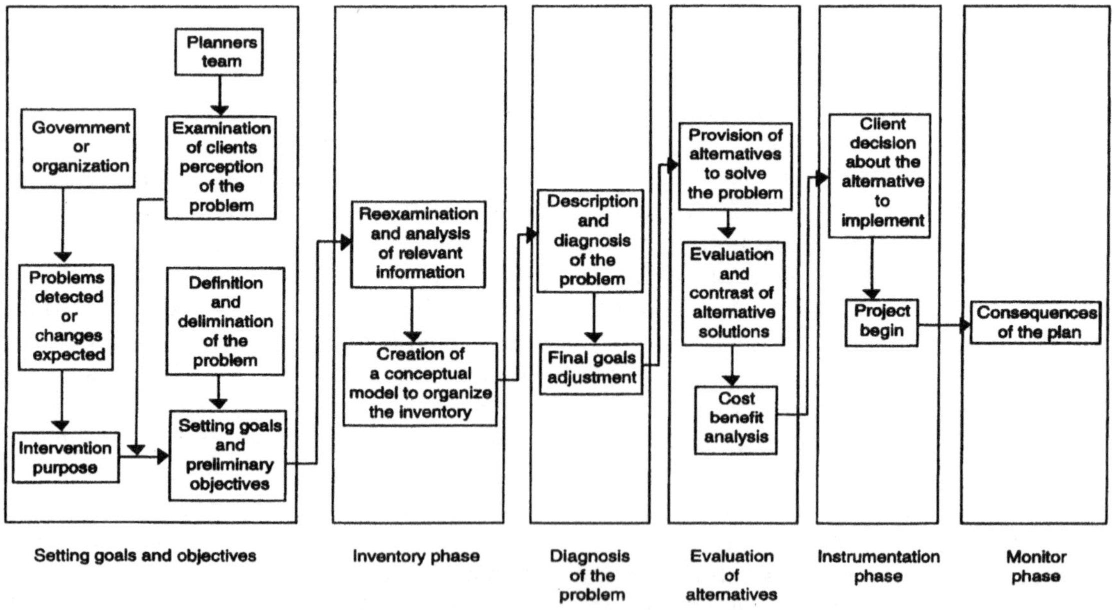

FIGURE 1.3. The sequence of steps required for rural planning as developed in this book.

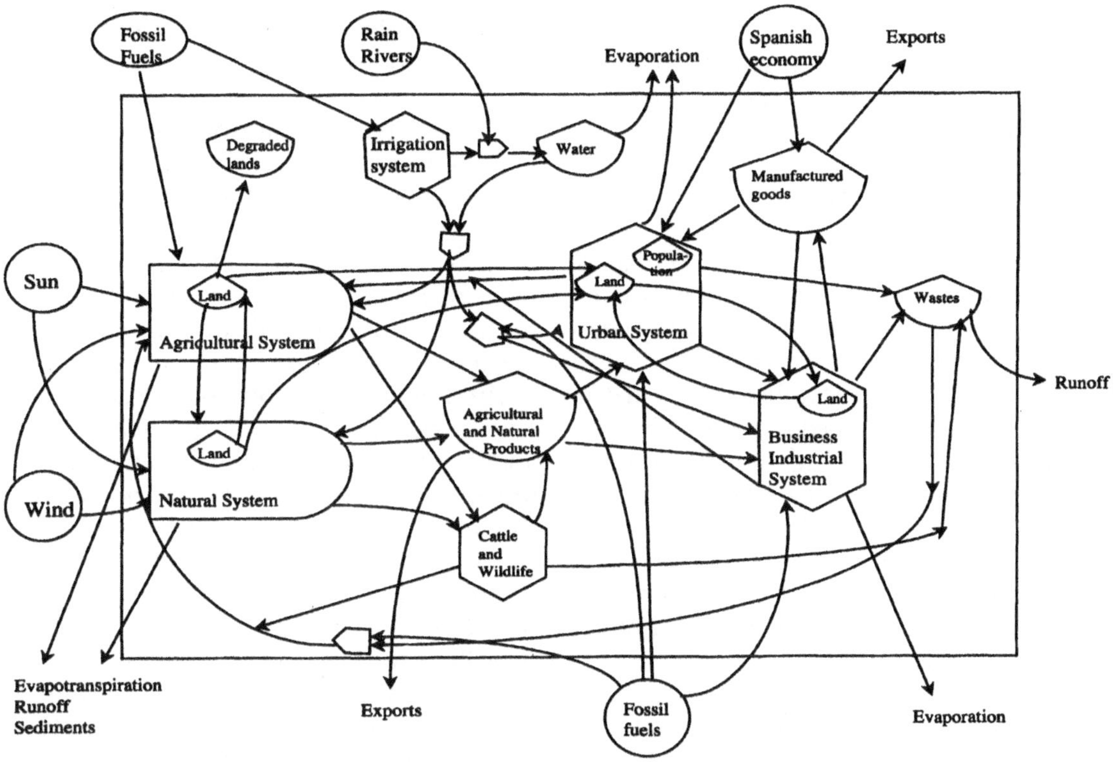

FIGURE 1.4. The water balance of the Ebro River watershed, Spain, illustrating a conceptual model of energy flows associated with water flows (modified from Lugo et al., 1983)

important to them and to locate specific interactions with water on maps. The resulting technical plan accommodated a majority of these sites. As a consequence, there was a wide public support of the plan. Time delays associated with protest and lawsuits against the plan were reduced. Involving the public at the beginning increased the efficiency and reduced the cost of the project.

1.4.2.2 Step 2: Inventory

Once the goals and objectives are established, the inventory phase of the project can begin. The inventory is guided by a systems-analytical framework. The system of interest is identified and bounded to fit the problem. The system has certain properties that are relevant to the problem and these properties must be inventoried. The analysis is concerned only with relevant information; no attempt is made to collect all the information about the system. A conceptual model (Figure 1.4) is created to organize the inventory. This model shows what topics will receive attention and how they are linked. A conceptual model differs from a mathematical model used to predict system behaviour because it is not necessarily quantitative. Predictive models will be discussed later (Chapter 16). They are not required at this point.

1.4.2.3 Step 3: Diagnosis

The inventory provides the basis for a diagnosis of the problem. At this stage the planner returns to the first stage of goal setting. Do the goals need adjustment? What is possible? Are resources adequate? What is the source of the problem? Can it be changed? These kinds of questions arise from the team's examination of both the inventory and the problem from multiple points of view. The authorities and local people will also be asked for opinions. The diagnosis establishes the form and structure of the plan. If the goals are unclear or contradictory, we must return to the goal-setting step and repeat the inventory and diagnosis. Once the diagnosis step is finished, the goals are established and we can proceed.

1.4.2.4 Step 4: Alternatives

One of the important elements in the diagnosis is the identification of alternatives. Usually a landscape planning problem is highly complex and involves multiple elements. Almost always we will identify multiple strings of causation that could have been involved in causing a problem. These strings of causation provide alternatives that might be changed to solve the problem. The planning team must lay out these alternatives and then select those with greatest chance of success, lowest cost, least disturbance, and so on. The plan might contrast alternative solutions.

1.4.2.5 Step 5: Evaluation of Alternatives

The planning activity begins by entering the landscape system and adjusting natural capital to obtain greater gains and solve the identified problems. Each alternative has a benefit and a cost and these must be determined as far as possible. Benefits and costs are economic, environmental, and social-and only some of them can be expressed in monetary terms. Frequently the nonmonetary values must be converted to monetary equivalents. Costs frequently feed back to other parts of the system and causes injury in several places. For example, soil provides many functions for production, one of which is to provide nutrients for plant growth. The exhaustion of soil nutrients can be overcome with fertilizer and yields on eroded land can even be increased above uneroded unfertilized land by using fertilizer. But continued erosion has many costs, including clogging stream channels with sediment, destroying the value of downstream water, and changing the rate of infiltration of surface water to groundwater. These costs must be charged against the increase in yield of the crop. Often a calculation of this type will show that increased costs exceed the increased value of the yield.

1.4.2.6 Steps 6 and 7: Implementation and Monitoring

Once the plans are finished, the client can make a decision about which alternative will be implemented and the project can then begin. At this point the work of the planning team is finished. However, if possible, planners should examine the implementation of the plan, using it to feed back into experience. We are operating in a continually changing environment. Conditions in the next case will always be different. For this reason, each analysis is a test of an overall method. Monitoring involves examination of the impact of the plan. Specialists often can be enlisted in monitoring programmes and their viewpoint can be especially helpful for identifying unexpected positive and negative impacts. In this way the practical work becomes an experiment that will result eventually in improvement of planning. We can ask that monitoring be formally included as part of the process. It seldom is!

1.5 Sources of Information

We have organized the information needed in rural planning into four units. These follow the logic developed above when we spoke about the inventory phase of the planning process. First, we consider the natural capital of land, soils, water, and climate. There exists an enormous literature on these subjects and we must restrict our analysis to a particular scale. In this book we have selected a scale of the landscape (Golley and Bellot, 1991; Marsh, 1991). That scale ranges from tens to thousands of square kilometres. Landscapes are parts of regions, which make up the great biomes or provinces of the earth, such as the tundra, grasslands, and deserts. Landscapes may be divided into watersheds or landscape elements, which can be further subdivided to the farm property or field (Figure 1.2). While we will not subdivide the analysis to this depth, because it would require too much detail, in planning it may be necessary to include individual properties.

Second, biological productivity is the fundamental source of rural value. Initially, we will determine the structure and species of the plant communities in the landscapes. Then, we will determine the biological production of the communities and how the environment influences natural rates of production. These give us a measure of the productive capacity of the land and water on a sustained basis, without subsidies. Finally, we will consider the process of agricultural, forest, and pasture production, which is subsidized by inputs such as fertilizer and pesticides. Then we can determine the benefits and costs of the subsidies.

Third, we are concerned with information on the social aspects of rural systems-the broad patterns of rural population and the capacity of rural social systems to adapt to and direct change. Traditionally, rural populations produced food, housed a large mobile population of landless poor, and produced enough surplus to populate the cities, provide the armies, navies, and labour gangs for civic projects and construction of railroads and other large projects.

Cities were stressful centres of disease. Cities did not typically produce enough surviving children to maintain their population. They were population sinks that depended upon a flow from the more healthy rural areas. This role was reversed with the development of modern medicine and public health at the end of the last century. Now the cities are full of excess rural population attracted to them by opportunities there or forced to them by economic change. Rural social systems are as diverse as are natural systems because they represent human adaptation to the natural environment, controlled by culture. Part of the interest of rural areas is its cultural diversity.

Rural populations are widely considered to be highly conservative and unresponsive to plans for change and improvement. It is important for the planner to understand how tradition and conservatism work for or against change. Capacity to deal with change depends upon ones control of resources and ones capacity to tolerate disturbance. Wealthy individuals have resources to control change so that they benefit from it. The poor farmer has almost no resources except access to land and water, his experience and his family's labour. He must use techniques and methods that have worked successfully in the past. He would be foolish to risk changing his technology on the advice of a stranger, who risks nothing. Therefore, to create change governments frequently resort to subsidies or threats to create the motivation to accept change. It is essential that these social costs be evaluated in a planning activity.

Further, the economic processes in rural landscape are examined. The producers use natural capital (land, air, and water), their labour and management, solar energy, and a variety of purchased inputs to produce products of economic and social value. With the income from these products they pay for the purchased inputs, pay for the capital, pay themselves a wage, and pay for the services and infrastructure that they require as citizens of a modern nation. Their wage establishes a standard of living, and if they are successful they will maintain a standard like that maintained by an industrial or service worker in the city. If they are unsuccessful they may live in rural poverty.

The rural producer produces for markets over which he or she has little or no control. Economic markets develop when individuals with something to sell meet individuals with money and desire to buy. The price paid represents the mutual decision to buy and sell. Agricultural markets are often global and are highly complex. The farmer with a perishable product is in a difficult position. Products are usually bulky and difficult to move. The farmer usually is not a marketer and does not have time to personally operate in the market. In that case, the farmer depends upon an agent or a middleman to move the product to the consumer at a suitable price, which often involves the price of storing and packaging the product. Further, markets are manipulated by governments who use agricultural products for political purposes. Manipulation is through direct payment to farmers to produce specific products, which often have an industrial use, though indirect subsidy of the inputs to production, to setting selling prices, and to providing access to public land and resources at below-market costs. Seldom is agricultural production a free market. The consequence is that the farmer faces a market that operates like the climate. It is crucial to his or her success, yet there is nothing he or she can do about it directly.

The farmer has been encouraged to organize into economic groups to cope with market changes. The cooperative movement is an example. In this instance the farmers combines money to purchase supplies at lower cost. Further, the cooperative stores the product so that the farmer receives the best price possible. It may even employ marketers to search out new markets for the products. In this way the individual farmer improves his or her competitive position at the cost of conforming to the overall objectives of the group.

1.6 The Technical Methods

These sources of information focus on subjects such as physical environment, productivity, social systems, and economics, which provide a context that guides the inventory and diagnosis. Next we have to consider the technical methods which can be used to organize and integrate the information into formats useful for identifying and evaluating alternatives. These include systems analysis which can be used to develop the concept of the plan and predict the consequences of alternatives, and landscape analysis techniques, such as remotely sensed information and GIS, to organize, weight, and evaluate ground surface data.

The systems-analytical approach requires that the system be identified explicitly. In planning, this means that the area of the Earth's surface being planned is delimited. It is highly desirable that these limits fit natural limits appropriate to the issues being addressed. For example if the plan includes the role of water, the areas must be watersheds, not political units.

Second, the system processes of relevance are identified and quantified, if possible. Processes involve impacts to the system and outputs from the system. Inputs and outputs must include natural and economic capital, biological and social costs and benefits, as well as economically assessed elements. These data help us to understand how the system performs and how it fits into larger-scale systems, of which it is a part.

Third, the factors within the system that cause the observed behaviours and the linkages between them are determined. There will be a family of direct and indirect impacts. We are seldom able to include many factors in an analysis. The expense is too great and the output becomes questionable when the model is too large. It is essential to restrict this step to relevant issues that are central to the goals of the plan.

At this stage a conceptual systems analysis is complete. The analysis identifies the boundary, the performance and the potential causes of the performance of the system of interest. However, the analysis is entirely mental. Each viewer of the model may derive different perspectives from it. If we have sufficient quantitative data, of adequate quality, we can use the conceptual model to predict outputs from manipulation of the inputs or the internal arrangement or performance of the parts of the system.

Mathematical system models usually incorporate feedback processes. Feedback means that two elements in a system are linked in such a way that the impact of one on the other is transmitted, as information, back to the initiating element. The flow of feedback is often transmitted through other system elements.

The linkages between system elements may be described by linear programming methods but these produce results that are usually too simple and often are unrealistic. Rather, systems are more likely to operate in multivariate modes, in nonlinear sequences, with direct and indirect feedback. Systems modelling is an active field of research and the planning team must include a specialist in this subject if an efficient use of this special methodology is realized.

A second technical subject that plays an important role in planning is the use of remote sensed photography and GIS. The ground surface reflects light in a spectrum of wavelengths which can be recorded by an airplane or a satellite. These pictures can be interpreted to show many features of the surface. The actual pictures need to be interpreted and linked to the objects on the ground. This is a highly technical and rapidly changing subject that requires special training to employ efficiently (Cloke and Park, 1985).

Finally, land planning methodology also uses a variety of less technical and mathematical methods, which are appropriate to many applications (FAO, 1993). These methods are frequently more useful than those employing complex technology, because they require mental agility and experience and do not require alteration of questions and data to fit the technology. Usually these methods balance favourable against unfavourable alternatives weight inputs and outputs using an arbitrary scale, evaluate weighted sums of impacts based on matrices of evaluation. The method called ELECTRE will be presented in Chapter 15.

1.7 The Organization of the Book

This book develops according to the previously described analysis of rural planning, dealing first with foundation concepts and continuing on with methodologies.

The first part concerns the analysis of the physical environment, including geographical information systems, land, climate, water, and soil. The second part focuses on the biological features of rural landscapes, including conservation of biological resources, organic production, and agricultural, forest, and range productivity. The third part involves social and cultural issues, and the fourth part involves economics.

The part devoted to methodology has two chapters. The first is concerned with land and soil planning and the second with systems analysis. The last part of the book is a case study which provide an overview of the information required and processes followed for rural land planning.

Each of the chapters indicates the relevance of the information to rural planning but also represents background information necessary to understand the application. Of course, it is impossible to anticipate every need of the reader or planner. Deeper knowledge about a topic can be gained from the references in each chapter.

1.8 Summary

Rural planning is a complicated and sensitive process. It involves the natural environment and all the forms of life that live in rural areas and in waters associated with rural areas.

As human populations have increased and rural development has pushed against natural limits, the rural region has assumed an important conservation, as well as a production, purpose. These new demands are added to those demands of urban populations for food, products, water, air, recreation, and scenery. Rural areas frequently have paid for their freedom from the stress and pollution associated with city life by lower standards of living, a poor image, and social neglect. Yet our food, water, and air come from or are strongly influenced by the rural regions surrounding us. There are many examples of economically viable, beautiful, diverse, and interesting rural landscapes which have been planned and managed to create sustainable societies based on agriculture, tourism, and rural industry.

Planning has many aspects that cannot be developed in a single volume. Yet with a hierarchical system approach it is possible to expand or contract the analysis to fit the need.

In summary, we can view rural planning as a search for answers to questions, with the planning process acting as a system to translate information and concepts into answers (Figure 1.5a). The purpose of the book is to unpack the black box labelled rural planning. In a general sense, rural planning as a problem-solving exercise employs analysis and synthesis, as does every other problem-solving subject (Figure 1.5b). Analysis and synthesis must be defined for the specific case. In rural planning we propose that we begin the analy-

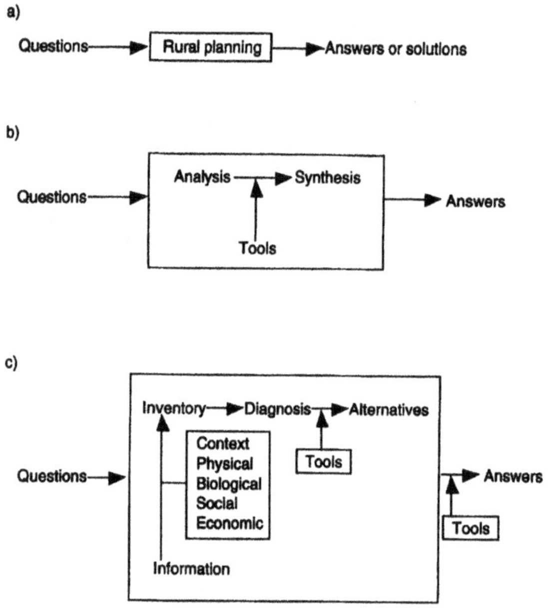

FIGURE 1.5. A hierarchical view of rural planning. The planning activity translates information on the questions or problems into answers (a). The method of rural planning involves analysis and synthesis (b). Inventories and diagnoses are developed in the analytical phase and alternatives are evaluated in the synthesis phase (c).

sis with the inventory and diagnosis phases (Figure 1.5c).

Inventory requires information, which is derived from a variety of sources but is local and specific to the question being posed. Information is also controlled by both concepts and theories derived from fields that are concerned with the physical environment, biological processes, social factors, and economic processes.

The process of inventory and diagnosis leads to the synthetic phase where alternatives are identified. Synthesis requires a variety of tools. Some are well known, such as statistics, and some are highly technical, such as GIS and remote-sensed images. Maps are usually produced at this phase of a project. These maps help us identify a series of alternative answers to our questions.

The ultimate choices among these alternatives requires other tools, such as system analysis, landscape evaluation systems, and so on. In this way we reach a final conclusion to the project and an answer to the question. The chapters of the book are designed to follow in sequence, focusing on the contextual concepts and on the theory and tools necessary to carry through the planning process.

References

Amir, S. A programme to improve the presentation of environmental impact assessments. *Landsc. Plan.* 2:229–241; 1976.

Bertrand, G. Paysage et géographie physique globale. Esquise methodologique. *Revue Géographique des Pyrénées et du Sud-Ouest* 39: 249–272; 1968.

Christian, C.S. The concept of land units and land systems. *Proceedings of the 9th Pacific Science Congress* 20:74–81; 1958.

Cloke, P.J.; Park, C. C. *Rural Resource Management*. New York: St. Martin's Press; 1985.

COPLACO. *Plan Especial de Protección del Medio Físico de la Provincia de Madrid*. Madrid: Ministerio de la Vivienda; 1975.

Food and Agriculture Organization of the United Nations (FAO). *Guidelines for Land-Use Planning*. FAO Development Series 1. Rome: FAO; 1993.

Golley, F.B.; Bellot, J. Interactions of landscape ecology, planning and design. *Landsc. Urban Plan.* 21:3–11; 1991.

Gómez Orea, D. Análisis de la calidad ambiental: Procedimiento para la formulación y representación de los valores paisajisticos de los espacios naturales. Ph.D. Thesis, ETSIA, Polithechnical University of Madrid; 1975.

Gómez Orea, D. *Planificación Rural*. Madrid: MAPA; 1992

González Bernáldez, F.; Martín de Agar, P.; Ródenas, M.; Sancho Royo, F. Un systeme d'enregistrement et minimization des impacts sur l'environnement: IRAMS. *Proceedings of the Colloque Internationale Informatique et Environement*. Arlon, Belgium: Fundat. Univ. Belge; 1974.

Hills, G. A. *The Ecological Basis of Land-Use Planning*. Research Report 46-57. Ontario: Department of Land Uses and Forests; 1961.

Johns, D. M. *Environmental Quality Evaluation and Impact Analysis Methodology.* Winnipeg: Center for Settlement Studies, University Manitoba; 1973.

Lassey, W.R. *Planning in Rural Environments*. New York: McGraw-Hill; 1977.

Lewis, Ph.H. Quality corridors for Wisconsin.

Lugo, A.; Patiño, M.P.; Mayoral, O.; Pizarro, D. et al. Models of energy flow for rural planning in the Ebro river watershed. *Options Méditerranéennes, Series Studies 1983-II*. Zaragoza: CIHEAM; 1983.

Marsh, W.M. *Landscape Planning: Environmental Applications*. 2nd ed. New York: John Wiley and Sons; 1991.

McHarg, I.L. *Design with Nature*. New York: Natural History Press; 1969.

Ramos, A., ed. *Planificación Física y Ecología: Modelos y Métodos*. Madrid: EMESA; 1979.

Roy, B. How outranking relations helps multiple criteria decision making. Cochrane, J.; Zeleny, M., eds. *Multiple Criteria Decision Making*. Columbia, SC: University of South Carolina Press; 1973:179–201.

Steinitz, C. Hand-drawn overlays: Their history and prospective uses. *Landsc. Archit.* 9:444–455; 1976.

Steinitz, C.; Rogers, P. *A Systems Analysis Model of Urbanisation and Change: An Experiment in Interdisciplinary Education*. Cambridge, MA: Graduate School of Design, Harvard University; 1968.

Steinitz, C.; Sinton, D. *Program GRID*. Cambridge, MA: Department of Landscape Architecture, Graduate School of Design, Harvard University; 1967.

Steinitz, C.; Sinton, D. *Program IMGRID*. Cambridge, MA: Department of Landscape Architecture, Graduate School of Design, Harvard University; 1975.

van der Maarel, E.; Vellena, K. *Towards an Ecological Model for Physical Planning in the Netherlands*. Seminar UN Economic Commission for Europe, Ecological Aspects of Economic Development Planning Report. Rotterdam: Ministry of Housing and Physical Planning; 1977.

Villarino, T. Modelo para la ordenación integral de espacios naturales y su aplicación a la Sierra Calderona. Ph.D. Thesis, ETSIM, Politechnical University of Madrid; 1978.

Part II
Bases for Resource Interpretation and Valuation in Land Planning

2

Information Systems for Rural Planning

Arwyn Jones, Alan Belward, and Marc Van Liedekerke

2.1 Introduction

The decision-making processes involved in rural planning require historical, contemporaneous, and predictive information on the state of the natural environment. Rural planners have in the past made extensive use of both aerial photography and topographic maps to provide such perspectives. Since 1972, rural planners have also had access to images taken from space. The photo interpretation and conventional map-reading skills employed in the analysis of these data continue to play an important role in rural planning.

However increasing use of digital technology, and the demand for the storage, analysis, and display of complex environmental datasets have led to increasing reliance on computers. It is often difficult to use a single processing strategy or software tool to extract the necessary information. Consequently, for certain applications, a range of computer-based tools are needed to answer specific questions. Such decision support systems aim to provide a planner with sufficient knowledge to effectively manage all available resources. The technologies outlined in this chapter also help the rural planner to face one of the most demanding challenges in the field, namely, the need to access, collate, assess, process, visualize, and understand a range of information from diverse sources.

This chapter deals with a simplified definition of an information system as an integrated chain of applications that starts with a question, and that then requires the collection of data, which are processed to generate new results for delivery to the appropriate decision-making environment. The bases for resource interpretation and valuation, the information on production systems, and the human use of rural space, as described in following chapters, could all use such an information system. The merits and disadvantages of integrating three distinct technologies to form an information system for rural planning applications are discussed.

2.2 Information Systems

The first element of our system concerns the use of networking technologies and in particular, the concept of information cyberspace known as the World Wide Web (or WWW). The WWW implements the concept of linking multimedia information held on computers located all over the world. An information system based on WWW technology can be used to locate and provide access to data. Since the introduction of the WWW and the advent of client/server technology, there has been a tremendous growth in the use of the Internet to receive and publish information. By the end of 1995, the Internet was composed of 50,000 individual networks, which served more than 6 million host computers. With this level of usage and the increasing commercialization of cyberspace, the future possibilities of this technology appear boundless. Benefits to the rural planner are threefold: (i) the combined possibilities to search for knowledge and the provision of virtually instantaneous access to information; (ii) the ability to receive data; and (iii) the real-time publication of results. All these factors should combine to provide the planner with faster decision making since the delays from information retrieval and third-party responses can be reduced.

The second element is a Geographic Information System (GIS) which is based on the concept of digital cartography. By definition, a map is a representation, to a certain scale and on a flat medium, of a selection of features on, or in relation to, the surface of the Earth (Valenzuela, 1991). To be useful, a map must be able to convey information in a clear and unambiguous fashion. A GIS is an information system that works with data located by geographic coordinates (also known as spatially referenced). In this sense, a GIS can be thought of as a high-order map product.

The term "geographic information system" has in recent years become synonymous with a rapidly emerging technology for processing spatial data. A GIS can be defined as a powerful set of computer-based tools for collecting, storing, retrieving, transforming, and displaying spatial data from the real world for a particular set of purposes (Burrough, 1986). In real terms, a GIS consists of a database system for holding information together with an integrated set of operations for working with the data. GISs are rapidly becoming a standard tool for the management of natural resources and the production of high-quality output (i.e., maps). GISs present the rural planner with the possibility to combine a range of data in a way not previously feasible. The analytical power afforded by spatial analysis, combined with the predictive capabilities of a GIS allows the planner to investigate a range of possible scenarios before implementing a decision.

The final element in our overall system concerns the provision of environmental indicators and thematic maps on the state of the Earth's surface. In the past, these data would invariably have been supplied through traditional field studies. However, over the past 25 years, such detail has been provided increasingly by sophisticated digitally based remote sensing systems. In essence, the term remote sensing is used to describe the acquisition of information about a target of interest without physical contact. In most cases the carrier of this information is electromagnetic energy which is acquired by some sensor on an airborne or spaceborne platform. In the context of rural planning, remote sensing provides an important and complementary technology to GIS by providing unique data on the state on the Earth's surface.

Rural planners may not be aware of the potential offered by these key technologies. The subsequent sections show the merits of each of these components to the rural planner and discuss how they can be integrated to provide an information system.

2.3 A Simple Guide to Networked Information Systems

There is a growing awareness by organizations and individuals that their computers are, or can be, part of a fully interconnected network (TIME, 1994). This knowledge has started the tendency to make information and services available for other networked users. The ability to search the globe for relevant information and datasets has tremendous implications for the rural planner. From a networked desktop computer it can be possible to examine existing studies, search for, examine and download relevant datasets, and use the electronic publishing capability of the network to rapidly disseminate results.

2.3.1 The World Wide Web (WWW)

The key to the availability of network resources is the provision of servers on computers all over the network. A server consists of special software that accepts requests (or queries or commands) and sends a response automatically. Requests received by the server may have originated from a user on the same computer as the server software, or from a user on a computer on the other side of the world. Software programs that ask for resources from servers are called client programs. Clients send requests to a server using a standardised format called a protocol. The server responds by supplying information, usually in the form of files containing data of various sorts. Such client software is being developed all the time, providing better and more convenient ways of interacting with servers. Different versions of a particular client may be developed for different computers. Thus different versions of a client may be provided for use on PCs, Apple Macintosh, or UNIX computers.

The World Wide Web is a collection of programs that can understand the various information retrieval protocols (e.g., FTP, Telnet, WAIS, Gopher) currently in use on the Internet as well as the data formats of those protocols (e.g., ASCII, GIF, JPEG, MPEG, PostScript, etc.) while providing a single consistent user interface for all of them. In addition, these programs understand a new communication protocol (HyperText Transfer Protocol, or HTTP) and a new data format (HyperText Markup Language, or HTML) which are both geared toward hypertext and hypermedia. Hypertext is a term that describes a computer interface to text that allows cross-references to be followed. In a graphical situation, the user can follow cross-references by clicking with their mouse on a cross-referenced phrase. This action would bring up the document at the other end of the cross-reference. Hypermedia is the extension of this to include graphics and audio as themes that can be selected or viewed. An example of a WWW page is displayed in Figure 2.1 where italics can identify the hypertext links.

Each document in the WWW information space has a unique identifier, called a Universal Resource Locator (URL), which identifies the location (machine and directory) of the document and the protocol to be used to transfer it.

For example, the URL

```
rea.ei.jrc.it/about_ei/ein.html
```

identifies the HTML document ein.html in the directory /about_ei which is on the machine rea.ei.jrc.it, which can be retrieved by the HTTP protocol. One of the more popular WWW clients is the Netscape Navigator. This evolved out the NCSA's Mosaic browser, which was developed as a single interface that could access all types of information (i.e., text, graphics, sound, or video). For an introduction to, and an overview of the WWW, HTML, and HTTP, the reader is directed to Web sites [1]–[5].

Common questions from users include: What information is available on the network? Where can the information be found? and How can the information be put to use?

In parallel to the emergence of networks and the increase in information of every kind, there has been a stream of development of tools that try to provide an answer to these questions. Only a few of these tools have been generally accepted to become de facto standards: information producers need to organize and process their information in a form compatible with these tools, and information consumers need to employ them to find the relevant information. These new applications, called Resource Discovery or Network Information Retrieval (NIR) tools, offer three types of information management: searching, browsing, and retrieving.

Browsing is the method whereby a user explores an information space without the use of automated tools. Instead, a user navigates through a space of information nodes by traversing from one node to another node; each

FIGURE 2.1. The result of a search of WWW pages containing the phrase "rural planning."

node could contain information and/or pointers to other information nodes. The effectiveness of browsing depends very much on the organisation of information into nodes and on the description of pointers that provide guidance through the space. Large volumes of data tend to make browsing impractical.

Searching on the other hand can be seen as the automated process, where the user provides some information about the resources being sought and the system provides some appropriate matches. In many cases, effective resource discovery is accomplished by a combination of both methods, and often a multistep procedure alternating between searching and browsing yields the best retrieval results. Information is physically retrieved from a server to the requesting client using a predefined communication protocol (e.g., FTP).

In the case of the WWW, the natural method of retrieving information is by browsing, carried out by following links within HTML pages. Consequently, the successful discovery of relevant information depends on the effective arrangement of the links. However, this method of information discovery is not so suitable when the possible number of resources is very large (as is the case now with the WWW) or if there are time constraints for the retrieval of the information.

Fortunately for the general user, there exists a number of organizations that provide search engines or catalogues of information organized according to subject matter, or organization or by service offered, or customize the query to specific domains (see, e.g., Web sites [6]–[9]).

The result of a query on "rural planning" on the Lycos search engine resulted in over 100 items being identified (Figure 2.1).

2.4 A GIS

In the same way that paper maps convey graphical information by representing features by points, lines, and areas or polygons, a GIS also uses the same convention to store data. Points are discrete locations of objects that are too small to be shown as a line or area at the scale of the maps.

A line feature is a set of ordered coordinates that, when connected, represent a linear feature. A line need not have any width (e.g., an isoline or contour). An area is a closed figure whose boundary encloses a homogeneous area such as a country or a lake. A GIS can also use a grid or raster to display information. In this case, the value of each cell in the grid represents a feature (e.g., elevation from a digital terrain model, land cover category derived from satellite and airborne sensors).

One major advantage of a GIS over conventional maps is the ability to combine and process different themes (e.g., elevation, urban areas, rivers, well locations, etc.). Certain complex spatial operations that would be very difficult, time consuming or impracticable become easy to perform with a GIS. A typical operation may involve locating a house within certain planning restrictions, on a south-facing slope, not more than 200 m from a major road. Such a task would take many hours if done manually using maps. However, a GIS system can identify suitable areas in a matter of minutes. In this context, GIS are used to assist planners by indicating alternatives in the way resources can be developed and the consequences of these actions on the environment.

There are two reasons why this type of operation works. Firstly, data structures within GIS are topologically related. Topology is a mathematical characteristic that allows for the explicit definition of spatial relationships. In practice, topology expresses different types of relationships as lists of features. For example, an area is defined by the lines comprising its border. Each segment of the border is aware of adjacent links and the identity of the polygon that it encompasses. Topology allows data to be stored efficiently which in turn allows faster processing.

Secondly, it is possible to attach knowledge or information to the data by means of attributes. Data can be coded with information about the real world (e.g., road class, river names, and rainfall values). Using GIS tools, it becomes possible to manipulate information not only on the basis of the location of the data but also on its associated information. A typical operation may involve selecting all sampling points that have recorded more than 10 mm of rainfall.

A GIS consists of three basic elements: data management, modelling, and visualization. The first component is undertaken by an associated database system, usually relational in nature but increasingly moving toward object-oriented data models. Once the information has been entered into a GIS, powerful analysis can be undertaken. Such techniques range from map algebra, proximity analysis, predictive assessments, reclassification, and surface generation to areal and volumetric analysis.

Tomlinson (1976) suggested that analytical capabilities should be open ended in order to allow the incorporation of new techniques that reflect changes in the needs of the user.

While users of early GIS might have been content to investigate the relationship between various datasets by means of simple operations such as overlay, buffering, reclassification, and corridor analysis, the current user is probably more concerned with investigating the variance of their data and thus requires increasingly sophisticated functionality (Journel, 1996).

For any study, a crucial element is the presentation of the final results. Data need to be combined with supporting cartographic datasets (e.g., roads, rivers, and political boundaries) to produce a series of graphic files for viewing on the screen or in order to generate hard copy maps.

Alternatively, a series of tables or graphs are required. A GIS provides the user with the tools to generate such information. All types of maps incorporating a variety of symbols and colours can be generated (see the URL [10]). This domain is the main focus of synergy between GIS and the WWW and will be discussed further.

For fuller accounts on the background and fundamentals of GIS, see Burrough (1986), Rhind and Mounsey (1989), Star and Estes (1990), Belward and Valenzuela (1991), or Mather (1993). For a comprehensive online view of GIS and associated developments, the attention of the user is drawn to Web site [11].

2.5 Remote Sensing Systems

2.5.1 Fundamentals

Most of the sensors used to provide information for rural planning record visible and near infrared electromagnetic radiation (EMR) emitted by the Sun and reflected back to the sensor by Earth surface features (Bird, 1991). Some sensors also measure middle infrared EMR, which may be emitted from very hot terrestrial features (such as fires or volcanoes) or may be sunlight reflected from very bright

TABLE 2.1. Regions of the electromagnetic spectrum used in remote sensing

Name	Wavelength range	Sensors (see text for details)
Visible	0.4–0.7 mm	Virtually all "passive" land resource satellites
Near infrared	0.7–1.3 mm	do.
Short-wave infrared	1.5–2.5 mm	Landsat TM, JERS–OPS, Meteosat, airborne sensors
Mid-infrared	3–5 mm	Landsat TM, NOAA AVHRR
Thermal infrared	8–14 mm	Landsat, AVHRR
Microwave	1–20 cm	Radar systems: RADARSAT, ERS, JERS, SEASAT

surfaces (e.g., sun glint from water). Most sensors also make at least one measurement at far infrared wavelengths. These measurements relate to the EMR emitted by the Earth's surface, and may be used to infer thermal properties of the target (Vogt, 1991).

A second important group of sensors providing data for rural planners are the so-called "active remote sensing sensors." These active systems provide their own EMR. The most common are the Radar systems that operate in the microwave regions of the electromagnetic spectrum. Table 2.1 shows the wavelength ranges for various parts of the electromagnetic spectrum most commonly used in remote sensing.

2.5.2 A Brief History

Photographs of the Earth, taken during the United States's Mercury, Gemini, and Apollo manned space missions prompted the U.S. National Atmospheric and Space Administration (NASA) to develop dedicated unmanned Earth resources monitoring satellites. The first of these was the Earth Resources Technology Satellite-1 (ERTS-1), launched in 1972. This carried two sensors: a four-waveband multispectral scanning system (MSS) and three television cameras. ERTS-1 was the first in a series of such satellites, and was renamed Landsat-1 in 1975. Landsat-1 was operational until January 1978 and was joined by Landsat-2 and Landsat-3, launched in 1975 and 1978, respectively. The Landsat MSS programme gave us more than 32 billion km^2 of images of the Earth's surface-each unique in time, though many images covered the same locations.

MSS images cover an area of 185 × 185 km at a spatial resolution of around 79 m. In common with other digital satellite images, those from the MSS are made up of picture elements, called pixels. Pixel dimensions to a large extent determine the spatial resolution of the sensor. Fine-resolution sensors can image small objects on the ground; coarse-resolution sensors can only image larger features, though this is also dependent on other factors.

Forshaw et al. (1983) review the issues behind spatial resolution of sensors. The sensors on all three Landsats provided data in four spectral bands; green, red, and two measurements in the near-infrared. Landsat-3 also had a thermal-infrared band. In 1982, Landsat-4 was launched, and the programme moved into a second phase. Landsat-4 (and subsequently Landsat-5, launched in 1984) carried a modified MSS and a new instrument, the Thematic Mapper which has improved spatial resolution (30 m), and seven spectral bands; blue, green, red, near infrared, two middle infrared, and thermal wavelengths. Unfortunately, the last satellite in the series, Landsat-6 failed to achieve orbit when it was launched in 1993, so more than ten years after launch, we are still using data from Landsat-5. However, Landsat-7 is to be launched in 1999. This will offer 15-m ground resolution. Information on all the Landsat satellites and sample images can be found using the URL [12].

In 1986, the French launched the first SPOT (Système Probatoire de l'Observation de la Terre) satellite. This has been followed by two subsequent satellites, giving a decade of unbroken access to data from the High-Resolution Visible instruments (HRV) that they carry. The SPOT satellites provide high spatial resolution data (10-m panchromatic and 20-m multispectral cover in green, red, and infrared wavebands) and image an area of 60 × 60 km. In addition to the higher spatial resolution offered by these satellites, SPOT represented a landmark in civilian remote sensing as these satellites were the first to provide stereoscopic imagery, thus allowing rural planners to use photogrammetric techniques previously only applicable to air photography. The latest satellite in the series, SPOT-4, which was launched in March 1998, also has on board a low spatial resolution sensor called Vegetation. This instrument was specifically designed to daily monitor land surface parameters on a global basis at a medium spatial resolution of one kilometre. For more information see URL [13].

The rural planner not only needs access to information on the terrestrial environment, but also meteorological conditions (Foot, 1994). Meteorology requires data on a very high temporal frequency. There are two options for satellites to provide such information. Firstly, if satellites are placed in geostationary orbit, at an altitude of 35,800 km above the equator, they have a fixed view of the Earth and can send back images at regular intervals. But no one geostationary satellite can view the whole Earth, and resolution is always poor in the polar regions.

Secondly, satellites carry sensors with swath widths (the area seen by the sensor) of thousands of kilometres, but wide swath widths can only be obtained by reducing spatial resolution to the order of kilometres. Meteorological satellites have a long history. The first experimental satellite was launched in 1960, and now Europe, the United States, Russia, Japan, India, and the People's Republic of China all contribute to the current global meteorological satellite system.

The first polar orbiting meteorological satellites operated from 1960 to July 1966. A third generation of these satellites (NOAA-6 to NOAA-14) is still in use, and the series is guaranteed into the future. The satellites share a similar orbital pattern to Landsat and SPOT, and the instrument they carry (the Advanced Very High Resolution Radiometer, or AVHRR) offers spectral measurements in the red and near infrared wavebands. Though designed for meteorological applications, AVHRR data have been extensively used by nonmeteorologists. Papers in D'Souza et al. (1996) review many AVHRR-based land applications ranging from agricultural production forecasting, continental scale land cover mapping, and locating wildfires through to evapotranspiration and water stress monitoring. For more information see the URLs and associated pointers at Web sites [14] and [15].

The second group of sensors the rural planner must consider are the active systems based on radar. The primary instruments are Synthetic Aperture Radars (SAR). The first civilian SAR-carrying satellite was Seasat-A, launched in 1978. The satellite was designed primarily with global oceanographic monitoring in mind, yet also provided first radar images over land. Unfortunately, the operational mission only lasted 106 days. Nevertheless the quality and uses of the data prompted the world's space agencies to continue the development of civilian radar systems. Thus, in 1991, ERS-1 was launched by the European Space Agency (ESA), and in 1992 Japan launched JERS-1 carrying a SAR and an Optical Sensor (URL [16]) which has a resolution of around 20 m in seven wavebands and, like the SPOT HRV sensors, offers stereoscopic data acquisition in a panchromatic channel.

ERS-1 provides measurements relating to sea state, sea surface winds, ocean circulation, and sea and ice levels, as well as all-weather imaging of ocean, ice, and land and sea surface temperatures. As with Seasat, the imagery has found many terrestrial applications, including the powerful techniques of SAR interferometry. ERS-2, launched in 1995, ensures the continued supply of good quality radar images. For images and information on the ERS system, see URL [17].

Perhaps the potential of SAR data can best be judged by the fact that the first commercial Earth observation venture, Radarsat, launched in 1995, was based on this technology. The SAR here offers many different spatial and temporal configurations in response to customers' requests. For information see URL [18].

Finally, the rural planner should be aware that digital satellite imagery from Russian RESURS satellites is increasingly available. Of particular interest are the archives of historical very high spatial resolution (2 m) data, and important new data from the MSU-SK sensors which offer 170-m resolution images over a swath width of 600 km. See Legg (1990) and URL [19].

2.5.3 Implications for the Rural Planner

It is evident from the above discussion that most satellite sensors acquire data in more than one waveband and are therefore referred to as multispectral systems. The multispectral characteristic of remotely sensed imagery is important because the way in which the Earth's surface features interact with EMR varies with wavelength region. At its simplest EMR from the Sun arriving at the Earth's surface will either be absorbed by the surface it hits, will be transmitted through the target, or will be reflected back from the target.

The relative amounts of energy transmitted, absorbed, or reflected depend on the nature of the target and the wavelength of the EMR. Thus, multispectral images can be used to identify and discriminate between different features of the Earth's surface (e.g., vegetation communities, soil types, water quality). Multispectral measurements of vegetation are often combined into a single parameter as vegetation indices. These are all highly correlated with the interception by vegetation of photosynthetically active radiation, and thus, indirectly, with primary production. The most widely used index to date has been the Normalized Difference Vegetation Index (NDVI), where the NDVI is defined as the difference between near-infrared reflectance and red reflectance divided by the sum of the two. Many variations on this exist (Perry and Lautenschlager, 1984), and new indices are being developed (Pinty and Verstraete, 1992). A more detailed discussion of soil, plant, and water reflectance characteristics can be found in Belward (1991).

In interpreting SAR imagery from ERS-1, ERS-2, RADARSAT, and JERS-1, the rural planner must be aware that the factors controlling the reflectance of EMR at both optical and microwave wavelengths are fundamentally different. For example, at near-infrared wavelengths EMR is interacting with micrometer-scale objects such as cell walls, cell organelles, and intercellular air spaces, yet at microwave wavelengths interactions are typically with leaves, stems, and branches.

However, microwave interactions with Earth surface targets also vary with the precise microwave wavelength used, and the nature of the target (N.B.: microwave measurements are usually expressed in terms of frequency rather than wavelength). Generally, surface roughness, structural variations at the centimetre and decimetre scales, and the electrical properties of the target (particularly the complex dielectric constant which is largely controlled by the moisture content of the target) affect the signal, such that rougher surfaces, surfaces facing the sensor, and increasing moisture content of a target all tend to increase radar backscatter. The complex nature of SAR imagery means that specialist software and sophisticated quantitative analysis methods frequently need to be employed in their interpretation.

Nevertheless, radar offers the rural planner a number of advantages. The wavelengths involved are such that the sensors can "see" through clouds. The images are generally unaffected by changes in atmospheric and weather conditions, and as the satellite provides the EMR, rather than the Sun, imagery can be obtained both day and night. This means that the supply of imagery, and therefore information, can be guaranteed to a greater extent than from optical systems.

In addition, the recently developed techniques of radar interferometry allows the rural planner to construct precise digital elevation models (DEM) from radar images, and the techniques of differential interferometry even allow precise measurements on the order of centimetres to be made. The underlying principle is the same as that governing stereo air photography. An interferogram is obtained by analysing two radar images of the same target obtained from two slightly different positions. Applications other than DEM construction include volcanology (for detection of preseismic swelling), crustal motion, fault slip, glaciology, motion detection, and for land subsidence. One of the first reported applications was the mapping of the Landers earthquake in Los Angeles, on June 28, 1992. Using this technique, surface deformations of some 10 cm were detected (Massonnet et al., 1994, and URL [20]).

Earthquake analysis has also been performed using SPOT data from Japan's Kobe earthquake of January 17, 1995. Postquake images from January 20 were used in disaster relief planning by combining colours added as plots of earthquake intensity based on the Mercalian scale (which takes into account an earthquake's destructive effect on surface features) with infrastructure information such as roads overlaid onto the earthquake damage map in a GIS. The stereoscopic capabilities of the SPOT sensors make these another good source of data for the creation of DEMs (Downman et al., 1987).

Satellites provide rural planners with information on a repeated basis. Depending on the orbital pattern of the satellite, the sensors on board will give repeated images of the same points, or all points on the Earth's surface. This can vary from an image every 30 minutes (of a full Earth disk, but not the whole planet) to complete global coverage once or twice per day. This is vital given that so many processes considered in rural planning are dynamic-soil erosion, or changes in a crop through a single growing season; development as a function of irrigation or nitrate application regime; effects of flooding, or the impact of drought; the risks of wildfires; and even changes in land use and land cover over the years are all examples. Satellites can provide a picture of conditions before, during, and after an event. As the first satellite dedicated to Earth resource monitoring (ERTS-1) was launched as long ago as 1972, rural planners have access to permanent records spanning around 25 years. This multitemporal aspect to satellite remote sensing is a key advantage over the temporally fixed information available from conventional topographic or thematic maps.

Finally, satellite remote sensing offers the rural planner information at many spatial scales (the area covered by a single image and the finest detail that can be seen in that image). The area of an image depends on the scanner geometry and the orbital characteristics; thus geostationary satellites such as Meteosat see half of the Earth as a single scene and polar orbiters like Landsat image a swath. The swath width is variable; the TM and MSS sensors cover 185 km, SPOT's HRVs cover 60 km, and the NOAA's AVHRRs cover 2500 km. Swath width dictates the maximum along-track dimensions of any image. The along-track dimensions are not fixed. These are determined by the forward motion of the satellite, though high-resolution satellite images (such as from SPOT or Landsat) are usually square. As a result, an imaginary grid covering the Earth's surface can be constructed using the orbit ground tracks and the along track image divisions. These grids can then be used to locate satellite images.

Satellite sensors have widely differing spatial resolutions. SPOT panchromatic data at 10-m resolution have been available for a decade, and civilian systems capable of delivering 1–3 m data will be available by the turn of the century (e.g., the Advanced Land Observing Satellite to be launched by Japan in 2002 will offer 2.5-m resolution optical data and L-band SAR). Balancing this, the AVHRR systems have provided daily global images at 1-km resolution for the last 15 years. Thus the rural planner can obtain a series of synoptic views of their area of interest at various scales.

The multispectral, multitemporal, multiscale nature of satellite imagery provide the rural planner with synoptic and repeated views of his or her area of interest, with a means to halt action, with a permanent record, with access to multispectral information that can be translated into geophysical measurements and with access to three-dimensional views of the Earth. These data can be processed to generate thematic maps (showing factors such as land use, land cover, soil types, water resources, etc.), topographic maps (showing drainage networks and relief) and provide a dynamic perspective (showing changes in vegetation status, land use, soil erosion patterns, etc.) both inter- and intraannually. These maps can be integrated directly into GIS, or indeed GIS may be used to help generate them from the remotely sensed data (Wilkinson, 1996).

However, the huge range of sources of data, types of data, processing algorithms, requirements for staff training, dedicated hardware, and customized software can lead the rural planner to suspect that remote sensing is a "closed shop"-specialists only need apply. This understandable view is highly unfortunate. New initiatives in Europe and the United States aim to put the potential beneficiaries of Earth observation in touch with the data and service providers who can help them. In Europe one first point of contact could be the help desk service offered by the Centre for Earth Observation programme. This service can be accessed using URL [21] or by contacting CEO, European Commission, Space Applications Institute, JRC-Ispra, 21020 Varese, Italy (Churchill, 1996).

Other contacts would be the URLs cited above, the national space agencies, or the European Space Agency's guide and directory service (see URL [22]) or by contacting ESA Users services, ESRIN, Via Galileo Galilei, 5, 00044 Frascati, Italy. Non-European users should contact their national space agency, or the Operations, Data, and Information Systems Division, EOS-DIS, NASA HQ, Code YD, 300 E Street SW, Washington D.C. 20546, USA. For information concerning SPOT data availability contact either URL [23] or SPOT Image, 5, rue des Satellites, BP 4359, F 31030 Toulouse cedex, France; or, in the United States, SPOT Image Corp., 1897 Preston White Drive, Reston, VA 22091.

2.6 Data and Information Flow in Land Planning Applications

To demonstrate how an integrated decision support system, based on the three information technologies discussed previously, can be used to answer specific land planning applications, we will present a generic case study on how one might utilize a set of software tools for planning and assessing the risks of fire in Mediterranean environments.

This section will outline the general objectives of such an information system, emphasizing the generation and limitations of suitable datasets and the processing required. Although the following discussion is not based on a single case study, the example draws heavily from a range of projects with which the authors have been involved. By using a demonstration scenario, many more elements can be covered than one would normally find in a true-life case study.

The role of the WWW in locating datasets has been discussed and a typical response is shown in Figure 2.1. The role of the WWW in disseminating spatial data and the mechanisms by which users can interact with GIS through the WWW is discussed in Section 2.6. Consequently, this part of the chapter will focus on the integration of remotely sensed data and GIS.

2.6.1 Introduction to the Case Study

The role and the effect of fire in environmental issues are the subject of much research and discussion (Crutzen and Goldammer, 1993; Chuvieco, 1995). On a global scale, fire is being used to radically transform natural landscapes to agricultural areas with potentially severe impacts on global atmospheric chemistry. On a regional level, fire can be a regulatory influence on the functioning of ecosystems by controlling vegetation composition, biodiversity, and soil erosion. From a human perspective, fire is perceived as a local threat only when bush fires threaten to destroy property and people.

The risk of fires is, to a degree, climate dependent. The Mediterranean region, where high summer temperatures and low rainfall lead to high levels of vegetation stress, is an area where the early identification of risk, the monitoring of ground and meteorological conditions, and the provision of effective fire-fighting tools is crucial to manage the effects of fire on the environment.

An integrated information system could be used to help the user to: (i) to identify areas most at risk from fire (preventive actions); (ii) to develop a management model for combating fire (detection and control actions); and (iii) to plan and manage postfire situations (recovery and impact). Because a GIS handles large, multilayered, and heterogeneous databases and can query the existence, location, and properties of a wide range of spatial objects in an interactive way, it is the obvious tool for managing and displaying the information that is required for fire risk assessment and management.

The integration of remote sensing and GIS is helpful in all the above stages, particularly for fire risk assessment where satellite and airborne sensors can be used to monitor and map the surface characteristics that control the probability of combustion.

Such a system offers planners the possibility to make strategic decisions on how the environment should be managed to reduce the risk of fire, in addition to tactical decisions to fight

Table 2.2. Key datasets required for fire risk assessment and the role of remote sensing in generating this information

Feature	Element	Source
Relief	Altitude/elevation Slope/gradient aspect	Aerial photography SPOT stereo SAE interferometry
Infrastructure	Roads Railways Footpaths	SPOT HRV panchromatic Aerial photography RESURS
Hydrological resources	Rivers drainage network stability of flow Lakes and reservoirs extents/volume water quality Soil moisture Aquifers	Radar systems SPOT HRV multispectral Landsat TM
Vegetation condition	Type Biomass Condition moisture content senescent Evapotranspiration Fuel loading	SPOT HRV multispectral Landsat TM AVHRR
Land use		Colour aerial photography
Past fire history	Burn patterns Regrowth	SPOT HRV multispectral Landsat TM AVHRR
Meterology	Current synopsis wind direction rainfall temperature Past situation Forecasts	AVHRR Meteosat
Fire event	Location monitoring	AVHRR ATSR

and control fires by: (i) acting as an enabling tool to bring disparate data sources together; (ii) undertaking data processing and modelling; and (iii) generating cartographically orientated output.

2.6.2 Managing Fire in the Rural Environment

Fire risk maps provide the basic information for fire management by illustrating the spatial distribution of possible risk. Such maps can be generated at a variety of scales. In the broad sense, such maps have been used to quantify the overall effects of fire on the landscape (in explaining why certain areas burn) while for the regional level, risk maps can be used to implement a strategic warning system and at a local level, to manage the allocation of resources to control individual burning events or identify possible causes (e.g., proximity to roads).

The estimation of fire risk involves calculating the probability of ignition, the behaviour of fire in a given ecosystem and the effects of the burning on that ecosystem. To derive these parameters, the planner needs access to datasets that describe vegetation type and condition, weather, terrain, fire occurrence, and human activity, and a suitable model that provides a reasonably accurate indication of risk (Chuvieco, 1995). Table 2.2 highlights the various parameters used in fire risk models and identifies the role of remote sensing in providing such detail.

It is evident from the above table that remote sensing provides the user with most of the variable information required by the fire model. Vegetation type and associated availability of fuel can be mapped in detail by several sensors. The condition of that fuel and the water stress can be assessed by various vegetation indices and through the use of thermal imagery (D'Souza et al., 1996). Current research suggests that soil moisture can be estimated by the use of radar systems and that the meteorological conditions before, during, and after fires can be monitored on a routine basis.

By combining all these elements into a GIS, the planner can begin to develop and implement the necessary analytical components that are required for the decision-making process. Additional information such as urban areas, population patterns, the location of wells, fire breaks, and anthropogenic activities can be included and integrated with the remotely sensed information and used by a particular model. However, any GIS-based analysis will only be as effective as the quality of the model being implemented as there is a degree of uncertainty in the accuracy and reliability of models being used by various organizations (Vasconcelos, 1995). As in the case of any model, it is impossible to take into account all the variability of the real world.

However, it should be noted that most of the information derived from remotely sensed imagery has been inferred and a degree of caution has to be borne in mind when using these data in actual models. Calibrating a biomass estimation derived from remotely sensed imagery against actual values in the field is a considerable exercise. Even though there is a considerable volume of work on the detection of vegetation structure (especially, the understory of woodlands) and vegetation water content, there seems to be disagreement on the quality of results and, hence, the quality of any fuel map is restricted

The user's attention should be drawn to the issues concerning the preprocessing of imagery prior to use (e.g., radiometric, atmospheric, and geometric correction) and the consequences these actions can have on the information content of remotely sensed datasets. This topic falls outside of the scope of this chapter but comprehensive details can be found in Ryerson (1996).

2.6.3 Uncertainties in Processing Spatially Organized Datasets for Rural Planning

One of the most common problems associated with GIS-based projects is the lack of attention to the limitations and the errors associated with the datasets being used. The user should never assume that a dataset, whether it is derived "in-house" or is purchased from a value-added company, is completely accurate. Unfortunately, not enough attention is given to the question of accuracy of digital data. The following paragraphs outline some of the problems with handling spatial data in a GIS.

The traditional advantages and disadvantages between raster and vector data are well documented (Kennedy and Meyers, 1977), but these issues are less important in current GIS implementations., The user should be wary, however when merging information based on differing data models because the representations of reality are often very different.

An assessment of a land cover map derived from SPOT HRV multispectral data (20-m pixels) against a digitized vectorial field by field survey showing footpaths and narrow lines of trees will give rise to spurious correlations between the two. A further example of this problem is the contrasting nature of information given by a DEM (an arithmetic interpolation of elevation for a given area) and an isoline map (a relatively precise contour between heights). Their usage should reflect such considerations.

Scale and resolution are other areas that can lead to error. As a result of the digital nature of the data within the GIS, scale can be regarded as infinite. There is no technological reason why the coastline of Europe cannot be digitized from a 1:10,000,000 source map and redisplayed or plotted at a much larger scale. However, problems will certainly arise if a more precise dataset is integrated with this larger-scale product.

Merging a dataset of coastal urban areas, digitized at 1:50,000, with a coastline of comparable scale but derived from the much coarser product, will probably result in towns being placed in the sea or too far inland. A compounding factor will be the generalization inherent in the smaller-scale product. The resampling of raster-based datasets can lead to equal problems.

One of the most common operations undertaken by users of spatial data is to transform raw information (e.g., imagery, digitizing, and field survey) to suitable map projection. This is known as Geocoding or Geo-referencing. Associated techniques modify the geometric relationship of the data into some other arrangement without substantively changing the contents of the data (Star and Estes, 1990).

The choice of map projection is an important area where the user has to consider the reason for holding the data. Latitude and longitude is a common reference system for recording international coordinates. However, the use of such an angular system does not allow the calculation of length and area.

This chapter has illustrated throughout how remotely sensed data can be processed by various means to provide thematic indications on the state of the Earth's surface. As in all digital datasets these products have a degree of uncertainty attached to them. For the various indices that have been presented, the majority assume a degree of linear relationship between reflectance and the surface properties being investigated.

However, this is not always the case. For example, the NDVI becomes unreliable where vegetation cover is sparse. The other main method for processing remotely sensed data is the use of various multivariate classification routines (Wilkinson, 1996). Image processing systems are very poor at providing the user with the facilities to assess the statistical structure of their data and to assess the integrity of the information being used to seed the classification. To this day, no one has managed to classify any remotely sensed image to 100 percent accuracy, because of the effects of boundary pixels, unknown categories which have been excluded from the list of target classes, as well as physiological variations in known land cover which will not fit spectral signatures of the "training area" (e.g., stress, shade, soil moisture conditions, etc.).

Users should not forget differences caused by the "snapshot" nature of many remotely sensed based projects. Phenological changes in vegetation cover can result in widely differing classification results and accuracy. The deciduous nature of Larch trees makes it easier to separate them from other conifers on winter rather than summer imagery.

In any information system, facilities must be provided to detect and correct errors in the data. This section is not a comprehensive overview of the errors associated with handling spatial data. For example, no mention has been made of the problems of data entry (from keyboards or digitizing tables), encoding or in building topology.

Some GIS systems provide the means to log the processing chain associated with a particular dataset so that it is possible to derive the associated accuracy. However, such functionality is only the bare minimum that is needed. GIS manufacturers have to go a long way to provide the user with comprehensive error trail and auditing facilities.

2.7 Accessing Mapped Information: The Internet and GIS Data

The previous section gave an example of how of GIS and remotely sensed data may be used together. An important next step in any rural planning system is the dissemination of information to potential users. The possible outputs from a GIS-based information system range from tabulations of information and statistics to the more graphic elements such as predictive and thematic maps. As the paper has already discussed, the hypertext model supported by the WWW easily provides the mechanisms to facilitate access to a range of data formats. In addition, there are a number of software products that allow the translation of text and common word processing formats to HTML (e.g., the HTML Assistant which can be found at [24]). However, the transfer of map-based data via the WWW, directly from the various GIS that are currently on the market, can be a little more involved. The following section outlines some of the current solutions that might be of interest to the rural planner.

2.7.1 Graphic Snapshots

Sometimes it is useful for a user to know that certain WWW servers publish spatial data in a graphic form which does not need further processing or the use of additional tools to be viewed. These static snapshots of pregenerated maps and images come in formats such as GIF and TIFF, which can be viewed by most WWW clients both as stand-alone images or embedded in HTML pages. Of particular interest here is the GeoTIFF [GEOTIFF] format which is emerging from a joint effort between a large number of GIS vendors and data. The advantages of producing static graphic snapshots are that these data become ready for widespread and easy access, and doing so is quite easy to implement. Any developer who knows how to convert GIS data to a widespread graphics format and who has elementary HTML knowledge is capable of setting up a site containing this kind of information. The disadvantages are that, apart from the GeoTIFF format, data in graphic format are of limited use for further elaboration; also the limited resolution obtained with such formats often pose a problem. Typical servers that carry useful information, updated in a timely fashion and useful for a wide public, including rural planners, are the many WWW weather information servers which produce and publish on a regular basis various static maps representing weather parameters in a number of different ways and at various scales. Often these maps are embedded in HTML documents that add textual information and pointers to the graphic contents of the maps.

2.7.2 Data in Native GIS Format

Through the WWW browser, a user could be given access to GIS data published in native format. Retrieving such data allows knowledgeable users to really utilize them using the appropriate tools and to integrate them with other data. A typical example is the provision of GIS ARC/INFO export files or ArcView shapefiles. The export format guarantees portability across platforms. Most WWW browsers can be configured in such a way that on retrieval of a data file with an agreed extension, automatically the right application is launched, which subsequently ingests and elaborates the data.

2.7.3 Spatial Database Catalogues

The key to finding relevant data in the WWW is to have the right tools to access this information space and to have adequate organization of the information for various levels of abstraction by means of lists, catalogues, and indices. Applied to spatial data, one could have at the lowest level, lists of datasets, each set described by a single line of text, augmented with pointers to a more extensive description of the data, derived (value-added) data such as images (e.g., graphic previews), and plots (e.g., time series). Locations of other related data, applications and databases could be provided. Such lists could be browsed (in an efficient manner if an adequate organization of the data is provided) or searched (provided that a suitable search engine or indexing mechanism is in place). An example of an ambitious project that is partly concerned with solving problems of this kind is the Alexandria Digital Library [25].

2.7.4 Map Generators

The integration of GIS and WWW could also be interpreted as a GIS computation component where a GIS is actively involved as part of the client-request/server-response mechanism, that is, where a user would have on its desktop a WWW embedded facility which provides GIS functionality to local or remote datasets.

There are many reasons why it could be worthwhile to integrate a GIS with the WWW environment:

(i) GIS computation power residing at the server side can be brought to the user without the need of buying expensive GIS software and associated hardware.

(ii) The WWW environment is a multimedia environment, allowing easy integration of many types of information, and one of its most essential features is the possibility to interlink documents in a variety of ways.

(iii) There is a cheap, (almost) device-independent standard user interface environment (e.g., the Netscape WWW browser is almost free and works on UNIX, Windows, and Mac platforms).

(iv) A GIS development integrated with the WWW has the potential to reach many people for sharing data and applications.

Alternatively, it can be expected from the outset that response times for users could be high due to the competition for limited computation resources at a limited number of servers, and also due to the potential high volume of data to be transmitted.

Currently, the possibilities within standard HTML and HTTP servers that allow a developer to integrate any application, including GIS components within the WWW environment in a client/server way are rather limited. At the client side, an HTML page can be filled with FORMs and Imagemaps to solicit input from the user.

Input from the user is routed through the server to the CGI-BIN, which allows the calling of external programs or scripts that elaborate an output then routed back to the client. Typically, the result of such an external application is an HTML document (e.g., containing an annotated map)

Programs, or scripts as they are called, that accept FORM input have been written in various languages: sh, csh, Perl, C, C++. Perl is a favourite among developers because of its strength in string handling during user input and formatting for output. In the case that a GIS computational component is required, the script should find a way to call external GIS routines. For example, by using a UNIX script, one could launch ARC, specify on the ARC command line a developer's AML routine, and pass to it the desired parameters. This process is illustrated in Figure 2.2.

2.7.4.1 An Example of a Map Generator

An example of a sophisticated WWW interface to select themes and generate customized maps using ARC/INFO is given in Figure 2.3. By means of a FORM, the user is given the possibility to select a number of themes that will be included on a map.

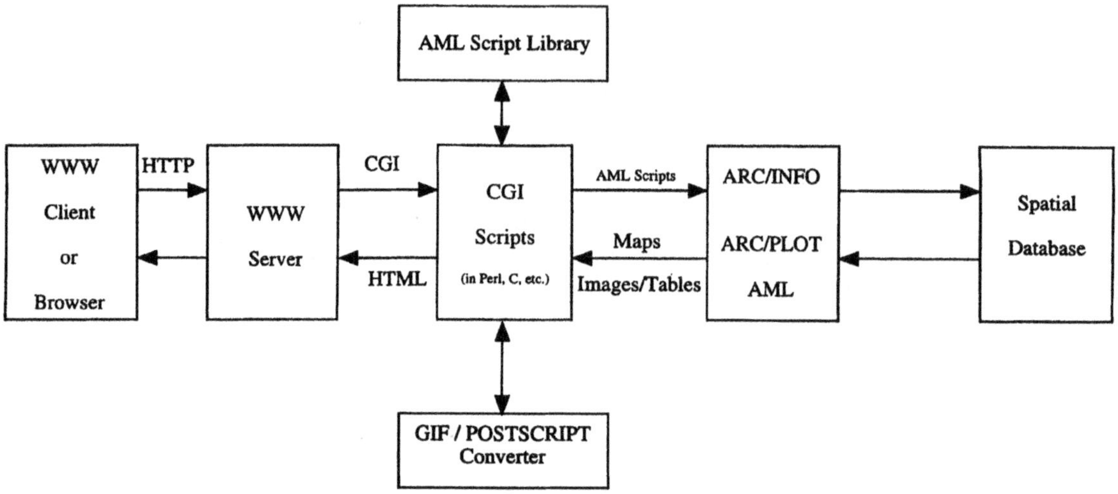

FIGURE 2.2. A mechanism for implementing GIS functionality within a WWW browser.

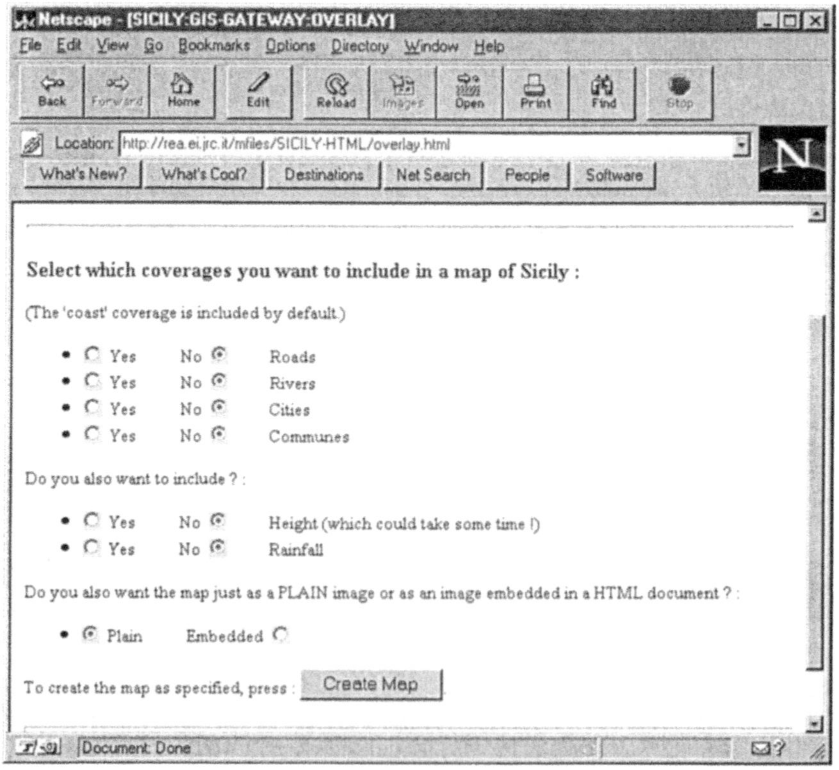

FIGURE 2.3. A FORMS interface that allows a user to select specific themes in order to produce a customized map.

On selection of the relevant parameters, the submit button launches a script, which in turn calls an ARC/INFO AML routine, which gets the desired data, feeds them to plot program, an image conversion program, and embeds the resulting image in an HTML document, which is routed back to the client. The user is presented with an appropriate image (Figure 2.4) (see [26]).

The advantage of such a system is that a user is offered GIS functionality to a degree of sophistication determined by the complexity of the HTML-based interface, which, taking into account the recent developments in the WWW world, is increasing.

The disadvantages from the user point of view are that there is no access to native data, access is potentially slow due to poor communication and slow server processing times, and there is no true interactivity with the data (as compared to, e.g., directly operating a GIS system which allows quick zooming, panning, feature finding, etc.). Solutions to all these problems are now potentially possible due to developments such as Java. Briefly, a Java applet is a piece of code that is embedded into an HTML page and implements a program that can run at the client side and with a graphical user interface embedded in the HTML page itself. The complexity of the embedded Java program is unlimited and, theoretically, one could envisage that all (GIS) computation could be done at the client side and not at the server, which implies that computation power would scale up with the number of clients wanting to generate maps on the fly.

2.7.5 Real Time Map Browsers

Many problems associated with real-time generation of maps by servers stem from the fact

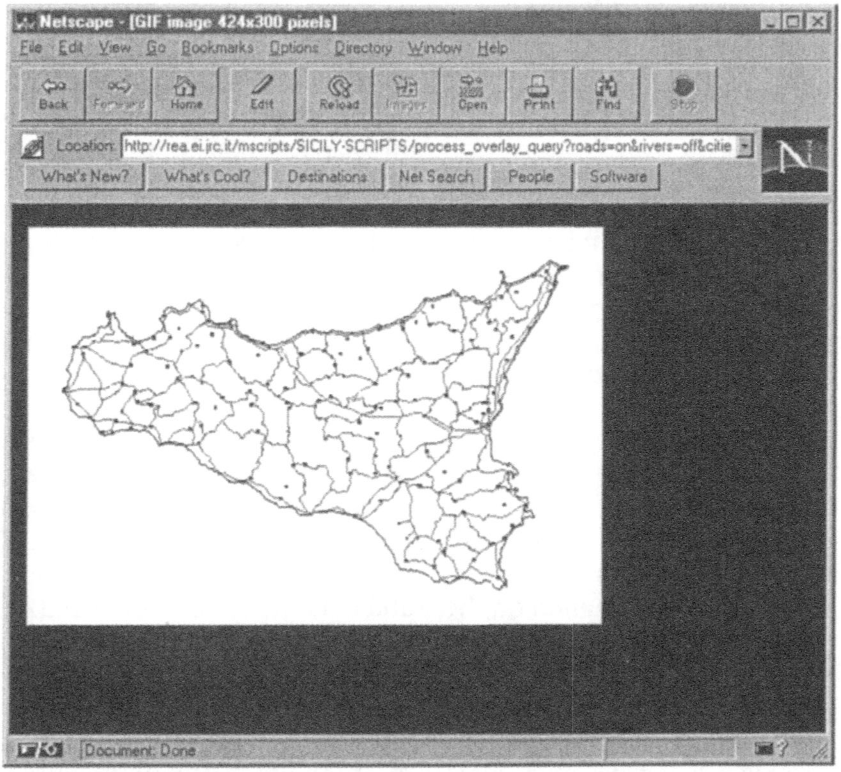

FIGURE 2.4. The result of generating a map in real-time from the interface shown in Figure 2.3.

that the programs that generate the maps are not built to operate in a client/server mode and are not really designed for fast performance.

One of the leading GIS providers, ESRI, has realized that, to meet critical demands, GIS computation should shift to the client/server paradigm and design at the server side should be such that map access to very large spatial databases is guaranteed for quick browsing, displaying, querying, and retrieving. This requires a powerful spatial data server.

One such example is the ORACLE-based Spatial Data Engine (SDE) from ESRI which performs spatial query and with subsecond response times for millions of features and hundreds of users. At the client side, a range of software products can be used to generate the requests (Netscape Navigator, ArcView, Java applications, C++, etc.) (see [27] for further details).

2.8 Conclusions and Future Developments in IS Technology: Their Effect on Land Planning Applications

In summary, an integrated information system holds tremendous benefits for the rural planner. The WWW is a means of intelligently linking various information and data sources, potentially searching for and retrieving information and providing user-friendly access to the final product.

GIS can be viewed as the integrating environment for the various spatial datasets and in providing the analytical and predictive elements required to manage the environment. Remote sensing is the crucial, unique and current provider of a range of information on the state of the environment.

The three information systems discussed in this chapter are all developing rapidly and there is always the danger that the technology described by texts such as this becomes dated very quickly. Users of remotely sensed data will see higher spatial resolution images, hyperspectral information, and improved data access within the next decade, while users of the Internet and the WWW will undoubtedly benefit from higher bandwidths, lower telecommunications costs, and enhanced functionality being offered as a result of novel client/server developments such as Java. Whilst this undoubtedly places increasing demands on users in terms of computing requirements and algorithm development, the rural planner should take comfort in the fact that these are also rapidly advancing areas; new users should be encouraged to take advantage of the unprecedented possibilities that these technologies provide.

However, as this chapter has demonstrated, there are a number of lessons to be learned. In particular, a user has to be aware constantly of actual and potential errors and validation issues. The ease by which existing and new datasets can be merged, integrated, or processed to a higher level often hides the cumulative uncertainties that accompany each step in the processing chain. To overcome these problems, users of spatial datasets and processing systems such as GIS need to have rigorous statistical understanding of their data and the techniques that they wish to employ. All new datasets need some degree of qualitative and quantitative validation to assure other users of their integrity.

The incorrect use of scale in digital maps can lead to spurious errors if data are taken beyond the limit for which they were created. The user must understand the errors and limitations inherent in all digital data. In the same context, sensor capability must be used responsibly. For example, there is no use trying to derive meaningful linear information from AVHRR or information on mineral composition from SPOT HRV data. The physical properties of the surface condition in question and radiometric resolution and characteristics of the sensor must be matchable. In the same light, an increase in the spatial resolution of imagery is not always beneficial for people attempting to derive land cover classifications in an automated manner as the number of boundary pixels, a major source of error, will increase dramatically.

It is worth noting, however, that a number of commercial ventures, based on "small satellite" technology have been proposed. These satellites run only a limited number of sensors, but constellations of them are placed in orbit. This means that such systems can offer very high temporal coverage. Most also offer very high spatial resolution. Space platforms offering 1-m resolution imagery, formerly very much the province of military observation satellites, will soon be available to the rural planner, leading to dramatic increases in satellite-based cartography and inputs to GIS.

In order to ease the transfer of information and data between users, there are a number of interesting studies being undertaken at the present which are concerned with the possible implementation of open technology between GIS producers and users. In particular, the European Umbrella Organization for Geographical Information (EUROGI) and the GISDATA initiative of the European Science Foundation Social Science Programme. The standardization of data structures, terminology, and operability between the multitude of GIS vendors can only be good news to the user.

As demonstrated above, the integration of GIS technology with remotely sensed datasets is an effective approach in the generation of maps associated with fire risk, propagation, and recovery. GIS is also useful for assisting in the classification of remotely sensed data.

However, a major limitation in the advancement of GIS for many environmental applications is the limited development of analytical and modelling capability. This includes tools for understanding the statistical and quality issues of datasets (e.g., the accuracy of the data, the variance of datasets, the validity of sampling schemes, the possible error in location, and the suitability and validity of a processing technique). The importance of geo-statistics within commercial GIS systems cannot be overstressed. Interested parties are directed to the increasingly relevant and lively discussions of the role of geostatistics at [28].

Finally, the developments of network-based technology and, in particular, the World Wide Web has truly revolutionized the way in which people can work. The publication and updating of information can now be undertaken on both a personal level and in real time. This has tremendous implications for those interested in exchanging or searching for specific knowledge. Functionality and the range of services being offered on the Web are increasing all the time.

The next book in this series might well be a multimedia affair, accessible only in digital form over the network or on a CD-ROM. In order to accommodate these uses, a number of fundamental technological changes have to take place. These include lower costs for telecommunications, improved networking infrastructure within countries (already happening is some areas), and electronic commerce.

List of Web Sites

[1] altavista.software.digital.com/inethistory/nfintro.htm
[2] www.yahoo.com/Computers_and_Internet/Internet/World_Wide_Web/
[3] www.yahoo.com/Computers_and_Internet/Software/Data_Formats/HTML/
[4] www.yahoo.com/Computers_and_Internet/Internet/World_Wide_Web
[5] home.netscape.com
[6] altavista.digital.com
[7] guide.infoseek.com
[8] a2z.lycos.com
[9] www.yahoo.com
[10] www.esri.com
[11] www.geo.ed.ac.uk/home/giswww.html
[12] geo.arc.nasa.gov/esd/esdstaff/landsat/lpchron.html
[13] www.spotimage.fr
[14] daac.gsfc.nasa.gov
[15] www.esrin.esa.it
[16] www.eoc.nasda.go.jp
[17] gds.esrin.esa.it
[18] www.ccrs.nrcan.gc.ca/ccrs/radarsat/rsate.html
[19] www.ssc.se/ssc/sb/resurs.html
[20] gds.esrin.esa.it:80
[21] ewse.ceo.org
[22] gds.esrin.esa.it
[23] www.spotimage.fr

[24] ftp.cs.dal.ca/htmlasst/htmlafaq.html
[25] alexandria.sdc.ucsb.edu:3366/ADL/index.html
[26] ultra.ei.jrc.it/projects/sicily
[27] www.esri.com
[28] java.ei.jrc.it/rem/gregoire

References

Bird, A.C. Principles of remote sensing. Belward, A.; Valenzuela, C., eds. *Remote Sensing and Geographical Information Systems for Resource Management in Developing Countries.* Dordrecht: Kluwer Academic; 1991:1–30.

Belward, A. Spectral characteristics of vegetation, soil and water. Belward, A.; Valenzuela, C., eds. *Remote Sensing and Geographical Information Systems for Resource Management in Developing Countries.* Dordrecht: Kluwer Academic; 1991:31–54.

Belward, A.; Valenzuela, C., eds. *Remote Sensing and Geographical Information Systems for Resource Management in Developing Countries.* Dordrecht: Kluwer Academic; 1991.

Burrough, P. *Principles of Geographic Information System for Land Resource Assessment.* Oxford: Clarendon Press; 1986.

Churchill, P.N. *The CEO Programme.* Special Publication SPI 96.28. Varese, Italy: European Commission, Space Applications Institute, JRC-Ispra; 1996.

Chuvieco, E., ed. *Remote Sensing and GIS Applications to Forest Fire Management.* Proceedings of the Workshop; 1995 September 7-9; University of Alcalá de Henares, Spain. Alcalá de Henares: ERSEL; 1995.

Crutzen, P.; Goldammer, J., eds. *The Ecological, Atmospheric, and Climatic Importance of Vegetation Fires.* Report of the Dahlem Workshop on Fire in the Environment. Chichester: John Wiley and Sons; 1993.

Downman, I.; Gugan, D.J.; Muller, J-P.; Peacegood, G. The use of SPOT data for mapping and DEM production. CNES, ed. *SPOT-1 Image Utilisation, Assessment and Results.* Toulouse: Cepadeus Editions; 1987:1213–1220.

D'Souza, G.; Belward, A.; Malingreau, J-P., eds. *Advances in the Use of AVHRR Data for Land Applications.* Dordrecht: Kluwer Academic; 1996.

Foot, J.S. Hazard warning in meteorology; the importance of remote sensing. Wadge, G., ed. *Proceedings of the Meeting on Natural Hazard Assessment and Mitigation: The Unique Role of Remote Sensing;* 1994 March 8–9; the Royal Society, London. London: Royal Society; 1994: 13–16.

Forshaw, M.; Haskell, A.; Miller, P.; Stanley, D.; Townshend, J.R.G. Spatial resolution of remotely sensed imagery. A review paper. *Int. J. Remote Sens.* 4:497-520; 1983.

GEOTIFF. www-mipl.jpl.nasa.gov/cartlab/geotiff/ geotiff.html.

Journel, A.G. Modelling uncertainty and spatial dependence: Stochastic imaging. *Int. J. Geograf. Inf. Syst.* 10:517–522; 1996.

Kennedy, M.; Meyers, S. *Spatial Information Systems: Introduction.* Louisville, KY: Urban Studies Centre; 1977.

Legg, C. Digital satellite imagery from the Soviet Union. *Proceedings of the 16 Annual Conference of the Remote Sensing Society;* 1990 September 19–21; Remote Sensing Society, Swansea, UK. Nottigham: Remote Sensing Society; 1990:65-74.

Massonnett, D.; Rossi, M.; Carmona, C.; Adragna, F.; Peltzer, G.; Feigl, K.; Rabaute, T. The displacement field of the Landers earthquake mapped by radar interferometry. *Nature* 364:138–142; 1994.

Mather, P.M., ed. *Geographic Information Handling: Research and Applications.* Chichester: John Wiley and Sons; 1993.

Perry, C.R.; Lautenschlager, L.F. Functional equivalence of spectral vegetation indices. *Remote Sens. Environ.* 14: 169–182; 1984.

Pinty, B.; Verstraete, M. GEMI: A nonlinear index to monitor global vegetation from satellites. *Vegetatio* 101:15–20; 1992.

Rhind, D.; Mounsey, H. *Understanding GIS.* London: Taylor & Francis; 1989.

Ryerson, R., ed. *Manual of Remote Sensing.* 3rd ed. Falls Church, VA: American Society of Photogrammetry; 1996.

Star, J.; Estes, J. *Geographic Information Systems: An Introduction.* Englewood Cliffs, NJ: Prentice-Hall; 1990.

TIME. *Time Magazine International,* 30 (July 25th):44–50; 1994.

Tomlinson, R.F. *Computer Handling of Geographical Data.* Paris: UNESCO; 1976.

Valenzuela, C. Basic principles of Geographic Information Systems. Belward, A.; Valenzuela, C., eds. *Remote Sensing and Geographical Information Systems for Resource Management in Developing Countries.* Dordrecht: Kluwer Academic; 1991:279–295.

Vasconcelos, M.J. Integration of Remote Sensing and Geographic Information System for fire risk management. Chuvieco, E., ed. *Remote Sensing and GIS Applications to Forest Fire Management.* Proceedings of the Workshop; 1995 September 7–9; University of Alcalá de Henares, Spain. Alcalá de Henares: ERSEL; 1995:129–147.

Vogt, J. Land surface temperature retrieval from NOAA AVHRR data. Belward, A.; Valenzuela, C., eds. *Remote Sensing and Geographical Information Systems for Resource Management in Developing Countries.* Dordrecht: Kluwer Academic; 1991:125–152.

Wilkinson, G.G. A review of current issues in the integration of GIS and remote sensing data. *Int. J. Geogr. Inf. Sys.* 10:85–101; 1996.

3

Role of Land Surface Relief in Land Use Allocation

Francisco Pellicer Corellano

3.1 Introduction

Human communities develop their activities in an environment or system that is the result of the interaction of physical, biological, and anthropological elements throughout history in the same geographical space.

The physical environment is merely a part of the total global system, but because it can be divided into separate elements, each with their own identities, it can be considered in itself within the process of environmental planning.

Planning a physical environment involves directing all actions toward ensuring the quality of all life systems, especially in those surroundings where humans are both the dominant species and the protagonist. It is precisely this principle that implies:

(i) the need to predict and correct both the undesired repercussions of human activity on the environment (negative impacts) and of the environment on human activities (natural hazards);

(ii) the study of the aptitudes and limitations of the territory for different uses; and

(iii) the research into potential resources.

In this chapter, geomorphological aspects of the land are considered, including their characteristics, capacities, aptitudes, natural hazards, and impacts provoked by human action.

Consideration of ecosystems from the geomorphological point of view is necessarily synthesized and complex. It is expressive of the characteristics of the physical environments and is useful for planning. Today, nobody doubts that the dynamics of the land surface, despite great technical possibilities and the capacity of humankind to transform relief, conditions the use made of each part of it.

3.2 Geomorphological Resources and Functions of the Rural Environment

Traditional rural activities are closely linked to relief. Features of relief, such as altitude, slope, orientation, resistance of constituent materials, and so on, define the main patterns of spatial distribution of energy and associated natural forces in the local and regional areas.

The rural environment is a system whose constituent elements express a physiognomy, structures, and various functions, which are strongly linked with one another by reciprocal causality relationships, in which relief always plays a significant role. In the Mediiterranean rural landscape, relief, considering all possible exceptions, still determines the distribution of many of the land uses, whether they be traditional or new.

Geomorphological organization:

(i) governs the topography which, in turn, determines the slopes, land division, exposure, and even certain features of the climate;

(ii) governs the relief shaping processes;

(iii) conditions the distribution of soils according to the rock outcrops and surface formations; and

(iv) directs hydrology, in the configuration of basins, river beds and flood plains and in the establishment of groundwater aquifers.

Sites chosen for small towns and villages as well as the location of traditional buildings (farmsteads, mills, etc.) depended largely on the conditions of geographical stability, exposure, and slope in order to obtain the best results. Furthermore, relief has often been a deciding factor when selecting a site for defence purposes.

In the case of new urban development for second homes and housing estates in areas near urban agglomerations, relatively abrupt relief means an increase in the costs of the urban infrastructure (supply networks, access routes, etc.) as well as higher transport costs. This effect is often compensated for by open views, closeness to nature, and a lower cost of land.

The current deep transformation of rural landscapes means that relief, as an outstanding element of the system, plays a different role from the traditional one. Today, apart from the traditional function of producing food and other materials (wood, fibres for textiles, etc.) the rural landscape must also provide open spaces, breaks between urban agglomerations, and a healthy environment (Gómez Orea, 1992).

3.3 Approaches to Geomorphology

Land surface relief results from the interaction of internal forces acting upon the lithosphere and external forces causing hydrological and atmospheric processes (Figure 3.1). Biological and cultural processes develop in this plane of contact between the solid Earth crust and the two fluid media surrounding it. These processes, in turn, contribute to the shaping of relief details.

Therefore, in the genesis of relief, two energy systems intervene—solar energy and geoenergy-in constant interaction of the great Earth spheres—lithosphere, hydrosphere, atmosphere, biosphere, and anthroposphere. Relief is the state, at a given moment, of the dynamic relationship of facts, forces, and processes found in the lithosphere—geological and tectonic structures—to those which take place in the external envelopments-shaping processes (Muñoz, 1992).

The lithospheric structure and dynamism are merely one component of relief. Rock outcrops, located according to the tectonically determined structures, are shaped through processes linked to the external forces. There are several shaping agents, including wind, temperature changes, sea, rivers or glaciers, activity of living organisms, and, in particular, humans. Active morphogenetic processes are extremely numerous, have infinite combinations, and reach high levels of complexity. They are articulated in morphogenetic systems according to environmental conditions in which the geological aspects—structure and tectonics—are just as important as the bioclimatic, hydrological, and human aspects (Muñoz, 1992).

Topographic contrasts are usually very marked. The available soils and vegetation have become poorer, since the impact of man on these spaces over thousands of years has made them more fragile. Hardly developed, subdesert-type soils are often found, due to the existence of unfavourable lithologies and arid conditions with scarce organic supply from the existing plant species. The great plant masses, on better developed soils, are found mainly in mountain areas with higher rainfall, in spite of the problem presented by steep slopes and constant human impact. As a consequence of these contrasts in the physical environment, the resulting landscapes, more or less humanized, present variable potential uses. But distribution of resources tends to set the regions off balance (Peña and Pellicer, 1991).

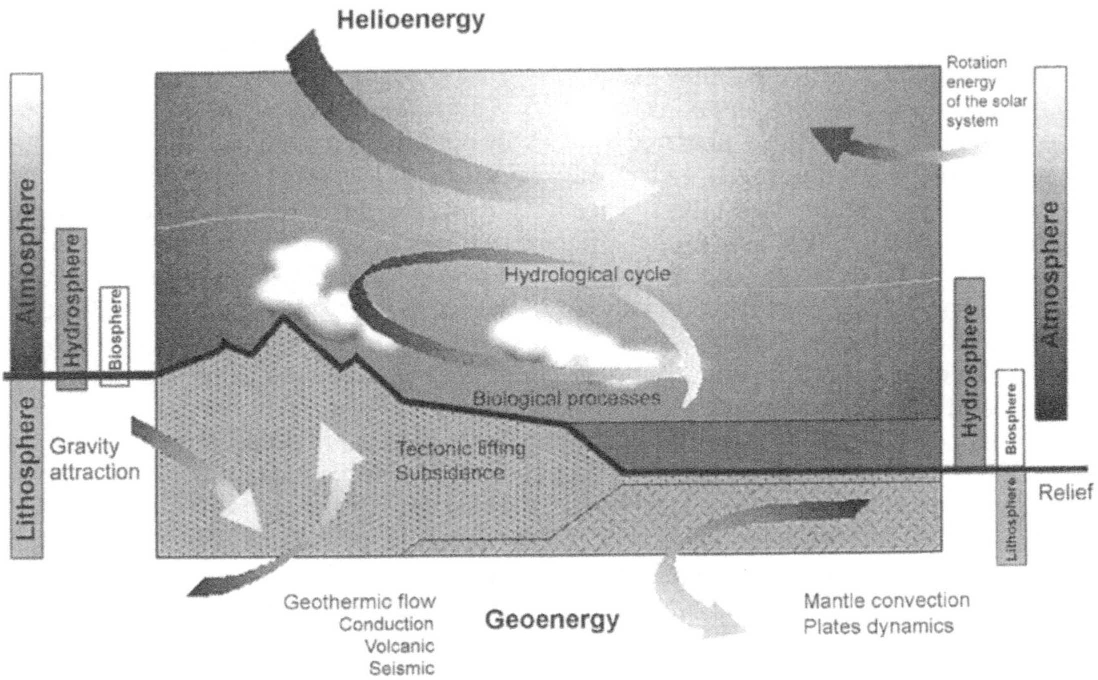

FIGURE 3.1. Relief is shaped by the internal and external forces that operate on the Earth's surface.

In the Mediterranean region, the climate is capable, in itself, of generating erosive processes. Furthermore, the actions of humans rapidly increases the erosion processes to irreversible levels, causing desertification. The degradation dynamics are greater in those areas where semi-arid climate and not very resistant lithologies (clay, loam, and gypsum) coincide with high population pressure. The LUCDEME Project (Fight Against Desertization in the Mediterranean) is developing an important research task in this respect.

The human impact on the plant cover over centuries to obtain areas suitable for crops, abusive felling and grazing, mining and public works have all accelerated over recent decades due to the mechanization of these activities and to the lack of due care to conserve the traditional walls in fields, tracks, and other adoptive structures of the rural landscape. The proliferation of forest fires and the advance of irrigated areas have led the environment to be increasingly vulnerable to the attack of erosion. Nor should it be forgotten that ecological degradation results in a loss of the productive capacity of the soil in the short run and later to an impoverishment of the agricultural sector of the affected region (Peña and Pellicer, 1991).

Geomorphology, as a science that studies the relief forms and the processes which explain their origin and evolution, can be considered as an important auxiliary science in planning. However, descriptive and analytical aspects which are irreplaceable for the understanding of the present configuration of the land, may hardly be significant for planning because they are presented on a geological time scale and do not respond directly to the planning objectives and aims. The importance of geomorphology as a factor of the physical environment and the need to give adequate answers to the planning demands have brought a new line of work called applied geomorphology to the forefront. Likewise, relief is considered jointly with other variables in the landscape and territorial investigation studies.

There are numerous methodological approaches:

(i) *Analytical approach.* This is based on the

concept that the land surface is the result of the integration of different elements and factors of the natural system. Each component is considered as a layer of information and is represented on a thematic map or is integrated into a mappable geocoded data bank. Using superimposition or aggregation procedures, each territorial area can then be characterized. For each use proposed, the significant elements are selected and weighted, and their analysis and valuation is then carried out.

(ii) *Synthesis approach.* This approach considers the land surface as a mosaic of homogeneous units or spatial units whose points show similar physiognomy, structure, and function on a given scale. The procedure normally begins with the definition of very obvious units and develops a sequential process of aggregation or disaggregation according to certain coherent steps and criteria, finally reaching the degree of detail required by the objectives proposed. The approach generally requires a professional team with experience and training, capable of defining the criteria and of establishing the precise limits objectively. This is normally a quicker and more economical procedure.

(iii) The most recent contributions place special emphasis on the *interface* areas where the imbalances are most apparent, where there is more information, and where action, when taken properly, is most productive. As interfaces are the field of action of two or more systems, this is where most conflicts and problems arise: caused by the problems of each system in itself, as well as the problems originating from the systems as a group. Relief lies in the interface of the lithosphere, the atmosphere, and the hydrosphere. It is in this interface, in an environment where the biosphere develops, that life is based (Tricart, 1994). An example of a simple interface is that of the fluvial point bars; this is where the nature of the substrate is recognized as well as the hydrological processes acting in the basin. The point bars thus become the environmental representation of a system of relationships. The rural area, from this perspective, is found on the border between natural, urban, and cultivated subsystems, and could be defined as an area of conflict or interface.

Any of the procedures leads to the elaboration of maps or numerical bases of georeferenced data—the only way that information can be operative in the planning process. It is very important to define the criteria of delimitation and the degree of necessary generalization to the objectives proposed, and to determine the adequate work scale. The information collected should be significant for planning. It is of no sense to include data, however precise and rigorous it may be, that does not contribute anything to the objective. Information should fit, wherever possible, to the administrative procedures upon which the decisions will be based.

In a broad-scale planning project, the geomorphological information should form a subgroup which explains and justifies, in itself, its analysis, appraisals, and proposals without the need to go to complementary documentation (Marker, 1990). Elaboration of maps means a first-order scientific exercise facilitating the ranking, structuring, definition, synthesis and communication of knowledge which is very often accumulated in an inappropriate order, without recognizing its interrelationships or valuating its significance within the system.

3.4 Basic Documentation for Relief Analysis in Planning

Cartography is the universal graphic language for transmitting spatial concepts. Cartographic information, aerial photography, satellite images, and those images obtained with other sensors—sonar, lidar, radar images, etc.—are extremely efficient for making inventories, and to manage, analyse, synthesize, and represent ideas, forms, and relationships that occur in a bi- or tridimensional space. A map is a graphic representation of forms and relationships that take place in a given space, drawn upon a plane with a relationship of similarity called scale, and a necessary degree of generalization and simplification which varies according to the scale and objectives of the map.

One of the great problems with cartography is projection. The unsolvable difficulties in representing the Earth's spherical surface on a bidimensional plane have been solved in very different ways. There are far more than a hundred projection systems with advantages and disadvantages depending on their application. Regarding planning, one must bear in mind these problems of projection when large regional spaces are covered, when pieces of cartographic information from different origins are crossed and, in particular, when automatic cartography is to be worked out.

A similarity relationship called scale is established between the map or obtained image of the land surface and reality. Scales are interpreted as a fraction and therefore those with the smallest denominator will be the largest. Choice of scale is very important in the planning studies because it determines the resolution to be used in the different relief analyses. The scale will determine the generalization, simplification, and sign symbol systems in the mapping of the results. Maps with smaller scales bring less number of symbols and concepts, and the simplified representation of many (Figure 3.2).

Any map will allow the definition of the relative situation of a point to a fixed reference system of geographical coordinates such as latitude and longitude.

3.4.1 Topographic Maps

The purpose of topographic maps is to represent the land surface. They are based on a positional reference system of coordinates, where relief is represented by contour lines. Thus, the topographical map permits the definition of the relative situation of a point on the land surface and its height compared to a geoid or plane of reference, traditionally identified with sea level.

Contour lines, hydrographic network, communication networks, population centres, and further topographical features of the land represented on conventional maps may also be contained on data carriers, ready for processing directly on the computer. They are known as Numerical Mapping Bases. In the digital territorial models (DTM), positional references and the third dimension, integrated in a data carrier as x, y, and z are ready for automatic analysis.

From topographical maps, digital territorial models, and numerical mapping bases, numerous maps can be derived following relatively simple processing procedures, including slopes, exposures, hypsometry, drainage networks, and diagram blocks. The necessary hardware and software are available to practically any user. The biggest limitation for the development of these techniques continues to be the lack of appropriate resolution of topographical databases of wide land surfaces.

3.4.2 Geological Maps

The geological map is the cartographical representation of the distribution of outcrops in a given area, as well as the chronological succession and the deformations they present. Traditionally, geology studies the internal processes that affect the Earth's crust, the properties and characteristics of the crust, and its capacity to supply resources and withstand human action. The general geological maps offer a lot of information, but in the majority of cases they are not adequate for the tasks of planning and need to be re-elaborated in order to obtain significant information.

The geoenvironmental or geoscientific studies (Cendrero et al., 1992) greatly surpassed the strict limits of classical geology and developed specific thematic maps—i.e., maps showing hazards, capacities, aptitudes, impacts, etc.—where particular geological features are also described, as well as external processes and the relationships between man and the geological components of ecosystems.

3.4.3 Aerial Photographs

The value of aerial photographs lies in the fact that they are a midway mark between the concrete

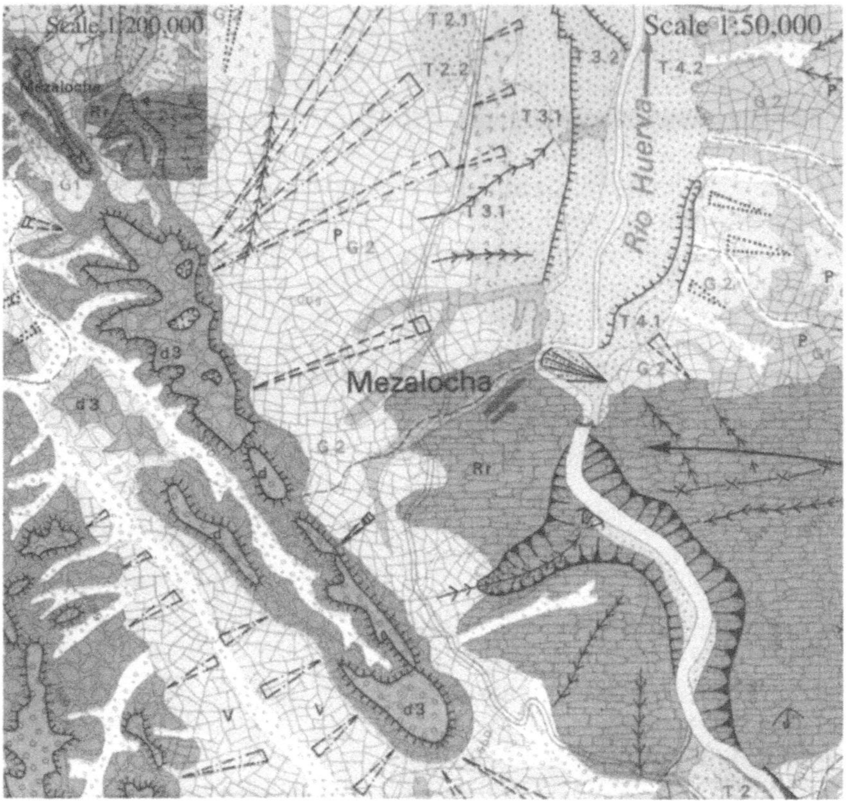

Geomorphological details (scale 1:50,000)

- Geomorphological boundary
- Fault, probable
- Joint
- Structural ridge (small elongated hill)
- Dip and strike
- Dipslope
- Mass movement
- Anticlinal/sinclinal axis, with plunge direction
- Facet
- Mesa (small plateau)
- Residual hill
- Main watershed
- Break of slope, concave
- Break of slope, convex
- Scarp, minor
- Scarp, major
- Gully/ravine
- Alluvial fan, active
- Glacis, accumulative; non or slightly dissected
- Glacis, accumulative; dissected
- Glacis, mixed; non or slightly dissected
- Glacis, mixed; dissected
- Glacis, erosional; non or slightly dissected
- Glacis, erosional; dissected

Geomorphological details (scale 1:200,000)

- Geomorphological boundary
- Fault, probable
- Structural ridge (small elongated hill)
- Anticlinal/sinclinal axis, with plunge direction
- Facet
- Main watershed
- Break of slope, concave
- Break of slope, convex
- Scarp
- Gully, ravine
- Glacis, accumulative
- Glacis, mixed
- Glacis, erosional

FIGURE 3.2. Generalization, simplification, and simbolization in geomorphological map design. (Central Huerva Valley, Spain.) (Mensua et al., 1981).

limited reality of the visual perceptions of nature at land level and maps that are elaborated, orthogonally projected cartographic representations.

Aerial photographs are images of the Earth's surface taken from the air, normally from a plane. According to the angle formed by the projection of the optical axis of the camera and the vertical projection over the nadir, one talks of vertical aerial photographs when deviations from nadir are less than 10 degrees and oblique photographs when the angle is greater than 10 degrees. If the horizon can be seen, they are called panoramic or high-oblique photographs. Aerial photographs used for research are taken during a line of flight, in successive exposures at regular intervals, so that there is a 60 percent coverage between successive images. Thus, each point of land is registered twice from different angles, a necessary condition for the obtention of a tridimensional or stereoscope image of the photographs through the stereoscope. A systematic sweeping is made of the land in strips that overlap 10–25 percent. The direction and height should remain constant, thus avoiding deviations because of nodding and warping.

As a rule, conventional aerial photographs are difficult to interpret because of both their unusual point of view and scale. However, the accumulation of concrete references is such that a close look and a little training are sufficient to be able to discover the most significant elements of the landscape: population centres, communication networks, hydrographic networks, and so on, which make up a precise grid of spatial references.

An aerial photograph, however, is not a map. It lacks the necessary degree of data selection and generalization for planning and what one would demand from a cartographic document. Landscapes have been compared with medieval "palimpsests," which are documents with different superimposed messages and symbols that offer a lot of information but require a laborious process of discreting and interpretation. The interpretation of photographs consists in unravelling the different messages of the complex natural and cultural systems, and translating them into an intelligible, clear, and operative language.

Aerial photography has been the most important technical contribution of the first half of this century to the elaboration of environmental studies. It has been just as important as the subsequent development of remote sensing. The reasons for its effectiveness (López Vergara, 1971) include the following: (i) the focus and panoramic amplitude; (ii) the quality of the photographic image, which surpasses direct human vision; (iii) permanence, immobility, and geometric fidelity; and (iv) the instantaneous register as a result of successive flights which permit an evolutionary analysis.

Despite its geometric fidelity, the relief and conic projection system cause modifications in scale and optical distortions which must be taken into account. In itself, relief alters scale—e.g., variation of flight height—being greater in the mountains and smaller in the depressed areas. But, furthermore, due to the effect of the conic projection, the image becomes more distorted the further it is from the central point and the nearer it gets to the edges. The combination of relief and conic projection leads to a radial displacement, centrifugal at the peaks and centripetal at the low parts. Its incidence grows with the increase of relief and decreases with flight height. Furthermore, one must consider that aerial photographs show an exaggerated relief, since the interpupilary distance of the observer is smaller—excepting the scale differences—than the separation between two photograph shots of the stereoscopic pair. Nevertheless, despite the sensation of unreality, the exaggeration is very useful because details of relief which would otherwise go unnoticed can be appreciated.

The keys for interpretation are the abc of the photointerpreter. They are reference elements intended to facilitate the identification of real objects using photographic images. They are reference models or samples, selected with careful and flexible criteria, to admit the multiple variations within a coherent and clear scheme. Each study requires its own interpretation keys to be written up according to the land characteristics and study objectives. The identification of the objects by their image and

TABLE 3.1. Information given by the Geomorphological Map of France (Tricart, 1965, 1972)

Location	Coordinates	
Morphometry	Heights	
Structural context	Structural region	Bedrock Folded range Sedimentary basin
	Lithological nature and resistance rate	Limestones Granites Sandstones Marbles, etc.
	Tectonic	Dips Faults Anticline
Morphological context	Glacial Periglacial Humid Subtropical Littoral Fluvial	
Dating	Forms age Later modifications	
Form nature	Crests Erosion surface Fluvial terraces Badlands Dolines	
Surface formations	Granulometry Consolidation type Thickness	

ground truth as well as on the map are essential procedures in making keys for interpretation.

Since the 1930s, the development of photogrammetry and photointerpretation have contributed effectively to the knowledge gained of the territory, particularly of the parts that are remote and not easily accessible. However, it is in the last two decades, with the development of computer science in combination with the new sensor systems—radar, infrared, etc.—that telecommunication and access to the results of remote satellite perception have meant a genuine technological revolution, the real dimension of which has yet to be discovered.

3.4.4 Geomorphological Maps

Geomorphological maps are complex cartographic elaborations upon which representations are shown of morphometric, morphographic, morgenetic, morphodynamic, and morphochronological aspects, including additional information such as topography, hydrography, and certain lithostructural elements.

Geomorphological cartography lacks a single realization and a presentation system. Different schools have elaborated their own methodologies. However, the Tricart and ITC systems have been widely used and have had a decisive influence over other mixed cartographic formulations.

The legend of the geomorphological map of France (Tricart, 1965, 1972) is based on a clear system that is easily read and open to numerous variants. The conceptual precision, classification and ranking of data in accordance with a particular scale, the agreement on the data coding, and the coherence between the information represented as well as the system of graphical transcription, are the keys to success and diffusion. The Geomorphological

Map of France offers information on localization and morphometry of each geomorphological unit, the structural context in which it is placed, lithology, degree of rock resistance, tectonic aspects, the morphogenetical context, the age of the initial forms and successive modifications, nature of the forms and on the surface formations (Table 3.1).

In Holland, the International Institute for Aerospace Survey and Earth Sciences (ITC) has developed a particular methodology that has been widely diffused abroad, whose principles have been specially described by Verstappen (1970) and Verstappen and van Zuidam (1991). In the general geomorphological cartography of the ITC, the morphogenetic aspects are the basic factors, followed by morphography, morphostructure, lithology, morphometry, and morphochronology. The basic morphogenetic units are structural, volcanic, denudative, fluvial, lacustrine/marine, glacial, periglacial, aeolian, karstic, and anthropological/biological, including transitional forms.

3.4.5 Cartography Based on Remote Sensing

As observed, conventional aerial photographs make use of the visible part of the spectrum. But from the 1960s onward, technological development have allowed images to be obtained in other wavelengths, particularly in the infrared and microwave range. Since then, the artificial satellites began to provide new information on the land surface. Currently, these platforms are those which are most often used in remote sensing although they often need the joint support of aerial photography and an appropriate field work (Curran, 1985; Sabins, 1986) to complete their full potential.

The new sensors and remote sensing systems provide abundant and extremely precise data on relief. The main advantage of the optical-electronic sensors compared to the photographic sensors is their possibility to obtain simultaneous information in several bands of the electromagnetic spectrum, which helps to distinguish between the different lithologies, surfaces, and soil formations, hydrological aspects, and land uses. Furthermore, they are on a digital carrier which is the most appropriate support for automatic analysis and mapping. Other advantages of remote sensing are in the width of the panorama, the information captured on the nonvisible regions of the electromagnetic spectrum, homogeneity in datataking, and global and periodical coverage of the land surface (Chueca and Julián, 1997). The image resolution has reached estimable levels in recent sensors and has provided very detailed maps.

The active sensors that are most often used in relief studies are the radars which operate in the microwave region. Among the space platforms using radars, the European satellite ERS-1 is worthy of mention. It is functional nowadays and equipped with a Synthetic Aperture Radar (SAR) and has been used on occasions for geomorphological mapping work (Chotin et al., 1995).

The Multispectral Scanner (MSS) is a line-scanning sensor and has been associated with the Landsat satellites from 1972. The second generation of Landsats are also equipped with the Thematic Mapper™ sensor with better spectral, spatial, and radiometric resolution. It has been used in structural geomorphology studies (Kayan and Klemas, 1978; Chueca, 1992).

Among the meteorological satellites of the Tiros-NOAA series, the Advanced Very High Resolution Radiometer (AVHRR) sensor has been occasionally used in geomorphology work which involves the study of the dynamics of the desertification processes or the movement of ice in polar or subpolar areas (Kergomard, 1989).

The French satellite SPOT, equipped with the High-Resolution Visible (HRV) sensor stands out among the pushbroom scanners used for observing the land surface. The SPOT images offer new possibilities in the field of geomorphology because they can offer stereoscope images (Nossin, 1989) and because they take images of the same points more frequently.

However, one must not forget that the use of remote sensors is an auxiliary technique, ideal on occasions but discrete on others. It allows work to be done more quickly but not more correctly. In geomorphological cartography, for instance, the listed advantages are considerable for the development of synoptic cartography of large areas, working on a small scale (Chueca and Julián, 1997). Nevertheless, in detailed observation and geomorphological cartography, use of conventional aerial photographs is still preferred.

3.5 Geomorphological Bases for Planning: Analytical Components

As has been observed before, geomorphology considers a group of diverse elements and the result of their multiple interactions. It is, therefore, a complex and synthesized concept. Considered as a whole, it sometimes presents difficulties in collecting, elaborating, and applying data. In order to avoid these problems and to analyse certain simple relationships between relief and planning, different morphological aspects, geological components closely linked to geomorphology, geomorphological processes and any of the information derived from them can all be studied separately. The analytical variables offer great versatility, they can be aggregated selectively, and the significance of each variable can be weighted with regard to the possible uses. There are numerous methodological approaches. The choice of the most adequate formula forms part of the research process which will lead to the solution of each problem.

3.5.1 Morphometric Bases

Morphometric bases correspond to the geometry of the land. They define the spatial parameters of altitude, surface, slope, and certain typological features of the relief forms which find an easy numerical expression.

3.5.1.1 Hypsometry

The consideration of certain elevation intervals is most significant for presenting the area of study and for expressing certain morpho- and bioclimatic limits in the studies of the physical environment. This process is very simple. It is obtained from the information given by the contour lines of the topographical map. The difficulty lies in the definition of the elevation intervals that really adjust to the objectives of the study. The most usual expression is that of the hypsometric maps, but they can also be represented by histograms.

3.5.1.2 Slope

Slope refers to the inclination of the land with respect to a horizontal plane. It is established when the inclination is related between two points and the distance on the horizontal which separates them. It is generally expressed in percentages [Slope percent (%) = differences in height (100/distance on the horizontal], or in degrees (angle formed between the plane of the land and the horizontal plane).This limited and precise concept of the slope becomes more significant when other features are integrated, such as the slope length which determines the degree of roughness of the land; the concave, convex, or mixed form on an inclined plane; the simple or broken form; and the position of each unit of slope with respect to the surrounding units (sectors of peaks, mountainsides, or depressions).

Slope is an important localization factor for numerous land uses: it determines accessibility, run-off, susceptibility to erosion, potential for the mechanization of agricultural activities, urbanization, and so on. The classification of slopes for planning purposes differs according to the uses allocated to each portion of land and to the degree of technification and energy consumption applied. It is very important, therefore, to define beforehand the precise objectives and aims of the study. In general terms, the classification included in Table 3.2, from the Soil Survey Staff (1951), can be considered.

3.5.2 Morphographic Bases

The morphographic bases refer to the attributes that define the physiognomy, position, and dis-

TABLE 3.2. Slopes categories, opportunities and limitations of use (based on the variation degrees proposed by the Soil Survey Staff, 1951)

Categories	Slopes	Hydric erosion	Agricultural use	Urban infrastructures
Flat	<3%	Unrestricted	Unrestricted	Unrestricted
Smooth slope	3–10%	On certain kinds of soil	Terraces required	Airports excluded Limitations on motorways, drains, etc.
Moderate slope	10–20%	Rapid over land flow. Rough erosion.	Mechanization problems	Excluded (except recreational areas)
Rough slope	20–30%	Very rough erosion	Exceptional growing on very resistant soils and appropriated methods	Excluded
Very rough slope	30–50%	Very rough erosion	Grass and woods	Excluded
Steep slope	>50%	Very rough erosion	Extremely limited	Excluded

tribution of the relief components. In planning studies, orientation, exposure, and geomorphological configuration are frequently used.

Land aspect and exposure are both important topoclimatic factors because they determine the sun conditions (sun or shade) and are subject in different ways to the action of the local winds. In the traditional Mediterranean landscapes, it is a highly appreciated factor as it determines conditions of energy efficiency for agricultural and urban uses. The significance of the aspect and exposure increases in combination with slope.

The physiographical description of the land can be made through the classification of the topographical forms in simple mappable types (Figure 3.3). They are easily recognized on the topographical map and on the aerial photograph. They are very useful in the initial recognition phases, the preliminary presentation of the study, and as a basis for subsequent studies. The typology can be enriched with many other specific forms (cirques, terraces, volcanic cones, dolines, etc.) and microforms (ridges, irrigation ditches, polygons, etc.)—as many as are considered necessary for the planning project.

3.6 Geological Characteristics

Result of the internal evolution and dynamics of the Earth's crust, the nature of the materials, and their structural arrangement are all elements that explain and, to a large extent, direct the physiognomy and relief-shaping processes, and in some way condition the land uses.

3.6.1 Lithology and Superficial Deposits

The rocks of the geological substrate and the superficial formations, derived from the in situ alteration of the substrate or from the erosion deposits (alluvions, moraines, dunes, pyroclasts, and volcanic ash, etc.) all determine a series of land features that must be taken into account when planning. Land morphology, susceptibility to erosion, permeability, acidity, or alkalinity of soils, the presence of specific processes (clay expansion, gypsum, and lime solution), the availability of mineral or building materials, and the carrying capacity of the soil as regards the construction of infrastructures and buildings, are all aspects related to this factor.

As occurs with other elements of study, the classifications according to classical criteria (genetic, mineralogical, textural, and chronological criteria) are hardly useful for planning. Once more, one must design and predefine this variable in each case, according to new criteria which have been adjusted to the planning objectives.

Plains	Top plains Pediplains Panplains

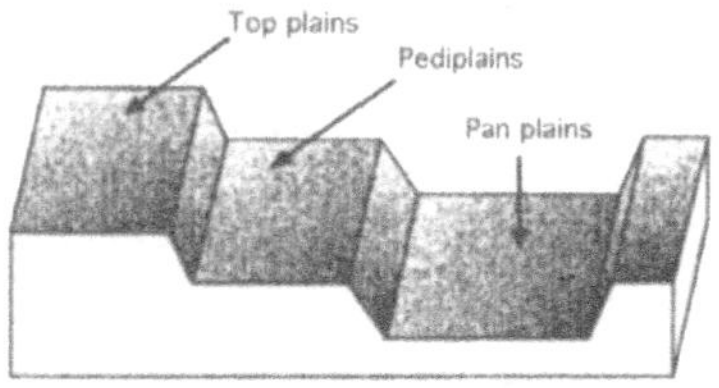

Slopes	Escarpments, >100% Taluses, 15–100% Glacis Fans

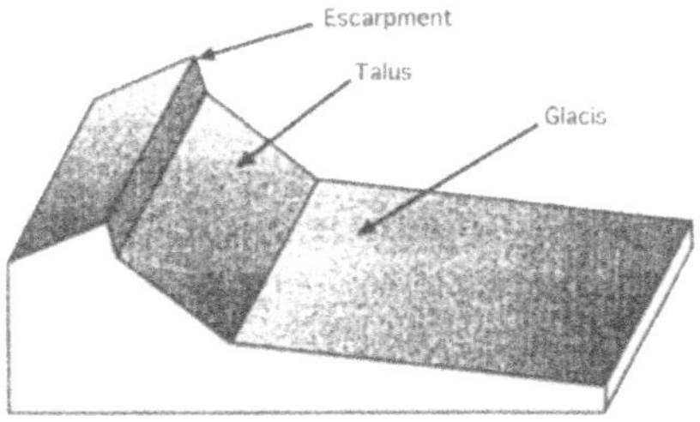

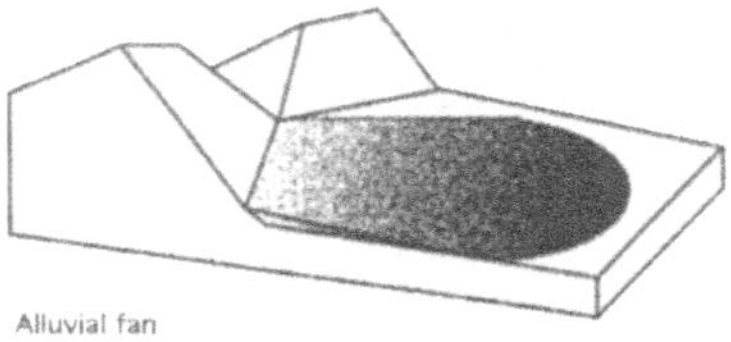

Summits	Edged Hillocked

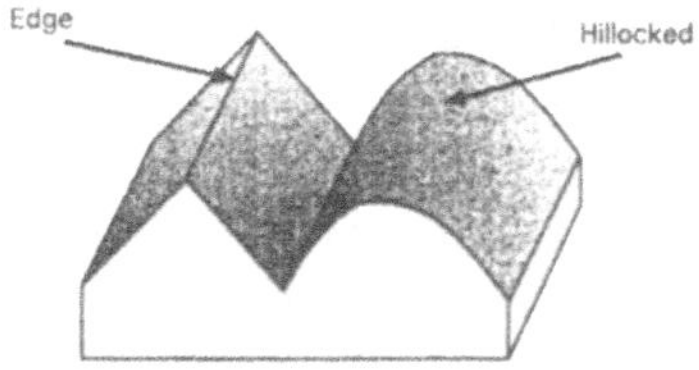

Valleys	Canyons Gullets V-shaped U-shaped Plains basins

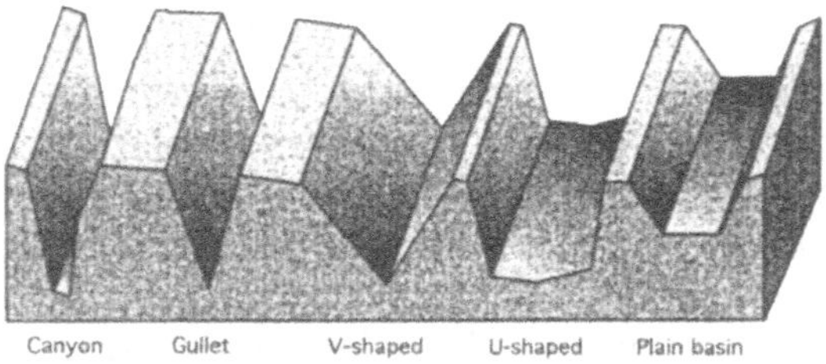

Depressions	Open Closed

Hills	

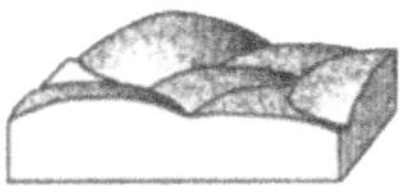

FIGURE 3.3. Frequent recognizable and exposed topographic types in 1:25,000 to 1:100,000 scales.

3.6.2 Tectonic and Structural Elements

In certain studies and places, it may be worthwhile to consider the structural arrangement of the material of the land indicating the dips, folding axes, or fractures which affect the materials. The presence of recent fractures or diapyrical deformations, for example, may be indicators of particular hazards or limitations. The obtention and valuation of these data is a very specialized task which, as is also the case with other variables, must be considered not as an aim in itself, but ancillary to the planning objectives.

3.7 Synthetic Information on Relief in Planning Studies

Numerous phenomena of the geomorphological landscape cannot be explained by simple univocal relationships, they are in fact the result of dynamic combinations of internal elements and factors (lithosphere) and external elements and factors (atmosphere, hydrosphere, biosphere, and anthroposphere), which are sometimes discontinuous and unstable, and which keep the relief in a permanent state of evolution. Synthetic studies are partially complex approaches that consider homogenous relief units which, on a given scale, offer similar physiognomies, structures, and functions, and whose points keep a similar response to possible uses.

Geomorphological tradition has established a rich typology of configurations which, more or less precisely, constitute the synthesis of physiographic, structural, lithological, chronological, dynamic, or climatic features.

The elemental geomorphological configurations can be associated according to their features (structural, denudative, etc.), their context (insular shelf, folding range, etc.), their lithological and sedimentary base (rocks and surface formations), their nature (specific definition of each form: crest, glacis, cirque glacier, etc.), their resistance (hardness, solubility, degree of rock weathering, etc.), the age of the evolutive forms and stages and acting morphoclimatic systems and processes, and so on.

This valuable information provides the keys to explain the configuration of the present relief, the recognition of the areas of greater geomorphological instability, the prediction of possible risks or impacts, the general dynamics of the system, and many other interesting aspects in planning studies. However, part of the information from the classical studies has very little significance in itself in planning studies and can easily be omitted (e.g., some morphochronological details).

In each geomorphological configuration, the distribution of the slopes, the resistance and nature of the rocks and surface formations, the type and intensity of the historical or functional genetic processes, etc., are factors which are associated, linked by multiple interactions, and structured in ranks according to some morphogenetic patterns.

The identification of the geomorphological configurations as complex systems, their recognition, description, and mapping allow a very expressive characterization to be made of the land, saving a considerable amount of time and financial resources. However, a sound training in geomorphology is necessary, as well as sound knowledge of the implications of relief in the planning processes.

Possible configurations on earth are extremely numerous. The most frequent and characteristic geomorphological configurations in the Mediterranean region, grouped according with morphogenetic criteria, are shown in Table 3.3.

3.8 Integrated Studies of the Natural Environment

The synthetic studies open to other environmental variables such as vegetation and land uses are considered in landscape (Bertrand, 1968; Sochava, 1972; de Bolós, 1992; Tricart, 1994) and in territorial studies (Hills, 1961; Christian and Stewart, 1968; van Zuidam and van Zuidam-Cancelado, 1979; Verstappen, 1983; Díaz de

TABLE 3.3. Frequent geomorphological configurations in the Mediterranean region. Concepts were gathered by morphogenetic criteria to exposed effects. In planning work criteria should be operative and tune with the proposed objectives.

Lithological reliefs	Karst	Poljes, dolines, uvalas; travertine forms; fluvial-karst canyons; blind valleys, sack bottom valleys; minor forms: bogaz; endokarst: caves, galleries, stalactites, stalagmites
	Volcanic	Shields, cones, lava domes, boilers, craters; lava plateau, lava blister, dykes; lava flows and lava fields
	Granitic	Bornhardt and crests; slab-like blocks or cheesewrings, granitic alveoles; arenization alveoles; micro-forms: piles, potholes, grooves, flat-flooded pans, parted blocks, polygonal cracks, perched boulders
	Other groups	Badlands on clay, pure-shaped conglomerates
Structural reliefs	Horizontal	Tabular reliefs (moors), witness hill, conical hill
	Folded	Slopes, crests, hog backs; monts, val, combe, hanging synclinal; cluses, ruzes
	Faulted	Fault-line scarps; host, graben
External dynamic processes modeling	Slope modelling by gravitational processes	Falls, rock avalanches, slumps, ground flows, mud flows, solifluction, creep
	Slope modelling by nonconcentrated hydric processes	Denudation forms: rills, gullies, pipes. Accumulation forms: alluvial cones, glacis
	Fluvial modelling	Straight river channel, meandering and braided; canyons, V-shaped valleys, U-shaped valleys; flood basin, terraces; alluvial fans
	Marine and littoral modelling	Denudation forms: cliffs, abrasion platforms, estuaries, "rias," fiords. Accumulation forms: bars or barrier-beaches, beaches, ripple marks and megaripple marks, lagoon type littorals, littoral dunes, intertidal plains, organic reefs, deltas
	Wind modelling	Denudation forms: polished faces, dreikanters, alveoles, grooves and flutes, mushroom forms. Accumulation forms: ripples, dunes, draas (huge accumulation of sand), loess
	Glacial modelling (mountains)	Denudation forms: horns, cirques, U-shaped valleys, polished faces, grooves, "roche mountonnée," overdeeped glacial basin, riegels. Accumulation forms: moraines (ground moraines, terminal moraines, lateral moraines)
	Periglacial	Denudation forms: crests, tors, cryokarst, snow hollows, gelifluction streams. Accumulation forms: protalus rampart, slope debris (cones, talus, fields, rivers, seas), tidy debris, rocky glaciers, mass-movement (gelifluctions, creeps), structured soils, cryoturbation
	Biological/ anthropogenic modelling	Excavations (open air mines, pits, ditches), accumulations (tips, fillings), impermeabilities (urban), road constructions, hydraulic infrastructure, ground conditioning for growing (cuttings, fires, nivelations, terraces, irrigation nets), global impacts on morphogenetic systems
Polygenic forms		Erosion surfaces, inselbergs

Terán, 1985; Sánchez Díaz, 1995).

These studies are based on three principles: (i) the recognition that lithology, morphogenetic processes, vegetation, soils, and the physical dimensions of human activities are the organizational principles of ecosystems and their expression in the landscape; (ii) those elements and factors, and their interactions, constitute, at a given scale, homogeneous spatial units as regards physiognomy, structure, and function, and therefore offer similar responses and potentialities to human use; and (iii) the environmental units can be grouped together in units of a higher rank, structured in different taxonomical orders according to the spatial resolution of the study and to the significance of the different factors in the different scales of recognition; in this way, environmental units form organized hierarchies.

The most advanced research tends to substitute dynamic approaches for those based on simple statistical inventories, which are useful for the preliminary exploration of large spaces about which little is known. Special emphasis is placed on the dynamic–temporal characteristics, and analysis is made of how evolution has given rise to a differentiation of the environment in units with their own problems. Remote sensing becomes a particularly appropriate instrument due to the possibility it offers for interpreting subsequent situations. Tricart insists on the need for dynamic integration and proposes the classification of environments according to their degrees of stability. The most visible form of stability can be seen in the topography (Tricart, 1994).

Spatial and dynamic recognition for planning purposes, inspired and supported by the experience from previous studies, gives way to numerous mixed methodological formulae. For example, applied geomorphological cartography that is designed in the University of Zaragoza (Peña et al., 1997) stems from the recognition of complex spatial units and from present and past geomorphological processes which configure each unit of relief and govern their dynamics. Normally the division into homogeneous units begins with aerial photography following some preestablished criteria in accordance with the previous knowledge of the land and with the aims of the research, which are contrasted on land with direct observations, sampling, laboratory analysis (González Hidalgo et al., 1991), etc. The following concepts are put forward:

(i) Structuring of the territory in geomorphological units, defined by the outcropping materials (lithology and superficial formations), geomorphological configurations (tabular reliefs, poljes, fluvial terraces, etc.), and morphogenetic systems. Scales 1:200,000–1:50,000.

(ii) Geomorphological processes, characterized by their genesis (fluvial, karstic, etc.) with subdivisions which are in keeping with the structural trait (study of neotectonics linked with active faults from recent dating and seismic epicentres), the denudation (factor of resistance to the weathering of the roquedo in comparison with the morphoclimatic conditions of the environment), transport (river, wind, etc.), and the type of sedimentation (cones, terraces, etc.). Scales 1:25,000–1:5000.

(iii) Morphodynamic units, defined by morphometric and morphographic criteria, outcropping materials, hydrographic network, present geomorphological processes, soil plant cover, and physical aspects of human activity. Plant cover is considered as soil protection and genesis factors (density, height, degree of rainfall interception, and other specific parameters are studied in experimental plots in order to make a quantitative definition of the protection factor), and is the basis for studies to be made of the interaction between geomorphological and edaphic processes. Scales 1:25,000–1:5000.

The thematic contents mentioned are the basis for the inventory and valuation studies of natural spaces. Use of these spaces can thus be optimized through the opportune plans of utilization, management, erosion prevention, prediction of natural and induced hazards and of the impacts of human action.

The need for approaching environmental problems as a whole has led some researchers (Chanlett, 1973; Hawley, 1974; Ramos Fernández, 1979; Gómez Orea, 1994; Recatala, 1995; Lugaresaresti Bilbao, 1996) to the integrated consideration of the physical and biological systems together with the sociopolitical and economic dimension, as a means to solve the conflicts which arise between human activities and natural surroundings.

3.9 Geomorphology Applied to Rural Planning and the Environment

Applied geomorphology (Tricart, 1978; Tricart and Kilian, 1982; Verstappen, 1983; Panizza, 1988) has made these three contributions to planning:

(i) identification, forecast and defence of the geomorphological processes (past, present, and future) that trigger off hazardous situations for man—geomorphological information is most significant in the study of environmental risks;

(ii) recognition and estimation of the consequences of human activity on the relief causing phenomena of degradation and imbalance in the environment—the applied geomorphological studies are a valuable tool in the studies of environmental impact; and

(iii) distribution and allocation of land uses depending on the danger and on the geomorphological resources.

3.9.1 Geomorphological Hazards

Consideration of the geomorphological as pects in territorial planning is also deeply significant in the study of natural hazards. Natural hazards often form part of current affairs due to the profusion, immediacy, sensationalism, and even morbidity of the news coverage of the natural catastrophes, whereas in fact this subject is as old as the history of man (e.g., the myth of the Great Flood). Unfortunately, once the tears have dried, the risks are forgotten and lack of foresight continues. The concept of hazard is defined by the extent of the effect on the human populations and their infrastructures, activities, and resources. Therefore, it is translated into loss of human lives and into economic terms. The consideration of time is highly significant, although difficult to estimate. Normally one understands the return period to be the lapse of time between events of a particular magnitude. The studies are based on the analysis of data from the past and on statistical extrapolation.

The greater or lesser probability of a given geomorphological process, whether natural or induced by man, to affect a space with a certain level of intensity, independently of it affecting human lives or goods, is called "geomorphological hazard." Many of the so-called "hazard maps" are in fact "natural danger" maps, since they do not take into account the vulnerability of the land. Geomorphological instability can be detected in three ways (Echeverría, 1997): (i) using monitoring and experimentation equipment (this approach is normally costly and needs long periods of observation); (ii) drawing up dyachronous cartography with the support of aerial photography and recent and historical documental sources; and (iii) integrating thematic maps correspondent with the risk causing factors and susceptibility of spatial units.

Normally, the elements that trigger hazards do not act on their own. Movement of mountain slopes are associated with floods and soil erosion in such a way that a flood is often accompanied by mudslides which are much more destructive than the action of the water itself.

Geomorphological hazard maps can be elaborated following several methodological procedures. Panizza (1988), proposes the differentiation between the factors of instability (geological, morphological, hydrogeological, climatic, anthropic, etc.) which give rise to a synthetic map, and the analysis of the effects (past and present forms and processes) which are shown on a map of morphological dynamics. The integration of both maps allows a geomorphological hazard map to be drawn up, indicating the actual and potential unstable areas.

Flood risk maps, because of the special spatial characteristics in which they evolve, take into consideration actual geomorphological and hydrological factors: types of plain, nature and morphometry of the sediments, microtopography, areas flooded in the past, estimation of floodable areas, flood return periods, flood duration, depth of stagnant water, flow energy, erosion, sedimentation, migration of the channels, flood defence constructions, etc. (Voskuil and van Zuidam, 1982; Ollero, 1996).

Hazard studies go beyond the actual geomorphological context. Apart from the studies of the hazard itself, the vulnerability of people, goods, infrastructures, and activities exposed to such a hazard must also be considered (Lugaresaresti Bilbao, 1996). Man's short historical memory, together with his trust in the defence constructions and in the great technological potential used to build them, favours very serious situations of lack of foresight: homes, shops, services, and industries are built and developed, and activities are practiced in areas that are naturally hazardous (flood plains, unstable slopes, active cliffs, seismic or subsident areas, etc.).

Prevention of hazards implies being aware of the danger, their magnitude and frequencies, dividing the territory into zones according to their degree of danger, taking a number of preventive actions, either structural or otherwise, and laying down administrative and judicial measures, besides informing the affected population (Figure 3.4).

3.9.2 Environmental Impacts

Man becomes increasingly more important as a factor of organization or disorganization of the geomorphological processes. As a consequence of the technological innovations and the motivations of the users, relief must fulfill numerous functions. It is subject to the ever greater number of interferences produced by the process by which human activity is introduced into wide sectors of the territory: built-up areas with high levels of space consumption, building up of new communication networks and expansion of existing ones, industrial facilities, highly technical agriculture, extraction of mineral resources, hydraulic infrastructures, and so on. The geomorphological processes are often interrupted (lost permeability, blocked surface water flows, compactation, etc.), geomorphological forms are destroyed (quarries, excavations, infilling, etc.), and landscapes lose their natural features.

Studies of geomorphological impact come within the framework of the studies of environmental impact. These studies try to predict the consequences of the implementation of a project on the environment and attempt to establish corrective measures. The basic methodology consists of the following:

(i) The description of the preoperational situation of the potential area to be transformed. Analysis of the morphology previous to the project, with the purpose of minimizing excavations or Earth movement and their effects as well as reducing the disfiguration of the landscape or attempting not to let the impact be seen. Analysis of the points of geological and geomorphological interest to be kept for their economic, scientific, or didactic value.

(ii) The definition and valuation of possible alterations, as well as the means to reduce or avoid them. Consideration of the actions of the project (excavations, infilling, interruption or transformation of run-off patterns, etc.) and its consequences on the geomorphological processes—increase of flash flood and flooding risk due to the barrier action of particular structures or the heightening of the swelling peak as a consequence of the impermeabilization of wide surface areas of the basins, provoking erosion, instability of shelves and slopes, and so on. Design of technical corrective measures and modifications of the project in order to avoid unwanted effects.

(iii) The establishment of a monitoring and recovery programme and the specification of the residual impacts.

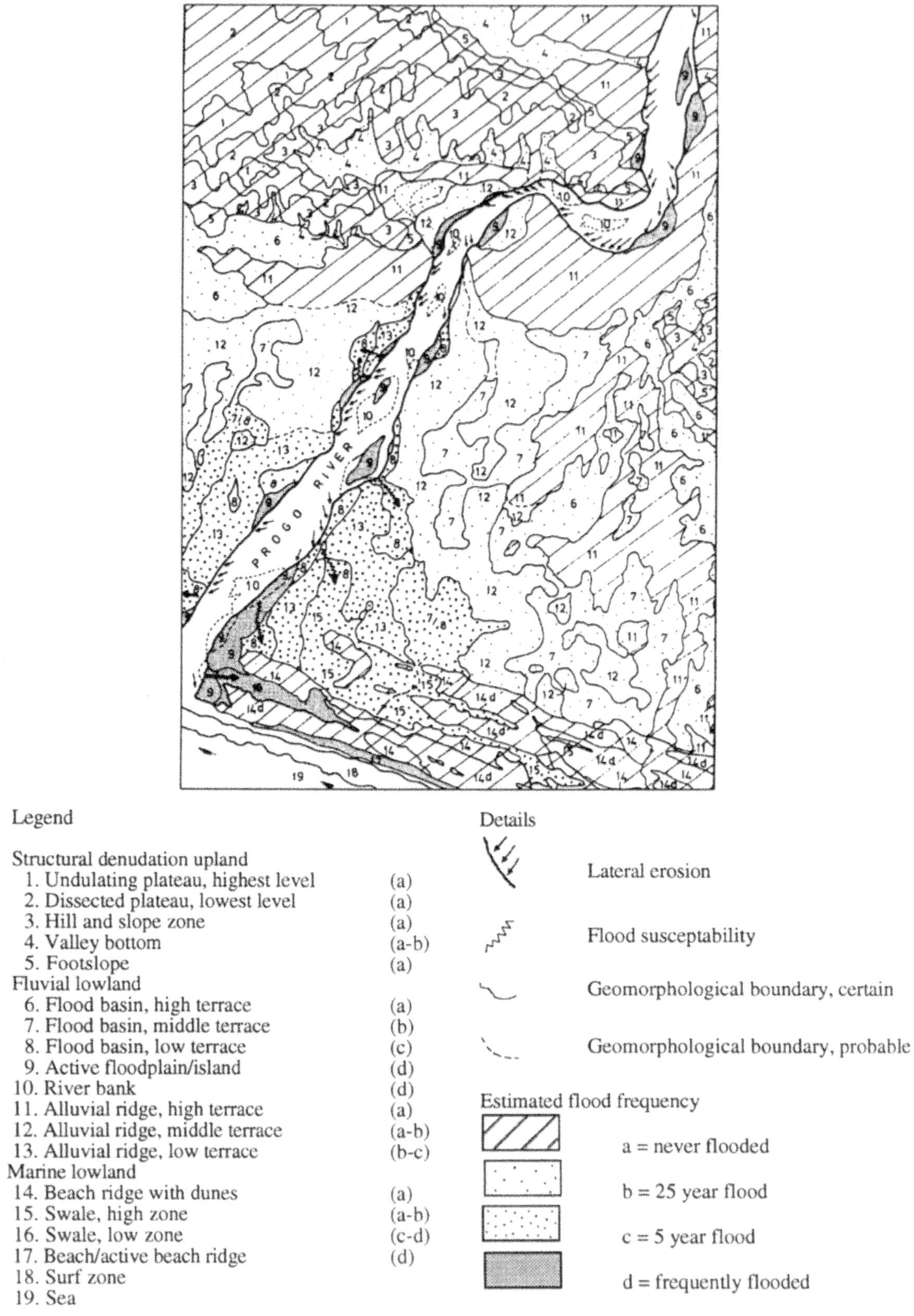

FIGURE 3.4. Flood hazard map of the lower Progo River Basin (Voskuil and van Zuidam, 1982).

3.9.3 Geomorphological Bases of Land Use Allocation

The consideration of geomorphological forms and processes in the allocation of land uses in rural planning requires the geomorphological zonification of the territory, with the purpose of determining the resources to be exploited or conserved, predicting hazard and risk situations connected to this type of process and giving priority to environmental improvement before allocating land for activities, thus avoiding having to take ineffective corrective action afterwards. The establishment of nonstructural measures—allocation of uses according to the potential and capacity of the land, creation of early-warning networks, and hazard monitoring—can mean considerable reductions in great defence infrastructures against natural elements and high costs of maintenance and correction of impacts.

However, rigid zonification must be avoided, as well as the division of the territory into highly specialized units, with strong physionomical and functional contrasts and frequent conflicts of uses. In the interface areas, gradual treatment should predominate as well as the interpenetration of the physical or social phases. Integrated planning is characterized by permeability, gradual transitions, and disappearance of the limits of the closed compartments in favour of a territorial organization in terms of networks that join living spaces (Bravard et al., 1995).

Rural planning with an ecological base is fundamental for the conservation of the natural patrimony of land. Cultivation of crops that are adequate for a greater soil protection, rationality in the rotivation of crop areas on a steep slope, planning urban areas, leisure spaces and communication networks in accordance with the aptitudes of the site, planning irrigated areas and irrigation techniques according to the response of the soils and the required profitability, fire prevention measures, reforestation according to the characteristics of the landscape to be restored, monitoring the positive or negative evolution of the fields abandoned in the mountain areas, and so on, are some of the topics with geomorphological roots or repercussions that ought to be considered in rural planning.

The geomorphological map is a main document in order to direct soils studies, because geomorphological configurations, surface formations, and modelling processes are important edaphogenetic factors. Other maps for specific objectives can be derived from the geomorphological map as shown in Figure 3.5.

The new functions of the rural areas are based in some way on the privileged environmental conditions they offer as compared to the strictly urban areas where most of the population live. The rural area is faced with a difficult challenge in itself: that of providing food and material for a world market and reducing the dependence on places elsewhere, maintaining the production potential for future societies, contributing to the recycling of the degraded urban ecosystems, offering quality landscape and supporting the numerous recreational, sport, and educational activities demanded by the new, progressively urban society (Gómez Orea, 1992).

The functions of the rural environment become especially valuable within the regional framework where analyses can be made of the interactive relationships that exist between the natural and rural environments with ecological functions of primary production, recreation, and recycling of the life systems, and the urban environments where the population is concentrated, and which are centres of industrial production, consumption, and generation of information.

As the urban societies evolve, there is a growing demand for natural surroundings for recreational, educational, and scientific uses as well as a preoccupation for the conservation of protected areas. Valleys and glacial cirques, karstic areas, river and coastal areas, endorrheic depressions, or dune systems are natural landscapes where relief is the structure factor, and are worthy of being considered in a conservation policy. Relief is fundamental to place numerous recreational, sport and educational activities demanded by the new society (Gómez Orea, 1992). The preservation of the ecological and cultural network of the structural axes of the traditional landscape, such as

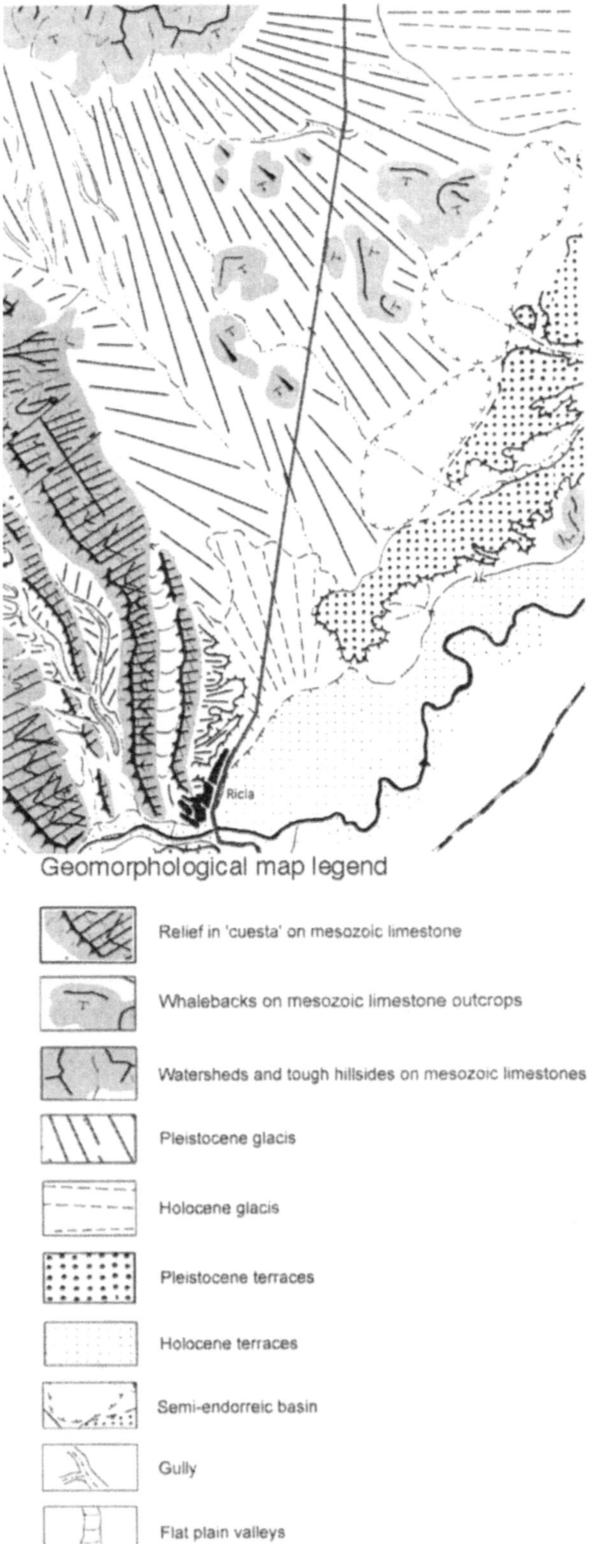

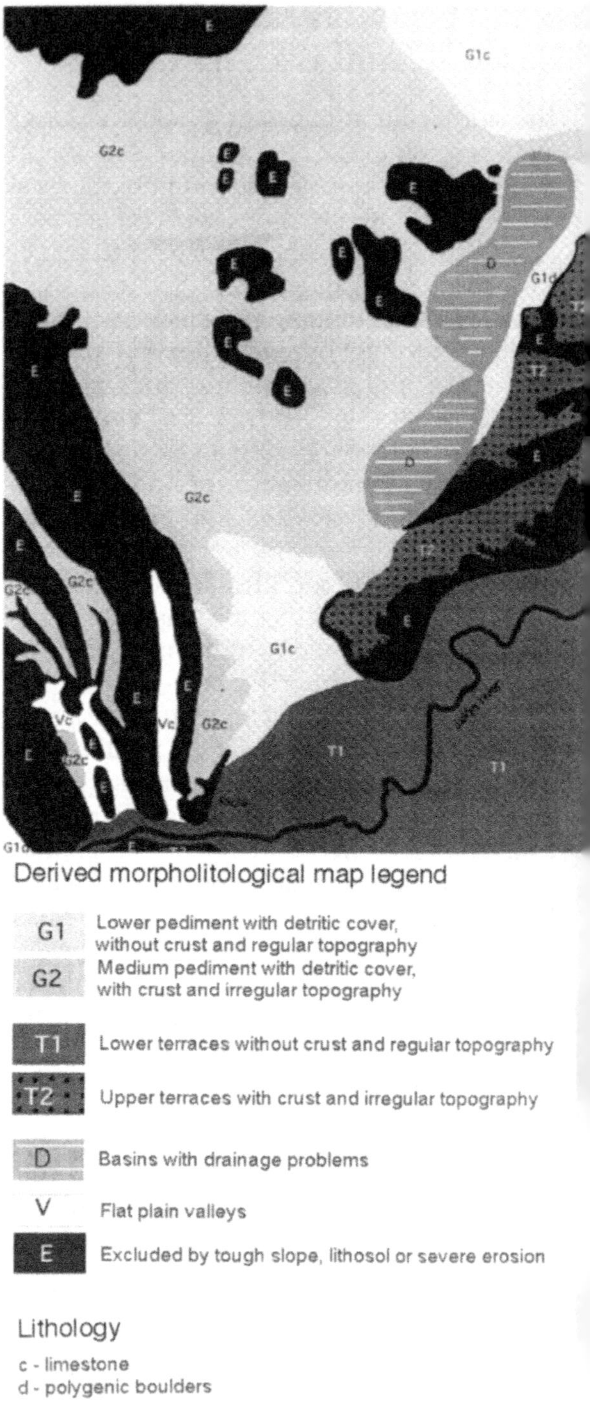

FIGURE 3.5. Geomorphological map (a) and derived morpholithological map (b) for soil studies applications (Central Jalón Valley, Spain).

hedges that act as biological corridors and retain erosion) is fundamental to maintain the identity, physiognomy, processes, and aesthetic and symbolic values of landscape in terms of sustainability.

The traditional landscapes offer an interesting fusion of geomorphology and culture. They express the knowledge and experimentation of centuries of cultures and, as such, constitute a valuable historic patrimony, hardly appreciated and about to disappear, where important ideas can be obtained for the management of the physical space. The observation of the inherited physiognomy of the land, of the agricultural practices, together with an appropriately worded questionnaire for farmers, can be of great interest for learning about the territory and for elaborating proposals.

3.10 Concluding Remarks

Relief plays an important role within the human life system. As an interface between the physical subsystems of the Earth and support for life, relief is a resource that cannot be substituted. It directs and contributes toward the diversification of numerous environments and can be considered as a significant factor of environmental diagnosis. Geomorphological organization is very rich in forms and processes in the Mediterranean world, where it conditions natural hazards and is situated in the interface between geomorphological processes and human activities. The capacity of man to modify the geomorphology of the land and the geomorphological processes, implies the generation of environmental impacts which must be foreseen and in certain cases, corrected. The consideration of the resources, potentialities and hazards of geomorphology cannot be substituted in the allocation of land uses in any process of economic and environmental planning.

References

Bertrand, G. Paysage et géographie phisique globale. Esquisse methodologique. *Revue Géographique des Pyrénées et du Sud-Ouest* 39:249–272; 1968.

Bravard, J.P.; Laurent, A.N.; Davallon, J.; Bethemont, J., eds. *Les Paysages de l'Eau aux Portes de la Ville.* Lyon: Programme Rhône-Alpes de Recherche en Sciences Humaines; 1995.

Cendrero, A.; Lütting, G.; Wolf, F., eds. *Planning the Use of the Earth's Surface.* Berlin: Springer-Verlag; 1992.

Chanlett, E.T. *Environmental Protection.* Berkeley, CA: McGraw-Hill; 1973.

Chotin, P.; Ait Brahim, L.; Deffontaines, B.; Rudant, J. P.; Chaouni, A. Apports des données ERS1-SAR sur la reconnaissance du réseau de failles dans la péninsule de Tanger, Maroc. *Photo-Interprétation* 2:146–148; 1995.

Christian, C.S.; Steward, G.A. *Summary of General Report on Survey of Katherine Darwin Region, 1946.* CSIRO, Australia, Land Research Series No. 1; 1968.

Chueca, J. Análisis estructural del área de Montalbán (Cordillera Ibérica turolense) a partir de la utilización de imágenes LANDSAT MSS. *Studium* 4:31–56; 1992.

Chueca, J.; Julián, A. La elaboración de cartografía geomorfológica a partir del empleo de imágenes satélite. Peña, J.L., ed. *Cartografía Geomorfológica Básica y Aplicada.* Logroño: Geoforma; 1997:49–66.

Curran, P. J. *Principles of Remote Sensing.* London: Longman; 1985.

De Bolós, M., director. *Manual de Ciencia del Paisaje.* Barcelona: Masson; 1992.

Díaz de Terán, J.R. Estudio geológico-ambiental de la franja costera Junquera-Castro Urdiales (Cantabria) y establecimiento de bases para su ordenación territorial. Ph.D. Thesis, University of Oviedo; 1985.

Echeverría, M.T. Cartografía geomorfológica temática y aplicada. Peña, J.L., ed. *Cartografía Geomorfogica Basica y Aplicada.* Logroño: Geoforma; 1997:147–182.

Gómez Orea, D. *Planificación Rural.* Madrid: Editorial Agrícola Española/MAPA; 1992.

Gómez Orea, D. *Ordenación del Territorio: Una Aproximación Desde el Medio Físico.* Madrid: Editorial Agrícola Española/Instituto Tecnológico Geominero de España; 1994.

González Hidalgo, C.; Echeverría, M.T.; Pellicer, F. Erosion and ecology in the middle Ebro Basin. Slope aspects as factor of erosional processes, an introduction approach. Sala, M.; Rubio, J.L.; García Ruiz, J.M., eds. *Soil Erosion in Spain.* Logroño: Geoforma; 1991:123–135.

Hawley, A.H. *Man and Environment*. New York: The New York Times Co.; 1974.

Hills, G.A. *The Ecological Bases for Land Use Planning*. Research Report 46. Ontario: Ontario Department of Lands and Forests; 1961.

Kayan, I.; Klemas, V. Application of LANDSAT imagery to studies of structural geology and geomorphology of the Mentese region southwestern Turkey. *Remote Sens. Environ.* 7:51–60; 1978.

Kergomard, C. Le mouvement des glaces de mer dans le détroit de Fram d'aprés les données NOAA-AVHRR. *Photo-Interprétation* 2:49–60; 1989.

López Vergara, M.L. *Manual de Fotogeología*. Madrid: Servicio Público de la Junta de Energía Nuclear; 1971.

Lugaresaresti Bilbao, J.I. Riesgos naturales e impactos ambientales en la cuenca de los ríos Gobela-Udondo y sector litoral. Ph.D. Thesis, University of Zaragoza; 1996.

Marker, B. Methods and approaches to environmental geology mapping: Meeting the planners requirement. Cendrero, A.; Lütting, G.; Wolf, F., eds. *Planning the Use of the Earth's Surface*. Berlin: Springer-Verlag; 1990.

Mensua, S.; van Zuidam, R.; van Zuidam-Cancelado, F.I. Consideraciones sobre el sistema de la cartografía geomorfológica del ITC, aplicado a un mapa geomorfológico sedimentario. *Geographicalia* 11-12:31–42; 1981.

Muñoz, J. *Geomorfología General*. Madrid: Síntesis; 1992.

Nossin, J.J. SPOT stereo interpretation in karst terrain, southern Turkey. *ITC J.* 2:79–91; 1989.

Ollero, A. *Los Meandros Libres del Río Ebro (Logroño–La Zaida): Geomorfología Fluvial, Ecogeografía y Riesgos*. Zaragoza: Consejo de Protección de la Naturaleza de Aragón; 1996.

Panizza, M. *Geomorfologia Applicata*. Rome: La Nuova Italia Scientifica; 1988.

Peña, J.L.; Pellicer, F. Medio ambiente y ordenación del territorio en Aragón. Bielza, V., coordinator. *Congreso Sobre Ordenación del Territorio en Aragón*; 1991 December 11-13; Zaragoza. Zaragoza: Departamento de Ordenación Territorial, Obras Públicas y Transportes del Gobierno de Aragón; 1991:1–24.

Peña, J.L.; Pellicer, F.; Chueca, J.; Julián, A. Leyendas para mapas geomorfológicos a escalas 1:50,000 y 1:200,000. Peña, J.L., ed. *Cartografía Geomorfológica Básica y Aplicada*. Logroño: Geoforma; 1997:129-146.

Ramos Fernández, A., director. *Planificación Física y Ecología*. Madrid: EMESA; 1979.

Recatala, L. Propuesta metodológica para planificación de usos del territorio y evaluación de impacto ambiental en el ámbito mediterráneo valenciano. Ph.D. Thesis, University of Valencia; 1995.

Sabins, F.F. *Remote Sensing: Principles and Interpretation*. San Francisco, CA: Freeman; 1986.

Sánchez Díaz, J. *Cartografía del Potencial del Medio Natural de La Isla de Gran Canaria*. Las Palmas de Gran Canaria: Cabildo Insular de Gran Canaria/ Universitat de Valencia/Universidad de las Palmas de Gran Canaria; 1995.

Soil Survey Staff. *Soil Survey Manual*. USDA Handbook 18. Washington, DC: USDA; 1951.

Sochava, V.B. L'étude des géosystèmes: Stade actuel de la géographie physique complexe. *Serita Geograficeskaja* 3:18–21; 1972.

Tricart, J. *Principes et Méthodes de la Géomorphologie*. Paris: Masson; 1965.

Tricart, J. Cartographie géomorphologique et classement des terres pour la conservation. *Mémoires et Documents* 12:215–222; 1972.

Tricart, J. *Géomorphologie Applicable*. Paris: Masson; 1978.

Tricart, J. *Écogéographie des Espaces Ruraux*. Paris: Nathan; 1994.

Tricart, J.; Kilian, J. *L'Eco-géographie et l'Aménagement du Milieu Naturel*. Paris: Librairie François Maspero; 1979.

Van Zuidam, R.A.; van Zuidam-Cancelado, F.I. *Terrain Analysis and Classification Using Aerial Photographs: A Geomorphological Approach*. ITC Textbook of Photo-interpretation; Vol. VII. Enschede: ITC; 1979.

Verstappen, H.Th. Introduction to the ITC system of geomorphological survey. *KNAG Geografisch Tijdschrift* 4(19):85–91; 1970.

Verstappen, H.Th. *Applied Geomorphology: Geomorphological Surveys for Environmental Development*. Amsterdam: Elsevier; 1983.

Verstappen, H.Th.; van Zuidam, R.A. *The ITC System of Geomorphological Survey*. ITC Publication No. 10. Enschede: ITC; 1991.

Voskuil, R.; van Zuidam, R.A. Examples of geomorphological and applied geomorphological mapping in central Java. *ITC J.* 3:290–302; 1982.

4

Climate As a Factor in Regional Planning

Vernon Meentemeyer

4.1 Introduction

4.1.1 Chapter Objectives

There may be no greater constraint on the biota and human societies of a region than climate. Climate can be viewed as a resource, offering immense benefits, but it also can produce episodic "events" that are hazardous to living things. Along with the topography (Chapter 3), it provides the foundation on which all land–water processes take place.

Over the past few decades the concept of climate has expanded beyond simple meteorological variables measured at weather stations to the broader concept of "environment" (Mather, 1978). Climatology as a discipline has become an Earth systems science in which systems concepts can be fully incorporated and, as such, is of central importance in land planning.

Since the pioneering work of Geiger (1927) in *Das Klima der bodennahen Luftschicht*, it has become evident that climate can be treated at many spatial scales, from macro- to meso- to microclimates. At each of these scales, climate near the Earth's surface varies greatly from place to place. Moreover certain elements of climate can be changed or even manipulated intentionally by human activities, especially at the finer scales. This chapter will review the most important elements of the regional climatic environment, attempt to cast this environment into a systems framework, and explore the elements of climate that can affect regional land use and activities. Examples of the usefulness of such information for planning will be presented.

4.1.2 Climate Versus Weather

Weather is often defined as the current state of the atmosphere, that is, the air temperature, humidity, sky cover, and barometric pressures at one instant in time. *Climate* is often defined as the long-term average of the hour-to-hour, day-to-day variation in these weather elements. Hence the climate can be described by the arithmetic means (or average) of meteorological variables observed over many years or decades. Meteorologists and climatologists call these long-term averages *normals*.

Unfortunately, such a definition restricts "climate" to the weather elements measured at weather stations. By modern thinking, however, climate is much more than records of temperature and precipitation. Rather climate involves flows (or fluxes) of energy and matter; forces and interactions, storage, release and exchange of energy and matter; and the transport of energy and matter across broad regions.

These forces and fluxes produce what we experience as weather or weather events, such

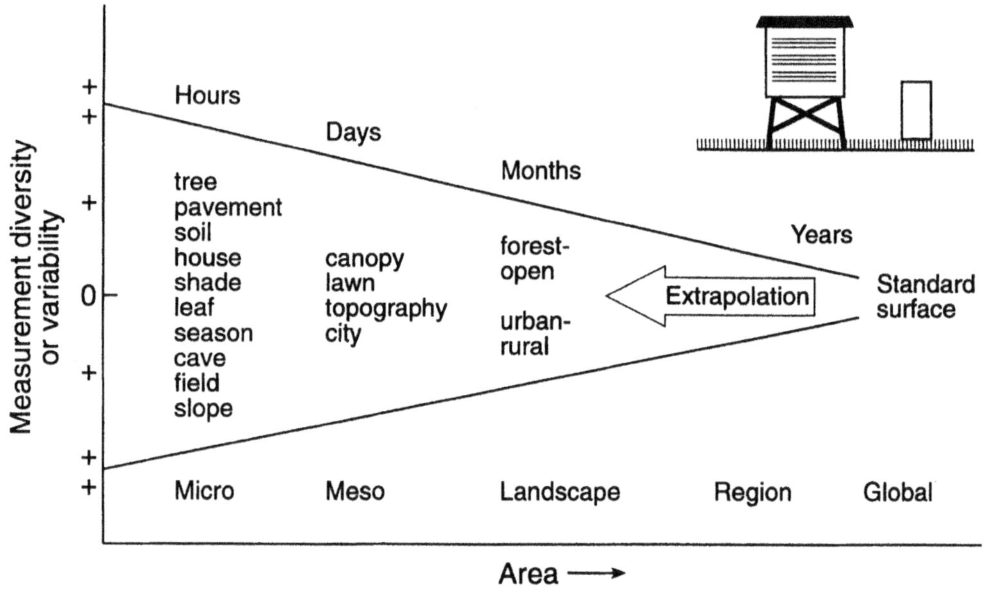

FIGURE 4.1. The relationships among spatial area and measurement duration and measurement diversity (after Meentemeyer, 1989).

experience as weather or weather events, such as thunderstorms, droughts, blizzards, and heat waves. Events range in duration from a second or less (lightning), to years (drought), or even millennia (ice ages), but in all cases events can be defined by their *frequency* of occurrence, *duration*, *intensity*, and *area* affected. If a sufficiently long record of observations of events is available, then the probabilities of occurrence of these events can be calculated for weather stations or regions. If a sufficient density of weather stations is available, maps of probabilities can be generated.

4.1.3 The Scales of Climate

In the latter portion of the nineteenth century an international agreement placed the instruments for observing the weather at 1.5 m above the surface of the Earth. This "station" was to be away from the influence of buildings and vegetation and on a level surface, generally a closely mown lawn. Hence, much of what is known about the climate of the land portions of the Earth comes from these stations. At a height of about 1.5 m, the influence of the surface around the station on absorbed radiation, evaporation, and air temperature is greatly diminished. Unfortunately the weather station may be a poor indictor of the general environment in places with much topographic variability and for the surface of the Earth. It is the surface environment that is important to most crops.

Although there is much disagreement regarding the exact dimensions of different scales of climate, generally three or four ranks are identified. Yoshino (1975, 1987) defined four scales of climate: macroclimate, mesoclimate, local climate, and microclimate (Table 4.1). At each category, the characteristic horizontal, vertical, and relevant time scales become smaller. In general, phenomena that

TABLE 4.1. Representative space and time scales of climate (after Yoshino, 1975)

Climate	Horizontal distribution (m)	Vertical distribution (m)	Lifetime of corresponding meteorological phenomena (sec)
Microclimate	10^{-2}–10^{2}	10^{-2}–10^{1}	10^{-1}–10^{1}
Local climate	10^{-2}–10^{4}	10^{-1}–10^{3}	10^{1}–10^{4}
Mesoclimate	10^{-2}–$(2 \cdot 10^{5})$	10^{0}–$(6 \cdot 10^{3})$	10^{4}–10^{5}
Macroclimate	$(2 \cdot 10^{5})$–$(5 \cdot 10^{7})$	10^{0}–10^{5}	10^{5}–10^{6}

appear to change slowly such that long observation periods are needed to understand and model them (Meentemeyer, 1989). Figure 4.1 shows that as climatic scales become finer (micro), measurements are seldom conducted for more than a few hours or days; but the number and variety of climatic flows and elements that can or will be measured increases greatly.

Similarly, at microclimatic scales the variables that appear to be important are more numerous. The matching of temporal and spatial scales is a practical problem that must be solved in any study or application involving climate. These practical problems are critical in planning activities. As temporal and spatial scales decrease, the value of measurements conducted for the macroclimate using standard surfaces and extrapolated to the meso local microclimate, become less and less useful (Figure 4.1).

We argue that climate has many parts, and many spatial/temporal scales which must be correctly matched. Climate is much more than the average weather. Furthermore, a better understanding of climate is facilitated by systems concepts. Indeed there appears to have been a convergence in the paradigms and resulting terminologies used by ecologists and by physical climatologists—the systems paradigm. Climate is one important aspect of the *abiotic* environment, often called the nonliving or physical environment by ecologists. As will be shown, climate from a systems perspective also opens opportunities for the incorporation of climate into policy and planning decisions.

4.2 Places and Regions in a Climatic Context

4.2.1 Macroclimatic Setting of a Region

As a first step in the examination of climatic influences, it is useful to examine the climatic records of weather stations in a region for a view of the macroclimatic patterns. Unfortunately many regions around the world have an extremely poor density of observing stations. Especially in Africa, the number of weather stations has declined in recent decades. Nevertheless, rudimentary records of temperature (usually daily maximum and minimum air temperatures) and daily precipitation totals are available, and even old observations from stations that are no longer in operation can have some utility.

The meteorological services of many countries have produced useful climatic summaries and/or a climatic atlas. Alternatively, a systematic appraisal of weather conditions around the world, such as the *World Survey of Climatology* by Landsberg (1969–1974), can be useful for understanding the climate of larger regions. Such data are, however, an incomplete depiction of climate, particularly for planning purposes at local scales.

Fortunately, meteorologists have learned a great deal about how different elements of climate interact. Meteorological data collected at weather stations have been analysed using statistical procedures (such as factor analysis and principal components analysis) that group variables tending to vary together in consistent ways. Such analyses typically identify four distinct types of variables: (i) an energy and/or heat factor; (ii) a moisture and/or humidity factor; (iii) a seasonal variability component; and (iv) a wind or atmospheric motion factor.

Thus, for example, air temperature, solar radiation, and soil temperature often vary in tandem, as do cloudiness, precipitation, and humidity.

For planning applications this means that one variable that is strongly correlated with another can often be substituted for the other. Often the variable for which detailed data are available can be substituted for one that requires special instrumentation or has missing observations.

4.2.2 The Climatic Data Sources

The availability of data at the macroclimatic scale has expanded recently for all portions of the world due to technological innovations. Many global and national data sets are made available on computer terminals via the Internet. These data range from historic station records dating from the nineteenth century (and earlier) to real-time weather observations. Observation times for the archived data range from hours (even minutes in some cases) to long-term averages or normals, for monthly and yearly intervals. The original data sources range from those collected by various National Weather Services, military services, research organizations, commercial organizations, and universities.

Data can be secured for weather stations, gridded and interpolated point patterns, the surface and upper air, forecasts and analyses, infrared temperatures, and current and archived satellite images among many products. It is possible to access, for example, the Climate Group's Data Library at Lamont-Doherty Earth Observatory, the National Climatic Data Center (NCDC), and the National Geophysical Data Center. All of the above are found in the United States, but they contain global databases and links to many other Internet sites with meteorological/climatic data. Many national weather services now have their own Internet sites which include sources of data and services. Some sites provide listings of all known sites with geophysical data, such as "Weather Data Available via the Internet" by Ilana Stern (Stern, 1997).

Vast quantities of data can also be secured on CD-ROM and 8-mm tape. Some data on CD can be secured from national archives and weather services, and some data can be secured commercially as from Earth Info Inc. (1997) in the United States or Chadwick-Healey, Ltd (1992) in the United Kingdom. Both produce CDs of basic monthly climatic data (temperature, precipitation, pressure, etc.) for locations globally. Much data can still be obtained via print. One of the most extensive compilations of world climatic data is found in Willmott et al. (1981). These volumes have data for 13,332 weather stations worldwide.

A crossroads has been reached: analyses for macroclimatic scales should no longer be hindered by the lack of data but by the volume and variety of available data. The data may, however, be of variable quality and incompletely checked for errors. This ease of access requires now more than ever filtering and checking of data to ensure that it is correct and appropriate for the applications at hand. The ancient problems of drift in weather station records caused by changes in station location, changes in instrumentation, and creeping urbanization persist; however, more data means that more attention needs to be directed to data quality. Of course to use these new data sources, the technological means to access and manipulate the data sets must be available.

4.2.3 Extreme and Rare Atmospheric Events

Knowledge of the normal temperature and precipitation of a region may be useful, but for many planning purposes knowledge of the probabilities of rare meteorological events, some of which may be uncommon to a region, is essential. Some events may represent extreme hazards to vegetation, fauna, crops, structures, and human systems, including health and mortality. Some common atmospheric hazards are shown in Table 4.2, broken down by the weather systems that generally cause them. Although these events may be occurring only infrequently, they have enormous implications for land use and land-use planning.

The categories and types of events shown in Table 4.2 are not mutually exclusive. High winds, rain, and flooding can be associated with more than one type of atmospheric system, and often occur together. Moreover, many hazards develop slowly and cumulatively by persistent weather patterns, as, for example, drought, forest fire risk, and air pollution levels. But all of these hazardous events can be expressed statistically and in probabilistic terms.

To do so, it is first necessary to determine the probability distribution of the event or phenomenon in question (e.g., wind speed,

TABLE 4.2. Meteorological hazards categorized by weather systems generally responsible for the events

Thunderstorms
Hail
Lightning
Heavy rain
Locally high winds
Downbursts
Wind-driven rain
Tornado
The midlatitude cyclone (low pressure system) (cold fronts/warm fronts/occluded fronts)
Severe thunderstorms/lightning
Early/late freeze events
High wind
Sleet and freezing rain
Blizzard
Snow and blizzard
The midlatitude anticyclone (high pressure system)
Heatwave
Drought
Stagnant air (inversions leading to forest fire, air pollution)
Tropical storm (tropical cyclone, hurricane, etc.)
High wind
Coastal flooding–storm surge
Heavy rain—flooding
Wind-driven rain
Tornado

or 24-hour rainfall totals). Many graphical and statistical techniques are available to express the probabilities of atmospheric events (see Dunne and Leopold, 1978). A simple frequency histogram of observed events for a specified time period, or a fitted normal probability curve can be used to estimate the probability of a specific event.

Alternatively, a cumulative frequency distribution plotted on a logarithmic scale represents a useful way to estimate the probability of various phenomena. The cumulative frequency diagram is constructed by ranking all values of the phenomenon of interest; and then plotting the values against the percentage of all events less than or equal to that value, as shown in Figure 4.2 for mean annual rainfall in Harare, Zimbabwe. The left and right axes, drawn on a logarithmic scale, show the probability of rainfall in a given year either greater than or less than the graphed value. Similar plots can be constructed for any phenomenon of interest, provided adequate data are available. Note too that the reciprocal of these probability values represent a predicted return interval (in the time units of the measured data) for the indicated phenomena.

Probabilities of occurrence can be calculated for individual weather stations as well as for an aggregated region. If a sufficiently dense network of stations is available, maps can be constructed and the data can be manipulated in a Geographic Information System (GIS). For example, a map of the two-year, 1-hour maximum rainfall (in inches) is shown in Figure 4.3. Station data and mapped data are particularly useful for planning purposes. Unfortunately, adequate probability analyses require longer periods of observation, especially to describe the probability of extremely rare events.

Some important resources such as streamflow yields are less directly related to weather and climate, but they can also be expressed probabilistically if measurements are made for a sufficiently long period of time. Often much information on such extreme occurrences has already been compiled by National Weather Services or other researchers, and may be in

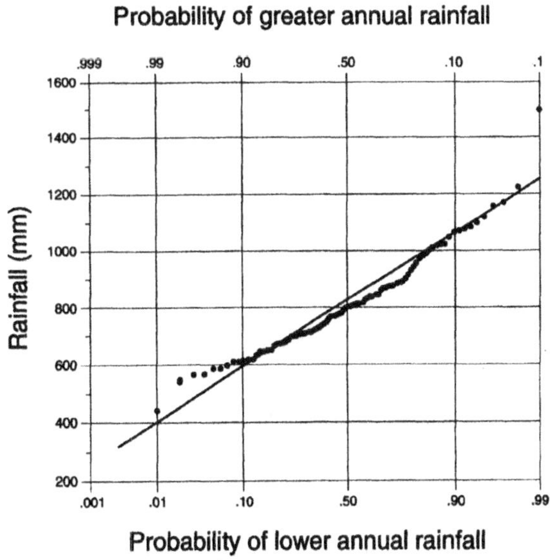

FIGURE 4.2. Percentage probabilities for annual rainfall totals for Harare, Zimbabwe.

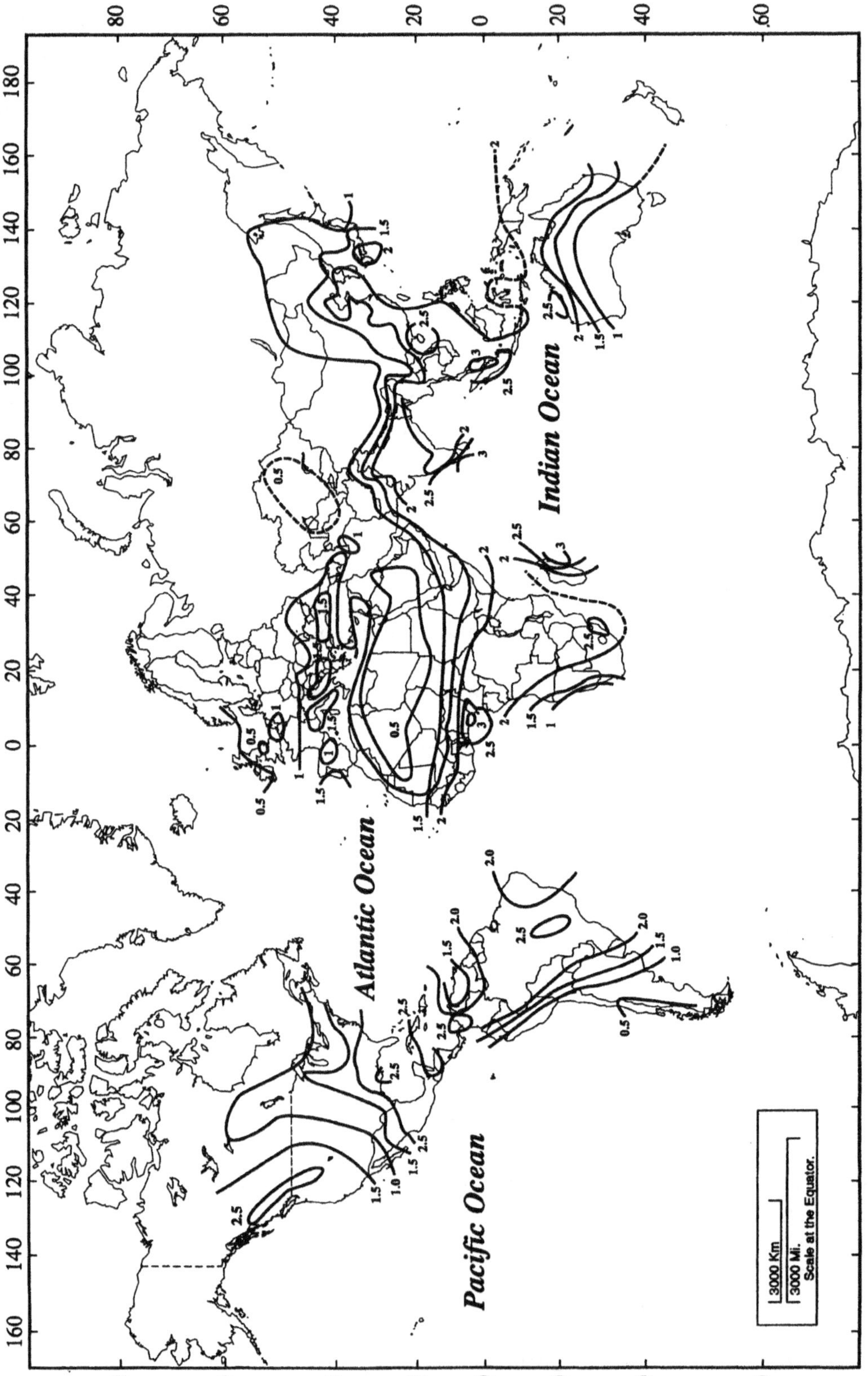

FIGURE 4.3. Spatial variation in the 2-year, 1-hour rainfall maximums (from U.S. Army Corps of Engineers, 1968).

the literature. It must be emphasized, however, that such analyses must be done for specific stations, sites, and regions. Moreover, the topographic and geographic setting can modify weather systems and produce additional hazards. For example, enhanced rainfall by mountains (orographic rainfall), heating of air on the leeward side of mountains (Foehn effects), cold air drainage from mountains (katabatic winds), land and sea breezes, and other locally important systems may not be recorded for individual weather stations or shown on regional maps. Such phenomena must be identified locally using basic knowledge of meteorological processes.

4.2.4 Conclusion

For large regions it is important to be able to describe the variability in macroclimate. Climate varies through time, especially seasonally, but it can also change over much longer time periods than one year: climate is not static. Climate also varies spatially across a region, especially in regions with much topographic relief. At local scales of tens of kilometres in distance, the nature of the surface of the Earth itself can modify the lowest layers of the atmosphere. At the microclimatic scales of decimetres and less, humans can in fact manage and manipulate the climatic environment near the Earth's surface (Sharpe, 1987).

It is my contention that it is necessary to see that climate comes in many scales and has many parts, of which only some parts may be relevant to an environmental or planning issue. Often the constraints for planning are the rare atmospheric events. The relevant parts of climate will generally change with the spatial scale and may only be relevant during a particular season or time of the year. The goal then is to find the relevant part(s) of climate and to determine how the part(s) enter mechanistically into a natural process or human activity. As will be shown, good planning requires good models of the climatic systems in operation.

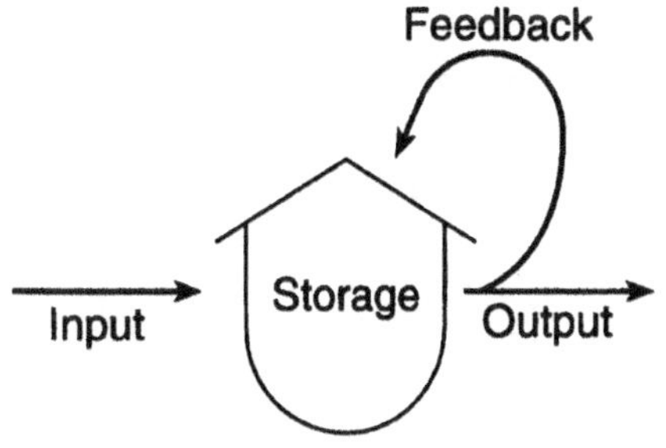

FIGURE 4.4. Basic structure of all systems.

4.3 Regional Climates from a Systems Perspective

4.3.1 The Input–Storage–Output Framework

In its most basic form, a system can be seen to have three critical parts: (i) an inflow of energy or matter; (ii) a storage component; and (iii) an output or loss component (Figure 4.4).

For all systems, a balance is eventually reached between the inputs and outputs. If for some period of time total inputs are less than total outputs the local storage (level) will go down. As the storage changes, the importance of each contributing pathway may change, but in the end the system will equilibrate.

The surface of the Earth, including the first few meters above the surface and approximately 1 m below the surface, represents the *Earth–atmosphere interface*, the place where most of the exchanges/transformations of energy and matter take place. The soil represents an important location for the storage of heat and water. To see the systems framework for climatic inputs–outputs, it is best to visualize flows upward from the soil (water) surface and downward to the soil (water) from the atmosphere. The most important fluxes include radiation (e.g., solar radiation), heat (e.g., a warm soil heating the air above it), and moisture (e.g., evaporation). It is also useful for some purposes to include fluxes of matter other than water, such as CO_2, and particulate matter, but flows of matter other than water will not be considered in this chapter.

4.3.1.1 Radiation Flux

All matter with a temperature greater than absolute zero (−273°C) will emit radiation. The amount or *flux density* of radiation is dependent on the surface temperature of the matter and its radiation efficiency (*emissivity coefficient*). For natural emission of radiation, heat in an object is being transformed into radiation and the emissivity coefficient represents the efficiency with which heat can be transformed. If emissivity coefficients are similar, the hotter the object the greater the amount of radiation emitted. At 0°C (+273°K), a perfect radiator emits about 316 W/m^2; at 15°C, 391 W/m^2; and at 30°C, 479 W/m^2 (Oke, 1987). Furthermore, the wavelength of radiation emitted depends on the temperature of the object. Thus, because of the great temperature difference between the sun (≈6000°K) and the Earth (≈300°K) the short wavelengths of the electromagnetic spectrum emitted by the Sun virtually do not overlap with the long wavelengths emitted by the Earth's ocean / land / atmosphere system. Radiation emitted by the Earth can therefore be termed longwave radiation (L) and solar radiation, short-wave radiation (K).

4.3.1.2 Heat Flux and Storage

Temperature is an indicator of the activity or motions of molecules within materials (average kinetic motion), but temperature itself cannot flow: temperature is a response to heat flow and storage. Heat flows by two basic mechanisms: (i) *conduction*, or the flow of heat from one molecule to another; and (ii) *convection*, flow due to the movement of a fluid such as air or water.

Climatologists often treat flows by convection and conduction together because it is difficult to distinguish exactly conduction from convection. Such flows of heat are termed *sensible heat flux* (Q_H), because the consequence of this heat flow can be sensed as a change in temperature with a thermometer. When surfaces such as soil, pavement, leaves, and so on, are warmer than the air temperature, sensible heat flows (Q_H) away from those surfaces, the amount being dependent on the temperature *gradient*.

A gradient is the degree of change for an entity per unit of distance. The understanding, calculation, and interpretation of gradients is critical in the use and application of climatic concepts. The greater the temperature difference per unit distance the steeper is the gradient, and the greater the potential for heat flux. This potential for heat flux is modified by the resistance to heat flow by the material.

4.3.1.3 Moisture Flux and Storage

Of great importance in climatic processes is the *phase change* of water from liquid to gas (evaporation), which consumes approximately 590 calories per gram of water (2.45 MJ/kg). When the phase change involves condensation of water, the *latent heat* within the molecules is released, and thus again becomes sensible heat. The transformation of long- and short-wave radiation into sensible heat and latent heat at the Earth's surface is largely responsible for "weather" and storms.

In places with dense crops or vegetation with a large leaf surface area, much the evaporation is due to water transport through the plant and leaf system. Such evaporation is termed *transpiration*. However, in terms of energy used, there is no difference between evaporation from water surfaces and bare soil and evaporation via transpiration. Hence, the term *evapotranspiration* is used to designate all possible water losses. The term evapotranspiration as used by climatologists and ecologists is directly equivalent to latent heat flux (Le) as used by meteorologists. Globally, evapotranspiration (Q_E) must equal precipitation (P). For local watersheds, Q_E is usually the second largest component of the hydrologic cycle. It is, however, the only surface climatic variable that is part of both the energy and water balance of a place. If Q_E is known, then one has some knowledge of the equivalent quantity of energy. Evapotranspiration is also powered by gradients, in this case, the difference in water vapour pressure between the surface and the atmosphere. If there is no vapour pressure gradient there can be no net evapotranspiration and no net latent heat flux.

Precipitation (P) in the form of rain, snow, sleet, and so on, is the primary moisture input to a region, except in the case of massive water imports via irrigation from streams or aquifers. Unlike Q_H, or Q_E, P occurs during "events" rather than as a continuous flow; however Q_H, Q_E, and P all can be highly seasonal. Water can be stored in the soil as soil water in the root zone of vegetation, groundwater as vadose and phreatic water, and free water surfaces such as ponds, lakes, and streams. The soil water balance and the progression of soil water storage through time and space is critical information needed to understand climatic fluxes. (The soil water balance and its applications will be covered in greater detail later in this chapter.) Water that cannot be added to storage or lost as Q_E must become runoff (R_0). A simple regional water balance may thus be written

$$P + I = Q_E - R_0 + \Delta St_1 + \Delta St_2 + \cdots + \Delta St_j \qquad (1)$$

where:

P = precipitation
I = imported water (irrigation, etc.)
R_0 = runoff
$\Delta St_{1\ldots j}$ = change in storage components

An expanded water balance for the soil at a place (microclimate) could be written as

$$P + I = (E + T) - (R_{suf} + R_{perc}) \qquad (2)$$

where P and I are as in equation (1) and

E = evaporation
T = transpiration
R_{suf} = surface water runoff
R_{perc} = runoff percolating through the soil

4.3.1.4 The All-Wave Radiation Balance

All of the Earth's surface weather systems are ultimately driven by solar radiation, which may arrive to an area in many different forms (Figure 4.5). Solar (shortwave) energy arriving at the atmosphere can either (i) pass through the atmosphere unaltered; (ii) be scattered; (iii) be reflected back to space; or (iv) be absorbed by molecules in the atmosphere. The short-wave radiation that passes directly through the atmosphere, unaltered, to the Earth's surface is called *direct beam* solar radiation ($K\downarrow_b$). Diffused (or scattered) solar radiation is called *diffuse beam* radiation ($K\downarrow_d$). Both direct beam ($K\downarrow_b$) and diffuse beam ($K\downarrow_d$) solar radiation can be reflected by a surface back into the atmosphere and into space ($K\uparrow$). The proportion of solar radiation reflected by the surface is called the surface *albedo* ($\propto$).

We saw in Section 4.3.1.1 that all matter emits radiation. The solar radiation absorbed

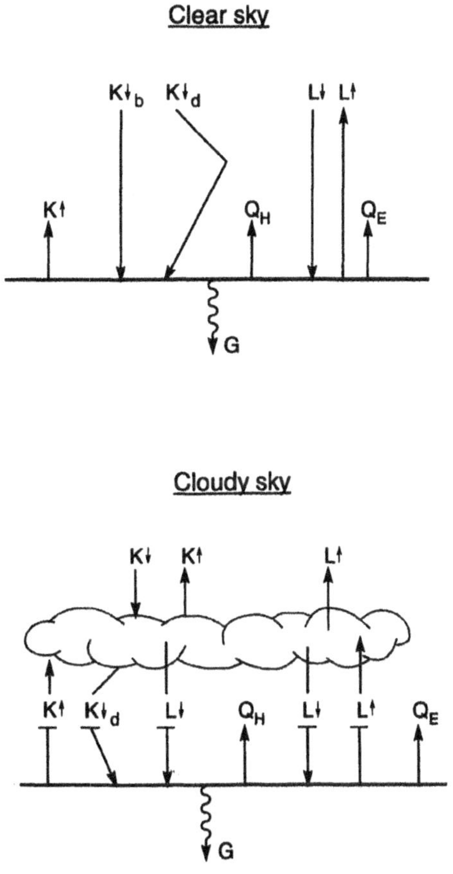

FIGURE 4.5. Radiation and energy fluxes at the surface of the Earth on a clear and on a cloudy day.

by the atmosphere is transformed into long-wave radiation (*atmospheric counterradiation*) and is emitted downward ($L\downarrow$) by matter in the atmosphere—that is, by gases, water droplets/crystals, or particles (aerosols) suspended in the atmosphere. Likewise, long-wave radiation ($L\uparrow$) is emitted upward from the Earth's surface (*terrestrial radiation*), of which some is absorbed by the atmosphere and some passes through the atmosphere and is lost to space. For the entire Earth, as a sphere in space, the long-term radiant gains at the outside of the atmosphere must exactly equal the losses or there would be a continuous build-up of heat with in the Earth–atmosphere system. At any particular place on the Earth's surface, however, the long-wave energy directed downward ($L\downarrow$) from the atmosphere versus upward ($L\uparrow$) from the Earth's surface may not be equal; this difference is called the long-wave radiation balance (L^*), and can be expressed as:

$$L\downarrow - L\uparrow = L^* \quad (3)$$

where

$L\downarrow$ = long-wave atmospheric radiation
$L\uparrow$ = long-wave terrestrial radiation
L^* = long-wave radiation balance

Generally, L^* for the Earth's surface is related to the difference in temperature between the atmosphere and the temperature of the Earth's surface.

Similarly, the difference between the short-wave radiation directed downward ($K\downarrow$) and that reflected back to space ($K\uparrow$) is called the short-wave radiation balance (K^*), which can be expressed as:

$$(K\downarrow_b + K\downarrow_d) - K\uparrow = K^* \quad (4)$$

where

$K\downarrow_d$ = diffuse (scattered) beam radiation
$K\downarrow_b$ = direct beam radiation
$K\uparrow$ = reflected radiation
K^* = short-wave radiation balance

Combining equations (3) and (4) yields the total, or *all-wave radiation balance* (Q^*), for any given location

$$K^* - L^* = Q^* \quad (5)$$

The net radiation balance (Q^*) can vary over a daily (or diurnal) cycle, or seasonally, and it can also vary greatly from place to place. Differences in Q^* from place to place, and from season to season, is the driving force behind weather and weather systems. Surpluses and deficits of radiation are compensated by flows of heat—both sensible heat and latent heat—as the overall energy balance tends toward equilibrium.

When Q^* is *negative* both short- and long-wave radiation are being lost from the near surface soil/water/atmosphere environment. To maintain equilibrium, sensible and latent heat must flow in to maintain overall energy equilibrium. Except for extremely high latitudes, Q^* is usually positive during daylight hours and negative at night. Q^* totalled for an entire season is typically positive in summer and negative during winter. Totalled for an entire year, it is typically positive at low latitudes and negative at high latitudes. The total Q^* can be negative at a place for an entire year, so long as heat is being imported to that place by ocean currents and winds. When Q^* is *positive*, the extra energy must be dissipated through a negative heat flux or directed to long-term heat storage (such as soil heat). The allocation (or use) of positive Q^* is one of the most important of all climatological processes—and is the root cause of virtually all weather. Cold fronts, warm fronts, tropical cyclones, and virtually all other weather phenomena are mechanisms to redistribute imbalances of radiation from place to place, and from season to season.

One of the most critical of all climatological concepts is the following equation, which represents the allocation of a positive radiation balance to sensible and latent heat flux and to changes in heat storage:

$$Q^* = Q_E + Q_H + \Delta G + \Delta M \quad (6)$$

where:

Q^* = all-wave radiation balance (or net radiation)
Q_E = evapotranspiration
Q_H = sensible heat flux
ΔG = change in heat in soil
ΔM = miscellaneous term for small losses and gains of energy (e.g., photosynthesis)

This equation is also termed the *energy balance*. The ratio of Q_H to Q_E is termed *Bowen's ratio* (β) and represents the relative allocation of positive net radiation to Q_H versus Q_E. Under fully wet (saturated) conditions and a deeply rooted vegetation with a full canopy, Bowen's ratio (β) may be less than unity, suggesting that latent heat flux exceeds sensible heat flux. For water bodies and for the world's oceans, this ratio is near 0.10–0.12 (Oke, 1987), that is, the bulk of the available energy is used for evaporation while sensible heat flux and temperatures may be kept low. In drier environments, increases and may eventually become greater than unity. Deserts can have values greater than 5, that is, sensible heat flux is more than five times larger than the latent heat flux. This partitioning of Q_E versus Q_H helps explains why all of the world's highest surface air temperature records have been set in deserts.

By adding water to the environment it is possible to manage in favour of more Q_E relative to Q_H. For small wet surfaces surrounded by dry air and much horizontal transfer (*advection*) of Q_H from surrounding areas by wind, Q_E can be greatly enhanced and Q_H consumed. This phenomenon has been termed the *oasis effect* as well as the *clothesline effect* (Oke, 1987).

When dry hot air moves from a surface with a positive over an irrigated field, steep vapour pressure gradients are produced at the leading edge of the field, which will therefore experience a large Q_E. Farmers have been known to mist agricultural plots to reduce surface temperatures during summer heat waves, though the wisdom of using scarce water resources in this way is questionable. Whenever small wet environments are surrounded by dry, warm air, the potential for cooling the air and surrounding surfaces at the expense of water loss is possible; Q_H is used to increase Q_E. Alteration of one part of the energy and water balance of a place invariably leads to changes of other parts; the thermodynamic and matter balance sheets must always balance. When adequate models of these exchanges, flows, storages, and related feedbacks are available, the changes can be predicted.

4.3.2 Feedback Controls by Surfaces

In the previous sections we have seen that climate can be depicted by: (i) the short-wave/long-wave radiation balance; (ii) radiation, sensible and latent heat flux lumped together in energy balance; and (iii) a moisture balance. Radiation, heat, and moisture all can flow, but only heat and moisture can be stored. Humans can impact many parts of these balances both intentionally and inadvertently, especially at microclimatic scales.

Attempts to increase regional precipitation by cloud seeding have a long history of variable success and variable benefits versus costs (Cotton and Pielke, 1995). Under most circumstances, precipitation and solar and long-wave atmospheric radiation in the open sky environment cannot be altered. Locally at the Earth's surface, interaction and feedbacks among moisture and energy flows are always present, however, and lead to the situation that when one part of the system is altered (as might result from irrigation, clearing, paving, etc.), other fluxes and storages are impacted (Sharpe, 1987). Effective planning requires an understanding of these feedbacks and their consequences. Here we examine different surfaces (local to microclimate) to see how energy and moisture balances can be altered by human activities.

4.3.2.1 Changes in Albedo

Changes in albedo of surfaces are relatively easy to make and can produce considerable

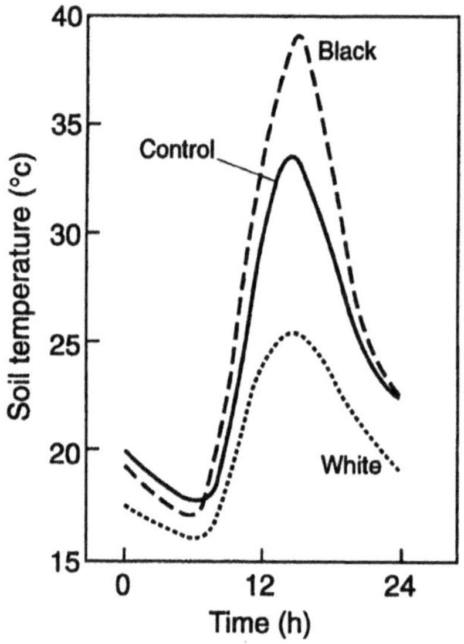

FIGURE 4.6. The impact of surface treatments on soil temperature at 10-mm depth during a sunny day (after Oke and Hannell, 1966; in Oke, 1987; reproduced with permission of Routledge).

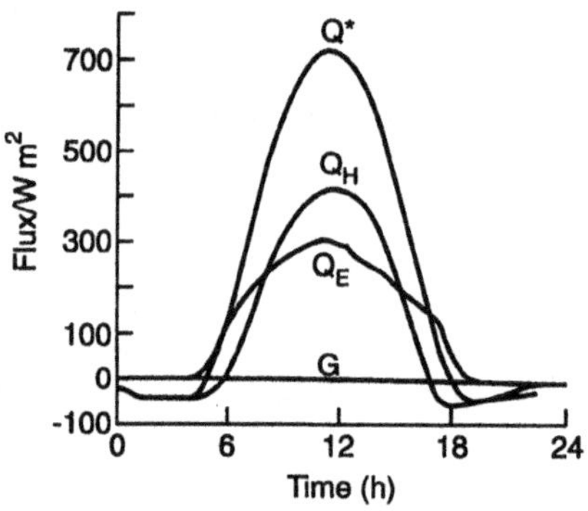

FIGURE 4.7. Energy balance of a cornfield (after André et al., 1989).

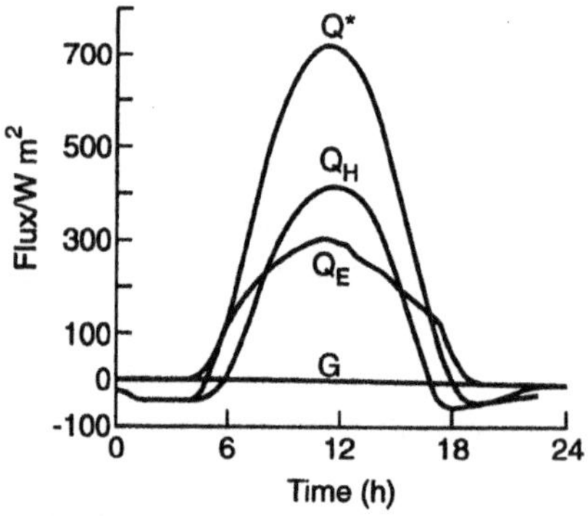

FIGURE 4.8. Energy balance of a pine forest (after André et al., 1989).

changes in surface temperature. Many experiments have been conducted in which surface albedos have been manipulated (see Geiger et al., 1995) using whitening and darkening agents. In one such experiment a short grass cover was lightened with talc and another plot was darkened by carbon black (Figure 4.6). A control plot was unchanged. The black surface was almost 6°C warmer in full sunlight than the control surface, and the white surface was about 8°C cooler. These experiments also showed that temperatures can be confounded by soil wetness and that during the night the temperatures of various surfaces tend to equalize (Geiger et al., 1995). Furthermore, all these results indicate that changes in one part of the system can cause feedbacks affecting other parts.

4.3.2.2 A Cornfield and a Pine Forest

Experiments conducted in France (André et al., 1989) revealed differences in net radiation and energy balance for a cornfield compared to a pine forest ("Landes" forest). Figures 4.7 and 4.8, which have been generalized from André et al. (1989), show that the cornfield had a higher Q_E, and a lower Q_H, than the pine forest. In addition, a small amount of heat (G) entered the soil in the cornfield, especially after 5:00 h. These data are for 16 June 1986. Even though the corn had a leaf area that covered only about 80 percent of the soil surface, it had a higher Q_E versus Q_H (numerically smaller Bowen's ratio, β).

The different Bowen's ratios for the different surfaces are caused by the low stomatal resistance (leaky nature) of corn plants versus the pine trees in the forest. Near sunset (17:00–18:00 h) sensible heat flux becomes negative indicating surface cooling by conduction and convection. Net radiation also tends to become negative at sunset as the heat accumulated in the vegetation and soil surface is emitted to space by terrestrial long-wave radiation ($L\uparrow$). If certain land uses cover relatively large areas, the temperature and humidity of the lower layers of the atmosphere can be impacted by the different allocation of $Q*$.

4.3.3 The Soil Water Balance

The soil water balance (see equations (1) and (2) above) is a model of water input, loss, and storage for a soil profile. It is a particularly powerful model because it can explain a great deal about a region's climate and has many practical applications (see Mather, 1978). The model incorporates climatic energy balance concepts as well as hydrologic concepts (Sanderson, 1990). Moreover, it is useful in simulations because its parts can be manipulated to determine the degrees of change in one or more parts of the system.

4.3.3.1 Soil Water

Basically the soil water balance is an account of the water in the root zone of a soil profile. Water is lost by evaporation from the soil and/or by transpiration through plants. Water below the root zone is essentially unavailable for transpiration. Water can also be lost via surface runoff and percolation; water is stored on and among the soil particles. The amount of water in a soil is usually defined by four markers or levels, from *saturation* (all soil space filled with water), to *field capacity*, the *permanent wilting point* (or percentage) and the soil's *hygroscopic limit*.

The field capacity is the quantity of water that the soil can hold against the force of gravity. Water in excess of field capacity is gravity water that generally percolates down, away from the root zone. Without further input, water is lost by evapotranspiration down to the permanent wilting point (PWP) which is the level of moisture below which plants can no longer transpire water. At this level, the small remaining quantity of water is held so tightly by soil particles that plants are unable to extract the water from the soil.

By slow, molecular diffusion a soil can dry further to the point that the soil is essentially in equilibrium with the humidity of air. Even the world's driest soil contains traces of water. For practical purposes the soil water of significance in the water balance is water held between the soil's field capacity and the permanent wilting point, for it is this water that is available for plants to use. This quantity is therefore termed the *plant-available soil water.*

Precipitation can add water to a soil up to the field capacity. Any water in the soil beyond field capacity quickly percolates down by the force of gravity. Evapotranspiration can reduce soil water to the permanent wilting point. The actual quantity of water stored by the soil at any given time depends on the texture, organic matter content, the rooting depth of vegetation, and a few minor other considerations. To some degree, the water-holding capacity of soil can be managed, mostly by vegetation (or crop selection) and by organic matter in and on the soil. The method does require, however, information about evapotranspiration rates.

4.3.3.2 Evapotranspiration

Evapotranspiration is the second largest component of the hydrologic cycle, but has been one of the most elusive climatological variables to measure and model (see, e.g., Brutsaert, 1982). Free water surfaces such as ponds and lakes are simpler to model than land surfaces, where transpiration through the plant is often the major pathway for water loss (output) from the soil to the atmosphere. In practice the measurement of evapotranspiration has been cumbersome, costly, and imprecise (Knapp, 1985).

The use of small wet surfaces to measure evapotranspiration leads to exaggerated estimates due to the input of sensible heat by advection. Evaporation pans measure only evaporation, and may also produce erroneous results due to advectional input of sensible heat. Lysimeters are among the most effective instruments to measure evapotranspiration. They consist of large containers or tanks of soil placed inside a slightly larger container, and which has its surface equal to the surrounding soil. They can contain growing vegetation, and evapotranspiration is measured by weight loss. Such "weighing" lysimers are costly to build and maintain. In practice evapotranspiration is usually estimated using a variety of methods ranging from simple empirical approaches to detailed mechanistic approaches which may require instrumentation and calibration.

4.3.3.3 A Water Balance Model

Since the early work of Thornthwaite (1948) and Penman (1948) on *potential evaporation (PET), the soil water balance has been mod*elled. The "effective" moisture available at a place is the degree to which precipitation is able to meet the potential for evapotranspiration. The concept of PET (or the reference evapotranspiration; see Knapp, 1985) has been variously defined and estimated. A generally accepted definition is that PET is the maximum amount of water that can be evapotranspired from a fully rooted vegetation with a full canopy cover of the ground and no shortage of soil moisture. Often this ideal environment is a shortly mown grass surrounded by a uniformly wet soil, and with no net advection of Q_H from surrounding areas. PET is then equivalent to the energy available to cause a phase change in water. It is highly correlated with positive net radiation Q^* and has been termed the environmental demand for water (Mather, 1978).

The actual amount of water evaporated and transpired, the *actual evapotranspiration* (AET), will equal PET under fully wetted soil conditions, and for free water surfaces; AET may fall below PET when shortages of soil water are present and when precipitation is insufficient to meet the PET. When AET is less than PET, a soil water shortage or *deficit* (*D*) will accumulate. When the soil is at its field capacity and precipitation exceeds PET, then the extra water cannot be stored or evapotranspired and then becomes *water surplus* (*S*). Water surplus can potentially become surface runoff or runoff as gravity water via percolation.

Table 4.3a–c shows the water balance "ledgers" for three locations (Zaragoza, Spain; Kiev, Ukraine; and Athens, Georgia, USA). The ledgers represent a monthly water balance for long-term climatic normals. The top line represents PET (based on Thornthwaite's estimate), the climatic demand for water by month which is matched against the precipitation (*P*), the monthly supply of water. The third line is the estimated soil moisture storage (*St*) followed by AET, water shortage (*D*) and water surplus (*S*). When *P* exceeds PET, then there is adequate water to meet PET, so AET is equal to PET (e.g., November through February at Zaragoza). When *P* is less than PET, soil moisture storage is used (soil moisture utilization) to meet the PET demand. In Table 4.3a–c, a maximum storage (*St*) of 300 mm has been used to represent a deeply rooted natural vegetation with full ground cover.

The depletion of water in the root zone can be complex and is generally not a linear function of PET. Rather this model (Table 4.3a–c) uses a function (decreasing availability) so that as soil water levels fall below field capacity, the moisture is ever more difficult to evaporate and transpire. Based on this model (Mather, 1978), some shortage of water, a deficit (*D*), may begin to accrue well before the permanent wilting percentage (PWP) is reached. When AET falls below PET a soil moisture deficit results, which in Zaragoza occurs on average from March through October. In seasons (months) that *P* is greater than PET, the extra *P* which cannot be stored or cannot be evapotranspired must become extra or surplus (*S*) water, as happens normally at Athens, Georgia, from late December into April.

The ledgers show the aridity of Zaragoza as indicated by the average annual total deficit of 478 mm. Kiev has an average annual deficit of 65 mm and Athens, 51 mm. When data for monthly averages are used in this model, Zaragoza does not generate a water surplus in any month. In reality, some short periods of wet weather will generate surpluses. Kiev has a typical temperate grassland climate in which *P* occurs mostly in the season of high PET. Athens, Georgia, is in a forested, humid subtropical environment that generates large amounts of surplus water (367 mm). It should also therefore be no surprise that the soils in this region are highly leached and acidic.

Figure 4.9a–c shows the three water balance ledgers represented as figures. The four seasons of the water balance year are soil moisture recharge, surplus, soil moisture utilization, and deficit. Zaragoza has on average only three of these seasons, because the surplus sea-

TABLE 4.3 (a) Waterbalance ledgers for Zaragoza, Spain. (b) Waterbalance ledgers for Kiev, Ukraine. (c) Waterbalance ledgers for Athens, Georgia, USA. PET = Potential evapotranspiration; *P* = Precipitation; *St* = Soil moisture storage; AET = Actual evapotranspiration; *D* = Soil moisture deficit; *S* = Soil moisture surplus

(a) Zaragoza, Spain. 41°39'N, 0°52'W, elevation 200 m

	J	F	M	A	M	J	J	A	S	O	N	D	Y
PET	11	17	33	49	88	118	143	137	93	55	25	14	783
P	15	20	22	28	40	28	19	14	29	36	31	23	305
St	24	27	26	24	21	15	10	7	5	5	11	20	
AET	11	17	23	30	43	34	24	17	31	36	25	14	305
D	0	0	10	19	45	84	119	120	62	19	0	0	478
S	0	0	0	0	0	0	0	0	0	0	0	0	0

(b) Kiev, Ukraine. 50°57'N, 30°30'E, elevation 183 m

	J	F	M	A	M	J	J	A	S	O	N	D	Y
PET	0	0	0	38	96	118	133	115	70	32	2	0	604
P	35	30	44	44	51	74	81	56	46	49	41	39	590
St	271	301	300	300	258	222	186	153	141	158	197	236	
AET	0	0	0	38	93	110	117	89	58	32	2	0	539
D	0	0	0	0	3	8	16	26	12	0	0	0	65
S	0	1	44	6	0	0	0	0	0	0	0	0	51

(c) Athens, Georgia, USA. 33°57'N, 83°19'W, elevation 243m

	J	F	M	A	M	J	J	A	S	O	N	D	Y
PET	11	15	31	65	108	148	161	150	108	58	23	10	888
P	122	117	126	99	100	85	121	76	98	70	90	100	1204
St	300	300	300	300	292	236	207	161	156	168	235	300	
AET	11	15	31	65	108	141	150	122	103	58	23	10	837
D	0	0	0	0	0	7	11	28	5	0	0	0	51
S	111	102	95	34	0	0	0	0	0	0	0	25	367

son is missing. Athens and Kiev have all four. This does not mean, however, that Zaragoza never has surplus water. Surpluses can be generated in unusual years and following short, episodic rainfall events.

The water balance calculations shown above in both ledger and graphical form are based on the simple empirical estimate of PET developed by Thornthwaite (1948) and the soil water balance model operationalized by Thornthwaite and Mather (1957). In Thornthwaite's estimate of PET, only simple weather data are needed. Monthly mean temperatures, monthly precipitation totals and the location's latitude, which is needed to calculate a day-length factor, are the only climatic data required. Additionally a measured or estimated value of soil moisture storage is needed. The soil moisture balance has important implications for agricultural and landscape planning, and for foundation and construction design. In practice, soils have variable levels of maximal soil water storage, and thus different soils within a region of similar climate may produce quite different seasonal soil water balance profiles.

4.3.3.4 Accuracy Versus Applicability

All methods that involve an estimate of AET involve trade-offs between ease of application and acceptance of errors, versus methods that provide greater accuracy but may be computationally cumbersome or may require data that are unavailable or difficult to gather, such as net radiation. Again this is a problem of

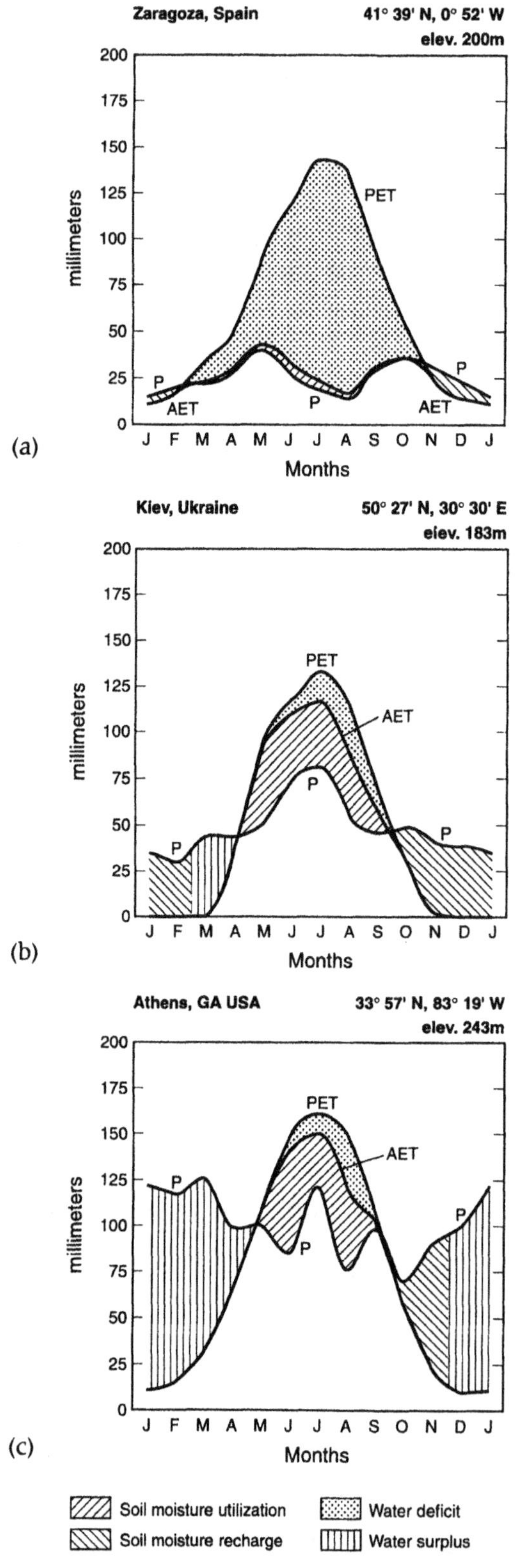

FIGURE 4.9. (a) The water balance for Zaragoza, Spain. (b) The water balance for Kiev, Ukraine. (c) The water balance for Athens, Georgia, USA.

matching the method with the application. In general, even the simple empirical methods provide relatively accurate estimations for longer time periods, such as months or years, while the more mechanistic approaches are needed for estimations over shorter time periods of hours to days (Knapp, 1985).

The water balance represents an environmental input–output system. It contains both meteorological–climatological concepts (i.e., AET is latent heat flux) and hydrological (i.e., precipitation, surplus, runoff, etc.) concepts. It can be used to classify climates (Thornthwaite, 1948; Sanderson, 1990), predict runoff (Ferguson, 1996), and make predictions about other environmental processes such as net primary production (Lieth and Box, 1972) and organic decay rates (Meentemeyer, 1978).

In its simplest form, the water balance is macroclimatic because it may not depict small place to place differences caused by land use, soil characteristics, or topography. The basic water balance framework has, however, great flexibility. It can be adjusted to more local, even site specific, situations by adjusting the parts. Soil storage capacity, rainfall infiltration rates, canopy closure, and frozen soil are examples of specific situations that have been modelled in detail.

Some of the more mechanistic approaches to the soil water balance have abandoned the PET concept to attempt a more deterministic evapotranspiration model (Wallace, 1995). Yet in spite of all the variations in the models, all known variations follow the basic input–output framework.

4.4 Applied Climatology, Management and Planning

4.4.1 Models

Effective planning and management from a scientific perspective require good understanding of how Earth and environmental systems operate. In other words, one needs good models of the systems. A model is a simplification and/or a representation of the system; a model can be a complex set of numerical equations, a computer program, a diagram, a map, a depiction using language, or any combination of these.

A good model is one that is able to make predictions. The predictions (or expected outcomes) can then be used to select management options. If the reflectivity of a surface is increased, then less solar radiation is absorbed leading to lower temperatures and perhaps less evapotranspiration. Paving an area leads to drastic changes in both energy and moisture balances; the Bowen's ratio is altered favouring Q_H over Q_E and precipitation is rapidly directed to runoff rather than soil storage. For many other manipulations the outcomes may not be so obvious. Climatology from a systems perspective does have the advantage, however, in that the accounts for energy and matter (mostly water) must balance.

Management and planning may involve periodic and rare events that require data and knowledge. Freeze dates can be important for many horticultural crops and can be expressed by statistical probability of occurrence, as we saw in Section 4.2.3. The duration and intensity of a freeze may also be important. However, the effect of any given phenomenon on a particular crop needs to be determined, usually empirically by observation. Freeze occurrences, duration, and intensity vary with topography and can be modelled or an existing model can be applied. Similarly, other hazards or risk factors such as hail, snow loads on buildings, temperature extremes, and fire risk can be quantified for their probability of occurrence, duration, and intensity. For local applications, topographic slope and aspect may need to be considered.

4.4.2 Impacts of Land Use on Local and Microclimates

It is not possible in this short chapter to review all of the work and resulting models that have been produced for local and microclimates. The volumes produced by Geiger et al. (1995) and Oke (1987) present excellent summaries of existing knowledge, and Oke uses a systems framework. It is important to understand that microclimates are generally imbedded within local climates that are imbedded within regional (meso) climates, and so on.

TABLE 4.4. A listing of landuse actions by humans which can lead to changes in local and microclimates

Albedo changes
Anthropogenic heat release
Shelterbelts
Flooding/draining
Clearing/reforestation
Pavements/structures
Fields
Furrows (height/orientation)
Mulching
Orientation
Air dams
Management of "edges"
Soil changes
Thermal capacity
Thermal conductivity
Surface roughness
Sprinklers (frost prevention)
Irrigation
Air pollution

Table 4.4 presents a listing of major examples by which human actions can lead to changes in the local and microclimates. Many items on this list represent land use changes, and all items are commonly reported in the literature (e.g., Geiger et al., 1995). They also suggest actions that planners might suggest which would have climatological significance.

At a larger scale, albedo changes were discussed earlier. Heat release can increase air temperatures locally, but is most pronounced in urban settings. The growing of shelter belts to control wind impacts has ancient origins and has often been studied. The flooding and draining of land leads to changes in surface thermal capacity and conductivity, albedo, and the Bowen's ratio. Clearing and reforestation alter all aspects of the energy and water balances as do pavements and structures.

At microclimatic scales the height and orientation of furrows in fields changes energy budgets. Mulching is a particularly powerful method of altering field microclimates and soil moisture balance. At night cold air drains down slopes and can be redirected by vegetation and/or structures. Soil thermal capacity and conductivity can be altered. Clearing and reforestation also alter the surface roughness and local wind patterns.

Irrigation greatly influences energy and water balances, and if the irrigated area is small, may cause oasis effects. Sprinkler systems are often used to prevent frost/freeze events harmful to crops via the release of heat by freezing water. Regional air pollution can alter to some degree the short-wave and long-wave radiation balance.

All of the climatic factors discussed above should be considered in the planning process. They are not only significant to the production of products but also are a foundation of human health and well being and the well being of the fauna and flora.

4.5 Summation

Climate has progressively become more synonymous with the term environment. Although the atmosphere is a critical component of the climatic environment, it is not the only component. Furthermore, climate at its finer scales is understood less and less by what is measured by weather stations, satellites and atmospheric soundings and more by the fluxes and storages of energy and matter near the surface of the Earth. Such a view relies less on the state of the atmosphere than on systems concepts.

The use of climate in regional planning is in reality a form of applied climatology in which climatic knowledge and models are used deductively—to predict specific cases from the generalizations. Such applications require a careful matching of time and spatial scales, and specific models to the specific problems at hand. Climate has many parts, which include thermal parts, moisture parts, seasonality, and atmospheric motion. The correct element of climate must be matched with the Earth surface process or phenomenon being considered. Atmospheric pressure, for example, has little direct bearing on soil temperature.

The systems concepts of input–storage–output are nicely employed in radiation and energy balance concepts, and in the water balance. These concepts are powerful building blocks for refined applications. The applications may require, however, decisions regarding the trade-offs between accuracy in prediction and the ease of application, which are regulated in part by climatic scale. In the progression from finer (microclimatic) scales to the macroclimatic, detail becomes less important. At fine scales a host of variables may appear important and can indeed be altered and manipulated.

At microclimatic and local scales climate can be managed to some degree. Land uses can then match the managed environments as well as the soil and topographic characteristics of the region. At these finer scales most spatial variability is related to land use and topography. Planning at the macroclimatic scales of climate involves mostly the anticipation of rare and extreme atmospheric events.

Weather events have the properties of frequency, intensity, duration, and areal extent (and also seasonality). Among these, frequency, intensity, and duration can be expressed probabilistically. Drought and freeze events are good examples. The macroclimatic and large regional climates cannot yet be manipulated intentionally, although we may now have entered that part of history of human habitation of the Earth in which cumulative alterations at the finer scales are leading to unintended, and perhaps inalterable, changes in the climate of the entire globe.

References

André, J.-C.; Bougeault, P.; Mahfouf, J.-F.; Mascart, P.; Noilhan, J.; Pinty, J.-P. Impacts of forests on mesoscale meteorology. *Phil. Trans. Roy. Soc. London B* 324:406–422; 1989.

Brutsaert, W.M. *Evaporation into the Atmosphere.* Dordrecht: D. Reidel; 1982.

Chadwick-Healey Ltd. *World Climate Disc: Global Climatic Change Data on CD-ROM*, with User Manual and Reference Guide. Price £595.00. Cambridge: Chadwick-Healey Ltd; 1992.

Cotton, W.R.; Pielke, R.A. *Human Impacts on Weather and Climate*. New York: Cambridge University Press; 1995.

Dunne, T.; Leopold, L.B. *Water in Environmental Planning*. San Francisco, CA: W.H. Freeman and Company; 1978.

Earth Info, Inc. *Environmental Databases on CD-ROMs*. Boulder, CO: Earth Info, Inc.; 1997.

Ferguson, B.K. Estimation of direct runoff in the Thornthwaite water balance. *Prof. Geogr.* 48:263–271; 1996.

Geiger, R. *Das Klima der Bodennahen Luftschicht*. Braunschweig: Vieweg; 1927.

Geiger, R.; Aron, R.H.; Todhunter, P. *The Climate Near the Ground*. 5th ed. based on the Hud ed. 1961. Braunschweig: Vieweg; 1995.

Knapp, H.V. Evapotranspiration and transpiration. Houghton, D.D. ed. *Handbook of Applied Meteorology*. New York: John Wiley and Sons; 1985:537–554.

Landsberg, H.E. *World Survey of Climatology*. Multiple volumes. Amsterdam, New York: Elsevier Scientific Publications; 1969–1974.

Lieth, H.; Box, E.O. Evapotranspiration and primary productivity. *Publ. Climatol.* 25:37–46; 1972.

Mather, J.R. *The Climatic Water Balance in Environmental Analysis*. Lexington, MA: Lexington Books; 1978.

Meentemeyer, V. Macroclimate and lignin control of litter decomposition rates. *Ecology* 59: 465–472; 1978.

Meentemeyer, V. Geographical perspectives of space, time and sale. *Landsc. Ecol.* 3:163–173; 1989.

Oke, T.R. *Boundary Layer Climates*. 2nd ed. London: Routledge; 1987.

Penman, H.L. Natural evaporation from open water, bare soil, and grass. *Proc. R. Soc. London Ser. A* 193:120–145; 1948.

Sanderson, M. ed. *UNESCO Sourcebook in Climatology for Hydrologists and Water Resource Engineers*. Paris: United Nations Educational, Scientific, and Cultural Organization; 1990.

Sharpe, D.M. Microclimatology. Oliver, J.E.; Fairbridge, R.W. eds. *The Encyclopedia of Climatology*. New York: Van Nostrand and Reinhold Co.; 1987:572–581.

Stern, I. Meteorology FAQ. Sources of weather data (7 parts). `ilana@kiowa.scd.ucar.edu`; 1997.

Thornthwaite, C.W. An approach toward a rational classification of climate. *Geogr. Rev.* 38: 55–94; 1948.

Thornthwaite, C.W.; Mather, J.R. Instructions and tables for computing potential evapotranspiration and the water balance. *Publ. Climatol.* 10:185–311; 1957.

U.S. Army Corps of Engineers. *Planning and Design of Roads, Air Bases, and Heliports in the Theater of Operations*. Technical Manual TM5-330. Washington, DC.: U.S. Army Corps of Engineers; 1968.

Wallace, J.S. Calculating evaporation: Resistance to factors. *Agr. Forest Meteorol.* 73:353–366; 1995.

Willmott, C.J.; Mather, J.R.; Rowe, C.M. Average monthly and annual surface air temperature and precipitation data for the world. *Publ. Climatol.* 34(1/2): 1–790; 1981.

Yoshino, M.M. *Climate in a Small Area*. Tokyo: University of Tokyo Press; 1975.

Yoshino, M.M. Local climatology. Oliver, J.E.; Fairbridge, R.W. eds. *The Encyclopedia of Climatology*. New York: Van Nostrand and Reinhold Co.; 1987:551–558.

Further Reading

Caldwell, W.J. Consideration of the environment: An approach for rural planning and development. *J. Soil Water Conserv.* 49:324–322; 1994.

Copeland, J.H., Pielke, R.A.; Kittel, T.G.F. Potential climatic impacts of vegetation change: A regional modelling study. *J. Geophys. Res. Atmos.* 101:7409–7418; 1996.

Gosz, J.R.; Sharpe, P.J.H. Broad-scale concepts for interaction of climate, topography, and biota at biome transitions. *Landsc. Ecol.* 3:229–244; 1989.

Host, G.E.; Polzer, P.L.; Mladenott, D.J. ; White, M.A.; Crow, T.R. A quantitative approach to developing regional ecosystem classifications. *Ecol. Appl.* 6:608–618; 1996.

Houghton, D.D., ed. *Handbook of Applied Meteorology*. New York: John Wiley and Sons; 1985.

Hulme, M. Using climate information in Africa: Some examples related to drought, rainfall forecasting and global warming. *IDS Bull.–Inst. Develop. Stud.* 25(2):59–68; 1994.

Jensen, M.E.; Burman, R.O.; Allen, R.G., eds. *Evapotranspiration and Irrigation Water Requirements*. ASCE Manual No. 70. New York: ASCE; 1989.

Kates, R.W.; Ausubel, J.H.; Berberian, M., eds. *Climate Impact Assessment: Studies of the Interaction of Climate and Society.* SCOPE; Vol. 27. Chichester: John Wiley and Sons; 1985.

Leenhardt, D.; Voltz, M.; Rambal, S. A survey of several agroclimatic soil-water balance models with reference to their spatial application. *Eur. J. Agron.* 4:1–14; 1995.

Rounsevell, M.D.A.; Jones, R.J.A. A soil and agroclimatic model for estimating machinery workdays: The basic model and climatic sensitivity. *Soil Till. Res.* 26:179–191; 1993.

Sheng, T.C. *Watershed Management Field Manual: Watershed Survey and Planning.* FAO Conservation Guide 1316. Rome: FAO; 1990.

Van Hamme, T. Rainfall and topography. *Hydrol. Cont.* 7:51–73; 1992.

Whilhite, D.A., ed. *Drought Assessment, Management, and Planning: Theory and Case Studies.* Dordrecht: Kluwer Academic Publishers; 1993.

White, D.H.; Howden, S.M.; Nix, H.A. Modelling agricultural and pastoral systems under environmental change. *Ecol. Model.* 86:213–217; 1996.

Wu, R.S.; Haith, D.A. Land use, climate, and water supply. *J. Water Resour. Plan. Man.–ASCE* 119(6): 685–704; 1993.

5

The Hydrological Cycle and the Role of Water in Mediterranean Environments

John B. Thornes

5.1 Introduction

The hydrological cycle is the movement of water from the atmosphere, through the soils and rocks, into river channels and to the sea. An important part of this cycle is the atmosphere and that is dealt with by Meentemeyer in Chapter 4 of this book. In this chapter we concentrate on water in the vegetation canopy, in the soil, as groundwater, and in rivers.

It would be difficult to overestimate the importance of water in rural planning, whether one is talking about the humid tropics, where water is abundant, or the near deserts, where water is scarce. Water is basic to human life and settlement cannot exist without it. Where water is most scarce or it is highly seasonal, it is a major factor in determin-ing agricultural activity and even long-distance water transfer may not offset the scarcity. Scarcity through time is as important as scarcity in space and many agricultural systems try to ensure that seasonal variations in rainfall are accommodated in the crop cycle. Plant growth is controlled by light and water, and in dry environments water is the limiting factor. Irrigation in drylands is therefore the main issue in water planning.

However irrigation is not the only concern. Water is needed for public consumption, for industrial use, and for a wide range of other purposes. Among these other aspects of water management, one of the most important in rural communities in Mediterranean environments is in controlling run-off from hillslopes and in rivers, where it may cause soil erosion and flooding. This is especially important in Mediterranean environments because of sparse vegetation cover and the intense character of rainfall. Moreover extensive erosion may lead to reservoir sedimentation, leading to a reduction in water resources availability. There is here a vicious circle (Ives and Messerli, 1988) that can lead to further degradation.

The huge demands that modern society, especially intensive agriculture, make on water resources mean that water resources are often too meagre to meet that demand. Reservoirs are often run down to dangerously low levels; rivers are run dry; and groundwater resources are depleted. Moreover, even in rural environments, water pollution by agricultural waste, residues from agricultural industry, rural villages and towns, and from toxic waste disposal add further to the burden of rural planning. They often pose formidable problems if the processes of degradation are carried too far, and the message has to be that prevention is better than cure. By careful rural planning some of the worst of these problems can be avoided.

Water resources are at the centre of desertification in the world's drylands. In great areas of Africa, India, and South America (especially the inner Andes and the northeast corner of Brazil) the problem is very grave. But the developed societies also have the problem, in Mediterranean regions worldwide, but especially in the Old World Mediterranean of southern Europe, of depletion of valuable resources due both to climatic and socioeconomic factors leading to regionally unstable development.

Desertification is now being addressed by the International Convention on Combating Desertification, signed in Paris in 1994. This Convention seeks international support for a major offensive on the problem. This attack must be based on a concept of rural planning that embodies a clear understanding of water in all its ramifications as a major element in the problem. Notwithstanding the debate on the political significance of desertification (Thomas and Middleton, 1993) it is the major issue for rural planning in large areas of the world today, because virtually any change to land use has an impact on the hydrological cycle. In addition, the effects may extend to the atmosphere at the global as well as at the local scale.

5.2 The Hydrological Cycle

Figure 5.1 shows that the majority of water is in the oceans. From a planning point of view, however, only the terrestrial part of the cycle is of direct concern. As rainfall reaches the Earth, it is intercepted by bare soil, natural vegetation, or an artificial surface produced by human activities. Interception depends on the nature of the surface. Some intercepted rainfall is detained and available for evaporation. With high rainfall intensities an overland flow is generated, which may reach rivers and streams and become runoff. The remaining water infiltrates into the ground and may move downslope laterally through the soil (as throughflow) or percolate through the soil and enter the bedrock as groundwater. Groundwater eventually returns to rivers or to the sea by complex pathways determined by the type and structure of the rocks (Strahkov, 1976).

5.2.1 Precipitation

Precipitation is water falling from clouds as rain, snow, or direct condensation, as in cloud forests. Rainfall is highly variable in space and time and there are many measures of this variability. Rainfall amount (mm) is expressed for a given unit of time, for example, per year (annual rainfall) and can be averaged over time as decadal or running means as shown in Figure 5.2, the rainfall at Murcia, southeast Spain for the past 120 years.

In this diagram the running mean is calculated over six years. It illustrates three features: (i) a high interannual variability, in dry Mediterranean regions rainfall may vary from 150–600 mm in succeeding years; (ii) periods of persistently high or low rainfall, as in the higher rainfall of the 1960s or the persistent droughts of the early 1980s; and (iii) longer period of change, in this case the fall from an average of about 500 mm in the 1890s to little over 200 mm in the 1940s. Rainfall is also strongly seasonal in the Mediterranean. Rainfall is dominantly in winter and spring, when evaporation is low and this is the main period of plant, and especially crop growth.

At a shorter period, daily and hourly rainfall is much less predictable. The rainfall intensity is critical in studies of erosion and overland flow. Intensities in Mediterranean regions are generally higher than in many other parts of the globe, though the highest intensities occur in tropical typhoons or hurricanes. In the Mediterranean the peak rainfall tend to be of the order of 150 mm/h. Although rainfall is usually reported in mm/h, they can be based on much shorter period, such as 10 minutes (Castillo and Beltrán, 1979). So it is crucial to refer to the recurrence interval of intensity, that is, the period over which such an intensity is likely to recur with a given probability. So we can speak of a 30 mm/h rainfall having a recurrence interval of, say, five years. Typically in Mediterranean Europe, a rainfall intensity of 60 mm/h might occur once about every five years.

Rainfall duration is also important and is usually determined by the weather type. Rainfall coming from cyclones is of relatively long

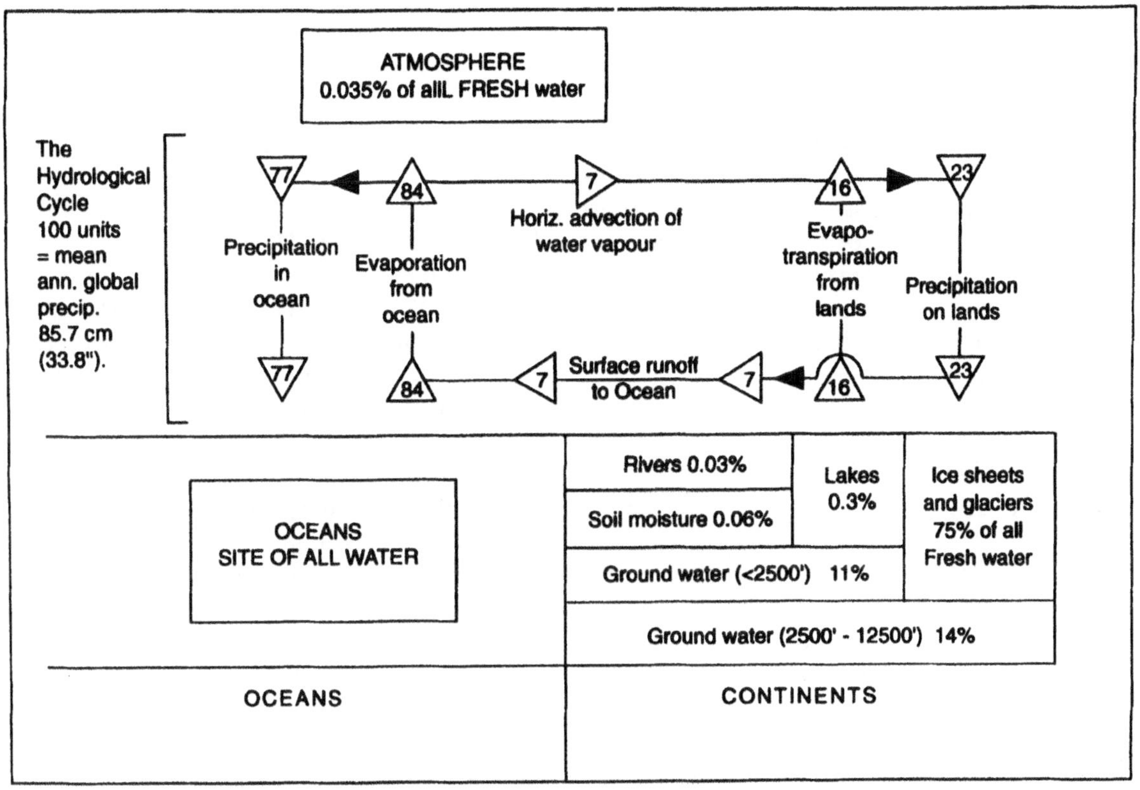

FIGURE 5.1. The world hydrological cycle (after Barry, 1969).

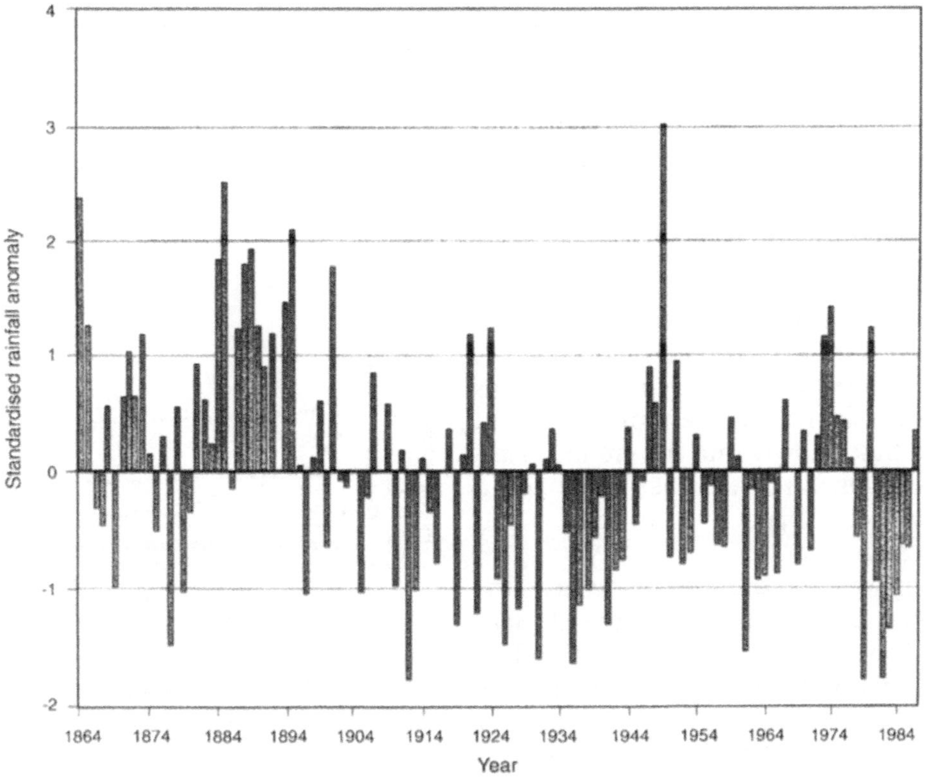

FIGURE 5.2. The long-term annual rainfall series for Murcia, Spain (after Thornes, 1995).

duration (e.g., 72 h). High intensities and short durations are more typical of convective storms. Ninety-five percent of all storms occurring in the dry southeast of Spain last less than 2 h. The distribution of storm duration provides key information on the capacity for soil water recharge, for plant growth, and for the likelihood that overland flow, flooding, and erosion will occur.

5.2.2 Interception and Detention

When water falls onto a plant cover some of it is captured on the leaf surface and some of it falls straight through the canopy as throughfall. The amount intercepted is available for evaporation back to the atmosphere and is lost for plant use. The canopy storage varies with the characteristics of the canopy and the rainfall. Dense canopies tend to hold more water and high intensity rain tends to produce less interception. Up to about 10 mm, storms with more rainfall tend to produce more storage. Above that amount, storage levels off and typically 95 percent falls through the canopy. The main role of interception is therefore in its breaking the impact of falling raindrops, though individual drops from canopies can be twice as large as that of rainfall (Brandt, 1988).

Rainfall can also be put into detention storage at ground level. Pools may develop on the surface where the soil infiltrability is low and storage tends to be high when the surface is rough, for example, just after ploughing. This water can evaporate to the atmosphere but some infiltrates slowly into the ground. Tractor implements can be designed to provide a large detention storage, to reduce runoff and hence erosion.

Keeping water on the surface increases the chance of infiltration and therefore of plant growth. Some water also runs as stemflow down plants and trees, and although the quantities may be small, typically 3–5 percent of total rainfall, it may be quite important because it is rich in organic matter and encourages plant growth. In agroforestry systems it can be especially important.

5.2.3 Water in the Soil

Water enters the soil surface by infiltration. The maximum rate is called the infiltration capacity. The infiltration rate is measured in millimetres per hour (mm/h) over a fixed area, so that it can easily be compared with rainfall. For example if the infiltration capacity is 25 mm/h and the rainfall is 15 mm/h, then all the rainfall can be infiltrated, and the infiltration rate equals the rainfall rate. Any interception has to be subtracted from the rainfall before infiltration is considered. Infiltration occurs rapidly at first (providing the rainfall intensity is high enough) and then tends to fall in an exponential manner (Figure 5.3) because, in the early stages, if the soil is not saturated, then the water is used to fill up the pore spaces in the soil. As the available space is occupied the rate diminishes. Although shown as a smooth process in Figure 5.3 it can be quite jerky. Water may enter large pores such as old root holes and then the rate increases. Generally, the process is well represented in Mediterranean soils by a simple equation of the form

$$f_t = A + B/t$$

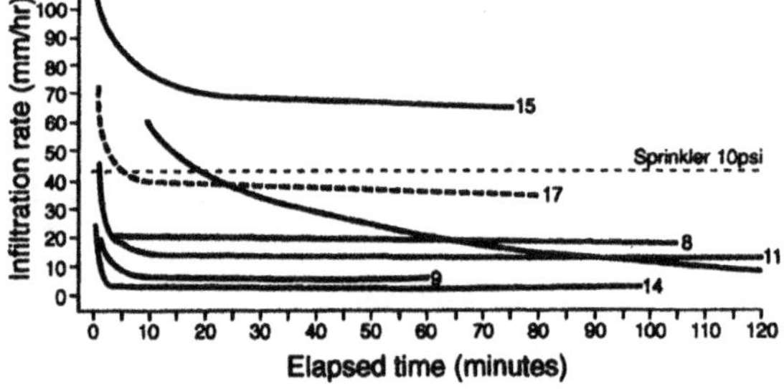

FIGURE 5.3. Infiltration curves for various lithologies in the Sorian area of Spain. The shaded area represents maximum precipitation rates for storms of 3-year recurrence and various durations (after Thornes, 1976).

where f_t is the infiltration (in mm/h), t hours after the start, and A and B are parameters that are determined from field experiments for typical soils. Coarse-textured soils have high infiltrabilities and fine textured soils low infiltrabilities. They are also high if the organic matter or stone content is high and if there are many macropores (Poesen et al., 1990).

Infiltration increases the water content of the soil. This is measured as the volume of water/volume of pore space in the soil and can be expressed as a fraction (e.g., 0.5, or as a percentage, 50 percent). When the pore space is filled, the soil is saturated. The hydraulic conductivity describes the rate of water movement in soils. The conductivity (cm/d) increases as the soil moisture content increases until saturated hydraulic conductivity is reached (Figure 5.4a).

Soil water suction is the force that the soil exerts on the water and slows down drainage. It is the opposite of pressure and is measured in Pascals or in metres of head. The drier the soil the more closely the water is bonded to the soil and the suction increases (Figure 5.4b). When the soil is very dry, water movement is difficult, which is why once the surface dries out, water loss by evaporation is negligible. The two curves determine the soil hydrological behaviour. Soil properties can be changed substantially by farming. Farming tends to increase soil infiltrability, and dry farming techniques encourage water entry. The bulk density (mass/volume), typically 1.0 g/cm^3 at the surface, increases with depth in most soils, and the conductivity decreases. If the soil surface is eroded, less infiltration means more surface overland flow.

Percolation (the movement from soil into the rock) operates under gravity, but the more water there is in the soil, the more percolates into the rock. This flux is determined by Darcy's law, expressed as

$$q = K_\theta (\varphi_2 - \varphi_1)/L$$

in which q is the water flux velocity (m/sec), K_θ is the conductivity dependent on water content θ, and φ is the pressure, comprising the height (gravitational head)—the soil suction. L is the distance between the two points (1 and 2) between which the height is measured. This law is appropriate for estimating flow when the soil or rock is saturated, but it is less effective in unsaturated soils especially when they have lots of large pores in them through which the flow obeys a different set of rules.

Darcian flow in hillsides depends on the relative role of gravitational effects, the soil properties, and the soil suction. In a single vertical

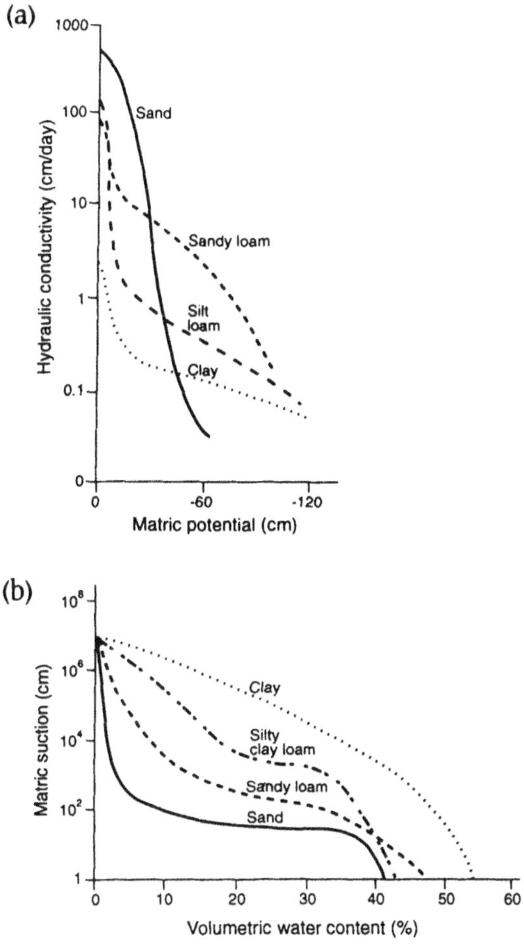

FIGURE 5.4. Relationship between (a) soil moisture and conductivity and (b) soil moisture and soil suction (after Ward and Robinson, 1990; reproduced with the kind permission of McGraw-Hill Publishing Company).

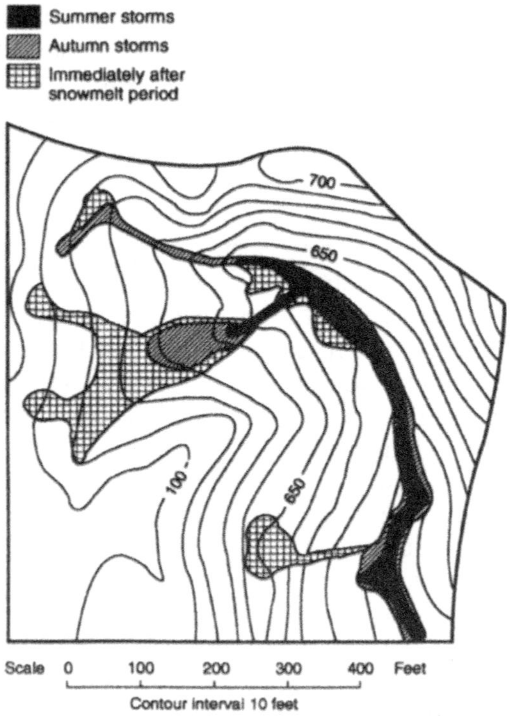

FIGURE 5.5. Partial contributing areas for stream flow in a small catchment. (After Dunne, 1978; copyright John Wiley & Sons Ltd. Reproduced with permission.)

profile, gravitational effects are vertical, and the rate of movement is controlled almost entirely by the conductivity. In a high-conductivity soil overlying one of low conductivity, the upper soil will tend to fill with water, while the lower may remain dry. When the soil is saturated overland flow occurs. With a similar arrangement on a hillside, water tends to move downslope through the upper soil as throughflow and may be forced back to the surface lower down as return flow. If the lower soil has higher conductivity, then percolation to lower levels occurs easily. In summer, with high suction near the surface, the flux may be upward as suction overcomes gravity. So long as the rainfall intensity is greater than the infiltration rate, then overland flow occurs after detention storage has been filled. This is called Hortonian overland flow (Horton, 1932).

When, by contrast, rapid infiltration occurs or the soil is very shallow, the storage capacity fills up and saturated overland flow occurs. The relative importance of saturated and Hortonian overland flow depends on rainfall, soil, and topographic position. With high annual rainfall, thin soils, and in hollows, saturated flow dominates. With high-intensity rains, low soil conductivity, and on the upper slopes of hillsides, Hortonian overland flow tends to prevail.

In very extreme storms Hortonian overland flow may occur on all hillslopes in a river basin. Usually, however, the areas contributing overland flow to river flow build up spatially throughout the storm, occurring first in the low areas near the river channels (Figure 5.5). If rainfall continues, wider and wider areas called partial contributing areas contribute to channel flow (Betson, 1964). Runoff (and hence flooding) can be controlled in the areas where runoff is generated. Overland flow typically has velocities of 1 m/sec, whereas subsurface water moves much more slowly (m/day) so overland flow is critical in causing flooding. Overland is called quickflow, whilst slower subsurface or groundwater flow is called base flow.

5.2.4 River Runoff

Unchanneled overland flow is called sheet flow. Concentration occurs in seepage lines or rills. The former are subsurface soil-filled depressions through which water can run easily, usually because the soil has a higher conductivity formed by chemical solution. Other subsurface features are actual tunnels through the soil called pipes. Rills are shallow intermittent surface drainage lines formed by erosion or sometimes by bad ploughing practice. They often give way to permanent steep-sided unvegetated features called gullies.

All these concentrated flow lines increase the rate of transfer of overland flow to the true river channels and therefore increase the probability of floods. The density of channels per unit area (km/km^2) is called the drainage density. Drainage densities (measured as km of channel per km^2) are high where the soils have a low hydraulic conductivity (i.e., very impermeable soils), and where rainfalls are very high, and/or where the vegetation has been removed by human action. Conversely,

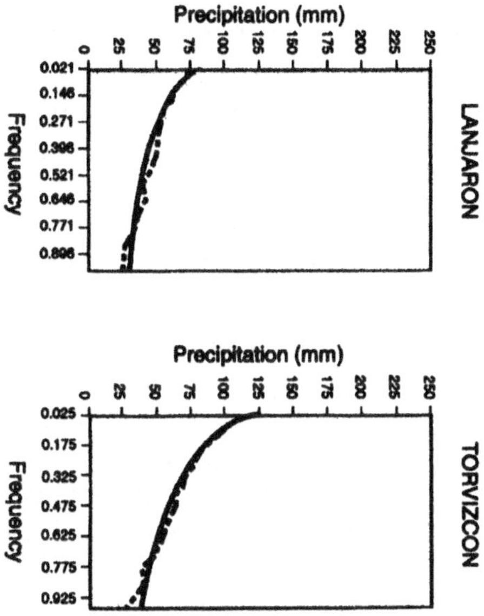

FIGURE 5.6. Recurrence interval of extreme rainfall events for two meteorological stations in the Alpujarras, Spain (after Heras, 1973).

drainage density is low where soils have high infiltrabilities and the rocks underneath have high permeabilities, soils have a high water storage capacity, rainfall is of low intensity, and/or there is a good vegetation cover.

Channel discharge is usually estimated from the depth using a rating curve. If depth is recorded continuously by a gauge, then data can be extracted for any time and the probability of flows of different magnitudes estimated. A given flow (event) can be specified by a

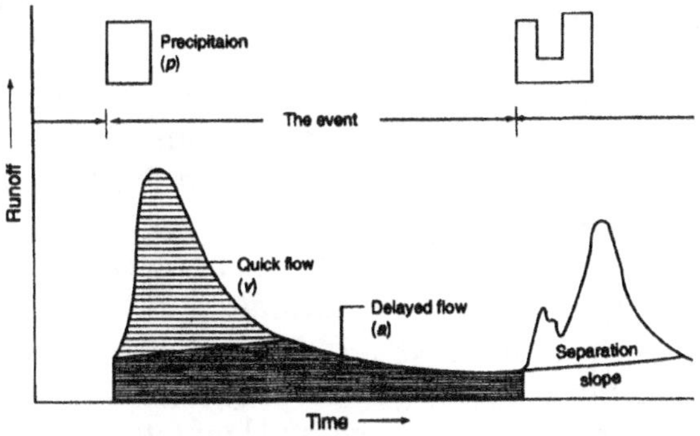

Figure 5.7. Event hydrograph. (Reprinted from Hewlett and Hibbert, 1967, with permission from Elsevier Science.).

given return period or probability (Figure 5.6), an important tool in the planning of rural land use where flooding is likely to be a problem. In problems of pollution and irrigation the low flows are of main interest, but the same prin-ciple applies.

The plot of discharge against time is called a hydrograph. This term may apply to a single event or to a season or to several years. For single events the main features are shown on Figure 5.7. After rain the water rises on the rising limb of the hydrograph. This is followed by the peak flow and the recession limb. Each rainfall event produces a separate hydrograph comprising both quickflow and base flow. Separating these may be difficult and is done using both chemical and graphical methods. The word "flood" usually refers to flows that are above the average low flow or base flow conditions. In ephemeral channels there is zero flow between flood events. A long-term hydrograph usually shows a decrease in base flow resulting from a progressive reduction of the groundwater inputs to the river as a result of low rainfalls.

5.2.5 Groundwater

Groundwater is the water in underground rocks that is available for use. It is usually measured by the depth to water in wells, though its position can also be inferred from springs. Water in a hillside is either in the unsaturated zone or in the saturated zone. These are separated by the water table or piezometric surface whose position varies from season to season and year to year and is generally related quite strongly to the overall shape of the hillside. Such a picture is simplified because the geology is, in reality, very complicated, and faults, folds, and other structural features tend to control the distribution of the saturated zone. Sometimes impermeable horizons hold saturated levels above the water table and these are called perched water tables. Moreover, water percolates from the river into the underlying gravels to provide water to suballuvial aquifers. These are especially important in water resource problems in semiarid environments.

Groundwater occurs in largest quantities in

permeable rocks. When the water level is unconstrained upwards the aquifer is said to be unconfined. In confined aquifers the upward movement is usually prevented by an overlying impermeable rock called an aquiclude when the water may be under considerable pressure and supply artesian wells. The exploitation of groundwater by wells is an ancient practice, and around the Mediterranean complicated systems were developed to tap water. Some of these, though over a 1000 years old, still play a part in water resources management. The construction of huge dams backing up large reservoirs is a relatively recent development and is still relatively unimportant compared with groundwater resources.

5.3 The Water Cycle in Different Environments

5.3.1 The Temperate Model

The processes described above occur with varying intensities in different environments. So much so that it is possible to think of distinctive assemblages of processes occurring in well defined zones. A set of such zones has been proposed by Eagleson (1970) on the basis of the essential processes operating to determine the annual water budget.

The description above assumed a temperate environment with annual rainfall in the range 100–2000 mm/year, with temperatures on average at about 12°C and rainfall distributed evenly throughout the year. Natural vegetation cover is more or less complete, soils are generally about 1 m thick, soil moisture typically ranges between 30 and 100 percent, and throughflow and deep percolation dominate the picture.

Overland flow is relatively rare and flood hydrographs may have double peaks representing overland and throughflow components as well as protracted recession limbs due to the large amount of base flow. Floods occur only with very extreme rainfalls or with catchment saturation after prolonged rain. Rainfall throughout the year means that severe droughts are relatively rare and soil suctions rarely reach values that are so high that crops cannot compete for water.

5.3.2 Tropical Forest Hydrology

In the humid tropics rainfall is usually in excess of 2000 mm and often 3000 or even 4000 mm. The soil is more or less permanently saturated, though even in the Amazon rai n forest there may be months when the soil is dry enough for leaves to be shed from trees. The high rainfall and temperatures yield a very high vegetation cover and biomass providing both high interception and evapotranspiration. There is also a strong ground level litter content, so that interception by the ground layer can be high (Ross et al., 1990). The soils are deep weathered so that the potential storage is very high.

In areas where the lithology is suitable, however, lateritic layers and hard horizons of iron and aluminium impede the flow, so that in heavy rain the soils above the laterite layers (normally at 1–2 m depth) become conduits for water. As a result overland flow hardly ever occurs under natural vegetation, but throughflow is very important.

Removal of the cover transforms the hydrological budget almost instantaneously and dramatically and may lead to severe erosion. The hydrographs of tropical forest streams may nevertheless be very "flashy," that is, they can rise very quickly. This is because with the soil already full of water the hillside acts like a pump, the later arriving water pushes out the water on to the slopes adjacent to the channels and causes the floodplains to become inundated with return flow as shown in Figure 5.8 (Nortcliff and Thornes, 1987).

Our own research shows that tropical forest floodplains may be saturated for several months in the year and that this can lead to waterlogging, so that the vegetation matter breaks down very slowly and large amounts of litter accumulate. This is another reason why changes to land use can have great impacts on the hydrological budget

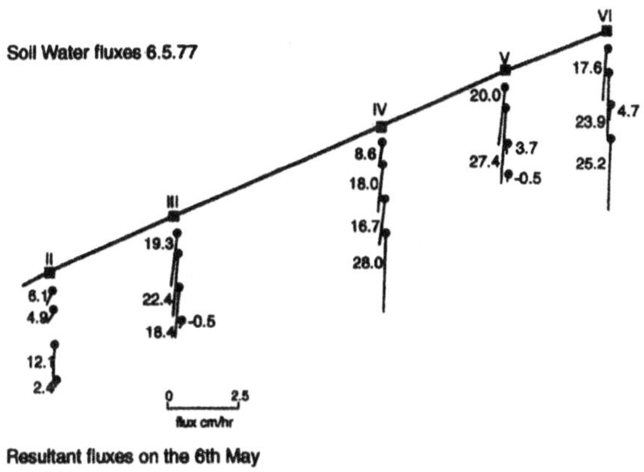

FIGURE 5.8. Tropical hill slope hydrology—fluxes at Maraca, Brazil (after Nortcliff et al., 1979).

5.3.3 Water in Drylands

The drylands are characterized by low precipitation amounts, between 10 and 400 mm/year, with sporadic or seasonal distribution of rainfall and with continually or seasonally high temperatures that lead to high evaporation and transpiration rates.

Vegetation cover is sparse and it increases linearly with rainfall up to about 500 mm, beyond which nutrients become the limiting factor. The sparse vegetation cover results from a competition for water and therefore evapotranspiration (and plant growth) are limited to periods of heavy rains. There is usually a seasonal change in cover that follows the rains.

The "greening" of deserts is a well-known phenomenon. The litter layer is highly restricted to beneath bushes and trees, and bare areas can occur between the woody species. The soils are thin, often less than 0.5 m, stony, and low in organic matter content. Rainfall is not only sporadic but also intense. As a result, Hortonian overland flow is much more common, and percolation, throughflow, and groundwater base flow are absent or rare. Channels therefore have ephemeral flows with extremely steep rising and recession limbs, as shown in Figure 5.9.

In these environments river flow for reservoir recharge is extremely difficult to predict, and as a result water resources based on river flows are often unreliable. Often the groundwater withdrawal reflects water that was stored in aquifers hundreds or even thousands of years ago and is not being renewed significantly under contemporary circumstance. Because of their high uncertainty ephemeral channels and their riparian environments pose severe problems for rural planning.

5.3.4 Snowmelt Regimes

Large areas of the globe are subject to snow for at least some part of the year. This includes mountains, such as those of the Mediter-

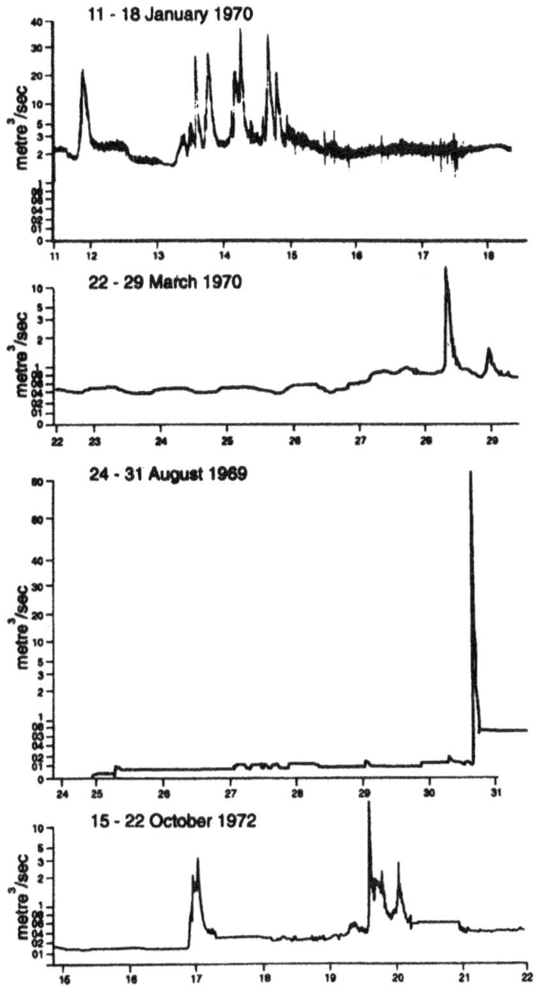

FIGURE 5.9. Selected traces of stage records from the Las Tosquillas station, Rio Ugijar, Spain, for different seasons of the year. These are typical of ephemeral channel flow hydrographs (after Thornes, 1976).

ranean, whose rivers supply water resources to the surrounding foothills and lowlands. In nival runoff regimes the soil itself may be frozen for large parts of the year or even permanently, as in the Canadian permafrost regions. The engineering of permafrost terrains is a special problem in its own right and is not treated here, but soil freezing in winter is much more common.

The effect is to reduce the infiltrability of the soil and hence encourage runoff at the time of snowmelt. Rivers dominated by snowmelt therefore tend to have very low winter runoff and high spring runoff when the snow melts. In the summer period, if the basins are glacierized as in the Alpine regions of Europe or the foothills of the Himalayas and the Rockies, there may also be a strong diurnal regime of snow melt. Under ice sheets, such as Vatnajokull in Iceland, water may build up in subglacial reservoirs that cause the ice to "float" and release large volumes of flood water in events called jokulaups.

Even short mountain snow seasons with relatively modest snow volumes can be critical in supplying irrigation water, and properly managed these can provide sustainable production for some rural environments, such as the foothills of the Sierra Nevada in Spain.

5.3.5 Water in Mediterranean Environments

Mediterranean environments cover a wide range of rainfalls, from about 200–1200 mm and have a highly variable annual potential evapotranspiration. The rainfall may vary from 250 to 1000 mm in as little as 10 km in the mountainous environments of the Betic cordillera in Spain or the mountains of central Crete, for example. Further south, the rainfall is less and the evapotranspiration potential higher, and the gradient steepens further as the Sahara is approached. Nevertheless, all Mediterranean environments are characterized by strong seasonal contrasts in rainfall, with the majority falling in the months of October–May and little if any in the months of June–September.

In some areas, as along the southeastern coast of Spain, the rainfall is polarized into two seasons, an early rainfall in September–October and a late rainfall in April and even early May. This pattern has important implications for all aspects of hydrological behaviour. The soil moisture regime reflects a combination of evaporation, plant water use, and rainfall, and typically in the drier areas of the Mediterranean soil moisture is around 10–25 percent in winter and 3–5 percent in the sum-

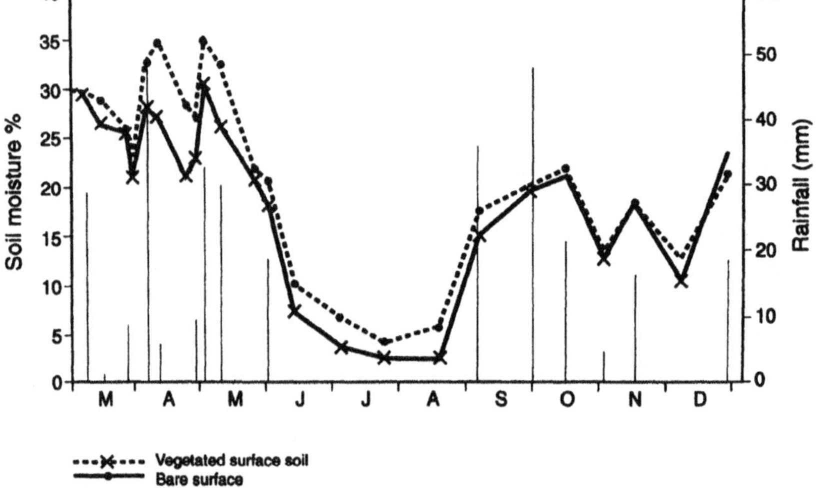

FIGURE 5.10. Seasonal soil moisture curve in Murcia, Spain (data courtesy of Professor Francisco López-Bermúdez).

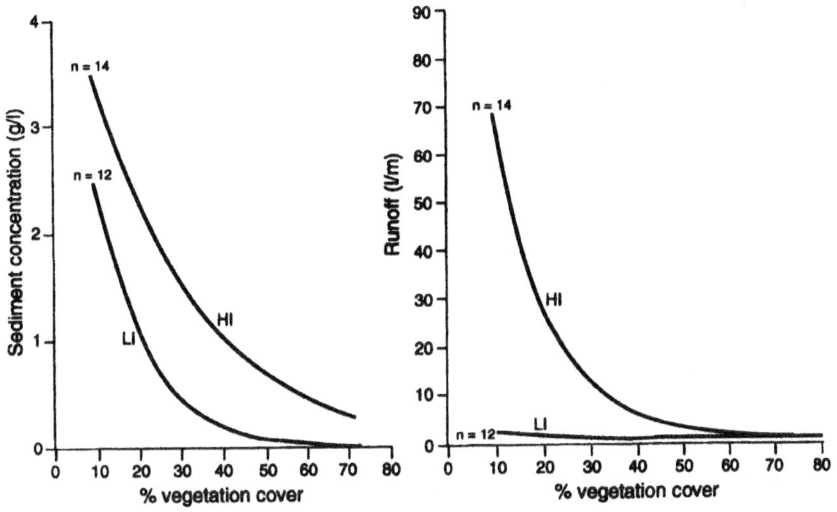

FIGURE 5.11. Runoff and sediment yield in relation to vegetation cover. HI = high intensity, LI = low intensity simulated rainfall (after Francis and Thornes, 1990).

mer season, with strong fluctuations for individual events (Figure 5.10).

As a result, soil suctions are very high in summer and plant physiology is adapted accordingly. Typically the hydric stress in plants is lowest about dawn and rises steeply to midday when the suction applied by the plant may be equivalent to 40 atmospheres in oaks such as Quercus rotundifolia. Unfortunately the hydrophysiology of Old World Mediterranean plant species is still quite poorly known.

Hillslope hydrology in Mediterranean environments is characterized by Hortonian overland flow and very little subsurface flow. Soils often are petrocalcic, that is, they have strong and sometimes weakly permeable layers of calcium at depths of only a few decimetres, and the soil above is often thin and highly porous. As a result the infiltration rates may be very high but storage is low so the soils quickly fill with water. Moreover, where the soils are very dry at the onset of rain, their hydraulic conductivities are low so the wetting front moves only slowly down through the soil.

Often after a heavy storm, the profile shows water penetration only to about 10 cm. When overland flow does occur, it is mainly in the form of rill or gully flow rather than sheet wash. The runoff is strongly influenced by the scattered vegetation cover. The flow may be structured by the vegetation to maximize water capture—as in the case of vegetation arranged in stripes normal to the contour—such as the "tiger bush" in the Sahel.

Wherever rilling and gullying do occur, the rate of runoff is accelerated and drainage densities are high, leading to rapid rise of the hill slope and channel hydrographs. Figure 5.11 shows that runoff decreases as vegetation cover increases, though less rapidly than for sediment yield, and this applies to a wide range of environments, especially in Mediterranean regions.

Recharge to aquifers is typically only about 10–15 percent of the annual rainfall. Recent work in Castilla–La Mancha shows that recharge is mainly the result of large events rather than a higher frequency of small events. Coupled with substantial groundwater withdrawals (see Section 5.6.2) this means that base flow in rivers is relatively rare and the only perennial channels are those fed from mountains and snow-fed streams.

All the major perennial rivers (such as the Rhone, the Po, and the Nile) of the Mediterranean fall into this class. Many of the smaller rivers are almost permanently dry, except where low flows have been engineered for amenity or to carry away pollution. Despite 400 years of engineering, floods in this environment still result from intense autumn storms such as the gotas frias that affected the southeast of Spain in 1973 and 1982, the largest floods in living memory. In these events the hydrographs are dominated by the passage of waves or "bores" of water that propagated rapidly down the channel, so that rising limbs are exceptionally steep and, because of transmission losses into the channel, the recession limb may be almost as steep.

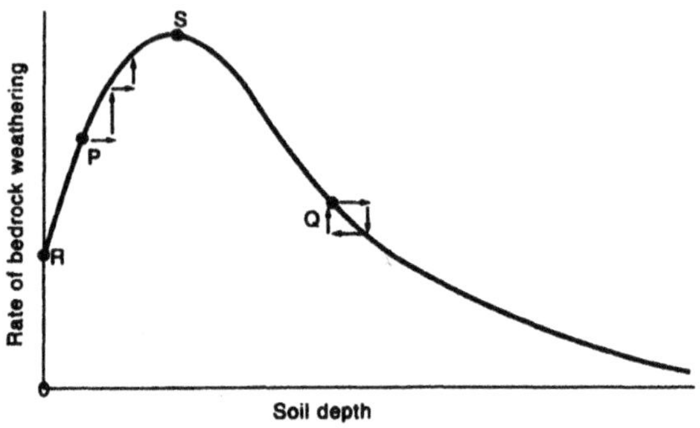

FIGURE 5.12. The relationship between soil depth and weathering rates (after Carson and Kirkby, 1972).

5.4 Water and Erosion

5.4.1 Erosional Processes

Denudation is the overall lowering of the land surface and is measured in mm/year of land surface loss. It comprises four processes: (i) weathering, which is the breakdown of rocks at the land surface by chemical processes; (ii) solute transport, the removal of materials in solution; (iii) detachment; and (iv) sediment transport.

The term erosion is normally used to indicate the mechanical detachment of weathered material and its transport away from the site of production. There is an intimate link between weathering and erosion. If the rate of production is faster than the rate of removal, eventually there is a build up of weathered material (soil). Sooner or later the soil may become so thick that the rate of weathering slows down because water movement slows down and solution transport is reduced. If the weathering slows down and the soil becomes thinner, water circulation rates are increased and weathering speeds up again. This negative feedback implies a steady state soil depth.

This relationship is shown graphically in Figure 5.12. When the weathering rate is always higher than the erosion rate, there is always material to be transported away, and therefore the rate of removal depends on the forces of transportation. This is called the transport-limited case. Conversely, when the erosion rate is persistently higher than the weathering rate, all the weathered material will be removed, and so erosion becomes weathering or production limited.

Which of these cases prevails depends on a number of factors outlined below. One of these is the rock type. If the rock has a very high resistance to weathering, such as quartzite, then only a small amount of erosion will be needed to remove all the available soil, and thereafter erosion is weathering limited. On the other hand, if rocks are easily eroded because they are very soft or very weathered, then the amount of erosion depends on the capacity of the flow to carry away all the material. This is called transport-limited removal.

Weathering is controlled by the mineral suite that makes up the rocks, the temperature, the available moisture, and the vegetation cover that produces organic acids that help the breakdown process. Generally rocks are made up of a mixture of minerals that have varying susceptibilities to chemical decomposition. The minerals fall into a scale of chemical resistance called the Bowen reaction series. Rocks comprising mainly minerals from the resistant end of the series, such as quartz, weather very slowly. Rocks with a large proportion of mica, especially the graphitic and albite micas, such as metamorphic schists, weather relatively quickly. In hot wet climates weathering of all rocks is faster than in cold or drier climates.

Hot, wet climates also have more plant production and levels of soil biological activity are high, and this also encourages high weathering rates. In the spectrum of weathering rates, the Mediterranean environments come on the low side. This is a result mainly of the persistent dryness of the soils under contemporary conditions, the low availability of circulating water, and the slow breakdown of plant organic matter. Note however that in the past, under different climatic and vegetation conditions, rates of weathering could have been higher.

Detachment and entrainment of particles is by wind or flowing water. The former produces aeolian erosion, the latter fluvial erosion. In fluvial erosion detachment occurs by splash and by flow detachment. Rain splash is the force produced by falling rain drops. The momentum of falling drops is a function of its mass and velocity, and the mass is a function of raindrop size, which in turn is related to rainfall intensity.

Intense rains tend to produce larger drop sizes than less intense rains, and throughfall tends to have larger drop sizes than rainfall. Velocity is determined by the fall height up to terminal velocity after which it is constant (Thornes and Brandt, 1987). The amount of splash (measured by splash cups, Morgan, 1995) is highly correlated with rainfall intensity and can be expressed as $G = kI^2$, where G is the sediment entrained by splash ($g/m^2/min$), I is the rainfall intensity (mm/min), and k is a coefficient representing the properties of the weathered materials that are being entrained.

Entrainment by flowing water is a function of the bed shear stress applied by the flow in relation to the size and sorting of the weathered material. Particles of silt or fine sand are most readily entrained. As the particles become smaller the cohesion between them makes them much more difficult to entrain. As they get larger, higher forces are needed to entrain the greater mass.

In fact, the situation is more complicated because weathered materials and deposits are usually a complex mixture of different materials, and vegetation makes transport more complicated. If there is no vegetation, then a simple erosion law is provided by $E = kq^m s^n$ (Musgrave, 1947), in which q is the discharge per unit width (m^2/sec), s is the sine of the slope gradient, and k, m, and n are empirical coefficients. The coefficient m is normally taken to be 1.66 and n about 1.0 (Carson and Kirkby, 1972); k is an empirical coefficient reflecting the soil erodibility.

The presence of vegetation results in a lowering of the amount of erosion entrained and transported relative to the bare soil (Figure 5.11). The amount of erosion decreases exponentially compared with the bare soil value. At first the fall is very rapid, so that with 30 percent vegetation cover, erosion is reduced to about 30 percent of its bare soil value. This remarkably useful result has been repeatedly justified for a range of environments since it was first described by Elwell and Stocking (1976), though recent work has shown it to depend on rainfall intensity (Francis and Thornes, 1990) and slope (Rose, 1994).

When the eroded sediments are transported across hillslopes, the shape and roughness of the hillslope as well as its vegetation cover and pattern, play an important role in determining how much will actually reach the channel. In particular, the processes depend on whether the flow is in the between-rills domain or in the rill domain. In the interrill domain splash and the slow downslope movement of a carpet of material are the dominant agents, whereas in the rills and gullies it is the tractive stress of the flowing water that plays the major role. Because rills are ephemeral features that are usually ploughed out by farmers, the factors leading to their initiation are important.

On different lithologies it seems that a critical area has to be reached before rills initiate. This is a reflection of the need to develop a critical overland flow, so that where area increases rapidly downslope (in converging, concave plan-forms) the distance from the divide (top of hillslope) required before rills can start is less. On diverging slopes (spurs) the rate of increase of discharge is slower and so the critical levels are reached later, if at all. In summary, vegetation and topography play a key role in determining not only the type but also the magnitude of the erosional processes involved. These are both attributes over which human activity exerts a great deal of control.

5.4.2 The Rate of Erosion

When attempts are made to decide on action to mitigate erosion, it is usual to attempt to estimate how severe the problem is at a particular place and time. For example, under the national plans required by the International Convention to Combat Desertification some expression of the regional severity of the problem is required if mitigation policies are implemented. This turns out to be remarkably difficult to achieve, partly because the objectives of measurement are often poorly specified.

Concern for agricultural potential production often emphasizes the loss of the soils productive capacity; concern for flooding emphasizes the loss of soil water storage capacity and changes in drainage density; while concern for offsite effects (such as downstream reservoir filling) requires the sediment yield from a hill slope or catchment. Another reason for the difficulty is that estimates are required for and provided at different scales. Thus at the local scale of the field, emphasis is on the soil types, the hill slope morphology and land use history. At the regional or national scale it will be on geological, climatic, and policy factors, while at the global scale it will be on climate, relief, and geology. For each of these scales a different estimation strategy is called for.

Visual identification of erosion is easy, but estimation of rates of erosion by this method is almost impossible. Often terrains that have been badly eroded may now be producing relatively little sediment, and sometimes terrains used for antierosion propaganda overstate the case for antierosion measures. This is because the landscape may have long-ago reached a state in which further erosion would be difficult if not impossible. Gilman and Thornes (1985), for example, showed that the badly eroded terrains of southeast Spain in some cases date back to the Bronze Age (4000 BP) and that contemporary erosion rates are, generally speaking, relatively modest except on especially sensitive lithologies such as the Tertiary marls.

The simplest and locally most useful tool is to estimate erosion directly from runoff plots designed to capture the water and sediment in troughs. This has the advantages that the controlling factors can easily be identified, the measurements are relatively easy to make and require low technology, and the technique is standardized and hence can be compared from one site to another (López Bermúdez and Rognon, 1996).

The disadvantage is that erosion rates tend to be overestimated, the effects of hillslope shape and distance from divide are difficult to quantify, and the boundary conditions in the trough may materially affect the results. This problem is called the sediment delivery ratio problem. As area gets greater, less sediment is produced for each square metre. The bigger the catchment, the lower the proportion of the eroded material that escapes from the catchment by the main river. The ratio of the sediment (tonnes) eroded within the catchment to the sediment carried out by the stream at the mouth of the catchment is called the sediment delivery ratio (SDR). The SDR decreases logarithmically as the catchment size increases.

A similar problem is the magnitude and frequency concept. This combines the amount of work produced by an event of a given magnitude with the frequency of occurrence of events of that magnitude. Large events of great magnitude have a lower frequency of occurrence, but do more work. High-frequency events occur regularly but individually do much less erosive work. Any estimate of erosion should really cover the entire spectrum of events. Thus observations need to cover the extreme cases and this may require many decades or even hundreds of years of data to provide a representative value of the real rates. Nevertheless this technique is still the most generally used and has provided, over the years, data that could be used to improve understanding of the process (Mutchler et al., 1994).

At the next level up, estimates can be provided from the sediment discharge of rivers. Sediment discharge is estimated by obtaining a curve that relates the discharge to the amount of sediment in transport. This is then used with flow record to calculate the total output from the basin.

In a now famous study, Fournier (1960) estimated erosion from sediment yield of a large number of rivers, empirically demonstrated that erosion was greatest in semiarid mountain environments, and provided equations to estimate these rates.

About the same time Langbein and Schumm (1958) provided an estimate of the erosion rates in the United States using sediment yield data. The graph is reproduced in Figure 5.13. They suggested that the peak sediment yield occured in the Mediterranean environments where rainfall was insufficient to provide a strong vegetation cover, but sufficient to provide significant erosion. As rainfall increasesd beyond this point, natural vegetation increased, and therefore reduced erosion. As rainfall decreased beyond this peak, there was insufficient runoff to generate erosion. One other important feature was that a small change in rainfall either side of this peak would produce a large reduction in rainfall. We should add that if the vegetation is cut down or restrained from growth, the erosion should increase further along the added dotted line.

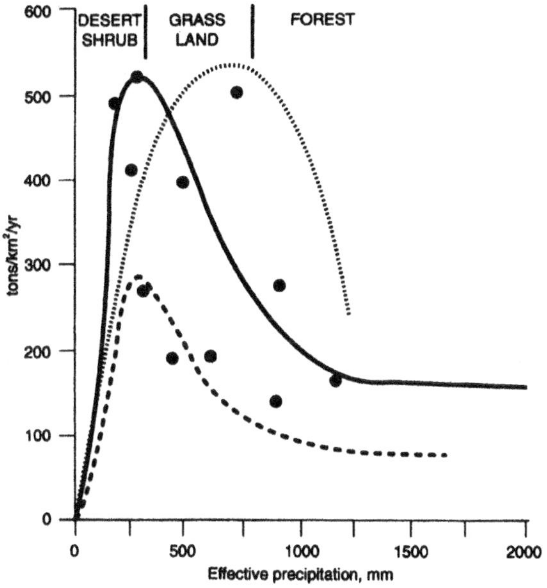

FIGURE 5.13. Variation of sediment yield with effective precipitation (after Langbein and Schumm, 1958). The dotted curve is based on both reservoir and sediment station records (after Rango, 1970).

A revision of this empirical result was provided by Walling and Webb (1981) in which they claimed that a further peak existed representing environments with a strong seasonal climate. A major problem with these empirical rules about erosion is that they depend on data that represent a relatively short period (though for a large area) and they are from catchments with a wide variety of human activity. Therefore the relationship to climatic parameters may be poorly established because the results are in fact related to human controlling actions. Nevertheless, other experiments and mathematical modelling (Kirkby and Neale, 1986) have tended to support these results.

Beyond this, estimates of erosion rates have to rely on mathematical modelling. At the simplest level, the approach adopts estimates from empirical fitted regression equations, such as the widely used Universal Soil Loss Equation (Wischmeir and Smith, 1965), that predicts soil erosion as a function of a set of variables including rainfall, relief and cover properties. This is extensively treated in standard texts (e.g., Hudson, 1995) and is not developed here, except to say that the equation and its derivatives tend to overestimate erosion, especially in mountain environments, where the soil may already have been lost to erosion hundreds of years ago. It deals with potential, not actual erosion.

A more sophisticated model for erosion was developed by Kirkby and Neale (1986) that provides for vegetation growth as well as erosion and its interactions with vegetation cover. Some of the results are shown in Figure 5.14, which gives the erosion estimates for Almeria. The strong role played by vegetation is indicated by the sustained high erosion rates predicted for Almeria. This model is at the regional scale of resolution and operates on a daily time-step. By contrast the MEDALUS model (Kirkby et al., 1995; Thornes et al., 1997) operates at the hillslope scale and is intended as a direct management tool. The model enables us to evaluate the effects of dif-

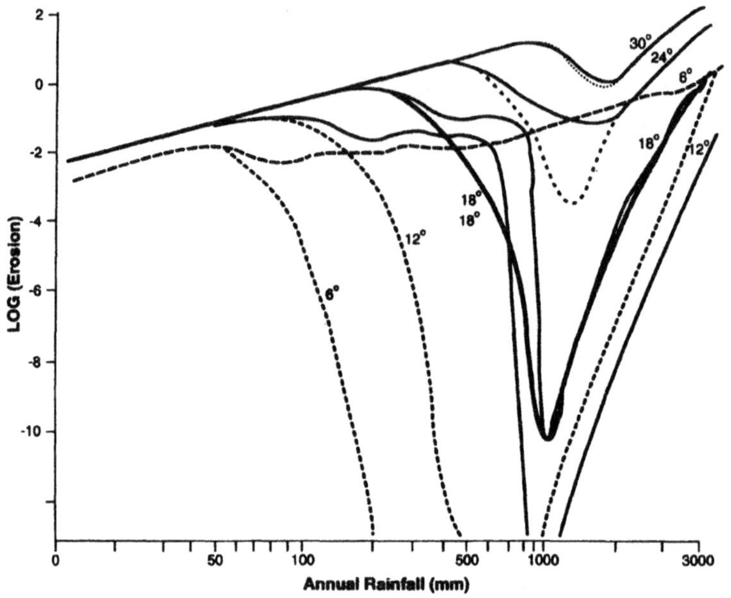

FIGURE 5.14. Rates of erosion in Almería. (After the model by Kirkby and Neale, 1986; reproduced with permission of John Wiley & Sons Ltd.)

ferent land-use treatments on erosion and also the impacts of different sequences of wet and dry years. Such computer-based models have the advantage of enabling experiments that would be totally unrealistic or impossible in the real world. At the same time, the models have to be verified against real-world experiences and are only as good as the quality of the input data and parameters against which they are calibrated.

5.4.3 The Impact of Human Activity on Erosion

Most of the world's environments are affected in some way by human activity. The major impact on land comes from rural land abandonment and engineering construction (Thornes and Brandt, 1994). Hydrology is dominated by changes in land use and, particularly, the loss of vegetation cover, which inevitably leads to increased erosion rates if sustained. This qualifier is important because in many, if not most, environments, provided there is sufficient water, warmth, and light, vegetation can quickly colonize areas where it has been removed.

Typically this growth and reestablishment follows a logistic type of curve, slow at first, accelerating after a few years, and then slowing down again as plants begin to compete for water, light, and nutrients. Typically, in dry Mediterranean conditions this new equilibrium cover will be reached after 12–20 years, being slower if there is a lack of rainfall (Brandt and Thornes, 1996). This assumes that recovery is taking place by reestablishment naturally.

Reafforestation, using nursery-grown plants, such as native pines may take about the same time, but may need extensive maintenance to reach comparable biomass recovery rates. One of the major causes of low vegetation recovery rates is the sustained use of abandoned or fallow land for grazing. Although light grazing may encourage growth and even reduce erosion, the combination of loss of cover and soil compaction by animal hooves is usual enough to increase soil loss dramatically. Abandonment as a soil conservation policy can only work if grazing is withdrawn.

The second most important type of human interference is the development of terraced landscapes. This tends to reduce erosion rates as steeply sloping terrain is replaced by gently sloping surface separated by walls or other retaining structures. However, a major problem can arise if land is cleared of vegetation for terracing without the construction of proper retaining walls and then abandoned.

The terraces hold and concentrate water which then may cause pipes and gullies to develop resulting in land that cannot be ploughed but produces high sediment yield. This occurred extensively on the marl lands in Murcia in the late 1980s, when the promise of water for irrigation, combined with government subsidies for land "improvement" and easily bulldozed marl lithologies led to disastrous erosion over large areas in the province, from which the land has not even begun to recover 10 years later. By contrast well-engineered and managed terraces can be a major factor in reducing erosion.

5.5 River Runoff and Flooding

5.5.1 Causes of Flooding

Rivers are concentrated flow paths for the water arriving from the hillslopes and channels for conducting sediment supplied by the catchment. The water in channels is from groundwater seeping into the channel from bed and banks, and quickflow which is the direct runoff from the slopes. Flooding occurs when the channel is insufficient to carry the water and sediment supplied to it and overbank discharge occurs causing loss of life and property. River channels also adjust their morphology at times of flood.

Floods are caused by intense rains that exceed the soil capacity to store the water by infiltration. Over time the probability of overbank flows can be changed, negatively (i.e., more floods) by removing vegetation from a catchment, carrying out drainage works that speed up the delivery of water to the channel, or by decreasing the infiltration through changing the catchment cover, for example, by urbanization. Even engineering works in the channels themselves may enhance flood events.

The impact of rainfall depends partly on the previous "wetness" of a catchment. Sustained rain, even at low intensities if of long duration, can wet-up the soil so much that a later small storm can be enough to cause a flood event. Predicting the occurrence of a flood is a well-developed art and is often approached through mathematical modelling (e.g., see Shaw, 1988). Even after a flood has been generated in a catchment, it may be possible to predict its size and speed down the channel, a process known as flood routing. This is again achieved through computer solutions to the complex Navier–Stokes flow equations.

5.5.2 Impacts of Flooding

The prediction of flow in ephemeral channels is made more difficult by the fact that the channel is already empty. At low stages the flow occupies the depressions left by the previous flow, then the different depressions are linked until the flow is deep enough to behave like a normal river. Even then the river may operate like a "bore" from a damburst, with a steep front that may move at great speed.

In the Santa Barbara catchment, Almeria, Spain, a bridge span, 18 m above the channel bed, was removed by an advancing flood wave, which further downstream killed over 100 people. Often these floods have a very high sediment content, approaching mud flows, and are of a pulsed character, with periodically higher waves behind the main front. Large amounts of damage are done by the sediments carried in the flows as huge boulders may be suspended in the flow and by the extensive deposition that occurs where the floods debauch onto lower, less sloping land. The lands subjected to flash floods, mudflows, and debris flows are some of the most hazardous on Earth (Cooke, 1984).

The channels themselves may be extensively modified, and major structures, such as irrigation canals, road and railway bridges, and public facilities and buildings, may be destroyed. Channels may change course or at least divert later flows into old channels. Construction in the hazard zones has to be controlled to mitigate disasters and to reduce damage losses, both in a spatial sense, but also vertically. Ground and lower floors of buildings can be used to store goods that have low susceptibility to water damage, or the flood plains have to be used for noncommercial uses such as recreation. In underdeveloped countries and even some developed ones ephemeral channels are used for dumping waste ranging from radioactive materials to organic waste from jam factories. This is extremely dangerous, because not only can plumes of pollutants develop in the suballuvial gravels, but such waste can be reentrained in flood events, with only a poor capacity to predict where they are likely to be redeposited.

5.5.3 Watershed and Channel Management

The traditional, easiest, and cheapest method of avoiding flood disasters has been to avoid the floodplain. Settlements throughout history have pursued this line of action, by locating on terraces or ridges well above the main channel. However as populations increased, land, especially with fertile soils and easy working conditions, came under pressure both for agricultural and industrial–residential use. The increased loss of life and the rise of insurance claims for flood disasters reflect the relentless increase in the commercial use of flood plains. Floodplain occupants are often too ready to blame nature or the upriver land uses rather than their own fallibility (Ives and Messerli, 1988; Wu and Thornes, 1994).

Nevertheless, people have come to accept the costs of major disasters, and so more and more has to be spent on flood protection. Equally, humankind has placed greater and greater confidence on the capacity of river managers to provide that protection. As a consequence the risks are often increased rather than decreased by mitigation efforts.

These efforts fall into three classes: (i) land management in the catchment (sometimes called watershed management); (ii) channel management; and (iii) flood warning systems. The first is based on the reduction of runoff and traditionally involved afforestation on contour-ploughed slopes. In many areas this has been a very effective policy. Trees intercept rainfall, reducing its volume and its intensity; they increase the organic matter content of the soil and hence the infiltration rate; and by increasing the detention storage and slope roughness, they increase infiltration opportunities. Afforestation is not, however, a universal solution. It may lead to water losses, especially with trees of high productivity (such as *Eucalyptus* species), the harvested crop may

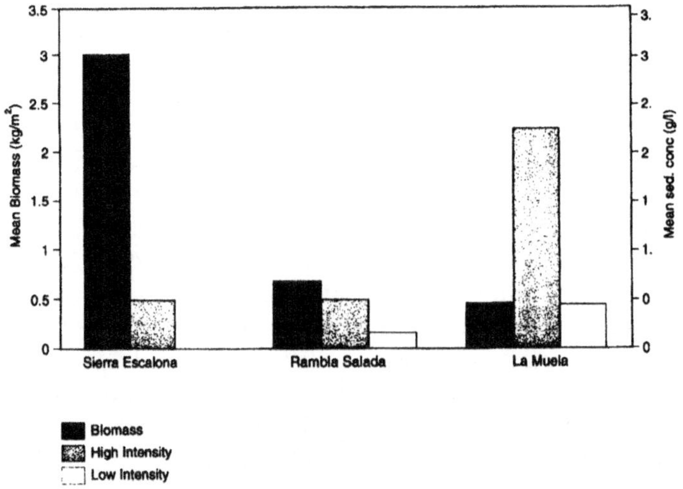

FIGURE 5.15. Runoff and sediment yield at three sites in South East Murcia, Spain. La Muela is a shrub site; Rambla Salada is an Anthyllis bush site; Sierra Esalona is under forest (after Francis and Thornes, 1990).

yield low economic returns and the trees may provide poor yields or simply die because of drought. Contour terracing may lead to increased erosional potential if tree growth is unsuccessful, native vegetation cleared, and the ground ploughed to make way for trees.

By contrast, provided that grazing can be limited, abandonment to natural regeneration may be a successful policy in environments where trees cannot flourish without costly maintenance (Francis and Thornes, 1990). Thus, as shown in Figure 5.15, Mediterranean scrub may provide as good a protection from runoff and sediment yield as do fully grown planted trees. Another problem is that afforestation often is carried out, by economic necessity, on the rough, poor, elevated ground at the upper end of catchments on lands of little economic value. By contrast, runoff and sediment yield reduction may be best achieved by reafforestation of the lower parts of catchments, especially of riparian zones, where trees can act as a filter to sediment and water, require a really less cover and lead to less water demand (Obando, 1997).

In the smaller channels check dams are frequently used to reduce the throughput of water and sediment to lower areas and to reduce the sedimentation of major storage reservoirs. This treatment is usually very effective provided that the check dams are well designed and built. They locally reduce channel gradients, induce sedimentation, and provide moisture for plant growth. In badly designed systems, down-channel scour may be increased so that isolated structures can do more harm than good and often are undermined and increase rather than decrease the sediment transport, as the bed and banks around the check dam are eroded.

Within major channels engineers have developed a complex armoury of techniques for reducing the impacts of major events ranging from large flood protection reservoirs to overflow lakes, levees, protective walls, and carefully engineered conveyance channels that conduct water and sediment most efficiently through the system with minimum channel instability. The design of such "regime" channels dates back to the civil engineering works of the Indian subcontinent at the beginning of the nineteenth century.

The third approach is to provide an adequate flood warning scheme that at least allows those threatened to move to safer places. These require capacity to forecast not only the likelihood of flooding, but also the peak flows and stages likely to be reached by major events. There is a well-developed procedure for estimating the level of water for different flood events and therefore the areas likely to experience different depths of flooding. This enables evacuations or remedial measures to be undertaken when floods are threatened or expected.

Experience shows that such systems have to be constantly updated, and the public constantly reminded of the procedures if they are to be effective. As economies shift away from public (the government pays for disasters) to private loss bearing (the individual or an insurance company pays), more attention is paid to both the value of land and its alternative uses. It is the job of the planner to ensure that scientifically based estimates of flood levels are available, that areas under risk have zoned land uses, and that warning systems are in place. All of these require a knowledge of the hydrogeomorphology of riparian environments.

5.6 Water As a Resource

Space only permits a brief discussion of the main issues and the reader is referred to specialist texts on water use by plants (Woodward, 1987), irrigation (Withers and Vipond, 1974), and groundwater (Freeze and Cherry, 1980), but a useful case study is Tobarra Ochoa (1995) who discusses economic as well as physical aspects of the problem.

5.6.1 Water in Plant Growth

Above all water is the main resource for plant production. To a rough approximation, plant growth (including crops) increases as the available soil moisture increases up to about 500 mm, beyond which nutrients control the

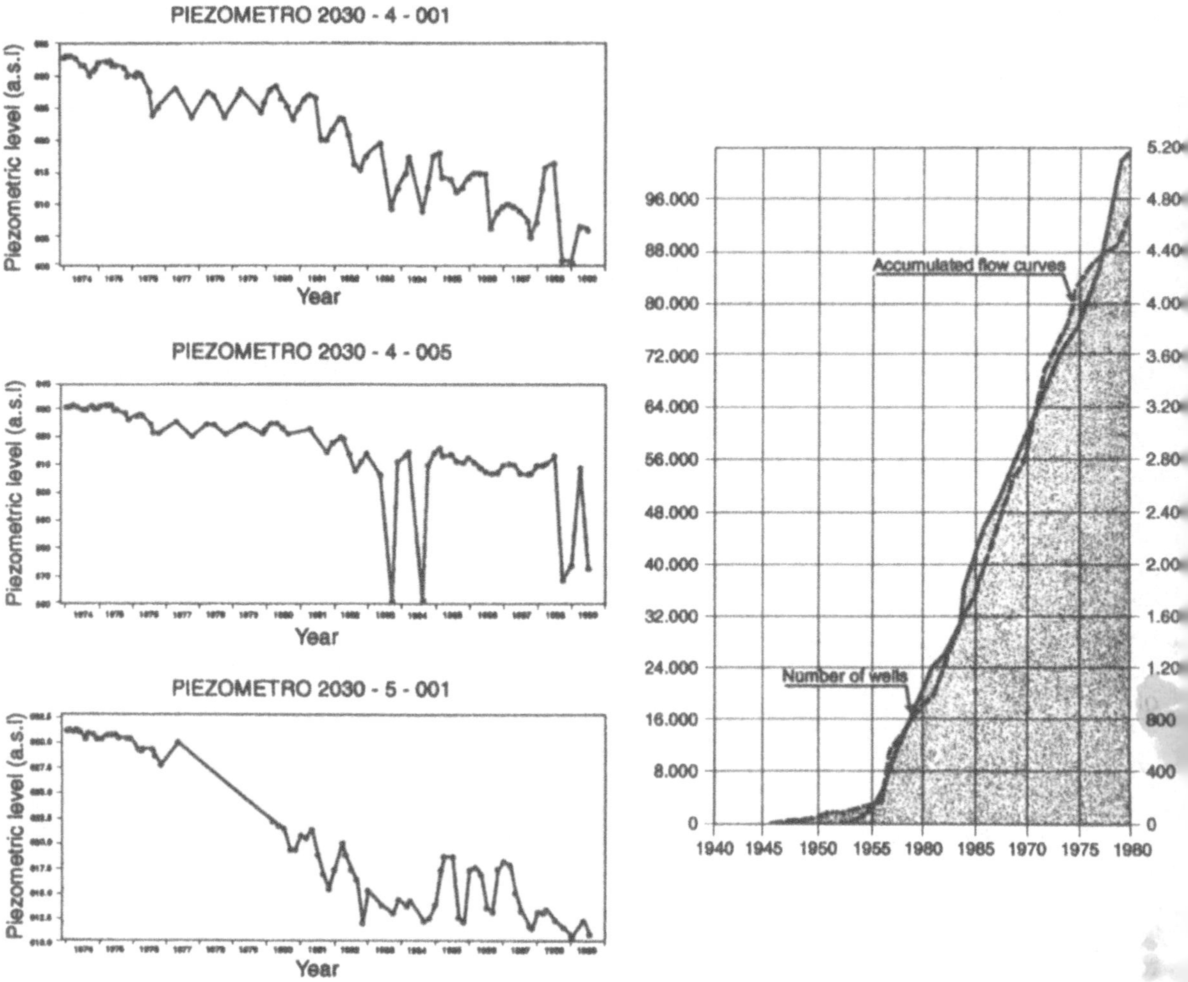

FIGURE 5.16. Groundwater resources depletion in: (a) Castilla La Mancha, Spain, showing after declining well levels (after ITGME, 1989); and (b) in Murcia, Spain, showing annual withdrawal (after Senent Alonso and Aragón Rueda, 1995).

growth. Since carbon production is directly related to evapotranspiration (Monteith, 1975) a common practise is to multiply the evapotranspiration rate with a coefficient representing the amount of carbon produced for each gram of water transpired. Not only is the total biomass production controlled by water but also the distribution of carbon to different parts of the plant. This enables plants to adapt to water stress by adjusting their physiognomy.

As stress increases, plants develop greater rooting capacity. Moreover many plants can adjust stress by reducing water loss through evapotranspiration. Sometimes this is by reduction of the stomatal openings in the leaves, sometimes by changing the plant surface characteristics (e.g., *Stipa tenacissima* rolls its leaves to reduce water loss) and sometimes by losing the leaves entirely. Because of this variety of controls, plants chosen for restorative work can have a large impact on the soil water budget.

5.6.2 Irrigation

Irrigation, after plant breeding and the "green revolution" remains the most important technical development for plant growth in the drylands. The technology dates back to preclassical times, but the major advances leading to very heavy water use postdate the submersible pump, spray-and-trickle irrigation techniques since the mid-nineteenth century. Modern computer-based systems involve massive financial investments and hence long-term strategic and financial planning. They may involve serious overexploitation of groundwater, as indicated for aquifers in the autonomous regions of Murcia and Castilla–La Mancha in Figure 5.16.

In coastal areas the withdrawal of freshwater for irrigation and the reduction in recharge of aquifers through dam construction has led to seawater intrusion, sometimes on a significant scale (Barrocou and Aru, 1993).

Throughout the Mediterranean the intensification of agriculture and the relaxation of the constraints of flat land on irrigation have led to the outward spread of irrigation from the traditional heartlands and the need for long-distance water transfers to make up for shortfalls at times of drought, and the crisis of water is now a major issue in rural resources management.

Water in drylands has always been accompanied by complex laws governing use, and these have been subject to revision in recent years as the pressures for sustainable development become more pressing and competing use, such as domestic consumption, the requirements of tourism, and industrial demands, also increase. Even now in most Mediterranean countries agriculture makes the heaviest demands even though this economic sector is declining in importance in the gross national product of European nations.

Traditionally water resource problems have been resolved by increasing supply by reservoir construction or increased groundwater pumping. As sites for reservoirs diminish and as the groundwater resources become more difficult and costly to extract, more effort will be needed in water conservation. There are at least three areas where action can be taken:

(i) Implementation of serious pricing structures for water. Most water for agriculture is still heavily subsidized even though the laws and mechanisms for fiscal control are in place. This continues to involve extravagant and wasteful use of water. Recent investigations of water fluxes in the irrigated lands of Castilla–La Mancha, for example reveal that both vertical drainage and evaporation are excessive.

(ii) Conveyance losses and poor timing of applications in relation to crop demands are still very high, especially in older systems.

(iii) Overland flow and channel flow in large storm events are still poorly managed and some 65 percent of available rainfall leaves catchment through direct evapotranspiration. Watershed management will provide one of the cheapest and politically most acceptable solutions to this problem of water sustainability as further intensification of production occurs, as methods of production increase in the promised "new green revolution" in genetics, and as land is abandoned on an increasing scale. These developments will impose heavy demands on the skills of rural planners.

References

Barrocou, G.; Aru, A. The Rio Santa Lucia catchment area. Brandt, C.J.; Thornes, J.B., eds. *Mediterranean Desertification and Land Use.* MEDALUS I final report. Chichester: John Wiley and Sons, 1993:534–559.

Barry, R.G. The world hydrological cycle. Chorley, R.J., ed. *Water, Earth, and Man.* London: Methuen; 1969:11–29.

Betson, R.P. What is watershed runoff? *J. Geophys. Res.* 69(8):1541–1552; 1964.

Brandt, C.J. The transformation of rainfall energy by a tropical rainforest canopy in relation to soil-erosion. *J. Biogeogr.* 15(1):41–48; 1988.

Brandt, C.J.; Thornes, J.B., eds. *Mediterranean Desertification and Land Use.* Chichester: John Wiley and Sons; 1996.

Carson, M.A.; Kirkby, M.J. *Hillslope Form and Process.* Cambridge: Cambridge University Press; 1972.

Castillo, F.E.; Beltrán, L.R. *Precipitaciones Máximas en España; Estimaciones Basadas en Métodos Estadísticos.* Madrid: Ministerio de Agricultura, Instituto Nacional para la Conservacion de la Naturaleza; 1979.

Cooke, R.U. *Geomorphological Hazards in Los Angeles*. London: Methuen; 1984.

Dunne, T. Field studies of hillslope processes. Kirkby, M.J., ed. *Hillslope Hydrology*. Chichester: John Wiley and Sons; 1978:227–289.

Eagleson, P.S. *Dynamic Hydrology*. New York: McGraw-Hill; 1970.

Elwell, H.A.; Stocking, M.A. Vegetative cover to estimate soil erosion hazard in Rhodesia. *Geoderma* 15:61–70; 1976.

Fournier, F. *Climat et Erosion*. Paris: Presses Universitaires Françaises; 1960.

Francis, C.F.; Thornes, J.B. Matorral: Erosion and reclamation. Abaledejo, J.; Stocking, M.A.; Diaz, E., eds. *Soil Degradation and Rehabilitation Under Mediterranean Environmental Conditions*. Madrid: Consejo Superior de Investigaciones Científicas, 1990:87–117.

Freeze, R.A.; Cherry, J.A. *Groundwater*. Englewood Cliffs, NJ: Prentice-Hall; 1980.

Gilman, A.; Thornes, J.B. *Land Use and Prehistory in South East Spain*. London: George Allen and Unwin; 1985.

Heras, R. *Estudio de Máximas Crecidas de la Zona de Alicante–Almería–Málaga y de las Lluvias Torrenciales de Octubre de 1973*. Madrid: Centro de Estudios Hidrográficos; 1973.

Hewlett, J.D.; Hibbert, A.R. Factors affecting the response of small watersheds to precipitation in humid areas. Soppe, W.E.; Lull, H.W., eds. *Forest Hydrology*. Oxford: Pergamon; 1967: 275–290.

Horton, R.E. The role of infiltration in the hydrological cycle. *Trans. Amer. Geophys. Union* 14:446–460; 1932.

Hudson, N. *Soil Conservation*. 3rd ed. London: Batsford; 1995.

Instituto Tecnológico GeoMinero de España (ITGME). *Sistema Aquífero No. 23, Mancha Occidental*. Madrid: ITGME, Dirección de Aguas Subterráneas; 1989.

Ives, J.D.; Messerli, B. *The Himalayan Dilemma*. London: Routledge; 1988.

Kirkby, M.J.; McMahon, M.L.; Abrahart, R.; Thornes, J.B.; Shao, J. The Medrush model for a single flow strip. MEDALUS 2nd Annual Report, covering the period 1st January to December 1994. London: MEDALUS, King's College; 1995:8–19.

Kirkby, M.J.; Neale, R.H. A soil erosion model incorporating seasonal factors. Gardiner, V., ed. *International Geomorphology*. Vol. II. Chichester: John Wiley and Sons; 1986:189–210.

Langbein, W.B.; Schumm, S.A. Yield of sediment in relation to mean annual precipitation. *Trans. Amer. Geophys. Union* 39:1076–1084; 1958.

López Bermúdez, F.; Rognon, P., eds. *Erosion Hydrique, Désertification et Aménagement dans L'environnment Méditerranéen Semi-Aride*. Murcia: Universidad de Murcia Publicaciones; 1996.

Monteith, J.L., ed. *Vegetation and the Atmosphere: Vol. 1, Principles*. New York: Academic Press; 1975.

Morgan, R.P.C. *Soil Erosion and Conservation*. London: Longman; 1995.

Musgrave, G.W. The quantitative evaluation of water erosion: A first approximation. *J. Soil Water Conserv.* 2:133–138; 1947.

Mutchler, C.K.; Murphree, C.E.; McGregor, K.C. Laboratory and field plots for erosion research. Lal, R., ed. *Soil Erosion Research Methods*. Ankeny, IA: Soil and Water Conservation Society; 1994:11–39.

Nortcliff, S.N.; Thornes, J.B. Flooding and organic matter accumulation in a tropical forest floodplain. *J. Biogeogr.* 15:49–59; 1987.

Nortcliff, S.N.; Thornes, J.B.; Waylen, M.J. Tropical forest systems: A hydrological approach. *Amazonia* 6(4):557–568; 1979.

Obando, J.A. Modelling the impact of land abandonment on runoff and soil erosion in a semi-arid catchment. Ph.D. Thesis, King's College, London; 1997.

Poesen, J.; Ingelmo-Sanchez, J.; Mucher, H. The hydrological response of soil surfaces to rainfall as affected by plant cover and position of rock fragments in the top layer. *Earth Surf. Process. Landf.* 15:653–672; 1990.

Rango, A. Possible effects of precipitation modification on stream channel geometry and sediment yield. *Water Resour. Res.* 6:1765–1770; 1970.

Rose, C.W. Research progress on soil erosion processes as a basis for soil conservation practices. Lal, R., ed. *Soil Erosion Research Methods*. Ankeny, IA: Soil and Water Conservation Society; 1994:159–180.

Ross, S.R.; Thornes, J.B.; Nortcliff, S.N. Soil hydrology, nutrient and erosional response to the clearance of terra firme forest, Maraca Island, Roraima, Northern Brazil. *Geogr. J.* 156(3):251–261; 1990.

Senent Alonso, M.; Aragón Rueda, R. Recursos hídricos subterráneos: Situación actual y gestión futura. Senent Alonso, M.; Calvo-Rubio, F.C., eds. *Agua y Futuro de la Región de Murcia*. Murcia: Asamblea Regional de Murcia; 1995: 105–127.

Shaw, E.M. *Hydrology in Practice*. 2nd ed. London: Chapman and Hall; 1988.

Strahkov, P. *Principles of Lithogenesis*. New York: John Wiley and Son; 1976.

Thomas, G.P.; Middleton, F. *Desertification: Myth or Reality?* Chichester: John Wiley and Sons; 1993.

Thornes, J.B. *Semi-Arid Erosional Systems*. Geography Papers No. 7. London: London School of Economics and Political Science; 1976.

Thornes, J.B. Mediterranean desertification and the vegetation cover. Fantechi, R.; Peter, D.; Balabanis, P.; Rubio, J.L., eds. *Desertification in a European Context: Physical and Socioeconomic Aspects*. EUR 15415EN. Brussels: European Commission; 1995: 169–179.

Thornes, J.B.; Brandt, C.J. Erosional energetics. Gregory, K.J., ed. *Energetics of Physical Environment: Energetic Approaches to Physical Geography*. Chichester: John Wiley and Sons; 1987.

Thornes, J.B.; Brandt, C.J. Erosion–vegetation competition in a stochastic environment undergoing climatic change. Millington, A.C.; Pye, K., eds. *Environmental Change in Drylands: Biogeography and Geomorphological Perspectives*. Chichester: John Wiley and Sons; 1994: 305–320.

Thornes, J.B.; Shannon, J.; Hooke, J. Transformation of ephemeral stream environments under desertification. *Proceedings of the International Conference on Mediterranean Desertification*. Brussels: European Commission; 1997.

Tobarra Ochoa, P. Derechos sobre el agua subterránea; una perspectiva económica. Senent Alonso, M.; Calvo-Rubio, F.C., eds. *Agua y Futuro en la Región de Murcia*. Murcia: Asamblea Regional de Murcia; 1995:365–379.

Walling, D.E.; Webb, B.W. Patterns of sediment yield. Gregory, K., ed. *Background to Paleohydrology*. Chichester: John Wiley and Sons; 1981: 69–100.

Ward, R.C.; Robinson, M. *Principles of Hydrology*. 3rd ed. London: McGraw-Hill; 1990.

Wischmeir, W.H.; Smith, D.D. *Predicting Rainfall Erosion Losses*. Agricultural Handbook 537. Washington, DC: U.S. Department of Agriculture; 1965.

Withers, B.; Vipond, S. Irrigation: *Design and Practice*. London: Batsford; 1974.

Woodward, F.I. *Climate and Plant Distribution*. Cambridge: Cambridge University Press; 1987.

Wu, K.; Thornes, J.B. Terrace irrigation of mountainous hillslopes in the Middle Hills of Nepal: Stability or instability. Chapman, G.; Thomas, M., eds. *The Quest for Sustainable Development in the Ganges Basin*. London: Mansell; 1994.

6

Evaluation of Soils for Land Use Allocation

Ramón Vallejo

6.1 Introduction

Soils support terrestrial ecosystems and human life. They largely determine the agricultural capacity of the land in spite of the increasing artificial inputs in cropping systems.

Soil regulates the exchange of energy, nutrients, and water between the atmosphere, lithosphere, and hydrosphere. In addition, soils are buffers and filters of pollutants, and are important sources and sinks for greenhouse gases. Therefore soil management influences water and nutrient cycles and can alleviate human impact on the environment.

Soil management faces three major challenges in the immediate future: (i) to meet the increasing demand for food; (ii) to reduce human involvement in the degradation of the soil resource; and (iii) to cope with the predicted effects of global change.

Soil management should be based on the ability of the soil resource to fulfil land use goals. Therefore, soil properties and their interactions with land use practices are a major component of rural planning.

6.1.1 Overview of Soil Genesis

The primary rocks on Earth were formed under conditions of high pressure and temperature, low oxygen concentration, and the absence of liquid water. When these rocks came into contact with the atmosphere, they were subjected to very different physico-chemical conditions than those prevailing during their genesis. These new conditions led to instability in the primary rocks, which in turn became disorganized and gave rise to new (secondary) minerals, which were in greater thermodynamic equilibrium with the conditions on Earth's surface. These products of weathered rocks form the raw material of soils and their solid skeleton. Clay minerals are characteristic products of rock weathering. Within the Earth's crust evolution time scale, soils have an ephemeral life span. An old saying defines soils as "a short rest of the sediments in their travel to the sea."

For a very long period during the Earth's history, soils developed under practically abiotic conditions, with the high production of sediments owing to the free expansion of the erosive processes. During this stage of soil evolution, the characteristic inorganic components of soils, the clay minerals, appeared. Since the Silurian, some 400 million years ago, the colonization of the terrestrial environment by vascular plants led to a period of "biological" evolution in soils, where the flow of organic matter supported the activity of a great diversity of soil organisms. The result of this flow of organic matter is the release of nutrients accumulated by plants, the transformation of part of the energy flow through the soil organisms into soil organization (soil structure), and the accumulation of a residual organic product of extreme importance to soil fertility, the humus.

As the consequence of these processes occurring in the Earth's history, it is possible to talk about the phylogeny of soils, in relation to changes in the composition of the atmosphere and in colonization of the land. It is also possible to define stages in soil development, such as juvenile, mature, or old, and even relict, and the eventual disappearance of the soil through erosion or integration into sedimentary structures. Soil age generally ranges be - tween hundreds and tens of thousands of years; exceptionally, on very stable land surfaces, soils may be more than 100,000 years old. Considering these time dimensions, anthropogenic soil erosion is practically irreversible and the soil is a nonrenewable resource.

Soil formation is a process driven by the action of climatic agents and the activity of organisms on rocks. Throughout this process, the nutrients previously stored in the rocks become available for plants within a more favourable soil environment. The soil has the capacity to store water and nutrients and the ability to buffer their flow. In addition, the soil environment provides a stable microhabitat for the roots. These properties are due to the presence of fine particles (clays and humus) with a high capacity for water and nutrient adsorption. The fine particles are arranged in aggregates with other larger particles, in such a way as to create interaggregate porosity which facilitates water and air flow.

6.1.2 The Soil Profile

Soil-forming agents have a preferential direction of energy flow, orthogonal to the land surface. Therefore, soil differentiation follows this direction in the form of distinguishable layers, called horizons. Weathering is more intense in the top soil or superficial bedrock, where rainwater is CO_2 enriched, biological activity is at its maximum, and temperatures show extreme fluctuations. Therefore weathering produces a gradient in bedrock transformation with a maximum at its upper boundary, in contact with the atmosphere. The organic matter flow follows the opposite direction of the mineral transformations. Litter fall is the main source of organic matter for the soil, and mostly enters the soil system through the soil surface. The organic matter is incorporated into the mineral soil through the action of soil fauna and leaching, in such a way that the deeper the soil layer the more transformed the organic matter and the lower its content. The organization of the soil profile into horizons is therefore a consequence of the combination of these two major flows.

In a soil formed *in situ*, under a forest cover, the sequence of horizons (denominations after the *Soil Taxonomy*, Soil Survey Staff, 1992) expected from the top soil down through the profile would be as follows:

O: litter horizons, or forest floor, with a low mineral particle content; slight transformation of plant remains (*Oi* layer) on top, then the "fermented" layer composed of recognizable plant remains (*Oe* layer), and the "humus" layer composed of amorphouslike organic matter (*Oa* layer) in contact with the mineral soil.

A: organomineral horizons, composed of highly weathered minerals and litter remains mixed with humidified organic matter.

B: mineral soil with low organic matter content, although mostly composed of deeply transformed and relatively old humus; the degree of mineral particle differentiation with respect to those of the parent material decreases in depth, both in composition and in spatial organization or soil structure. Often, *B* horizons are formed by accumulation of soil particles from the above horizons through leaching.

C: regolith or parent material with little mineral weathering and mostly conserving the rock structure.

Figure 6.1 shows soil-forming processes and the sequence of horizons in the soil profile.

The migration (leaching) of soil components in solution or as particles in suspension

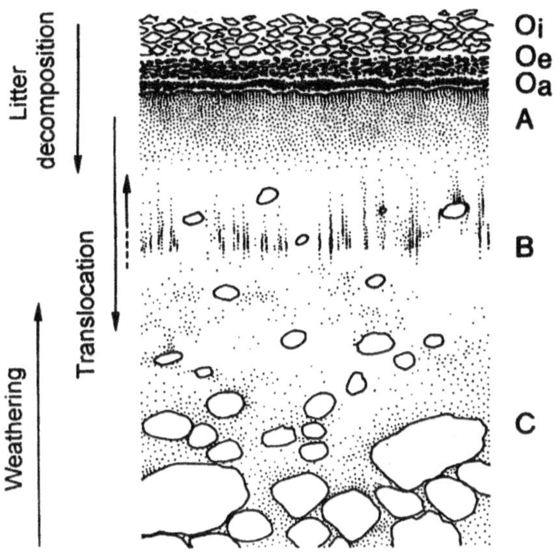

FIGURE 6.1. Soil-forming processes and the sequence of horzons in the soil profile.

can produce strong differentiation in the soil profile.

On steep slopes, the lateral flow of water and solutes from top to bottom introduces topographic soil patterns where the soils along the slope have genetic and functional relationships. The parent material of the bottom slope is colluvial material detached from the upper slope, so the soil in this position is not genetically related to the underlying bedrock.

6.1.3 Soil Fertility

Natural soil almost always presents limitations to plant growth, although vascular plants are able to colonize most existing soils. The most general soil resource limitation for plants is nutrient scarcity, especially N and P, in both absolute terms and/or in their availability for root uptake or spatial distribution. This may explain why most natural plant communities respond to fertilization (Etherington, 1982) and also the widespread development of mycorrhizal associations that facilitate soil nutrient uptake.

Mycorrhizae were already present and widespread at the onset of terrestrial colonization by plants. Plants have developed many other mechanisms to improve soil nutrient acquisition, such as root production and the release of enzymes, organic acids, complexing agents, mechanisms of active transport, and the above-mentioned nitrogen fixation symbiosis and the rhizosphere transformations (Marschner, 1986). At least, plants can adapt their growth rates in order to moderate their nutrient requirements. Therefore, lack of nutrients seldom prevents the colonization of soil by plants.

More drastic than nutrient shortcomings are water limitations, very low temperatures, or the presence of toxic elements (e.g., very saline or sulphuric environments) which can prevent colonization by vascular plants. In these particular soil conditions, plants have more specific adaptations, and clear causal relationships between soil types and plant communities can be established.

6.2 Today's Challenges in Soil Use for Food Production: Alternatives and Risks

6.2.1 Alternatives

To respond to the increased world demand for food and fibre in the next decades, three major strategies can be envisaged (Brady, 1990): (i) expansion of agricultural lands; (ii) increase in the productivity of currently cropped lands; and (iii) increase of cropping intensity (number of crops per year) on existing agricultural land.

At present, the distribution of land capabilities for agriculture in the world is (Brady, 1990): 50 percent of nonagricultural lands due to extreme cold or dry climate, or mountainous landscape; 25 percent nonarable lands used for grazing animals; and 25 percent potentially arable lands, of which 40 percent is already cropped. However, these categories are far from evenly distributed around the world. In Asia and Europe, for example, almost 100 percent of the arable land is already used in agriculture, and soils that would not be classed as arable are cropped by applying extra inputs of energy (e.g., by terracing steep slopes). In contrast, in the tropical and subtropical regions of Africa and Latin America only a small fraction of potentially arable land is cultivated at present. However, ploughing tropical forests in developing countries has many constraints: (i) costs associated with expanding new cropping fields, such as infrastructure or importing fertilizers; (ii) nondirect costs produced by the ecological risks of soil erosion, loss of biological diversity, loss of forest resources, and the release of huge quantities of CO_2 to the atmosphere, contributing to the greenhouse effect; (iii) sanitary problems existing in tropical regions, making their exploitation difficult; and (iv) low fertility of tropical soils and often poor development of management techniques adapted to tropical conditions.

These factors reduce the possibilities of expansion of arable lands and actually support the conversion of crop fields into forests or pastures, as has already happened in Europe.

From the estimates of the Food and Agriculture Organization, FAO (1987), the expected increase in food production will come mostly from the increased yield of cropped land. The increase in yields per hectare is expected to be achieved through the use of fertilizers, selected seeds and irrigation.

The "green revolution" dating from the 1960s has resulted in a large increase in yields in extensive areas of many countries. The principles of the green revolution were massive use of fertilizers and pesticides, seed/variety selection, agricultural machinery, and monoculture. These methods were directly exported from the developed countries to the rest of the world.

In relation to the use of fertilizers, industrial countries are generally using them in excess, causing contamination problems, whereas in most developing countries, yields are limited by lack of fertilizers. Data from 1982 showed that the industrial countries applied more than 110 kg/ha of fertilizers, whereas developing countries (excluding China and India) less than 40, and Sub-Saharan Africa 4.2 (Brady, 1986).

According to Marschner (1986), in the past, problems of soil fertility related to crop productivity where mostly tackled by modifying soil properties to suit plant requirements, for highly demanding genotypes selected according to their productivity. This approach functioned well for soils of medium to good fertility in the temperate region. However, these genotypes were poorly adapted to extreme soil properties, such as water shortage, salinity, or acidity. During the last three decades, this strategy has shifted to searching for plant genotypes that are better adapted to soil constraints, and into making more efficient use of soil fertility under limiting conditions. This strategy applies especially to the soils of tropical and subtropical regions. Plant productivity with these genotypes only causes a slight reduction of maximum crop yield (Marschner, 1986).

If the energy budgets of the cropping systems are considered (Lowrance et al., 1984), the power density (i.e., energy flow/surface unit) of preindustrial agriculture is not very different from that of natural ecosystems. The power density of industrial agriculture is ten times or more higher than natural ecosystems, due to the high energy (e.g., machinery) and chemical inputs used. As a consequence of these inputs, the agricultural land is more like urban–industrial areas than natural ecosystems. Modern agricultural systems are controlled by external factors (subsidies, prizes, exportations policy) whereas the opposite is true for natural ecosystems.

6.2.2 The Threat of Desertification

The increase in land use by humans has produced the side effect of soil degradation, leaving increased surfaces abandoned because of their loss of fertility after human mismanagement. This phenomenon has been especially dramatic in arid lands, where the risk of desertification is high.

Over the last 20 years, the total area of arable land has remained fairly constant, at around 1500 million ha. This has been achieved at the expense of reducing the surface area of forests and pastures and of increasing the area of "other lands" (Table 6.1). This includes badlands, deserts, watersheds, oil fields, open quarries, and roads and cities—all of which represent destroyed soil.

At present, there are some 4500 million ha of nonproductive land in the world, including mountain soils, of which around 2000 million ha have been degraded by human use at rates of 6 million ha/year during the last 50 years (Rozanov, 1990). The present rate of soil degradation due to human activity is thirty times higher than the average rate during the history of agriculture. These figures (Table 6.1) show that the historical balance between productive arable land and anthropogenically degraded land is negative (1500 less 2000 ha). Clearly this trend cannot be maintained in the future.

Considering soil degradation problems on a worldwide scale, soil salinization is and has been one of the most widespread problems. Since ancient civilizations such as Mesopotamia and the Indo Valley, the salinization of irrigated soils has threatened agriculture in the dry regions of the world. Nowadays, 50 percent of the total irrigated cropland is affected by salinization (Rozanov, 1990), and predictions of climate change indicate an increase in this process of soil degradation.

TABLE 6.1. Worldwide land use changes 1966–1980* (USDA, 1984, in Rozanov, 1990)

	1966	1980	Change
Croplands	1381	1452	+71
Pastures	3122	3117	−5
Forests	4236	4093	−143
Other lands	4336	4413	+77

*Data in million ha.

6.3 The Soil Resource: Overview of the Main Soil Properties That Characterize Soil Productivity

The soil is the support of most of plant life, but it is also the physical basis for buildings and other anthropogenic structures, it may be used for waste disposal, and it provides raw materials. In this chapter the main emphasis will be on soil as the support for plant growth, which is its main role for human use.

Soil productivity is defined as the ability of the soil to provide oxygen, water, and nutrients for the root system. This ability depends primarily on the organization of soil particles, that is soil structure, and, as a consequence, on soil porosity. Porosity determines water and gas flows and root penetration; therefore soil porosity mediates the effects of all soil properties on the roots. The soil volume, often limited in depth, defines the total capacity to store and supply water and nutrients to vegetation. The soil fine earth characteristics define the quality of soil productivity per unit soil weight (concentration). Both the quality of the soil fine earth and its quantity (or volume) define soil productivity. However, the soil properties have to be expressed in capacity terms (related to surface area) to have ecological significance, for example, to be compared with crop growth also expressed in surface area basis units.

6.3.1 Soil Depth

The primary physical property is soil volume, that is, the quantity of soil that can store water and nutrients. Soil volume is limited either in depth and/or laterally by rock outcrops in mountain areas (e.g., karstic systems). Consequently, vegetation cover on these landscapes is poor. The soil volume is composed of the solid phase and the pore space. The solid phase often constitutes around 50 percent of the soil volume, and is mostly mineral except in peaty and similar soils. The pore space is filled by water, which fluctuates in quantity, and the space remaining is occupied by air. Total soil porosity ranges between 30 and 70 percent.

6.3.2 Soil Texture

Soil texture defines the proportion of particle sizes grouped in classes that correspond roughly to different soil behaviour observed empirically (Figure 6.2). Texture has been used as the property that summarizes the characteristics of a type of soil. It is one of the few soil properties known in some way by farmers everywhere. The reasons for the importance given to texture are:

(i) practical difficulties in modifying texture in the field (i.e., texture is permanent feature of the soil);

(ii) texture defines the elemental units of the soil matrix that limit development of possible structures and porosity, that is, the limitation imposed by soil texture on soil structure is higher the coarser the texture; and

(iii) although texture does not directly affect the plant, most of the properties defining soil productivity are dependent upon it.

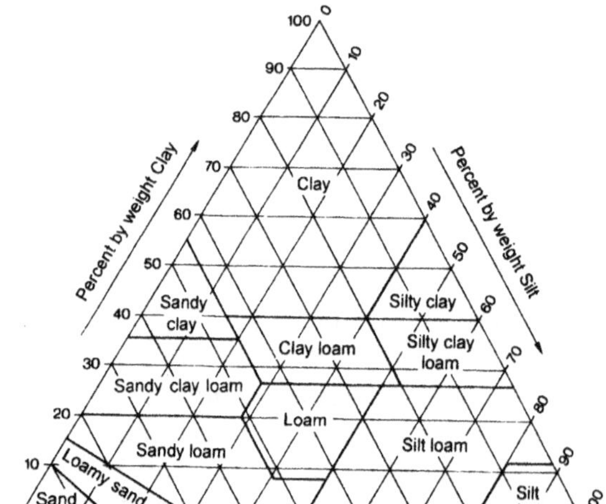

FIGURE 6.2. Soil texture diagram according to the USDA (Soil Survey Staff, 1975).

The most important factor dependent upon texture is porosity. Coarse-textured soils have a lower total porosity and the pores are of a larger diameter than in finer soils. As fine pores (micropores, diameter < 6 μm) hold water with greater energy than coarse pores (see below), clayey soils tend to be moister and less aerated than sandy soils. Crops in temperate and boreal climates are earlier in spring on sandy than on clayey soils, owing to the lower specific heat associated with the lower moisture content of sandy soils.

6.3.3 Soil Structure

This property defines the soil particle organization, that is, how the soil particles are packed and arranged. Two major features of soil structure are its variability in space and time. Structure is affected by climate changes (wetting and drying cycles), biological activity, and land use practices (e.g., ploughing, manuring). In contrast with texture, structure is a property that can be largely modified by soil management practices.

Soil structure directly determines the total soil porosity and the form and size distribution of single pores. Therefore structure ultimately controls the retention and flow of soil water and aeration. Soil structure also determines the mechanical soil properties affecting seed germination, root growth, mechanical resistance to ploughing, and erosion risk.

There are three major types of soil structure: particulate, massive, and fragmented. In particulate structure, the soil particles remain unattached and loose, so soil porosity is only interparticular. This is the case in very sandy soils where no aggregating compounds are present. In the massive structure, soil particles are closely packed and no aggregates can be distinguished; this soil structure is produced by compaction, or where sedimentary soil parent material is not worked by biological activity, especially in fine textured soils. These are poorly aerated and difficult to colonize by roots.

Between these extremes, soils have a fragmented structure where the elemental particles are bound by cementing compounds, like clay, sesquioxides, lime, polysaccharides from microbial activity (especially in the rhizosphere), and humic substances. In addition, swelling and shrinking of clays owing to changes in soil moisture, the burrowing activity of soil fauna, and root permeation of the soil matrix contribute to the fragmentation and arrangement of the soil into aggregates of different size. The fragmented structure creates interaggregate macroporosity that facilitates root penetration and water and air flow. This is especially relevant in clayey soils, in which fertility is dramatically modified by structure.

Soil structure usually shows regular patterns in the soil profile (Callot et al., 1982). The top soil layers (*A* horizons) are rich in organic matter and well aggregated, concentrating most of the roots which themselves contribute to the soil aggregation and loosening. In arable soils, ploughing approximately the top 20 cm soil aims to mechanically create this structure. Both soil mixing by faunal activity and/or ploughing tend to homogenize the *A* horizons. The structure of the intermediate soil layers (*B* horizons) is strongly affected by root penetration and by the characteristics of the mineral components. These are very anisotropic layers, because of the modifications produced around the roots and the poor mixing activity. In depth, the *C* horizons structure is inherited from the parent material, and it is slightly modified by the root permeation that produce pathways for water circulation and bedrock weathering.

Roots cause modifications in soil structure by mechanical pressure produced by root growth and through the physiological activity of the root and rhizosphere, such as the release of root exudates, respiration, and sorption (Callot et al., 1982). Frequently, the soil around the rhizocylinder shows strong anisotropy with respect to the bulk soil.

Soil structure may break down in bare soils that are exposed to the dispersive kinetic energy of rain drops, as happens in annual crops where the soil remains seasonally unprotected. The stability of soil structure is related to the organic content of the soil aggregates which reduces the wettability and swelling ability.

The soil bulk density (D_b) integrates the soil components density (D_p) and their structure, namely, the soil dry mass of a soil volume under field conditions (Mg/m^3). The particle density of the common soil minerals is around 2.6 Mg/m^3, with little variation between soils; in contrast, soil organic matter has a lower and more variable particle density of between 1 and 1.8 Mg/m^3, depending on the degree of humification. Bulk density varies between 0.8 (organic soils) and 2.0 Mg/m^3 (very compacted or stony soils). Bulk density allows the calculation of soil properties on a surface basis from laboratory data currently referred to as soil (fine earth) mass.

6.3.4 Soil Moisture

Water relations between the soil and the plant (and the atmosphere) are expressed in terms of water potential (ψ), that is, the difference in free energy of water in the soil or plant with respect to that of pure water in a standard reference state. Water moves along gradients of water potential in the soil–plant–atmosphere system.

Depending on the forces acting on water, different components of water potential are considered: (i) gravitational (ψ_g), positive potential due to the elevation with respect to a reference level; (ii) matric (ψ_m), negative potential due to the matricial forces of adhesion and capillarity of solid surfaces, which are especially important in the soils; and (iii) osmotic (ψ_o), negative potential produced by the solutes concentration, which is only effective in causing water movement through a semipermeable membrane (such as the cell membrane), and which determines water relations within the plant and at the soil–root interphase. In soils it is only relevant at high solute concentrations (e.g., in saline soils). The water potential of the soil or the plant integrates all these components.

Water flow from the soil to the plant is produced when a gradient of increasing potential (less negative) is established from the leaf surface to the root and to the soil nearby. When the soil is very dry, the plant cells cannot sup-

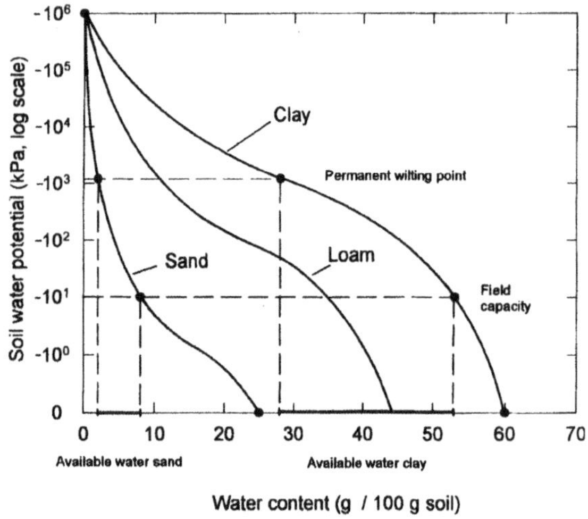

FIGURE 6.3. Characteristic soil moisture curves for different textures.

port enough dehydration to obtain a more negative potential than the soil that is necessary for water uptake, and plant wilting or stomata closure occurs. Two standard water potentials are defined:

(i) *field capacity*, that is, soil water content after wetting and allowing drainage for two days, giving values for ψ between −10 and −33 kPa; this roughly represents the maximum soil water content after fast drainage, with water retained only in the micropores and with the macropores filled by air; and

(ii) *permanent wilting point*, for which the plant is no longer able to extract the soil water, fixed at −1500 kPa.

The water content at potentials between these two limits is then considered to be the available soil water for plants.

The relationship between soil moisture content and water potential is described by curves characteristic of each soil type (Figure 6.3).

6.3.5 Nutrient Availability in Soils

Soil nutrient status can be divided in three main pools according to their degree of availability for root uptake.

The reserve nutrient store is made up of soil minerals and soil organic matter. Soil minerals, depending on their mineralogical nature and the climatic conditions are more or less easily weathered, slowly releasing nutrients, especially cations (Ca, Mg, K, Fe, etc.) and P. This is a long-term reserve, but is the basis of natural soil fertility.

Soil organic matter is much more labile than the mineral component, with turnover rates with a scale in years; through mineralization it releases its mineral content at different rates, depending on the specific chemical bond of the element with the organic compounds and the microbial activity interactions. In general, along with litter decomposition, most K is released very quickly, only a few months after litterfall, followed by Ca and Mg, and finally P and N. Soil organic matter is the main source of soil N, and is an important source of P, both nutrients being associated with the organic cycling within the soil–plant system.

The second nutrient store is the exchange complex which constitutes the short-term pool. This originates in the solid phase surfaces, especially of clay minerals and humic substances, together with their complexes. These fine particles have a chemical charge due to the presence of free radicals, mostly with a negative charge, which generate electrical fields around the particles surfaces—the so-called exchangeable complex.

These negatively charged surfaces attract cations that show a concentration gradient from the particle surface to the free soil solution where there is no influence of the exchange complex (Figure 6.4). Anions show the opposite pattern of distribution. The cations are bound reversibly in dynamic equilibrium with the soil solution. Due to the chemical attraction by the soil particles, cation concentrations are many times higher (one order of magnitude) in the exchange complex than in the soil solution.

The cation exchange capacity (CEC) defines the total soil capacity for reversible adsorption of cations. This depends on the nature and content of the fine particles. The CEC of humic substances is in the range 200–500 cmol/kg and for clay minerals is in the range 10–150 cmol/kg.

The percent base saturation defines the proportion of the CEC that is compensated by basic cations, usually accounting for the most abundant elements: Ca^{2+}, Mg^{2+}, K^+ and Na^+. The acidic cations H_3O^+ and Al^{3+} (including the cationic hydroxides) make up the rest of the charges.

The soil solution contains the nutrients that are readily available for root uptake. This pool is continuously modified by losses through root uptake and leaching, and inputs through rainwater, dissolution, and organic matter mineralization. These fluctuations are regulated to some extent through the exchange equilibrium with the solid phase, acting element by element. The most abundant cations are Ca^{2+} in neutral to basic soils, and H_3O^+ and Al^{3+} (and their hydroxides) in acidic soils. The most abundant anion is HCO_3^-.

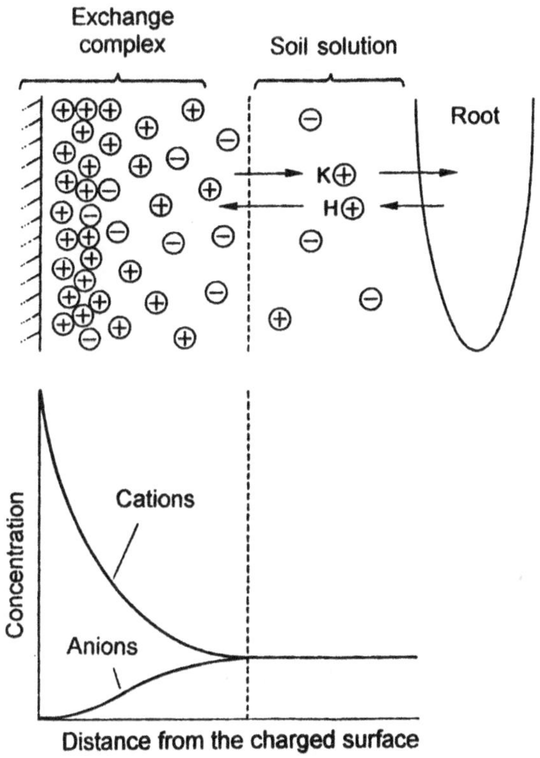

FIGURE 6.4. The exchange complex and the distribution of cations and anions. Schematic view of the exchange mechanism.

6.4 The Main Characteristics of Soil Groups in Relation to Their Management

The major groups of soils are distributed around the world according to climate, which is the main driving force of soil development, and the distribution of the nutrient stores in the natural ecosystems.

In very cold climates (arctic to boreal), plant growth, weathering, and soil formation are limited by lack of heat. Physical weathering predominates and soils are shallow and stony. The main nutrient pool within the ecosystem is stored in the bedrock and is unavailable to plants. At lower latitudes, where climatic conditions improve (i.e., in boreal forests) the forest floor accumulates high quantities of nutrients, especially nitrogen.

Temperate regions have deeper soils, with intermediate weathering and leaching, and large proportions of clay minerals which provide high CEC and natural soil fertility. Nutrient pools in mature ecosystems are mostly in the soil, with relevant amounts of some nutrients stored in the vegetation.

Deserts limit plant growth and hence biological cycling, and little weathering occurs. In many cases soils are remnants of moister past climatic conditions and show a considerable degree of weathering with respect to the bedrock. Salt accumulation and precipitation is much higher than leaching in these soils, frequently resulting in soils with an excessive salt concentration (saline soils). Under these conditions, nutrients are mostly stored in the soil. In dry areas, soil erosion generally exceeds soil formation, and therefore soil development is limited.

In tropical rain forests, soils have been suffering from high rates of weathering for a very long time. As a consequence, soils are poor in weatherable minerals and in clay minerals, thus developing poor natural soil fertility. Litter turnover is extremely fast owing to the optimal conditions for soil biological activity. In these ecosystems, nutrients are stored mostly in the vegetation, and therefore forest clearance can dramatically modify the whole ecosystem nutrient budget.

6.4.1 Calcareous Soils

Limestones cover around 10 percent of the world land surface (Pedro, 1972). The defining characteristic of calcareous soils is the presence of lime, the high base saturation of the exchangeable complex, mostly by calcium, and a slightly alkaline pH (7–8.5). These soils are formed from limestone or other lime-containing rocks with insufficient $CaCO_3$ leaching, or by lime precipitation from base-rich parent materials in poorly leached environments. Calcareous soils are characteristic of dry climates, from Mediterranean climates to drier conditions. Often, the leached $Ca(HCO_3)_2$ from the top soil profile reprecipitates as lime in the deeper soil horizons, or downstream at the bottom of the slope, thus forming hard pans (calcic and petrocalcic horizons) that reduce the volume of soil available for root growth. Soils derived from limestones in wet climates are carbonate-free, although highly saturated with Ca.

The presence of lime has several consequences for soil fertility:

(i) relatively high pH due to the $CaCO_3$–CO_2–H_2O buffer system;

(ii) low chemical availability of some macro- and micronutrients (P, Fe, Mn, Pb, and B);

(iii) flocculation of soil colloids through lime and Ca^{+2}, that promotes soil structure;

(iv) decreased litter decomposition rates in the presence of abundant finely divided lime, which covers and partially protects decaying litter from microbial attack.

Native plants living on calcareous soil are adapted to their chemical environment. Some of them live only on these soils—the calcicole species. In agriculture, calcifuge species are often grown on calcareous soils and they suffer physiological disorders. The most widespread of these is lime-induced chlorosis, that is, iron deficiency which causes yellowing of leaves due to a reduction in chlorophyll synthesis. Poor soil aeration worsens this physiological imbalance.

Crops that are particularly sensitive to chlorosis are vines, peaches, oranges, cherries, kiwis, soybeans, and peanuts. Calcicoles and lime-tolerant species have the ability to release chelating, acidifying, and reducing agents from the root to the surrounding soil. These improve availability of iron and other metallic micronutrients (Marschner, 1986). The chelating agents form soluble complexes with these metals which facilitate the metal absorption by the root. In the same way, root exudates from adapted plant species, including phosphates, improve P availability in calcareous soils.

Lime-induced chlorosis is overcome by applying an artificial iron-chelating agent (most usually Fe–EDDHA). Soil organic matter naturally improves iron availability for calcifuge crops because low molecular humic substances (fulvic acids) also act as chelating agents for these metals.

6.4.2 Gypsum Soils

These are soils of arid regions that contain gypsum even in small quantities ($CaSO_4$ $2H_2O$). This salt is much more soluble than lime (2.6 g/l at 25°C vs. 0.06 g/l at 25°C for pCO_2 $3 \cdot 10^{-4}$ atm) and releases a high Ca concentration into the soil solution. Gypsum is commonly found together with lime in the soil; in these situations, lime buffers the soil pH since gypsum is a neutral salt.

The elevated concentration of Ca in the soil solution of gypsum soils produces nutritional disorders in nonadapted plants, particularly disorders related to K uptake. There is a specialized flora, the gypsicoles, for these soils.

A frequent problem for the management of these soils under irrigation is the disintegration of the subsoil which is rich in gypsum (usually the C horizon), followed by subsidence.

6.4.3 Saline Soils

Soil salinization as a result of human activity is an ancient and important land-degradation process. The flourishing ancient Mesopotamian civilizations declined once their irrigated soils became saline, after developing sophisticated but inappropriate irrigation systems (Hillel, 1991). The FAO estimates that of the 237 million ha currently irrigated, around 40 percent is affected by salinity. Today some 1.5 million ha of irrigated land have been lost by salinization and waterlogging around the world (UNEP, 1992). Predictions of global change suggest a worsening of this process.

Saline soils are defined by the accumulation of soluble salts. These are salts that are soluble enough to produce osmotic problems for the plant cell membranes. Water flow to the cells is driven by a higher solute concentration in the plant cell with respect to the soil solution. These compartments are separated by the plasma membrane which behaves almost as a semipermeable membrane, allowing the passage of water, but few solutes. When the soil solution has a high solute concentration, water will tend to flow from the cell into the soil solution, thus producing what is called "physiological drought," that is, soil moisture can be high but the plant cannot absorb water. To reequilibrate solute differences, the plant has to accumulate salts in the cytoplasm. However these may be toxic for nonadapted plant species.

The most common soluble salts in soil are chlorides. NaCl is by far the most abundant, but Na_2SO_4 and $MgSO_4$ are also common. These are neutral salts, although they frequently appear together with $CaCO_3$, which buffers the pH below 8.5.

The halophytes have mechanisms that allow absorption of high quantities of salts in response to salinity in soils without suffering physiological disorders (e.g., *Suaeda, Salicornia*). These are native plants, although some cultivated plants show a similar behaviour. Another strategy, found in the glycophytes, consists of eliminate the salts through the leaves (some *Limonium*).

Among crop species, apples, peaches, oranges, beans, and carrots are sensitive to salinity, and beets, rice, cotton, and barley are tolerant.

Saline soils can arise in several ways: (i) through direct invasion of seawater in coastal strips; this salinization is widespread all over the world and the dominant salt is NaCl; (ii) through a rise in a shallow saline water table, either near the coast or inland; (iii) from rock weathering in sedimentary saline rocks; (iv) from irrigation using saline water; (v) from runoff and seepage in endorreic areas; (vi) due to "cyclic salt" carried by the wind near the coast; this has a low residence time in well-drained soils but directly affects the plant canopy; and (vii) due to the use of sewage sludge as a fertilizer; this often contains high levels of soluble salts that can cause salinization.

These salinization processes can be combined in different ways. With the exception of coastal areas, salinity is restricted to arid lands with insufficient drainage and where the intense evaporation demand make the salts move into the plant rooting zone and the top soil. Irrigating poorly drained soils enhances the latter process, mobilizing higher quantities of soluble salts from the deep soil layers into the soil surface.

Figure 6.5 shows a typical land salinization produced after the transformation of rain-fed into irrigated land in a semiarid climate. In this example, the source of salinity was the parent material (marls). The irrigation of the slopes mobilized salts down to the bottom slope and to the alluvial soils, and then produced a water table upraise.

Reclamation of saline soils requires the salt removal from the root zone. This is achieved by leaching or flushing. Leaching requires a well-drained soil or the installation of artificial drainage systems. Both reclamation methods require the use of large quantities of nonsaline water.

6.4.4 Sodic Soils

This term designates soils in which the exchange complex is highly saturated by Na+—that is, the exchangeable sodium percentage [ESP = (exch. Na^+/CEC) × 100] is higher than 15 percent of total CEC. In addition, sodium dominates in the soil solution with respect to divalent Mg^{2+} and Ca^{2+}. This is expressed in analytical terms by a Sodium

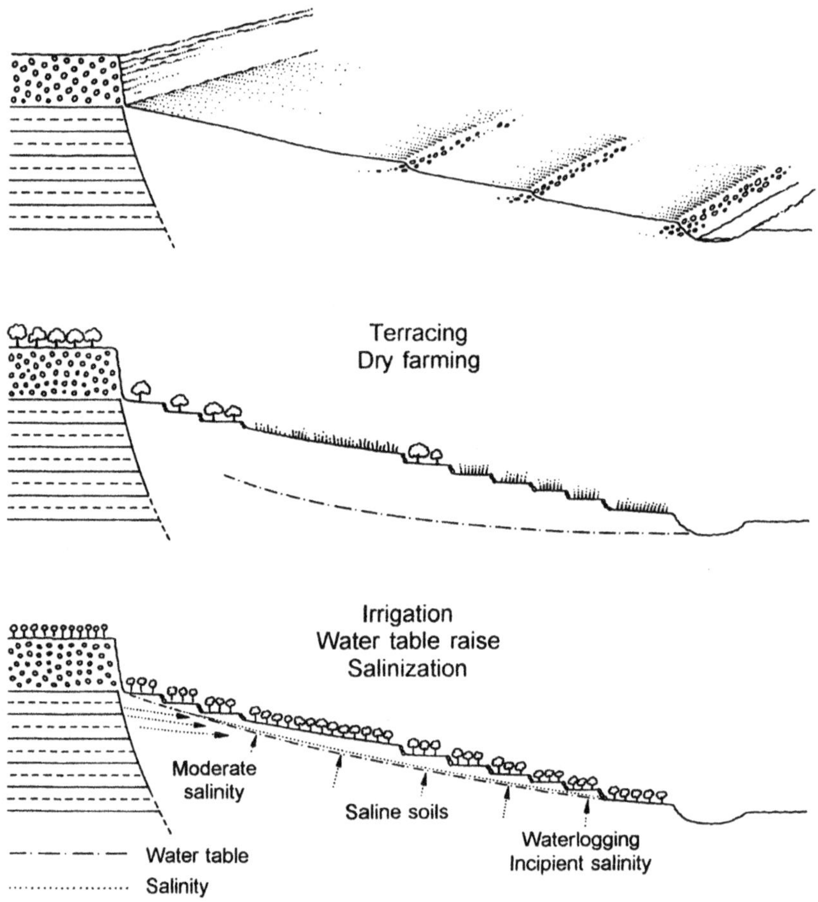

FIGURE 6.5. Salinization processes triggered by irrigating dry farming lands. Example from the Ebro Basin, North Eastern Spain.

Absorption Ration (SAR) in the soil solution which is higher than 13, for SAR = $[Na^+]/(1/2)([Ca^{2+}] + [Mg^{2+}])^{3/2}$.

In alkali sodic soils, pH is buffered by the Na_2CO_3 system and is strongly alkaline (i.e., higher than 8.5). This may produce physiological disequilibrium through sodium-induced calcium deficiency (Cramer and Lauchli, 1986). An even more dramatic effect of sodicity on soil fertility is the dispersion of soil colloids (clay and humus) owing to the deflocculating influence of sodium. This causes a breakdown of soil aggregates, giving way to a specific columnar structure that makes dry soil very compact and the wet soil swampy and poorly aerated. These soils show extremely adverse properties for plant growth.

Reclamation of sodic soils requires their conservation into saline soils. It is necesssary to substitute Na^+ by Ca^{2+} in the exchange complex and to neutralize the pH. Gypsum is commonly used as a source of relatively soluble calcium, although S or H_2SO_4 have also been applied when lime is abundant in the soil. The neutralizing reaction for gypsum application is as follows:

$$Na_2CO_3 + CaSO_4 \rightarrow CaCO_3 + CO_2 + Na_2SO_4$$
(neutral leachable salt)

$$2\ Na^+\text{–Exchangeable} + CaSO_4 \rightarrow Ca^{2+}\text{–Exchangeable} + Na_2SO_4 \text{ (leachable)}$$

After converting from sodic to saline, the reclamation involves leaching of the soluble salts, as carried out in saline soils.

Sodic soils are produced in dry climates

from sodium-rich parent materials, and from leaching of saline soils when concentrations of calcium or magnesium salts are low in the soil or in the irrigation water.

6.4.5 Acid Soils

Soils having a pH below 5.5 have specific properties that are associated with acidity. The most significant properties for crops are the low content of cationic macronutrients (Ca, Mg, and K), and the low solubility of P and aluminium toxicity for pH < 5.0. Acid soils are mostly distributed in temperate to boreal climates, especially on silicate rocks with a low level of basic cations, and in tropical climates, where leaching exceeds weathering rates.

Acidophilic plant species are specifically adapted to highly acidic soils. In the field, acidophilic and calcifuge species may be considered as the same, because acid soils are poor in calcium. However, the terms are not physiologically equivalent (Kinzel, 1983). Acidophilic species are tolerant to high Al^{3+} concentrations in the soil solution and are able to absorb NH^{+}_{4} as the main form of N in acid soils. Some of them show toxicity symptoms when NO^{-}_{3} is applied (e.g., some *Ericaceae* species). Many of these species show optimum physiological conditions at a pH close to neutral, but in natural environments are not very competitive in eutrophic soils; thus they are displaced to acid soils where more productive species cannot tolerate aluminium. In addition to the specific problems of toxicity associated with acid soils, they have a general low fertility status. Correction of acid soils by liming is a common agricultural practice in many areas of the world.

6.4.6 Tropical Soils

The most characteristic tropical soils are deeply weathered *ferralsols* where only the very resistant primary minerals are present (especially quartz), and secondary minerals are constituted by poor clay minerals, kaolinites, and iron and aluminium oxides. These minerals have a poor exchange capacity. In addition soil organic matter content is low due to the high rates of decomposition by the soil biota.

Some common soil features of tropical soils that are relevant to soil management are the following (Primavesi, 1984): (i) sandy layer on the top soil; (ii) compacted layer under the sandy top soil, often produced by the action of Al/Fe oxides; (iii) acidity; (iv) low CEC because of the soil mineralogy, CEC dependence on soil organic matter; (v) soil surface compaction by the soluble silica, either through the desiccation of the bare soil (in arable lands) or by the use of fire for crop residue management or in shifting cultivation; (vi) flocculation of soil colloids by Fe and Al oxides; (vii) Al^{3+} toxicity; and (viii) high P fixation capacity reducing its availability to plants.

The soil organic matter is especially important for maintaining soil fertility in highly weathered tropical soils where it may be almost the only source of nutrients. This is especially relevant in developing countries where low fertilization practices make crop yield mostly dependent on the intrinsic soil fertility. Given the socioeconomic conditions in many of these countries, soil management for these soil types and climate should promote the biological activity and structure and reduce exposure of the bare soil to surface degradation and erosion. Organic farming is especially suited for these conditions (Primavesi, 1984; Mulongoy and Merckx, 1993).

6.5 Constraints of Soil Management Intensity in Relation to Land Use and the Associated Benefits

The question to be discussed here is the following: How much can we alter a soil for a given land use, or what limitations does the soil impose on land use planning?

Clearly such a question cannot be answered without taking into consideration the economic framework of this land use. In a scale from high to low economical value of land use, involving a high to low input requirement, different soil transformations are possible: (i) creating/transporting a soil; (ii) soil reclamation; and (iii) soil fertility sustenance.

6.5.1 Creating a Soil

The soil is a nonrenewable resource on a human and ecological scale because soil formation is a long-term process. In very specific and exceptional circumstances (i.e., small areas of 1 ha in size) it is feasible to create an artificial soil, for example, sport fields made from pure sand with some amendments and sophisticated drainage systems that regulate the water table. Building and maintaining these artificial soils have practically no economic limitations.

On a larger scale, the most prominent example is that of the polders, built in the sixteenth and seventeenth centuries in Holland to gain land from the sea. The sea sediments produced sulphuric soils that required heavy liming in order to be productive.

In the Mediterranean, traditional examples of manmade agricultural soils are those made from soils or sediments moved from lowlands to uplands in mild climates, usually in order to grow high-value irrigated fruit crops (e.g., oranges, bananas).

In warm areas with poor soils, intensive crops are grown out of season on sand under simple plastic greenhouses, using fertiirrigation. The sand creates a well-aerated root environment and diminishes the salinization risk in the rooting zone. Usually a thin layer of peat or manure underlies the sand in order to provide a permanent source of nutrients.

6.5.2 Soil Reclamation

There are widespread practices aimed at improving or correcting soils for agriculture. The most energy-demanding soil reclamation technique is terracing, which involves building slopes that divert runoff into the fields. This practice is as old as agriculture in China and Indonesia, and it shapes the hilly landscapes all along the Mediterranean basin. As terracing requires a high-energy investment either both in building and maintenance, nowadays it is only practised for very valuable crops in developed countries. In the Mediterranean basin, most of the terraces are abandoned and are gradually returning to the natural equilibrium slope.

In dry areas, the soil rooting volume is often limited by the presence of a pedogenic hardpan (calcareous, gypsic, ferric, etc.), which can be removed using powerful machinery.

Drainage system installation in order to improve waterlogged soils and as a method of leaching saline and alkaline soils is another common practice.

Intrinsic soil properties limit modifications. Soil texture is almost a permanent soil property that in common cropping systems is not modifiable. On the contrary, soil organic matter is very sensitive to soil management either by addition of organic manures, crop residues management, or tillage practices. As a consequence, the properties that depend to some extent on soil organic matter can be modified (i.e., soil structure, CEC, soil biological activity, and N forms). Manipulation of chemical soil properties is more feasible.

6.5.3 Soil Productivity Maintenance

There are general practices of soil management aimed at maintaining the physical and chemical fertility of the soil. The most widespread are ploughing in different forms and depths, fertilizing, irrigating dry lands, pesticide application, and crop rotation. Organic farming is gaining increasing popularity, especially since the petrol crisis in the 1970s. Soil organic matter is a key factor in the natural fertility of the soil. It contributes to soil structure, either directly or through the enhanced biological activity, and therefore to the water infiltration capacity and soil aeration. It also affects the CEC (especially in sandy soils), and acts as a buffer and source of biologically regulated N and P.

Crop residues used as mulch have a major role in soil conservation and improving water supply for the crops. This is especially true in warm and dry climates, in soils with low organic matter content, and in areas with a high risk of erosion. Minimum or nontillage practices minimize mechanical soil work, thus reducing compaction, increasing soil organic matter, and facilitating the build up of biological soil structure. This is often combined with mulching. The technique usually requires an increase in herbicide application.

6.6 Methodologies for Soil Evaluation in Rural Planning

Soil classification allows soil characterization and mapping for land use planning. The systems of soil classification aim to group the soils according to their genesis, using diagnostic horizons and criteria. This is the case for the most widespread soil taxonomies at present: those of the FAO (1988) and the U.S. Soil Taxonomy (Soil Survey Staff, 1992). Soil classification systems lay on different criteria; therefore, only approximate equivalences can be defined between the larger soil groups in the different systems (Table 6.2).

Soil maps should contribute to rural planning at two levels:

(i) To determine broad land use allocation, i.e., agriculture, forestry, urbanization, natural parks, etc. For this purpose, synthesis maps are used at scales of between 1:250,000 and 1:100,000. Lower scales only have cultural use, such as the UNESCO/FAO soil map of the world at 1:5,000,000, or the Commission of the European Union soil map of the European Union at 1:1,000,000.

(ii) To develop land use projects, such as the establishment of regional plans, irrigation projects, marginal lands conservation, degraded land restoration, sanitary landfill allocation, and so on. These management projects require soil maps at scales larger than 1:50,000, usually around 1:20,000.

6.6.1 Soil Evaluation

A "good soil" for plant growth is not necessarily a pedogenetically developed or mature soil. Considering soil formation from bedrock, the degree of soil evolution and differentiation is related to its capability to supply nutrients and water. When the soil is formed from recent loose sediments, however, as on river banks, the pedogenetically nondeveloped soils can be extremely fertile for plant growth. In addition, soil evolution in many cases leads the accumulation of toxic elements (such as podsolization) or loss of nutrients.

The soil profile as described according to the pedogenetic criteria and the soil units, specified in the soil taxonomy systems, do not take into account sufficiently the physical properties that are fundamental for plant growth, such as the properties governing water flow. In order to understand soil–plant relationships, the soil profile analysis has to take into account soil structure, as this largely determines water flow and root penetration (Callot et al., 1982).

Therefore, land use planning often requires completion of the basic information provided by soil maps with the elaboration of thematic maps to define land use capabilities. This is the approach of the U.S. Soil Survey, which produces soil maps on a county basis at scales around 1:20,000. These soil surveys include the soil map at the level of soil phase, including soil properties relevant to soil management in addition to the pedogenetic soil units. For each soil unit, land use recommendations are derived for crop, pasture, and woodland management, recreation areas, engineering, and wildlife habitats.

There are some systems for soil evaluation that focus on the aptitudes and limitations of the soil for agricultural use. One widespread system is the Land Capability Classification of the U.S. Department of Agriculture. It is based on the soil properties, the climatic conditions,

TABLE 6.2. Major soil groups in the world according to FAO system (1988), including approximate *Soil Taxonomy* equivalents (Soil Survey Staff, 1992)

Soil unit	Soil order (Suborder)	Main features	
FAO	Soil taxonomy	Of the soil unit	Of the soil unit group
Histosols	Histosols	Organic soil, e.g., peat	
Anthrosols	Entisols	Profound anthropic modifications	
Fluvisols	Entisols (Fluvents)	Young alluvial soils	Soils not bound to specific zonal climatic conditions
Gleysols	Entisols (Aquents) Inceptisols (Aquepts)	Waterlogged soils	
Regosols	Entisols (Orthents)	Low soil development on consolidated parent material	
Leptosols	Entisols	Shallow soils, less than 30 cm depth	
Arenosols	Entisols (Psamments)	Very sandy	Soil formation conditioned by the parent material
Andosols	Andisols	Rich in amorphous minerals derived from volcanic materials	
Vertisols	Vertisols	Clayey swelling soils	
Cambisols	Inceptisols	Soils of initial soil formation	
Calcisols	Inceptisols (Ochrepts)		Soils showing salts accumulation characteristic of arid areas
Gypsisols	Inceptisols (Ochrepts)		
Solonetz	Aridisols, Alfisols and Mollisols		
Solonchaks	Aridisols (Orthids)		

Kastanozems	Mollisols	Deep dark-brown *A* horizon; calcareous	Soils with marked accumulation of base-saturated organic matter, characteristic of steppe or forest–steppe environments
Chernozems	Mollisols	Deep dark-grey *A* horizon; calcareous	
Phaeozems	Mollisols	Deep dark *A* horizon; rich in bases	
Greyzems	Mollisols	Dark-grey *A* horizon over argic *B**	
Luvisols	Alfisols	*B* argic with high base saturation; high CEC clays	Soils showing accumulation of clay, or of sesquioxides and organic matter in subsurface horizons
Planosols	Alfisols and Ultisols	*E* horizon over slowly permeable layer	
Podzoluvisols	Alfisols	Deep tonguing of *E* into argic *B* horizon	
Podzols	Spodosols	Accumulation of organic forms of Al and Fe in *B* horizon	
Lixisols	Alfisols	*B* argic with high base saturation; low CEC clays	Soil with intense weathering typical in tropical and subtropical regions
Acrisols	Ultisols	*B* argic with low base saturation; low CEC clays	
Alisols	Ultisols	*B* argic with low base saturation; high CEC clays	
Nitisols	Alfisols and Ultisols	*B* argic with blocky structure	
Ferralsols	Oxisols	Highly weathered *B* horizon	
Plinthosols	Oxisols	Reticulated iron, quartz and clay pan	

*Argic *B* horizon: subsurface horizon with higher clay content than the overlying horizon.

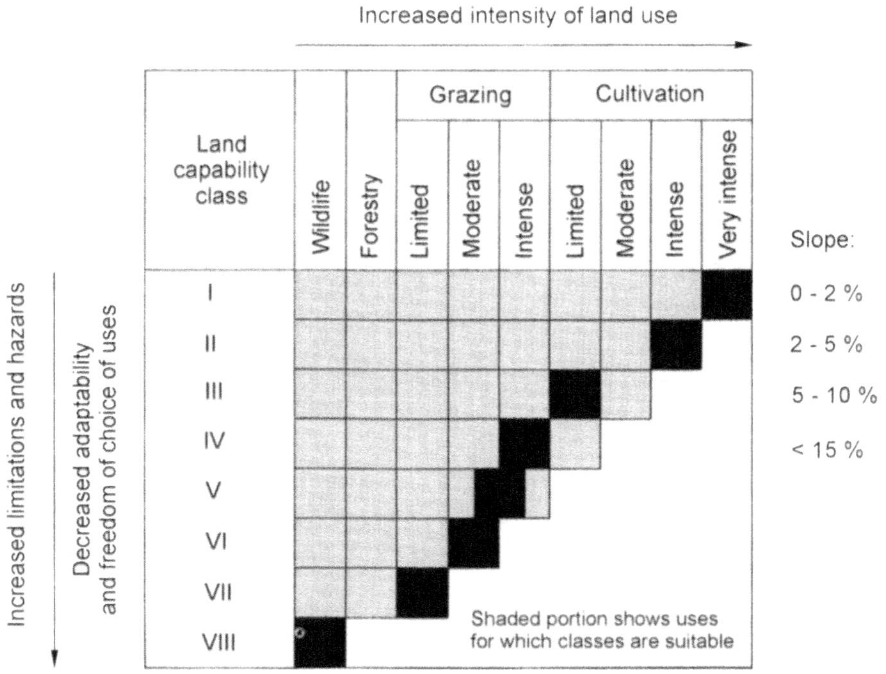

FIGURE 6.6. USDA Land Capability Classification. In black the best use for the land class (from Brady and Weil, 1996, adapted by permission of Prentice-Hall, Inc.).

and the common crops of a given region. The land capability units describe the limitations for cropping, the risks of soil loss or degradation and the response of the soil to management. Figure 6.6 summarizes the criteria to classify the USDA land capability units.

Forest soil evaluation requires consideration of the vegetation together with soil properties, because many forest soil attributes are dependent on the organic matter cycling (e.g., the forest floor). The approach commonly used is based on defining habitat types or forest sites, that is, sites where the most important ecological factors are homogeneous: soil and bedrock, climate and microclimate, and vegetation (usually referred as to their potential). In this methodology, the humus form types takes a relevant diagnostic role as it combines forest composition and productivity with the effects of soil and climate on decomposition (see, for example, the biogeoclimatic ecosystems classification from Klinka et al., 1984 and Sims et al., 1996).

6.7 Final Remark

The physical environment and the soil resource are not always dominant in decision-making processes in rural planning, but it is impossible to manage the land without taking into consideration the aptitudes and constraints of the physical environment.

Many examples in human history illustrate the disasters produced by misuse of soils. The adaptation of management techniques to optimize soil capabilities is one of the key points to ensure sustainable land use in the future.

References

Brady, N.C. Soils and world food supplies. Hintze, B., ed. *Transactions 13th Congress of the ISSS: Vol. I.* 1986 August 13–20, Congress Centrum, Hamburg. Hamburg: ISSS; 1986:61–79.

Brady, N.C. *The Nature and Properties of Soils.* New York: Macmillan; 1990.

Brady, N.C.; Weill, R.R. *The Nature and Properties of*

Soils. Upper Saddle River, NJ: Prentice-Hall, Inc.; 1996.

Callot, G.; Chamayou, H.; Maertens, C.; Salsac, L. *Mieux Comprendre les Interactions Solracine*. Paris: INRA; 1982.

Cramer, G.R.; Lauchli, A. Ion activities in solution in relation to Na^+–Ca^{2+} interactions at the plasma lemma. *J. Exp. Bot.* 37:320–330; 1986.

Etherington, J.R. *Environmental Plant Ecology*. Chichester: John Wiley and Sons; 1982.

FAO. *Agriculture: Toward 2000*. Rome: FAO; 1987.

FAO. *Soil Map of the World*. Revised legend. Rome: FAO; 1988.

Hillel, D. *Out of the Earth*. London: Aurum Press; 1991.

Kinzel, H. Influence of limestone, silicates, and soil pH on vegetation. Lange, O.L.; Nobel, P.S.; Osmond, C.B.; Ziegler, H., eds. *Physiological Plant Ecology III*. Berlin: Springer-Verlag; 1983:201–244.

Klinka, K.; Green, R.N.; Courtin, P.J.; Nuszdorfer, F.C. *Site Diagnosis, Tree Species Selection, and Slashburning Guidelines for the Vancouver Forest Region, British Columbia*. Victoria, BC: Ministry of Forests, Province of British Columbia; 1984.

Lowrance, R.; Stinner, B.R.; House, G.J. *Agricultural Ecosystems*. New York: John Wiley and Sons; 1984.

Marschner, H. *The Mineral Nutrition of Higher Plants*. New York: Academic Press; 1986.

Mulongoy, K.; Merckx, R. *Soil Organic Matter Dynamics and Sustainability of Tropical Agriculture*. Chichester: John Wiley and Sons; 1993.

Pedro, G. Les sols développés sur roches-mères calcaires. *Sci. Sol* 1:6–18; 1972.

Primavesi, A. *Manejo Ecológico del Suelo. La Agricultura en Regiones Tropicales*. Buenos Aires: El Ateneo; 1984.

Rozanov, B.G. Human impacts on evolution of soils under various ecological conditions of the world. Koshino, M.; Morooka, M.; Morishita, J.; Owa, N.; Marada, Y.; Yoneyama, T.; Ozari, Y., eds. *Transactions 14th ISSS. Plenary Papers*, 1990 August 12–18; Kyoto. Kyoto: ISSS; 1990:53–62.

Sims, R.A.; Corns, I.G.W.; Klinka, K. *Global to Local: Ecological Land Classification*. Dordrecht: Kluwer Academic Publishers; 1996.

Soil Survey Staff. *Soil Taxonomy*. Agriculture handbook No. 436. Washington, DC: Soil Survey Staff; 1975.

Soil Survey Staff. *Keys of Soil Taxonomy*. SMSS Technical Monograph No. 6. 5th. ed. Blacksburg: Soil Survey Staff; 1992.

UNEP. *Saving Our Planet*. Nairobi: UNEP; 1992.

Part III
Production Systems

7

Habitat Conservation

Robert G.H. Bunce

7.1 Introduction

Throughout the world it is now generally realized that pressure is increasing on the rural environment from factors such as pollution, agriculture, and urbanization. Many countryside habitats that were formerly widespread are now becoming scarce, as widely recognized by conservation organizations such as the Royal Society for the Protection of Birds and the British Trust for Nature Conservation. These voluntary organizations, in addition to government agencies, reflect an international consensus that it is necessary to maintain and enhance habitats for scientific, aesthetic, genetic, and cultural reasons.

Whilst the habitats are often defined subjectively and their extent inadequately recorded, the quantitative methods described in the present chapter can greatly assist rural planning by providing data on the extent and distribution of habitats and a basis for monitoring them. The introduction therefore covers the type of data required, before describing some conservation initiatives in Europe.

A worked example from Great Britain (GB)—that is, England, Scotland, and Wales but not Northern Ireland—is then presented of how data on habitats can be obtained at a national level from the wider countryside. This approach complements the information on special sites containing rare individual species or assemblages. Rural planning needs to take account of both, because they reflect different types of habitats.

Whilst this approach has enabled an integrated, consistent approach throughout GB the same is not true of Europe, so a similar classification is described at this level so that readers can appreciate the potential of the approach at a broader scale. By contrast, many small regions would benefit from having a similar framework so a further example at a level of approximately 600 km^2 is presented to show how local studies can be carried out relatively quickly.

It is important to point out here that the rural planner has a potentially important role in the conservation of biodiversity. The plan can often protect organisms and their habitats without reducing productive lands and waters in a region. However, to incorporate conservation goals into rural plans the planner must understand the national or regional conservation objectives. These objectives are usually expressed at higher levels, even at international scales, because the geographic range of distribution of species is usually larger than the area of concern to the planner. For this topic, as well as in economics and technology, we must have different geographical scales of reference so that we understand how local decisions act positively or negatively on regional or national conservation objectives.

Finally, it is always necessary for rural planners to have ready access to appropriate data, so the final section describes such an information system. This system is based on a 1 km^2 unit, but could be on a larger scale if required. It can incorporate any available data on habitats and species, regardless of the level of detail involved. The present chapter therefore primarily deals with a statistical approach to the development of a framework for strategic planning. It is complemented by an extensive literature on many aspects of habitat conservation, for example, description (Ratcliffe, 1977), evaluation (Usher, 1986), policy (DOE, 1995) and threats (Holdgate, 1994).

Except for the highest mountains in Europe the majority of habitats for which conservation measures are required are set in a matrix of land in various uses. There is usually, therefore, a degree of competition from different land uses, such as agriculture and forestry, for the available land. It is the role of the rural planning agencies to draw up policies to control development on these areas. At one extreme nature reserves or national parks may be set up with complete protection, but at the other extreme policies may be drawn up with only limited control over development. Whatever the legal structure within a given state, the first priority is always to identify the sites that require protection. The present chapter first outlines the requirements for defining habitats within a region or state, and then describes procedures for determining the extent of the resources present as a basis for planning.

Firstly, there is axiomatic that habitats must be classified into relatively homogeneous classes, before values are attached to them. Otherwise, inherently different sites with fundamentally different characteristics and controlling factors may be compared invalidly. For example limestone vegetation is usually species rich, whereas acid vegetation is species poor, so they cannot be compared directly. It is therefore necessary to construct a framework within which valid comparisons may be made. In many European countries vegetation is used to provide such a structure, often through the tradition of phytosociological analysis. Problems of consistency and definition inevitably arise between countries, but within countries such frameworks have been successfully employed. There is also a requirement to integrate data from groups of biota other than vegetation (e.g., birds and mammals) onto a common basis, but this is rarely done other than in an informal way. Ratcliffe (1977) laid out the criteria that should be used for nature conservation evaluation.

In GB, the Institute of Terrestrial Ecology (ITE) has worked with various planning authorities in setting up frameworks for identifying priorities with regions. During the course of such a project in the Highland Region of Scotland, an area of approximately 23,000 km^2 in the north of Scotland, the following criteria were used to set up an appropriate data base:

(i) *The acquisition of an inventory of basic regional statistics.* An integrated assessment of the overall resources of a region is essential, particularly of features that are not recorded on a systematic basis by other agencies (e.g., boundary types and amenity woodland), because the wider countryside often contains important wildlife resources. These need to be coordinated with information on individual sites that contain special habitats.

(ii) *An indication of the broad regional distribution of resources.* A knowledge of the distribution of habitats throughout the region is essential for planning, whereas the distribution of broad categories such as woodlands or water bodies is generally known from satellite imagery or maps. Other major land uses and land covers, however, may never have been quantified.

(iii) *The need to monitor land use changes.* Many changes in land use are small by themselves, but together have additive effects on habitat composition and extent. The expense of monitoring has usually restricted the availability of information needed for forward planning, but as described below sampling procedures are now available.

(iv) *The need to acquire information on biological species.* There is a need to determine the range of variation in species composition and assemblages in order to adequately define the habitats present, although some information is usually available from literature sources. This is done by recording plant species that are important in their own right, as well as being the most convenient way of defining habitats. Data from animals are more fragmented, but mammals and birds are often adequately covered, whereas insects are not.

(v) *The need to assess land use potential.* Many planning authorities are particularly concerned with retaining their rural population and therefore in identifying further employment opportunities. Apart from the physical suitability of land for different activities (e.g., increased forestry or improved management of marginal agricultural land or the development of peat resources) there is a need to be able to assess regularly the socioeconomic and environmental consequences of land use changes. Increasingly there are employment opportunities in conservation and in ecotourism.

(vi) *The need to test policy options.* It is important to assess the implications of proposals that may modify current usage on existing land use and settlement patterns. It is also desirable to consider the impacts of land use options—e.g., the maximization of agricultural production or increased afforestation—without having to make major surveys on each and every occasion. Certainly, all plans should include considerations of conservation impacts on land use opportunities.

7.2 Conservation Initiatives in Europe

Currently in Europe there are many initiatives concerning habitat conservation at regional, national, and supranational levels. In order to set these into context it is useful to examine the summary provided by the European Union (EU) Habitats Directive for the Internet. This not only indicates the policy context, but also the general requirements in terms of the type of information required.

In order to conserve biodiversity in the EU, legislation has been adopted in the form of the Birds and Habitats Directives, together with a specific financial regulation (LIFE). The Habitats Directive is the most important EU instrument for nature protection. It is due to be implemented over a number of years and member states have concentrated so far on the identification and mapping of sites for inclusion in the network which the Directive requires. This European network of protected areas—NATURA 2000—is a key element of the mechanism. In the Fifth Programme the creation of this network, the maintenance or restoration of natural habitats, and the protection of species of wild fauna and flora are the main targets for 2000.

Despite the major effort that was made for the integration of nature conservation concerns into the revised regional plans, there is still room for improvement. This is partly due to the fact that the environmental appraisals did not attract the same attention in the regional plans of different member states. In some cases this could be explained by the fact that nature characteristics were not included amongst the important features of a region, while in others there were objective difficulties concerning participation of regional or national environmental authorities in the planning procedure.

The EU and the member states face four challenges:

(i) Public authorities in the member states need to ensure that sensitive areas are identified. They need to be properly chosen to protect threatened species and habitats and sufficiently well represented across the territory of the Union to create a network. At the EU level this network will have to be established by the selection of the sites identified by the member states.

(ii) Member states need to determine the appropriate management methods to be applied to each type of selected site and to take into account those methods in their procedures, in particular for actions cofunded by the Union.

(iii) Member states and the EU will need to find adequate resources to ensure the sustainable conservation of the areas identified.

(iv) A European strategy needs to be developed to encourage local initiatives for development and employment supported by the structural funds, which can also contribute to the conservation of natural areas.

Within Britain, the U.K. Biodiversity Action Plan (which includes Northern Ireland) has been produced in response to the commitment made in Article 6 of the Biodiversity Convention adopted at the Earth Summit in Rio de Janeiro in 1992. The Biodiversity Action Plan considers the variability among living organisms from all sources, including inter alia terrestrial, marine, and other aquatic ecosystems and the ecological complexes of which they are part. This includes diversity within species, between species, and of ecosystems.

Three levels of biodiversity are apparent from this definition: (i) diversity between and within ecosystems and habitats; (ii) diversity of species; and (iii) genetic variation within individual species.

No organisms live in isolation from other living things and each has its own way of life, which contributes to the balance of nature. In general, this needs to be considered at the habitat level as a whole.

Habitat conservation is thus at the centre of maintaining diversity in its own right, but also because organisms are dependent upon them. Habitat conservation involves knowledge of:

(i) the overall populations and natural ranges of native species and the quality and range of wildlife habitats and ecosystems;

(ii) internationally important and threatened species, habitats, and ecosystems;

(iii) species, habitats, and natural and managed ecosystems that are characteristic of local areas; and

(iv) the biodiversity of natural and seminatural habitats where this has been diminished over recent past decades.

There is a very great deal known about the number and types of the species of the flora and fauna that live in the European countries. Nevertheless considerably more is known about some groups than others. The algae, fungi, ferns, bryophytes, and lichens are less well known than the flowering plants. For flowering plants, the location, and even population size, are known for the most threatened species. The abundance of more widespread plants is harder to assess, though in Britain the Countryside Survey (Barr et al., 1993) gives such information.

The methodology described below enables base lines to be drawn and for the composition of the general countryside to be defined. Additional information from other biological records, especially of rare species, is often available and can be added to the basic information described below. These data can then be incorporated into information or decision support systems, an example of which is described in the final section.

It is essential to set habitats into that appropriate level within a hierarchy:

(i) *Level 1*: Biogeographic regions (e.g., Scandinavia or any other appropriate level for which summaries are required at a broad level.

(ii) *Level 2*: Europe may be divided into relatively homogenous classes representing major geographical divisions; these correspond to what are often referred to as ecoregions, and in many cases occur in large blocks (this is the level described below in the section on European classification).

(iii) *Level 3*: *Landscape*. Within these units different component landscapes can be identified, for example the Bocage of Brittany within an ecoregion comprising northwest Spain, western France, and southwest Britain.

(iv) *Level 4*: *Habitats*. A given landscape will be made up of different combinations of habitats (or vegetation types, or biotopes); for example the Bocage may contain heaths, meadows, or forests.

(v) *Level 5*: *Species groups*. Within a given habitat different species may form groupings reflecting their ecological affinities, as expressed in phytosociological tables; for example, heaths may contain dry and wet groups of species but still form the same habitat.

(vi) *Level 6*: *Species*. These are often restricted to individual habitats, but are not mutually exclusive, as a given species may participate in several habitats.

(vii) *Level 7*: *Ecotypes*. Within certain species

selection has led to the evolution of ecotypes, which may have widely different ecological characteristics (e.g., forms of *Puccinellia maritima* on a salt marsh).

Levels 2–4 correspond to those described by Noss (1990). It should be noted that only the species level is constantly recognized, which is probably why so much emphasis has been placed on biodiversity at this level. None of the categories explicitly include measures of biodiversity (e.g., structure and genetics).

In addition to the above broad initiatives, many countries have conservation agencies whose purpose is to promote habitat conservation. However, there are many other sources of environmental information on soils, land cover, and individual species that need to be integrated on a common basis to provide adequate information for rural planners to assist habitat conservation. The approaches described below were designed for integration at international, national, or regional levels.

7.3 The Great Britain (GB) Land Classification: A Framework for Rural Planning

The traditional way to create sampling strata in ecology has been to make intuitive divisions of habitat types in the field, and then to describe each type from samples located subjectively. Such approaches—e.g., in phytosociology (Ellenberg, 1978) and soil science (Beckett, 1971)—are primarily intended for mapping and include no independent tests to validate the strata. Further, because of their subjective nature, valid statistical estimates of extent cannot be made, and though the categories are usually mapped and can be measured, no estimate of variability can be assigned to the measurements. In addition, those areas that cannot easily be assigned to a group may be ignored, especially where they fall between the standard definitions.

At the landscape scale, units are usually defined subjectively (e.g., Hills, 1974), or the system of classification is constructed from experience, e.g., the Agricultural Land Classification (MAFF, 1966). Landscape evaluation has presented similar, if not more difficult, problems for quantitative analysis, and there was much discussion in the 1970s about the validity of classification, which has now largely been resolved by the use of traditional methods.

The system is designed to provide a set of general integrated strata for assessing ecologically related parameters in GB and to coordinate information from other available sources such as bird or mammal surveys.

The ITE system had its origin in the expanding use of multivariate analysis for ecological data in the 1960s and 1970s. With advances in computing technology and the development of more efficient algorithms, it became possible to analyse progressively larger data sets with relative ease. Such methods were widely accepted in vegetation science (Greig-Smith, 1964), but were less readily accepted in other disciplines such as soil science (Beckett, 1971), although there was much discussion of their applicability (e.g., Ivimey-Cook and Proctor, 1966).

In the early 1970s, ITE (Bunce, 1984) applied a multivariate classification technique to environmental parameters derived from published maps, rather than to ecological field data. A series of sample units were classified into groups according to a set of environmental parameters that were interrelated in complex ways. The procedure formalized and mimicked objectively the recognition processes previously used to identify landscape types. Extremes of landform are readily identified, but it is more difficult to classify uniform landscapes that had no single dominant feature. For example, the Cairngorms, as the highest mountain plateau in GB, may be recognized as a landscape unit, but it was difficult to classify the southern uplands of Scotland which have no instantly recognizable boundaries. The method developed by ITE divided the whole land surface into nonoverlapping spatial sampling units, which were classified into groups according to their measured environmental features. These groups could then be used as strata from which to draw samples of ecological parameters as required.

The initial classification in 1977 was based on a sample of 1212 × 1 km squares at the intersections of a 15 × 15 km grid placed over GB. Four types of environmental parameters were recorded for each 1-km square: (i) climate (e.g., minimum January temperature and hours of sunshine in July); (ii) topography (e.g., altitude and slope); (iii) human geography (e.g., urban area and roads); and (iv) geology and drift (e.g., carboniferous limestone and peat).

The selection of variables was primarily according to availability, but the climate variables were screened by an initial statistical analysis, as described by Bunce et al. (1996). In other countries other information may be available, as described in the present volume, and variables such as soil depth, soil type, or geomorphology could be included. The problem is always concerned with the availability of consistent information across the whole of the region concerned, because detailed variables are often of limited availability. Experience suggests that variables such as climate and topography are usually available and that even if they are rather general in nature, the analyses can still provide a framework that is adequate to achieve most of the objectives described below.

A total of 40 environmental variables was recorded, details of which are given by Bunce (1996) which converted into attributes and classified using a multivariate procedure TWINSPAN (Hill, 1979) into 32 groups of 1-km squares, termed land classes. The 32 land classes were then used as strata to select sites for a field survey of GB.

It was neither practical, nor desirable, to classify all 240,000 squares in GB using the initial procedure. It would have taken tens of man-years to record manually the key attributes for every square, and, even then, there were potential errors in using a procedural dichotomous key. It was therefore necessary to use automated data capture methods to reduce the weighting of data by further multivariate techniques. However, the data and techniques prevented the same level of detail being recorded for all variables and the results did not perfectly match the original classes produced by the classification process. The procedure that was eventually adopted is described in Section 7.5.

The original 32 classes derived from the classification of the 1212 squares were used as strata for the random selection of eight 1-km squares from each class (i.e., 256 in total) for a field survey of vegetation, soils and land cover in 1977–1978. A summary of the information is given by Bunce and Heal (1984).

These data were then used to assess the effectiveness of the whole field sampling scheme at a national level and showed that, for the main land cover categories in Britain, the figures were comparable to those produced by independent official sources. In 1984, the same 256 squares were resurveyed, but the sample size was increased by a further four 1-km squares from each class to give a total of 384 squares. In the second survey, more detailed information on land cover was collected, and the Countryside Survey described below extended this to 508.

It is also possible to estimate the amount and distribution of features within regions (or subpopulations) of GB using the land class composition. There are two methods of producing estimates of cover for regions from features surveyed at a national scale: (i) the samples can be filtered and only those occurring in the region used to produce the land class means; or (ii) the total sample can be used to estimate the mean for each land class in the region.

Method (ii) involves the assumption that, for each land class, the true regional mean is the same as the GB mean. The regions are unlikely to be random samples from the whole GB population because they are spatially coherent subpopulations. If the average cover of a particular land feature within a land class is different in the region than over GB as a whole, then the resulting method (ii) estimate may be biased. However, a large region may contain the complete population of some land classes, and the populations of squares in the other land classes present are often small, and hence have little influence on the regional estimates.

As the land classes are spatially cohesive, they are more likely to produce a dispersed sample than a simple random procedure. The dispersion of samples was further assisted by the use of the grid on which the original 1-km squares were placed. Moss (1985) identified spatial cohesion as an advantage in both interpretation and presentation, especially where the strata are recognizable geographically. The initial sample-based classification showed a number of geographical outliers that were reduced when every square in GB was classified.

7.4 Results from the Countryside Survey

Countryside Survey 1990 (CS1990) set out to provide some of this information in the form of an overview of the countryside of Great Britain. Its main objectives were (i) to record the stock of countryside features in 1990, including information on land cover, landscape features, habitats, and species; (ii) to determine change by reference to earlier surveys in 1978 and 1984; and (iii) to provide a firm baseline, in the form of a database of countryside information, against which future changes could be assessed. It was a survey of the countryside in its widest sense, which concentrated on the common features and habitats that are most likely to influence the public's perception of rural Britain.

As a way of expressing regional variation in the results from CS1990, the land classes have been aggregated into four landscape types, each of which is dominated by certain land cover types and which simplify reporting:

(i) *Arable landscapes* (34 percent of GB)—land dominated by cereals and other arable crops, as well as intensively managed grassland. Concentrated in East Anglia and the eastern Midlands, but also in the central valley and eastern lowlands of Scotland. Present but less widespread in northeastern England, the Midlands and southeast Scotland.

(ii) *Pastural landscapes* (29 percent of GB)—mainly grasslands. Widely distributed in southwest England, west Wales, the west Midlands, and northwest England. Also in northeast England and scattered through the lowlands of Scotland and coastal areas throughout GB.

(iii) *Marginal upland landscapes* (16 percent of GB)—areas dominated by mixtures of low-intensity agriculture, forestry, and seminatural vegetation. Areas that are on the periphery of the uplands of much of north and west Britain, especially Wales.

(iv) *Upland landscapes* (21 percent of GB)—land generally above a height suitable for mechanized farming and frequently dominated by sheep farming and seminatural vegetation. Distributed in central, western, and southern Scotland, and the Pennine and Cumbrian mountains of northern England.

7.4.1 Field Survey

A stratified random sample of the 508 1-km squares was visited by survey teams. The squares were taken from all rural areas of GB, excluding any square with over 75 percent urban land. Within each square, survey teams: (i) mapped land cover on a field-by-field, or patch basis; (ii) recorded landscape features, such as walls, hedges and individual trees; (iii) recorded plant species in random plots located in fields, woodlands and linear features (hedgerows, roadside verges and stream banks); and (iv) sampled freshwater animals (macroinvertebrates) in streams and watercourses. The field survey repeated 256 1-km squares previously surveyed by ITE in 1977–1978, and 384 squares surveyed in 1984.

7.4.2 Land Cover Change from Field Survey

Overall, 87 percent of GB stayed in the same broad land cover type, and most of the large changes were typical of agricultural rotations between crops and grassland. In many seminatural vegetation types, including broadleaved woodlands, there was a rough balance between losses and gains and the totals remain unchanged.

In summary, the net changes between 1984 and 1990 were: (i) an increase in urban land of 10,500 ha/year; (ii) a 4 percent decrease in tilled land, mostly in barley which decreased by 33 percent; but there was also an increase in minor crops such as maize and linseed; (iii) a decrease of intensively managed types of grassland and an increase in weedier, unmanaged grasslands; (iv) a doubling of noncropped arable land (which would be the type of land typically resulting from land set aside from cereal production); (v) little change in the area of many seminatural vegetation types; (vi) a 4 percent net increase of built-up land, mostly in the countryside; and (vii) a 5 percent net increase in coniferous woodland.

The results contrast with the well-documented losses of seminatural habitats in Britain during the postwar period. Instead of a wholesale loss of these habitats in the wider countryside, there was, in the period 1984–1990, more of a balance between loss and gain.

7.4.3 Field Boundaries

The linear features that criss-cross the countryside form an important part of the British landscape. Many of these features are field boundaries, such as hedges, lines of trees, walls, fences, banks, or just grass strips. In 1990 the total estimated length of field boundaries was nearly 1.5 million km. On average, there were about 6.3 km of field boundaries in each 1-km square. Many of these boundaries (30 percent) contained more than one type of element, such as a hedge with a fence. Over two-thirds of boundaries (72 percent) contained a fence, and almost a third (31 percent) contained a hedge. Between 1984 and 1990, the total length of boundaries declined slightly but individual boundary types showed substantial losses or gains. Thus, hedges and walls decreased in total length, while fences and relict hedges (woody boundaries that had once been hedges) became more common.

7.4.4 Plant Diversity

In addition to major and obvious stepwise shifts from one type of land cover, habitat, or linear feature to another, change in the countryside can take the form of more gradual, subtle changes in the balance of species within habitats. The loss or gain of flowering plants in meadows is a typical example. Using the detailed records of species composition from the same plots in 1978 and 1990, it was possible to distinguish between those losses and gains in plant diversity due to shifts between vegetation types and those due to changes in the quality, that is, the species composition, within vegetation types.

The vegetation plots from open fields, woods, and moors were grouped into six types of vegetation corresponding to habitats—arable fields, improved grassland, semiimproved grassland, woodland, upland grass, and moorland. In Britain as a whole, three of these six major vegetation types (arable fields, semiimproved grass and woodland) showed significant losses of species between 1978 and 1990. Only one vegetation type, moorland, showed a significant increase in species diversity but this vegetation type is inherently species poor.

All the four landscapes showed a change in species composition with an increase of species more characteristic of intensively managed vegetation, and this change was pronounced in the lowlands. In the arable landscape, there was a net shift toward more intensive types of vegetation in 29 percent of all plots. In pastural, marginal upland, and upland landscapes, the shifts in the same direction were 27 percent, 11 percent, and 9 percent, respectively.

Hedges, verges, and watersides contained many species that were absent or rare in the surrounding landscape and represented an important reservoir of botanical diversity. This was particularly true in arable landscapes in which most types of species were more frequently found in the linear features than in the fields or woodlands.

Although meadow species were declining in hedgerows, verges, and watersides, these linear features still contained more of the total resource of meadow species than was found in the open countryside. Linear features are important for biodiversity not only because of their contribution to overall plant diversity in the countryside, but also because they can act as a source of locally native seed. In years to come, given the right conditions, it may be possible for species-rich habitats to regenerate from these seed banks. Any further loss of linear features, or the meadow species they often contain, may limit the scope to conserve biodiversity in the lowlands.

7.5 A European Environmental Classification

7.5.1 The Potential for Integration Across Europe

At the European scale data sets are more variable in quality than for GB. It was, therefore, decided that a comparable classification to achieve most of the objectives required at the European scale could be developed using climatic and altitude data alone. It was subsequently suggested that some of the classes could be combined or subdivided using non-climatic data, such as potential natural vegetation or geology.

The classification (described below) was first produced at a 0.5 × 0.5 degree resolution, and the major groups correspond to recognizable divisions of European climate, such as Mediterranean or continental, as described by Kendrew (1953). These cells are of different areas but are in most cases comparable within a country or region. Further resources could enable transfer to standard units of size, but the information content of the cells is likely to override size differences. The statistical procedures provides objective rules by which the classes are determined, as discussed by Jones and Bunce (1985). The classification has been applied successfully at a higher resolution to a 10 × 10 km data set for GB; subsequently, the classification can be extended to 1 × 1 km units, and the latter can then be nested into the ITE GB land classes.

Data derived from the Climate Research Unit (CRU) at the University of East Anglia (Hulme et al., 1995) were used as the basis of the classification. They contain the 1961–1990 mean climate figures for a "Greater European Window" region, extending from 32° West to 66° East and from 21° and 68° North. The data for air temperature, precipitation totals, sunshine hours, vapour pressure, wind speed (ground), frost day, and rain day frequencies are expressed at a resolution of 0.50° latitude/longitude as monthly minimum, maximum, and mean values. The interpolation of meteorological station data to the national grid used elevation as one of the predictor variables, and this enabled climate surfaces to be constructed for each variable.

Only a subset of the data was used in the classification, chosen to represent dominant trends in variation in the variables concerned; otherwise, the classification may be weighted according to the most common variables. The mean climate data were processed using Principal Components Analysis (PCA) and those variables with the highest component scores were used to produce the classification. The analysis was carried out in four stages:

(i) separate PCAs for each of the eight climate variables were carried out on the monthly data;

(ii) principal components with an eigenvalue greater than one were selected from each of the eight analyses and used in a ranked correlation analysis to select the months that correlated most for each climate variable;

(iii) the selected months were pooled together and a PCA was performed on this dataset; and

(iv) principal components with an eigenvalue greater than one were selected, and the monthly climate variables that correlated most with these components were selected for classification.

The variables most highly correlated to the principal components were selected, regardless of whether they were positive or negative. This selection produced six principal components in the PCA of the pooled variables with eigenvalues greater than one, resulting in twelve different monthly climate variables. The three altitude variables—high, low, and mean—were also added. Easting and northing were used as variables in the GB land classification to provide geographic position and surrogates for climate and oceanity to the data set.

At the European level, however, northing correlated well with the climate variables such as temperature and sunshine hours, but because of the complex climatic patterns in Europe, easting did not. Neither correlated strongly with oceanity, which is widely regarded as an essential variable to incorporate into climate classification exercises. Thus, northing was retained but easting replaced with an oceanity variable calculated as the annual mean temperature range divided by the sine of latitude. Oceanity and northing (latitude) were added to make the final classification data set of seventeen variables. These variables were then converted into attributes for TWINSPAN analysis (Hill, 1979). The 68 attributes were classified by TWINSPAN to produce 64 classes for the 5209 squares of the European window.

The geographic distribution of the classes was smoothed using the discriminant function, which resulted in outliers being removed. The results from this analysis have now been used as the final classification for the 5209 squares, rather than the TWINSPAN class, as described by Bunce et al. (1996). The discriminant function procedure has also been applied at 10 × 10 km, and can subsequently be used at 1 × 1 km, to model climate perturbations.

This classification is now being developed to provide a framework for describing European landscapes and for determining conservation priorities. It also has potential for achieving many of the objectives set for the GB land classification (e.g., stratification and integration).

7.5.2 An Example of a Regional Classification: The Picos de Europa

An example of a rapid local classification produced in seven days was carried out for the Picos de Europa in northwest Spain. An area, 20 × 33 km, encompassing the main mountain massif, was delimited for study. Environmental data were recorded for 91 attributes from each 1-km square, adopting the procedure described by Bunce (1984). These data included features such as altitude, lakes, cliffs, geology, climate zones, and river basins. TWINSPAN (Hill, 1979) was then used to divide the data into eight land classes (Bunce, 1984) reflecting natural divisions.

Within this environmental series a survey of land cover was carried out in 65 × 1 km squares in 1990. Seven major categories of ground cover were used to define the classes: cliff, scree, mountain grass/heath, meadow/herb-rich grassland, scrub, forest, and, finally, miscellaneous categories such as built-up land. These categories were then used as the basis for a classification which enabled the principal habitats of the region to be summarized in three zones described below.

7.5.2.1 Zone 1: High Mountain

This zone occupies 33 percent of the Picos and is composed of high mountain vegetation, lightly grazed by rebeccos (Pyrenean chamoix) and goats. Otherwise the land cover is mainly rock and contains many of the distinctive karstik features of the Picos. Species that require isolation, such as the wallcreeper, live in this zone as well as many species with alpine and subalpine affinities. It contains the specialized mountain, subalpine habitats, and limestone rock habitats. This zone has now formed the basis of the designation of the core area of the new national park and is not threatened by change other than by mountain recreation.

7.5.2.2 Zone 2: Scrub/Grazed Grasslands

This zone contains the scrub and grazed mountain grasslands and some of the highest meadows (up to 1400 m) of the high valleys and mountain sides. Screes and cliffs are locally important and form refuges for alpine and calcicolous species, as well as for nesting birds. Agricultural management is of critical importance, as much of this zone was originally tree covered and patches of shrub can rapidly expand without grazing. The mosaics of small meadows and lines and shrubs of trees are characteristic, but there are also extensive areas of heathlands and forests. This zone is threatened by decline in agriculture and local development.

7.5.2.3 Zone 3: Forest/Meadow

This zone occupies 36 percent of the region, and contains a mosaic of forest and managed grasslands, depending upon topography and aspect. It comprises the main valleys of the region and a large area on the northern fringe. The east/southern fringe has a higher proportion of forest due partly to its different geology and warmer climate. There is much variation because of geomorphological differences between valleys. The habitat in this zone is threatened by a variety of impacts from urban development to agriculture intensification and even locally by abandonment.

This classification is therefore an example of the construction of a rapid framework for identifying threats to habitats and could be developed further for planning purposes.

7.6 The Countryside Information System: A Delivery System for Rural Data

As discussed above, many countries have extensive data on habitats at a variety of levels but these are rarely coordinated in other than a descriptive way. Rural planning needs access to such data so that threats to habitats can be identified at an early stage and necessary policies developed for conservation. ITE has worked with the Department of the Environment over the last 10 years to develop such a system for Britain. It is essential that the researchers work closely with the planners; otherwise the system design will not necessarily be able to respond to their requirements. The system described below was developed in this way, and shows the potential for such an approach elsewhere.

The Countryside Information System (CIS) provides policy advisors and rural planners with easy access to strategic information about the GB countryside. There are major design issues involved. Three aspects of the design problem are considered in detail. First, the types of data and modelling tools that are required in such a system. Second, the data management issues that arise with system use. Third, the implications which the system has for education and training.

The CIS, which is designed to provide strategic information for GB, must provide access to both census and sample type data. The data can be stored and displayed at a relatively coarse scale of spatial resolution, but the errors and limitations of data at these scales must be flagged for the user. The system provides a good framework in which a wide range of information about the UK countryside could be brought together. However, in addition to this inventory function, access to more sophisticated modelling tools are needed in the longer term. Despite this limitation, however, CIS offered an important opportunity to develop a coordinated approach to the formulation and evaluation of policy.

7.6.1 CIS Design

The stimulus to the current phase of development of CIS was the need by the Department of the Environment (DOE) and other government agencies to gain easy access to the results of the 1990 Countryside Survey (CS90) (Barr et al., 1993). CS90 consisted primarily of the survey of the Countryside (as described above) undertaken by ITE. CS90 was partly funded by DOE, and builds on the two earlier surveys described above. Information available from disparate sources of information are generally difficult to use by policy customers when presented in the form of paper reports and tables. For future projects, access to results via some computer-based system was therefore considered essential.

The potential of expert systems technology as a means of providing policymakers with access to data and the skills necessary for its interpretation was explored in the late 1980s. The demonstrator systems developed dealt with the issues of identifying the environmental impacts of forestry at the national level, and the establishment and management of herb-rich meadows. However, the systems proved to be time consuming and costly to develop, and too specific in their focus to be of general use to policy advisers. Instead, a more flexible and general information system seemed to be required, which could link easily to other systems available to policy advisers. A simple prototype, written for Microsoft Windows, was developed to illustrate the concept. This system has grown into what is now known as the CIS.

The first phase of development of the CIS took place between September 1991 and March 1993. The approach adopted during the work has been one of successive prototyping and periodic review. In this way users had effective control on system development and shaped the design of the system as their level of understanding of the technology grew. During the development work the goal has been to create a system that:

(i) provides the rural planner with access to a data model which represents the key elements of the rural environment in a way relevant to national policy concerns;

(ii) is capable of being used within a time frame that is realistic in terms of the normal process of policy appraisal (i.e., hours or days, rather than weeks or months); and

(iii) requires no higher level of Information Technology (IT) skill for operation than is now commonplace in the modern office environment.

7.6.2 Data

In the ideal world, the policy adviser might desire access to data for the whole country for every environmental variable that is of interest. However, such census data are often impossible or too expensive to collect. Instead, rural planners have to satisfy themselves with sample-based information of some kind. Such data need not be less reliable than census information, but the user must be aware of the different assumptions that apply in using them. The requirement to provide access to both census- and sample-based data has been a fundamental factor affecting the design of the CIS.

The sample-based information contained within the CIS is derived from CS90 and includes estimates of change for the periods covered by the earlier ITE surveys of 1978– 1984 and 1984–1990 as described in Section 7.3.

Although the CIS was conceived originally as a means of providing access to CS90 data, as the project has developed its potential as a tool for making a whole range of other environmental data available has also been recognized. Thus, the requirement to hold additional information on the system has been added to the original specification. The system has been extended to include a range of census-based information at national scales. Once again, the 1 × 1 km cells of the national grid are used as the basic spatial unit.

Other census-type data held on the CIS include records showing which grid squares include any land falling under one of the different statutory designations (e.g., national parks, areas of outstanding natural beauty, and environmental sensitive areas), together with partial census data which apply to only specific regions or areas. The system has been designed so that users can add new data easily and hold different combinations of information. This aspect of modularity in database design will be discussed below.

7.7 Summary and Conclusions

Experience in developing the CIS has shown that at national scales, sample- and census-based information, a 1-km resolution can be used to inform rural planners about the environmental characteristics of regions, or the geographical patterns which a particular environmental parameter exhibits. For many purposes a raster-based system, with this coarse spatial resolution is adequate for policy appraisal at national scales, although in any further development of the system, the ability to model the sensitivity of landscapes to change may need to be added. In the short term, the most important factor affecting the value of the system is whether the range of data held on the system is sufficient to represent the essential aspects of the rural environment in ways that are relevant to the process of policy evaluation.

The CIS provides a framework for the integration of a broad range of environmental data. If the system is to be used widely, however, the general issue of data availability must be addressed. The advantage of the CIS is that it may help data suppliers to define a range of information products that can be generated from their own, more detailed corporate holdings. The availability of the system can assist them in making their information more widely available, without undermining confidentiality or the commercial value of their information resources.

The CIS may also help those parts of central government that deal with countryside policy, together with the associated statutory agencies concerned with the rural environment, to define the sorts of data flows that need to occur between them, if a coordinated and responsive approach to policy development is to be achieved. It could readily be adapted for other larger or smaller regions and has the potential for integrating the diverse types of data likely to be available in different areas.

7.8 Summary of Procedure

It is possible to summarize the steps that would need to be taken, whatever the scale of the region concerned, as follows:

(i) Determine the available information in land cover, land use, landscape, habitats, and species from a literature search and consultation with appropriate agencies.

(ii) Convert the available data onto a consistent format at an appropriate scale and develop a delivery system for ready access.

(iii) Identify gaps and then determine the resources that are available to carry out the necessary survey. Consider complete habitat survey or satellite image analysis.

(iv) If complete coverage is not possible, then construct a stratification system using statistical procedures from available consistent map information.

(v) Use the classes to coordinate sample information, and if resources are available, carry out a sample field survey as a basis for assessing ecological resources and provide a basis for monitoring.

(vi) Develop a coordinated data base into a format suitable for developing policies for rural planning and for determining the potential impacts of different scenarios of change.

References

Barr, C.J.; Bunce, R.G.H.; Clark, R.T.; Fuller, R.M.; Gillespie, M.K.; Groom, G.B.; Hallam, C.J.; Hornung, M.; Howards, D.C.; Ness, M.J. *Countryside Survey 1990: Main Report. Countryside 1990: Vol. 2.* London: Department of the Environment; 1993.

Beckett, P.H.T. The cost-effectiveness of soil survey. *Outlook Agric.* 6:191–198; 1971.

Bunce, R.G.H. The use of simple data in the production of strategic sampling systems. 1st International Seminar of the International Association of Landscape Ecology. *Methodology in Landscape Ecological Research and Planning Conference 4*; Roskilde University Centre; 1984.

Bunce, R.G.H.; Barr, C.J.; Clarke, R.T.; Howard, D.C.; Lane, A.M.J. Land classification for strategic ecological survey. *J. Environ. Manage.* 47:37–60; 1996.

Bunce, R.G.H.; Heal, O.W. Landscape evaluation and the impact of changing land use on the rural environment: The problem and an approach. Roberts, R.D.; Roberts, T.M., eds. *Planning and Ecology.* London: Chapman and Hall; 1984:164–188.

Department of the Environment (DOE). *Biodiversity: The UK Steering Group Report.* London: HMSO; 1995.

Ellenberg, H. *Vegetation Mittleuropas mit den Aloen in Okologischer Sicht.* Stuttgart: Ulmer; 1978.

Greig-Smith, P. *Quantitative Plant Ecology.* London: Butterworth; 1964.

Hill, M.O. TWINSPAN: *A* FORTRAN *Programme for Arranging Multivariate Data in an Ordered Two-Way Table by Classification of the Individuals and Attributes.* Ithaca, NY: Cornell University; 1979.

Hills, G.A. A philosophical approach to landscape planning. *Landsc. Plan.* 1:339–371; 1974.

Holdgate, M.W. Protected areas in the future: The implications of change and the need for new policies. *Biodivers. Conserv.* 3:406–10; 1994.

Hulme, M.; Conway, D.; Jones, P.D.; Jiang, T.; Barrow, E.M.; Turney, C. Construction of a 1961–90 European climatology for climate change modelling and impact applications. *Int. J. Climatol.* 15(12):1333–63; 1995.

Ivimey-Cook, R.R.; Proctor, M.C.F. Association analysis and phytosociology. *J. Ecol.* 54: 179–192; 1966.

Jones, H.E.; Bunce, R.G.H. A preliminary classification of the climate of Europe from temperature and precipitation records. *J. Environ. Manage.* 20:17–29; 1985.

Kendrew, W.G. *The Climate of the Continents.* 4th ed. London: Oxford University Press; 1953.

Ministry of Agriculture, Fisheries, and Food (MAFF). *Agricultural Land Classification.* Agricultural Land Service, Technical Report No. 11. London: Ministry of Agriculture, Fisheries, and Food; 1966.

Moss, D. An initial classification of 10-km squares in Great Britain from a land characteristic data bank. *Appl. Geogr.* 5:13–150; 1985.

Noss, R.F. Indicators for monitoring biodiversity: A hierarchical approach. *Conserv. Biol.* 4:355–364; 1990.

Ratcliffe, D.A. *The Nature Conservation Review.* London: HMSO; 1977.

Usher, M. *Wildlife Conservation Evaluation.* London: Chapman & Hall; 1986.

8

The Natural Systems: Potential Production and Forest Management

Carlos A. Gracia

8.1 Environmental and Biotic Controls on Production

8.1.1 Photosynthesis: Introduction, Units, Principles

Life of plants and animals is based on a constant flux of energy. This input of energy is used in me-tabolism, growth, and reproduction. From this perspective, individuals can be regarded as complex metabolic machines that process energy and materials. Green plants, which account for more than 99 percent of the total biomass present in the planet, fix a fraction of the electromagnetic solar radiation. This energy is stored in organic compounds of very high energy which, in turn, are the basis of the networks that maintain life on earth.

Photosynthesis is the process used by plants to transform the energy from electromagnetic radia-tion to chemical energy. This chemical energy will then be partially used by plants to compensate their metabolism and the remaining energy, if any, will be accumulated as new biomass. Energy fixa-tion, its use in plant metabolism, and the forma-tion of new plant biomass are core concepts dealt with in this chapter.

Before discussing the basic processes related to photosynthesis and primary production it will be useful to review the definition of some terms of general use. *Biomass* (*B*) is the total weight of the organism or community we are analysing. In plants, this concept involves above-ground and be-low-ground biomass. Biomass is referred to as units of mass per unit of area, and so the biomass of a grassland is expressed as g/m^2 and the biomass of a forest as kg/ha or t/ha, a classical unit in for-est studies which, to fit the recommendations of the International System of Units, should be expressed as Mg/ha.

All these units refer to the dry weight of the organisms once the water—which is an important and rather variable part of living tissues—has been discounted. Because the chemical composition of organisms is roughly constant with an almost constant fraction of carbon of around 50 percent of dry matter, biomass is often referred to as g C/m^2 or kg C/m^2.

The new biomass that plants grow per unit of land area and unit of time is the primary production (PP). It is important to distinguish between gross primary production (GPP), which is the total amount of carbon taken up by plants through the stomata, and the net primary production (NPP), which is GPP minus the fraction of this carbon used by plants to maintain their own metabolism or respiration. The NPP can be stored as starch or other compounds or can be used to make new wood or other plant tissues. It represents the biomass increase of the community, and it is the fraction of PP that can be used by animals.

Photosynthesis is the cornerstone of life. In a very simple way the process can be summarized as follows:

$$6CO_2 + 6H_2O + 709\ \text{kcal} \rightarrow C_6H_{12}O_6 + 6O_2 \quad (1)$$

but photosynthesis is not the only process that takes place in the plant. Thus, to maintain their own structures, plants respire a fraction of this energy. From a general point of view, these respiration activities can be summarized as

$$C_6H_{12}O_6 + 6O_2 \rightarrow 6CO_2 + 6H_2O + 709\ \text{kcal}$$

Comparing both equations, it is evident that photosynthesis and respiration are opposite pro-cesses. The conditions in which photosynthesis equals respiration giving a net balance of zero is referred to as the compensation point.

8.1.2 Factors Affecting and Limiting Photosynthesis

Important questions related to PP include the following: What factors affect photosynthesis? How do they affect it? From the point of view of natural resources managers, these questions can be reformulated as: Which factors can be modified to increase PP?

The increase of primary production is not, of course, always the aim of managers. For example, in areas with a high risk of erosion an alternative question could be: How can the foliage cover be increased to improve soil protection against heavy rains? This question illustrates the universe of basic and applied problems related to primary production. The analysis of factors affecting photosynthesis helps us to understand the basic mechanisms of primary production.

8.1.2.1 Light and Chlorophyll Content

Incoming solar radiation is absorbed through canopies. This absorption depends on the leaf area present in canopies, the shape and disposition of leaves, and the optical properties of leaf surfaces (colour, roughness, hairs, etc.). A fraction of the light absorbed by leaves is used by the chlorophyll to carry out the synthesis of organic matter.

Chlorophyll cannot be indefinitely accumulated in leaves. The upper limit depends on the amount of incident radiation that can be absorbed by the chlorophyll. An optical density of 2 corresponds to an absorption of 99 percent of light. This is the optical density of a solution of about 35 mg of chlorophyll per liter measured in 1-cm cuvettes (Margalef, 1974). This concentration represents about 350 mg of chlorophyll per m^2. Light into canopies is partially absorbed and partially reflected in all directions so, both faces of the leaf can absorb light. Adding the 350 mg/m^2 of both faces, the maximum active chlorophyll in leaves should be about 700 mg/m^2 of leaf area.

The ratio between the milligrams of carbon assimilated per hour and the chlorophyll concentration in absence of other limiting factors tends to be 5. Because plants have about 700 mg of chlorophyll per m^2 of leaf, they do not fix more than 3500 mg C/m^2 per hour, which is about 35,000 cal/m^2 per hour. The maximum photosynthetically active radiation reaching the plants can be about one-half of the solar constant (46 percent of solar radiation is available to plants) so we obtain the following: $2 * (46 * 60 * 104) = 552{,}000$ cal/m^2 per hour. Comparing the energy fixed and the input of energy on leaves we conclude that, even under optimal con-ditions, plants do not use more than 5–6 percent of incoming solar radiation, and very often the observed values are less than 1 percent.

Incident light on plants is often saturated. At around 500 $\mu E/m^2$ per second, the photosystems show a degree of saturation. This fact is not too surprising because in dense canopies most of the leaves are exposed to an environment of diffuse and attenuated light, which is why leaves have adapted to work optimally at such low or medium intensities. Only the leaves from the upper part of the canopy are exposed to the full solar intensity. These leaves

TABLE 8.1. Percentage of light reaching the soil in canopies with different Leaf Area Index (LAI) and light extinction coefficient values

	Extinction coefficient (μ)			
LAI	0.4	0.7	1.0	1.2
1	67.0	49.7	36.8	30.1
2	44.9	24.7	13.5	9.1
3	30.1	12.3	5.0	2.7
4	20.2	6.1	1.8	0.8
5	13.5	3.0	0.7	0.3

have modified morphology; they are usually smaller, thicker, and show a greater density of trichomes and stomata. These modifications allow them to successfully face the excess of radiation to which they are exposed.

Light intensity in the canopy can be easily measured using suitable light meters. The light intensity decreases with depth into the canopies. In diffuse light conditions (overcast sky) light intensity decreases exponentially with the leaf area index (LAI)—defined as the ratio between leaf area and ground area—measured from the top of the canopy downward according to the expression of Monsi and Saeki (1953).

Most values of LAI in natural communities are about 4. With a light extinction coefficient of 0.7, which as we have seen is the extinction coefficient of a canopy with an spherical distribution, a leaf area index of 4 absorbs 97 percent of incident light (Table 8.1). In most plant species, the light compensation point is about 1 percent of the incident radiation at noon on a clear day. Light is not usually a limiting factor in natural systems perhaps because this regulation between LAI and the light extinction coefficient is a constant in all types of communities.

8.1.2.2 Water

The water involved in the biochemical reactions of photosynthesis is only a very small fraction of the water transpired by the plant. Water is one of the most important ecological resources and is a key point to understand the organization of plants. Water is a vehicle of transport of the nutrients from the soil solution, where they are absorbed by the fine roots, very often with mycorrhiza, to the leaves. Without this vehicle of trans-port, nutrients are not accessible to the plant. In addition, water transpired through the stomata act as a coolant and help to maintain the leaf energy balance compensating for the increase of temperature produced by the absorbed energy.

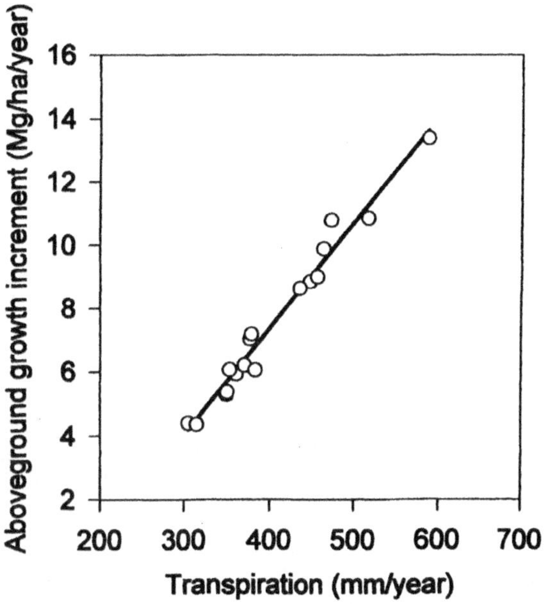

FIGURE 8.1. Above-ground growth increment in a holm oak forest as a function of the annual transpiration.

At the community level, the total leaf area per unit of ground, the LAI, is closely related to the water availability (Figure 8.1). The fact that a grassland, a shrubland, or a forest living in the same area differs very little in LAI is due to the fact that LAI does not depend on the species composition of the community nor on the structure of the population (density and size of the plants), as it is shown in Table 8.2. LAI is a relevant parameter because carbon assimilation is a function of the water transpired by the leaves.

It is easy to understand that if water is available to plants, the stomata remain open and gas exchange takes place in such a way that the output of an amount of water vapour is compensated by the input of an equivalent

TABLE 8.2. Leaf Area Index (LAI) of four different closed coniferous forests living in the same area in Catalonia (Spain). Despite the different species or tree densities, LAI remains basically constant reflecting its dependence on climate (water availability)

	Density (trees/ha)	LAI
Pinus sylvestris	382	1.31
	605	1.11
Pinus nigra	732	1.31
	1560	1.32

volume of air which contains the CO_2 that can be assimilated by the plant. The mmols of CO_2 absorbed by each mol of water transpirated is a measure of the water use efficiency (WUE). The WUE of most plants does not differ much from 5 mmol/mol, although plants living in humid environments are less efficient, and plants from arid lands can use water two or three times more efficiently.

The amount of rainfall is not the only meteorological factor that determines the amount of water available to the plant community. Some of the precipitation is lost to the ecosystem because it is intercepted and retained as a film on the vegetation, on the litter layer under the plants, or on the surface of the soil. This film is evaporated without passing through the living components of the ecosystem.

Some of the precipitation also may be lost to the ecosystem if it does not enter the soil but runs off the area into streams, or if having entered the soil it drains through saturated soil beyond the limit of plant roots. The rest of the precipitation is temporarily stored in the soil in the capillary pores between the soil particles. If a water potential gradient is created, water will flow from the soil pores into the plant and eventually be transpired from the leaves. The fraction of available water transpired by plants depends on the structure of the community (density, and architecture of plants) and the potential evapotranspiration of the atmosphere. Figure 8.2 summarizes the water flux in a *Quercus ilex* forest in Spain. On the average, for a rainfall of 548 mm, 81 percent is transpired by plants. This value increases to 92 percent if evapotranspiration is taken into account.

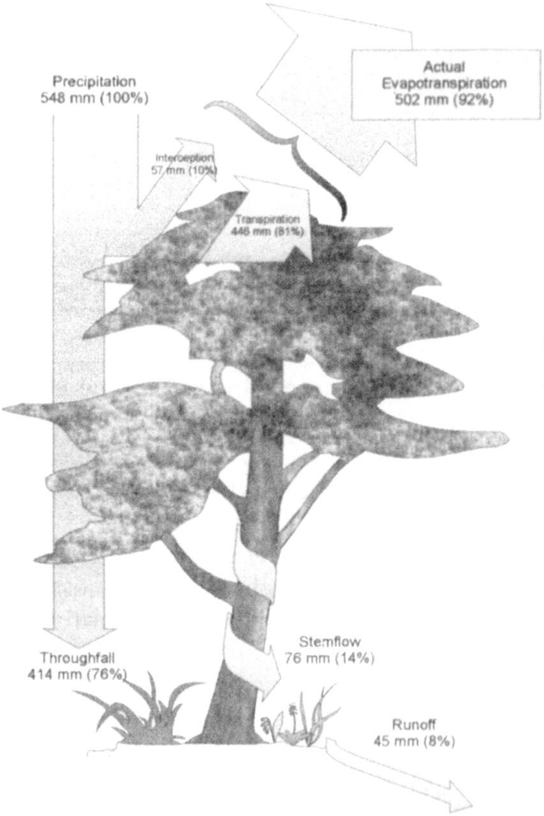

FIGURE 8.2. Water fluxes in a *Quercus ilex* forest in Prades (Spain). Data are the average of several years of observations (data from Bellot, 1989; Lledó, 1990; and Piñol, 1990).

8.1.2.3 Temperature

Temperature plays an important role in the control of chemical reactions. Van't Hoff in 1884 made the observation that chemical reactions increase their velocity by a factor ranging from 2 to 3 when temperature increases 10°C. The ratio of increments of velocity is designed as:

$$Q_{10} = V_{t+10}/V_t$$

A consequence of the Q_{10} factor is that increasing temperature increases the metabolism of plants. Nevertheless, not all processes increase at the same rate. Respiration increases much faster than does assimilation. Most plant species have a limit of about 5°C. Under this threshold the plant activity stops. It is not very difficult to understand this threshold temperature which is, in part, related to the physical properties of the water. Because water has a maximum density and viscosity at 4°C, the friction of water on the wall vessels increases the difficulty of transport of water through the plant.

Figure 8.3 describes the occurrence of the major terrestrial ecosystems types or biomes identified by the growth form of the dominant vegetation in relation to mean precipitation and mean temperature values on the surface of the earth. Primary production ranges from 1 Mg/ha per year in arctic regions with low mean annual temperature to 20 Mg/ha per year in tropical areas with high rainfall and temperatures.

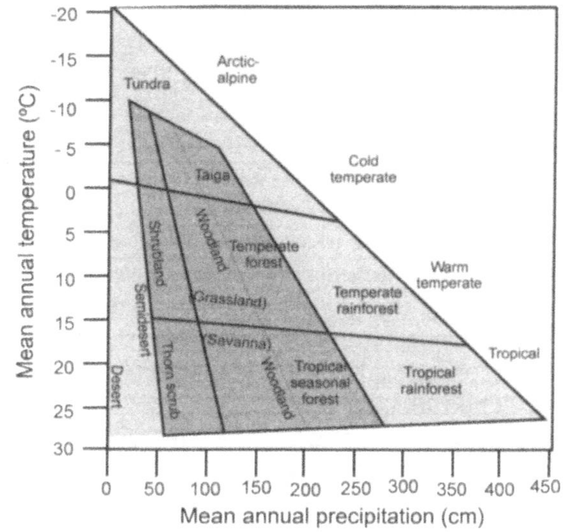

FIGURE 8.3. Occurrence of the major terrestrial ecosystems in relation to precipitation and temperature, the major climate factor for plant growth. The shaded area represents the range of precipitation and temperature that actually occurs in the surface of the Earth.

8.1.2.4 Nutrients

In presenting the basic traits of photosynthesis, it is common to use a simplification as we did in equation (8.1). It is evident that the formation of photosynthate not only depends on the availability of carbon, but other elements are needed as well. The presence of these elements in the soil and the availability of water to transport them to the leaves is often a limiting factor to the primary production in terrestrial ecosystems. Carbon is about 48 percent of dry weight; nitrogen in leaves is about 1–1.5 percent; and phosphorus is a bit less than 0.1 percent. The concentration of these elements is higher in the leaves and lower in bark or wood.

Despite the fact that nitrogen is very abundant in nature, and represents more than 78 percent in volume of atmospheric air, it is one of the most limiting elements to plant growth. This apparent paradox is due to the fact that plants cannot assimilate the diatomic nitrogen (N_2) present in the atmosphere. Specific bacteria and some blue-green algae are the only organisms able to fix atmospheric nitrogen and transform it into forms that can be used by plants. This nitrogen fixed in the ecosystem is recycled and used again by plants. The decomposition of the organic matter produces reduced nitrogen (ammonia), which can then be oxidized to form nitrate—two chemical forms that can be used by plants. All of the biochemical and photobiological processes of photosynthesis require nitrogen compounds. Fertilization, a practice used profusely to increase the productivity of crops, is partly a consequence of nitrogen limitation.

8.1.3 Predicting Gross Photosynthesis

The estimation of gross primary production can be carried out using gas analysers based on the absorption of CO_2. The assimilation rate varies over a wide range of values depending on the species and the environmental conditions. Maximum photosynthetic rates range from 3 to 80 mg CO_2/dm^2 per hour although the most frequent values are around 10–20 mg CO_2/dm^2 per hour.

The main environmental factors affecting the rate of carbon dioxide fixation are the light intensity and temperature, and, in Mediterranean-type ecosystems, water availability. Measuring the carbon uptake rate under different light intensities leads to the construction of the so-called light intensity–assimilation curves. At very low light intensities, CO_2 uptake is negative until the light compensation point—the light intensity at which CO_2 uptake is null—is reached. At these very low light intensities there is no photosynthesis, but the CO_2 assimilation is negative because of the respiration of the leaf. At higher light intensities, CO_2 uptake increases very quickly. At medium light intensities—in practice, at light intensities of about 500 w/m^2—CO_2 assimilation becomes saturated and the assimilation rate approaches a level that is independent of the light intensity.

Temperature affects the photosynthetic rate in a different way. As temperature increases, the CO_2 assimilation rate at maximum light intensity (A_{max}) increases to a maximum that depends on the thermopreferendum of the species. As temperature increases over that point, the assimilation rate decreases again. The lack of water during some periods forces the closure of stomata and, as a consequence, the carbon dioxide uptake is reduced. The water use efficiency—the amount of CO_2 taken up per unit of water transpired—is a good indicator of the way the plant uses water. Plants living in xeric environments have higher values of WUE than the plants living in environments in which nutrients, light, or other factors have a more limiting effect than water.

8.1.4 Carbon and Energy Budget: Cost of Formation and Maintenance of Plant Tissues

The chemical reduction of CO_2 taken up in photosynthesis is the way the plant converts the electromagnetic energy into a chemical form. The formation of a molecule of glucose requires the reduction of 38 molecules of ATP. In an opposite way, the oxidation of a molecule of glucose, whatever the biochemical pathway used, produces 38 molecules of ATP.

The sugars formed are not the final product. With the exception of structural polysaccharides, cellulose and lignin, which are the main components of the cell walls and wood and which can be regarded as final products, the remaining sugars are a source of energy for maintaining processes, for transport in the plant, and for the formation of new structural components.

Atmospheric CO_2 is taken up by the leaves through stomata while water and nutrients are taken up by the fine roots. The process of photosynthesis requires the transport of these nutrients to the leaves and the transport of assimilates from the leaves to the storage organs. Energy is used to take up the nutrients and to transport them and, as a result, less than 1 g of structural components is formed per gram of sugar.

In a practical analysis it is useful to distinguish between the cost of maintenance of plant tissues and the cost of formation of new materials. The maintenance respiration is considered to be proportional to the dry weight of the plant or part of the plant and means no increase of biomass, while the growth respiration is involved in the process of formation of new materials.

Because the chemical composition of a plant is relatively constant it is possible for most practical applications to use an integrated value of the cost of formation and maintenance of plant tissues. On the average, 1 g of carbohydrates gives about 0.68 g of new tissues, the difference being consumed during the growth process due to the transport of carbon and ions between organs and across cell membranes.

For practical applications, the caloric content of most plant tissues can be regarded as being practically constant and having a value very close to 5 kcal/g of dry weight. Because the formation of 0.68 g of plant tissue requires the consumption of 1 g of carbohydrates, the average cost of formation of plant tissues can be estimated as $5/0.68 = 7.35$ kcal/g of dry matter, the difference between the caloric content of the new plant tissue and the energy required for their formation being the growth respiration.

Similarly, an average value for the cost of maintenance can be estimated. On the basis of the plant composition and the cost of maintenance of each compound, an average value of 70 cal/g of dry matter/day can be estimated for most plants growing at 20°C. For other temperatures, this value should be modified because the cost of maintenance depends strongly on the metabolic activities that are temperature dependent. For other temperatures, the cost of maintenance can be estimated, using the Q_{10} factor, as:

$$M_t = M_{20°} \times Q_{10}^{(t-20)/10}$$

where M_t is the cost of maintenance at t°C.

The cost of formation and maintenance of plant tissues can be used to analyse the contribution of each part of the plant to their metabolic balance. Table 8.3 summarizes the metabolic balance of a Quercus ilex forest from the structural data obtained from field measurements.

TABLE 8.3. Components of the cost-benefit balance in an evergreen oak forest (*Quercus ilex*) growing in the Prades mountains (Tarragona, Spain). Leaf metabolic cost is referred to 1 m^2 of leaf area. Stem and roots metabolic cost are referred to 1 m^2 of ground.

Leaf specific weight	200	g/m^2
Mean leaf life span	730	days
First year		
Formation	1470*	$kcal/m^2$
Maintenance	5110**	$kcal/m^2$ per year
Second year		
Formation	0	$kcal/m^2$
Maintenance	5110**	$kcal/m^2$ per year
Mean annual metabolic cost of 1 m^2 of leaves (formation + maintenance)	5845	$kcal/m^2$ per year
Leaf Area Index	3	m^2/m^2
Leaf biomass	600	g/m^2 of ground
Stem biomass	11,000	g/m^2 of ground
Root biomass	12,000	g/m^2 of ground
Stem production	850	g/m^2 of ground per year
Root production	930	g/m^2 of ground per year
Cost of stem formation*	6248	$kcal/m^2$ of ground per year
Cost of stem maintenance**	1180	$kcal/m^2$ of ground per year
Cost of root formation*	6835	$kcal/m^2$ of ground per year
Cost of root maintenance**	1287	$kcal/m^2$ of ground per year
Total cost of nonphotosynthetic tissues	15,550	$kcal/m^2$ of ground per year
Total cost of nonphotosynthetic tissues per m^2 of leaf area	5183	$kcal/m^2$ of leaf per year
Total metabolic cost (leaf, stem and root) per m^2 of leaf area	11,028	kcal/m2 of leaf per year

*Assuming a formation cost of 7.35 kcal/g

**Assuming a maintenance cost of 7.35 kcal/g per day, a proportion of sapwood to total wood of 0.12. and a proportion of living parts in the sapwood of 0.035.

8.1.5 Predicting Net Photosynthesis

Gross photosynthesis minus respiration is the net photosynthesis or, in other words, the biomass increment responsible for the plant's growth. It is evident that knowing the amount of photosynthetic tissue, which is the leaf biomass or leaf area, and the biomass of the live tissues, is necessary to estimate both components of net photosynthesis.

The GPP of a community is limited by the total leaf area that can be supported. The GPP increases as the community matures and the LAI increases to a maximum value. At this point GPP remains constant or tends to decrease slightly. During all this process the biomass increases, which means a progressive increase of respiration. Necessarily, both values approach each other in time until the respiration equals the GPP in the mature communities.

This ideal situation—the climax—is never reached because natural disturbances do not allow it. For example, herbivores eat a fraction of the biomass and this reduction favours a certain increase of net production. McNaughton (1979) has studied in the savannahs of the Serengeti the effect of herbivores on net production, showing their importance on the maintenance of grasslands production (Figure 8.4). A high rate of grazing depresses production. If we take into account the relationships between GPP and R over time, we can see that the NPP values must be more variable than the

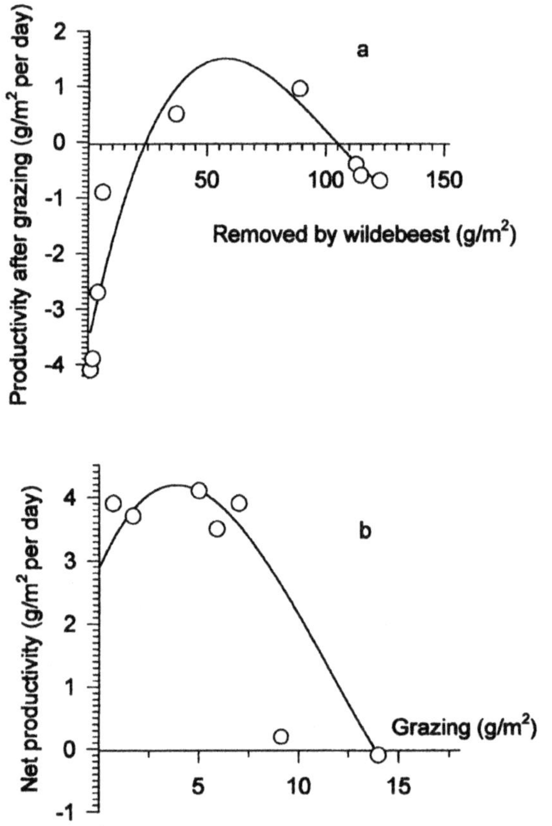

FIGURE 8.4. Grazing can stimulate the primary production of a savannah in the Serengeti. (a) Primary production of the grass after following the pass of a wildebeest flock. (b) Net production as a function of grazing intensity. In both cases production reaches a maximum at medium grazing rates (based on data from McNaughton, 1979).

GPP values. In fact, NPP relies not only on the type of community but also on the degree of maturity and on the structure. Many NPP data are available.

The measurement of the primary production of a natural community is not an easy task. The production invested in leaves and in other assimilatory parts and in wood must be taken into account. In addition to the above-ground fraction, the below-ground proportion, which is far more difficult to study, is not negligible. The below-ground fraction becomes quantitatively more important the more arid the environment is.

In Mediterranean conditions, the below-ground biomass is equal to or greater than the above-ground biomass. Carbon allocation to the different structures leads us to define the turnover time of plant structures which is the B/P_n or their inverse, the productivity P_n/B. The leaf turnover is low, one year, for deciduous leaves, 2 or 3 years for some evergreens (*Quercus*), from 2 to 10 years in conifers, and from decades to hundred years in the wood. Obviously, the renovation rate varies with the maturity of the community because a progressive accumulation of biomass leads to a smaller net production and to an increase of B/P_n. The value of the B/P_n ratio reflects some properties of the ecosystems.

8.1.6 Total Water Consumption by Whole Plants

Precipitation reaching the plants has a complex interaction in the plant's canopies which depends on the amount and intensity of each particular rainfall event as well as on the architecture of the canopy. A fraction of the water reaching the canopy is intercepted. Dry surfaces of stems, branches, and leaves retain this water. A fraction is directly evaporated without reaching the soil. The water intercepted can be as much as 90 percent for small rainfall episodes but, on the average it accounts for about 10–15 percent of annual rainfall.

When these surfaces are wet, a fraction of the water is conducted by the leaves and branches to the stem. This water washes the stem and reaches the soil as stemflow. In dry areas stemflow plays an important role because the water captured by the canopy is concentrated in the soil around the stem and a fraction of the nutrients lost as exudates are also concentrated in this area increasing the water and nutrients available to the fine roots. The remaining water not intercepted or conducted to soil as stemflow passes through the canopy and reaches the soil directly as throughfall. Details of water dynamics from a climatological point of view are available in Chapter 4.

Based on these considerations, Specht (1972) developed an indirect method to esti-

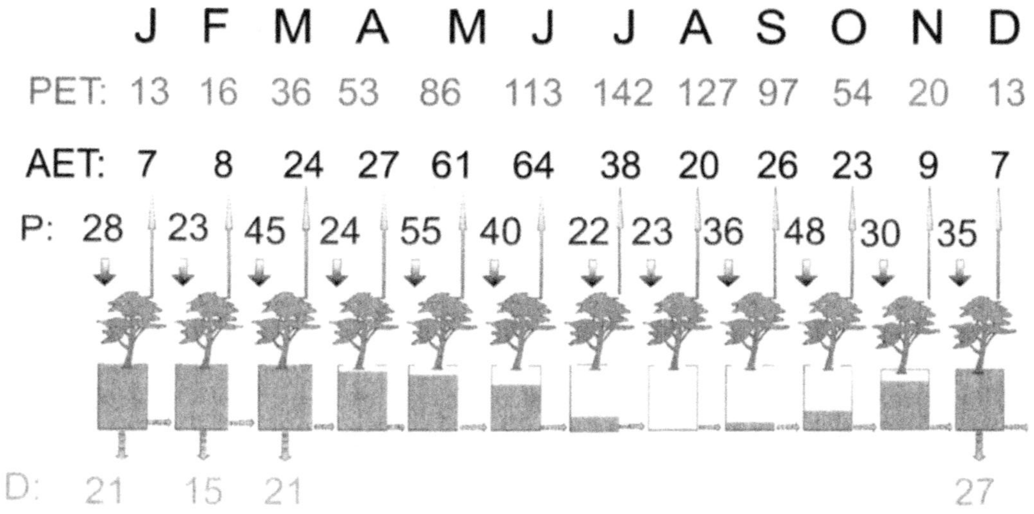

FIGURE 8.5. Monthly actual evapotranspiration (AET), water stored in soil, and drainage in plots near Barcelona, estimated from rainfall (P) and potential evapotranspiration (PET) using the evaporative coefficient *k* (Specht, 1972). Soil S_{max} = 5 cm assumed for calculations.

mate the water balance and actual evaporation of a whole plant community. The water input and output of the ecosystem over any period of time can be summarized in the following water-balance equation:

$$P + S_1 + B_1 = E_a + S_2 + B_2 + R + D$$

where P is precipitation + dew; S_1 and S_2 are the soil moisture storage at the beginning and end of the observation period, respectively; B_1 and B_2 represent the water held in the plants of the ecosystem at the beginning and the end of the observation period, respectively (as the change in B is very small, the value is usually ignored); R is the surface runoff; D is the internal drainage; and E_a is the actual evapotranspiration which is the sum of transpiration (E_t) + evaporation of intercepted water (E_i) + evaporation of soil water (E_s).

As precipitation is conventionally recorded in terms of depth of rain falling into a standard rain-gauge rather than as volume or weight of water per unit area, all terms in the water balance equation are expressed as depth of water entering or leaving a unit area of land as millimetres. After heavy rains, the surface soil will become saturated with water and excess water will move downward under gravity, gradually saturating deeper and deeper layers of the soil profile. Drainage (D) to subsoil reservoirs may eventually occur. Evaporation (E_a) is continually occurring and, during periods of limiting rainfall, soil moisture (S) will be utilized, at first largely from the surface where most of the roots are concentrated but, if the drought period is prolonged, soil moisture is drawn from deeper and deeper layers of the soil.

A typical annual sequence of soil moisture recharge and utilization observed under perennial evergreen vegetation, in the Med- iterranean-type climate of Barcelona, Spain, is shown in Figure 8.5. It will be noted that maximum and minimum values of soil moisture occur in the soil profile in winter and summer, respectively. The maximum value approaches the field capacity of the soil, a value that is found when the soil moisture has reached equilibrium after excess rain has drained under gravity to lower levels in the soil profile. The minimum value approaches the level at which no further water will flow from soil into the plant (i.e., the water potentials of the soil and root are then equal).

The daily variations in the evaporative

power of the atmosphere (expressed as the evaporation (E_o) from a water surface such as an evapor-imeter), may be integrated in a simple relationship if monthly instead of daily time intervals are used. A term such as available water (*W*)—the sum of the water extractable (S_{ext}) from the soil profile at the beginning of the month plus the water added by precipitation (*P*) during the month—may be used.

Provided that the leaf area index of the plant community remains relatively constant, a monthly index of evapotranspiration (E_a/E_o) is related to available water (*W*). When $W = (P + S_{ext})$ is limiting, $E_a/E_o = k\,(P + S_{ext})$; the index ($E_a/E_o$) tends to approach a value of unity when the ecosystem received an unlimited supply of available water (Penman, 1962; Rutter, 1964; Evans, 1971). As it is probably that the leaf area index of most perennial, evergreen plant communities remains relatively constant throughout the year, it can be assumed that the value of the coefficient *k* will also remain constant.

The twelve mean monthly values of precipitation (P_1 to P_{12}) and of pan-evaporation (E_{o1} to E_{o12}) for a locality provide the basic data for computation. If necessary, the monthly values of *P* can be adjusted to take into account irrigation, run-on, and run-off. The balance (precipitation + irrigation + run-on − run-off) enters the ecosystem.

The computations involve the following five steps repeated in sequence for the twelve months of the year:

$$W_1 = P_1 + S_{ext1}$$
$$E_{a1}/E_{o1} = kW_1 \rightarrow E_{a1} = kW_1\,E_{o1}$$

$$\text{if } E_{a1}/E_{o1} > 1 \rightarrow E_{a1} = E_{o1};$$

$$S_{ext2} = W_1 - E_{a1}$$

if $S_{ext2} > S_{max} \rightarrow D_1 = S_{ext2} - S_{max}$ and $S_{ext2} = S_{max}$. S_{max} is the maximum amount of extractable water (S_{ext}) that can be stored in the soil profile. Maximum soil storage is achieved when the whole soil profile is at field capacity; unextractable soil water is the minimum level of soil water to which the particular ecosystem is capable of depleting the soil profile.

Because no mean value of S_{ext} is known, the computations can be commenced by using an arbitrary value of S_{ext1} (e.g., $S_{ext1} = 0$). A series of erroneous values of S_{ext}, E_a, and *D* will be generated for the first few months of the year, but these will be corrected by repeat-

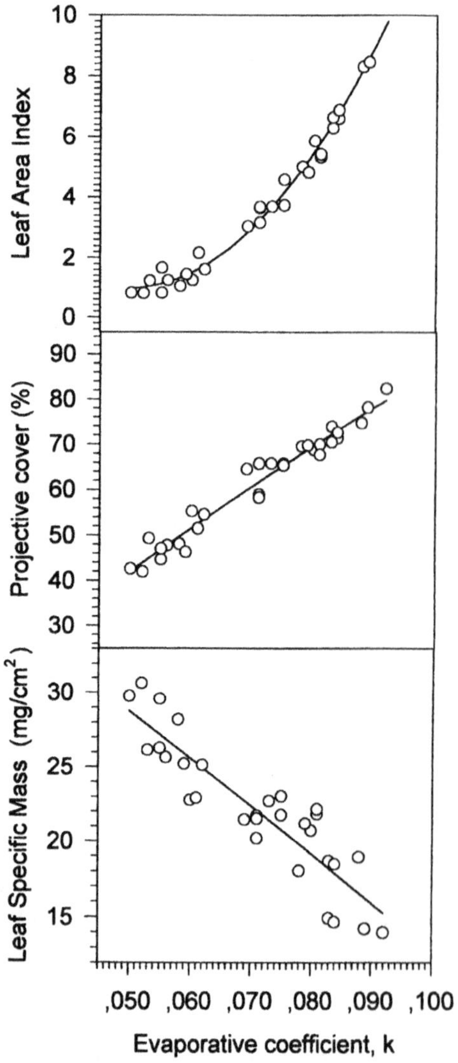

FIGURE 8.6. The evaporative coefficient reflects the structure of the plant community, in particular the LAI, canopy projective cover, and the mean leaf specific weight. Points in the plot represent different plant communities sampled following a gradient from Alicante (southeastern Spain) to the Pyrenees.

ing the series of twelve monthly computations, now using the December estimate of extractable soil water (S_{ext13}) to initiate the new series of calculations (S_{ext1} now equals S_{ext13}).

Observations reported by Specht (1957), Martin and Specht (1962), and Specht and Jones (1971) indicate that the climax vegetation of a particular area fully exploits the soil moisture available for plant growth. Also, the structure of the plant community (in particular the distribution of its leaf area) and the stomatal resistances of individual species must be such that actual evaporation (E_a) is controlled so that no deficit of extractable soil water (S_{ext}) can occur in any month of the year. If this is achieved, transpiration, although small, will continue throughout the year and, concomitantly, carbon dioxide fixation will result. The equation:

$$E_a = kE_oW$$

can be easily interpreted. Water actually evapotranspirated results from the combination of three main components: (i) the available water (W); (ii) the available energy in the atmosphere (E_o); and (iii) the active role of plants in transpiration, which is related to the value of k. The value of the evaporative coefficient reflects the structure of the canopy (Figure 8.6) provided that the minimum value obtained for the twelve months is retained. In fact, this means that plant canopies (leaf area, projective cover, etc.) are adjusted to a maximum value, which will cause the water stored in soil to approach zero during the driest period of the year.

A different approach was developed by Bormann and Likens (1979) using a whole catchment approach. The basic idea of this method is very simple. In a small experimental watershed it is possible to record the water input (using rainfall gauges) and output (using limnigraphs). If the bedrock of the watershed is unpermeable—the main limitation for the applicability of this method—one can assume that the difference between water input and stream flow output is lost by evapotranspiration. This method has proved to be very useful for ecological studies due mainly to the integration of many different environmental factors at a watershed scale. Nevertheless the distinction between transpiration and evaporation is not feasible in this approach, which gives an integrated value of actual evapotranspiration.

8.2 Regional Patterns of Biomass and Production

8.2.1 Water Availability and Temperature in the Biosphere

On a global scale, climate plays the largest role in determining the structure of both vegetation and soils in ecosystems. Temperature, precipitation, and potential evapotranspiration are important because they determine the rate at which biological and chemical reactions take place. Both the production of new organic mat-

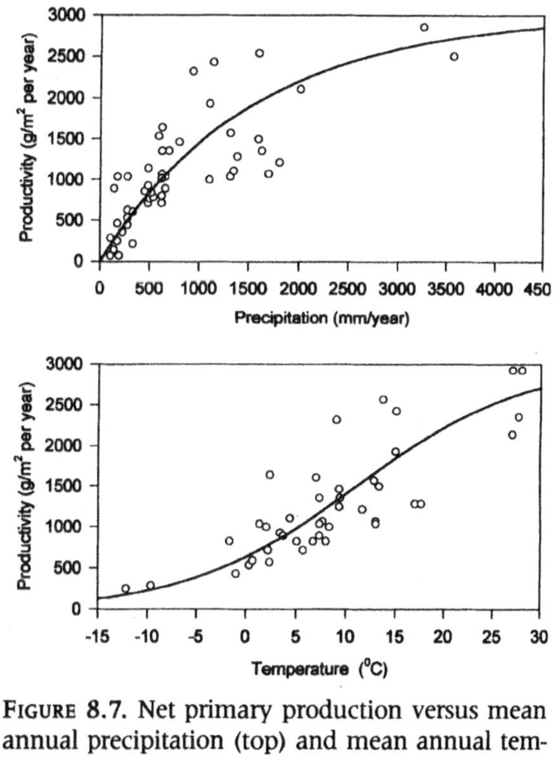

FIGURE 8.7. Net primary production versus mean annual precipitation (top) and mean annual temperature (bottom) (from Lieth and Whittaker, 1975).

TABLE 8.4. Net primary production and related characteristics of the biosphere (from Lieth and Whittaker, 1975)

		Net primary production (dry matter)			Biomass (dry matter)			Chlorophyll		Leaf surface area	
Ecosystem type	Area (10^6 km)	Normal range (g/m^2 per year)	Mean (g/m^2 per year)	Total (10^9 Mg per year)	Normal range (g/m^2)	Mean (g/m^2)	Total (10^9/Mg)	Mean (g/m^2)	Total (10^6 Mg)	Mean (m^2/m^2)	Total (10^6 Mg)
Tropical rain forest	17.0	1000–3500	2200	37.4	6–80	45	765	3.0	51.0	8	136
Tropical seasonal forest	7.5	1000–2500	1600	12.0	6–60	35	260	2.5	18.8	5	38
Temperate evergreen forest	5.0	600–2500	1300	6.5	6–200	35	175	3.5	17.5	12	60
Temperate deciduous forest	7.0	600–2500	1200	8.4	6–60	30	210	2.0	14.0	5	35
Boreal forest	12.0	400–2000	800	9.6	6–40	20	240	3.0	36.0	12	144
Woodland and shrubland	8.5	250–1200	700	6.0	2–20	6	50	1.6	13.6	4	34
Savannah	15.0	200–2000	900	13.5	0.2–15	4	60	1.5	22.5	4	60
Temperate grassland	9.0	200–1500	600	5.4	0.2–5	1.6	14	1.3	11.7	3.6	32
Tundra and alpine	8.0	10–400	140	1.1	0.1–3	0.6	5	0.5	4.0	2	16
Desert and semidesert scrub	18.0	10–250	90	1.6	0.1–4	0.7	13	0.5	9.0	1	18
Extreme desert: rock, sand, ice	24.0	0–10	3	0.07	0–0.2	0.02	0.5	0.02	0.5	0.05	1.2
Cultivated land	14.0	100–4000	650	9.1	0.4–12	1	14	1.5	21.0	4	56
Swamp and marsh	2.0	800–6000	3000	6.0	3–50	15	30	3.0	6.0	7	14
Lake and stream	2.0	100–1500	400	0.8	0–0.1	0.02	0.05	0.2	0.5	—	—
Total continental	149		782	117.5		12.2	1837	1.5	226.0	4.3	644
Open ocean	332.0	2–400	125	41.5	0–0.005	0.003	1.0	0.03	10.0		
Upwelling zones	0.4	400–1000	500	0.2	0.005–0.1	0.02	0.008	0.3	0.1		
Continental shelf	26.6	200–600	360	9.6	0.001–0.04	0.001	0.27	0.2	5.3		
Algal beds and reefs	0.6	500–4000	2500	1.6	0.04–4	2	1.2	2.0	1.2		
Estuaries (excluding marsh)	1.4	200–4000	1500	2.1	0.01–4	1	1.4	1.0	1.4		
Total marine	361	—	155	55.0	—	0.01	3.9	0.05	18.0		
Full total	510	—	336	172.5	—	3.6	1841	0.48	243		

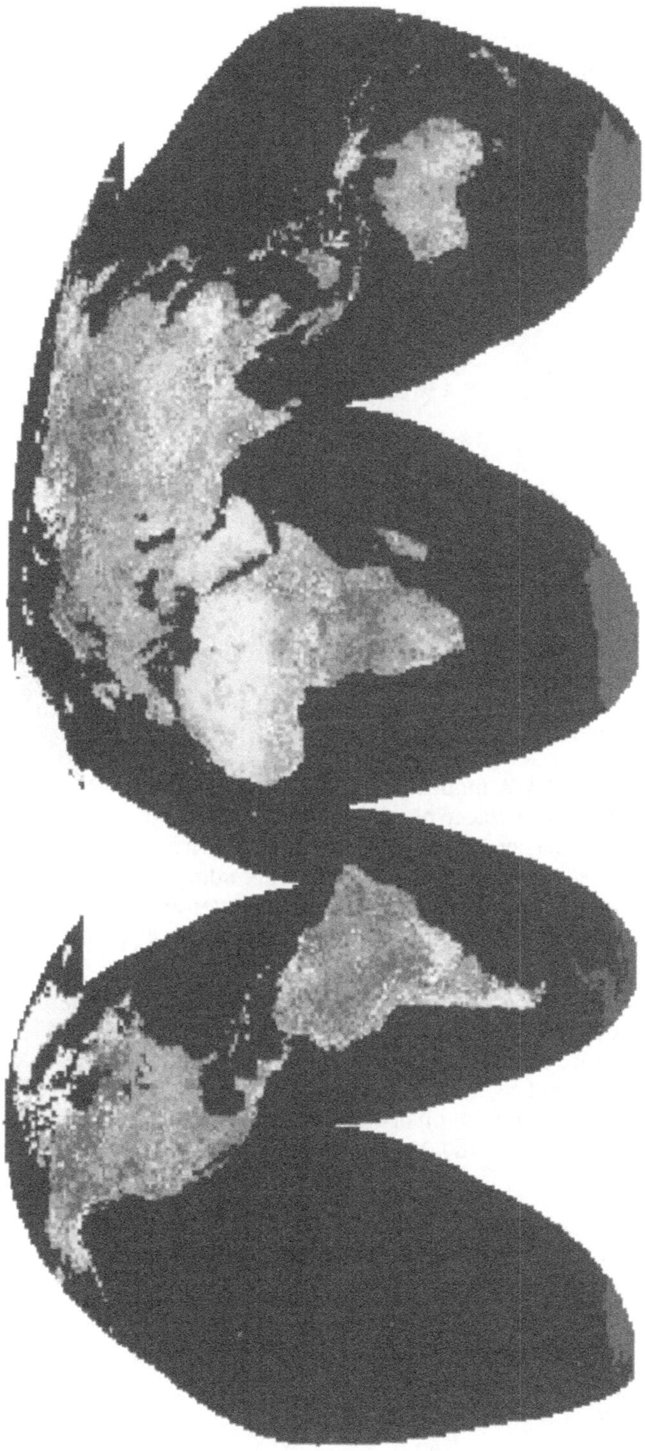

FIGURE 8.8. World greenness map as stated in June 1992 by remote sensing. The greenness is related to the carbon stored in the biomass. (Courtesy of the U.S. Geological Survey, National Mapping Division, EROS Data Center, as producers of the image, and the U.S. National Oceanic and Atmospheric Administration, as the source of the data.)

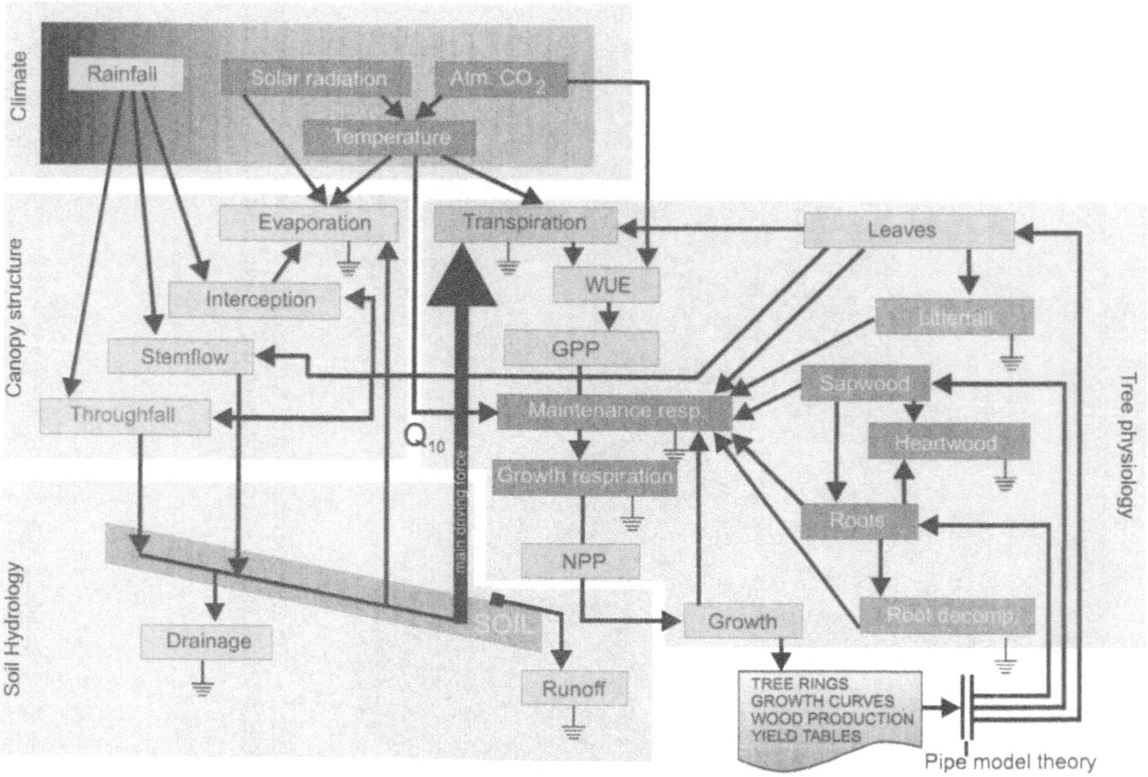

FIGURE 8.9. Flow diagram of the GOTILWA model. Rainfall not intercepted by the canopies reaches the soil as stemflow or throughfall. Water available to plants—once runoff, drainage, and water flux in soil are taken into account—can be transpired if energy is available in the atmosphere. From the transpired water—the main driving force in the model—carbon uptake is computed according to the water use efficiency of the plants. Maintenance and growth respiration of roots, stems, and leaves are calculated, and the net primary production, if any, is allocated to the different parts of the plant using an algorithm based on the pipe model theory. The time step used for calculations is one day.

TABLE 8.5. Monthly values of actual evapotranspiration (AET), gross primary production (GPP), respiration (R) and net primary production (NPP) of a *Quercus ilex* forest simulated by GOTILWA model using the actual daily weather data recorded during the year 1983 in Prades (Spain). The forest has a basal area of 32 m^2/ha, an aboveground biomass of 185 Mg/ha. Belowground biomass was 203 Mg/ha. Leaf biomass is 9 Mg/ha, LAI was about 5 with a minimum of 4.8 in August. In 1983, rainfall was 588 mm, PET 1025 mm and total incoming radiation was 4040 MJ

Month	AET (mm)	GPP (gC/m^2 leaf)	GPP (gC/m^2 ground)	R (gC/m^2 ground)	NPP (gC/m^2 leaf)	NPP (gC/m^2 ground)
January	6.1	4.2	23.9	12.9	2.0	11.1
February	24.2	16.5	94.4	35.1	10.4	59.3
March	31.2	21.0	121.8	45.4	13.2	76.4
April	40.9	27.2	159.8	62.6	16.5	97.2
May	66.8	43.3	261.3	106.3	25.7	155.0
June	0.8	0.6	3.2	98.8	−14.8	−84.6
July	5.1	3.7	19.8	103.3	−14.1	−75.3
August	19.5	15.3	75.6	87.3	−1.4	−7.0
September	20.5	16.6	79.1	42.3	8.7	41.5
October	50.7	35.8	194.8	33.6	29.6	161.3
November	26.5	18.3	102.9	34.3	12.2	68.6
December	20.4	13.9	79.6	30.9	8.5	48.7
Year	312.8	216.4	1216.2	692.9	96.4	552.2

ter by plants and microbial decomposition of dead organic matter are temperature- and moisture-dependent processes. The amount of water available to plants, relative to the potential to evaporate that water, determines the intensity of water stress and controls the length of the growing season. The amount of water percolating down through the soil profile and the chemistry of that water plays a large role in determining the soil structure and nutrient content.

In Figure 8.7 the relationships between water and temperature and the world biomass distribution (Table 8.4) are shown. Lieth (1975) has related productivity to climate variables. Having established consistent relationships from literature values, Lieth utilizes the worldwide climate to estimate geographic productivity patterns (Figure 8.8).

8.2.2 Modelling the Primary Production of the Biosphere

The ecophysiological processes related to primary production act together, at the same time, in natural communities. Some act at the molecular level, as in photosynthesis; others are leaf-level processes, for example, the leaf energy balance; still others act at a canopy level, like the light extinction; and, finally, some are community-level processes, for example, competition between individual trees. The integration of all this information from a microenvironmental scale to upper levels is needed to predict and explain the responses of plant communities. This integration is only possible with the help of relatively complex process-based models.

Figure 8.9 is a schematic representation of the GOTILWA model. It consists of a series of submodels that describe the operation of the various ecological processes involved in the growth of trees. GOTILWA focuses on the main features of Mediterranean forests. Nevertheless, the model is general enough to be applied to most forest types.

The name GOTILWA is an acronym for Growth Of Trees Is Limited by WAter. The name itself defines the main characteristic of the model. Water constitutes a key factor in the model. Rainfall, incoming solar radiation, maxi-mum and minimum temperature, and potential evapotranspiration (PET) are analysed daily. From the interaction between daily rainfall and the forest structure—density and size of the trees—the amount of intercepted water by the canopy layer can be estimated. Effective rainfall—the addition of throughfall, and stemflow—can also be estimated. The effective daily rainfall increases the water stored in the soil until the field capacity is reached. Any additional water is drained to the stream. Trees use the water stored in the soil.

The ratio between water availability and the atmosphere evaporative demand gradually determines the structure of the trees. The proportion of sapwood to heartwood, the leaf area of each tree, and, consequently, the leaf area index of the forest—all depend greatly on water availability. The carbon uptake by the trees is computed by using the water use efficiency and the water transpired by each tree on a daily basis.

The NPP is allocated to different parts of the tree according to a set of rules that define the carbon allocation priority criteria. The balance between the maintenance respiration of the different parts of the tree, the NPP, and the metabolic cost associated with the formation of new components determines the processes of leaf formation and leaf fall, tree ring formation, the rate of change of sapwood into heartwood, and, consequently, the change in tree structure within each size class. These changes will affect in turn the hydrological fluxes that will subsequently take place in what might be described as a feed-back process. Table 8.5 summarizes some results obtained simulating a *Quercus ilex* stand using the actual daily weather data recorded in 1983. GPP and NPP both on a leaf area and on a ground area basis are shown. The results show a negative net production during the months of June to August. This stop in summer growth is typical in most Mediterranean-type ecosystems and in this particular case is supported by field observations.

8.3 Production and Biomass of Forest Ecosystems

8.3.1 Structural Characteristics of Forest Mass

Forests have much higher biomass than other ecosystems. The total biomass (above-ground and below-ground) of a mature forest can be greater than 500 Mg/ha. At the same time, forests are the most productive ecosystems, with values of net production that go from 3 to 20 Mg/ha depending on the biome. As a whole, forests produce about 60 percent of the carbon fixed annually and store over 80 percent of the world's organic carbon.

The frequency distribution of stem diameters (measured at the conventional height of 1.3 m) is one of the basic aspects of the stand structure. This distribution reflects important traces of history. The diameter composition can be used, by means of dimensional techniques, to calculate the biomass and the production of the forest. Furthermore, the distribution of size classes is the result of a combination of dynamic processes in which both mortality and reproduction are important.

These distributions are not correlated, however, with the age of the tree. Even among trees from the same stand, those with the same age can show very different diameters depending on their genetic background, the competition with neighbours, or the characteristics of the environment. Forests that are planted by man are formed by trees of approximately the same age, their diameters are distributed around a mean value, and the variance of the diameters can be very reduced. These are the so-called even-aged forests. In contrast, in the uneven-aged forests we find trees with very different ages leading to a less regular distribution. We will see later on that these distributions are very different.

The information about tree diameter distribution in a forest is of the utmost importance. Because a great part of a forest is inaccessible from the ground, the foresters have analysed in detail the structural relationships of trees. The majority of the dimensional variables of a tree correlate with their diameter. Thus, diameter is one of the most important variables to estimate the leaf and branch biomass, the canopy diameter, wood volume, and even below-ground variables as root biomass (Djema, 1995).

The allometric equation:

$$Y = aX^b$$

is a good description of the relationship between an independent variable X, usually diameter, and a dependent variable Y. The values of the a and b constants can be estimated using linear regression of the log-transformed original data. Unfortunately the values of such *constants* are species-dependent and are less constant than could be expected.

Obviously, the carbon that the tree uses to produce its different parts (roots, stem, wood, bark, branches, leaves) depends on the balance between the production and the respiration as it has been previously described. As this balance relies on the environmental conditions, the carbon allocation to the different

TABLE 8.6. Alometric relationships of *Pinus halepensis* in 1460 stands in Catalonia (Spain) change according to the altitude (m a.s.l.) of the stands

	Leaf biomass vs DBH*				Aboveground biomass vs DBH			
Altitude (m)	α	β	r^2	n	α	β	r^2	n
0-200	0.034	1.62	97	238	0.160	2.10	96	71
200-400	0.041	1.58	97	696	0.110	2.20	97	181
400-600	0.046	1.55	99	413	0.141	2.10	98	118
600-800	0.054	1.52	95	113	0.134	2.17	96	43

*Diameter at breast height.

parts of the tree will be affected by the environment and so will the allometric coefficients. Table 8.6 shows the variation of *Pinus halepensis* allometric coefficients for leaf and total above-ground biomass in Catalonia, Spain. *Pinus halepensis* is a native species from the Mediterranean area that can be found at different elevations but it is more abundant below 500 m. The β allometric coefficient decreases from 1.62 in plots under 200 m a.s.l. to 1.52 in plots over 600 m. Comparing trees of similar diameter, those living at a higher elevation have a smaller leaf biomass.

8.3.2 Basal Area, DBH Classes, and Height

As has been pointed out, one unique measure plays a very large part in the description of forest stands: the diameter at breast height (DBH) which is usually measured over bark. There is generally a good correlation between the diameter of the trunk and the height, crown diameter, stem volume, or total biomass of trees; thus, DBH summarizes a lot of the information that ecologists and foresters need to guide the management of a forest.

Basal area is the surface of cross-sectional area of a tree or trees, usually expressed in m^2/ha, measured at breast height. It is a structural variable, and its value increases with the increasing diameter. Although it is not independent of the tree density, the basal area is scarcely related. A young even-aged forest cohort is characterized by a tree diameter distribution in which the most abundant trees are small trees and in which trees are less abundant, leading to inverted J distributions.

In natural conditions, as the mature mass and the mean tree size increases, and thus so does the variance of the distribution, a reduction of the population density is observed. Nevertheless, this decrease in density does not mean a diminution of basal area, which continues to increase during the life of the trees. At the beginning, an increase of leaf area parallels the increase of basal area. Later on, depending on the water availability, the maximum leaf area index is reached.

The sapwood cross-sectional area and the leaf area of each tree are two stretched variables. Thus, the forest leaf area–i.e., the sum of the leaf area of each tree–and the forest sapwood cross-sectional area are also related. When the forest reaches this maximum leaf area index, the accumulation of new xylem in the last formed rings is compensated by new heartwood. A fraction of the old xylem looses its functionality. From then, any increase of forest basal area is based on an increase of heartwood. However, the sapwood area as well as the leaf area index remains constant.

8.3.3 Some Values of Production and Biomass

Some patterns of primary production in different natural systems are evident. Primary production is higher in the tropical rain forest and decreases toward the poles. Grasslands and tundra areas are less productive than forests in the same areas. Forest biomass is very high but the green fraction (leaves, understorey grasses and herbs, etc.) is very low and so, the productivity ratio (P/B) is very low and decreases with the forest age. The diminution of productivity can be deduced from the dynamics of the community as described above.

8.4 Forest Demography

8.4.1 Growth of Individual Trees

The growth in size reflects the net production of individual trees. Their knowledge constitutes a way to evaluate the net primary production and becomes necessary to plan the management of a natural resource of such a low turnover as forests. Nevertheless, the life of the trees covers several hundreds of years and it is obviously incompatible with direct observation by man.

The problem of such a long life can be easily overcome, at least in some species, because the seasonal growth leaves permanent marks

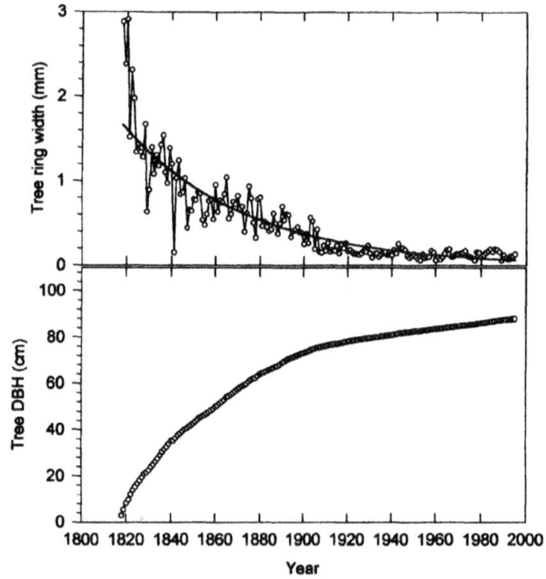

FIGURE 8.10. Annual tree rings of a *Pinus sylvestris* tree growing in the Sierra de la Demanda (Soria). The solid line represents the trend fitted by means of Richard's function (top). The annual DBH increments are represented accumulating the values of the annual ring width (bottom).

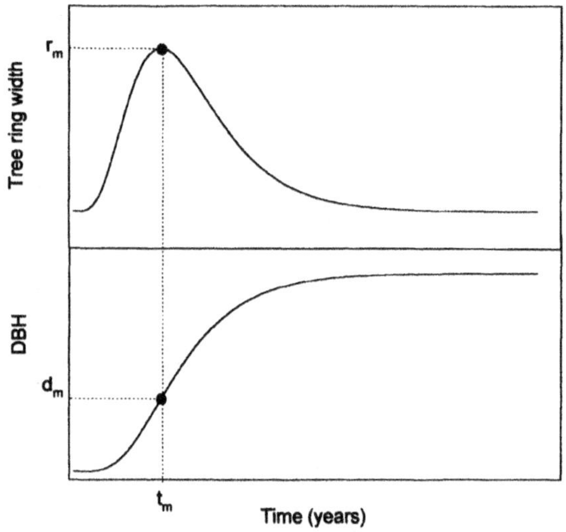

FIGURE 8.11. Growth curves can be represented as annual increments—tree rings (top)—or as cumulative growth—DBH (bottom). Both types of curves have the same information. The time at which the maximum growth rate occurs is shown as t_m. At that time, the tree has a DBH of d_m and produces their maximum ring (r_m).

in the xylem: tree rings that can be observed in cut trees or from cores that can be extracted using the Pressler borer. Rain forest species extend their vegetative period over the year, exhibiting continuous growth. In those trees, rings cannot be observed, and the growth should be analysed by installing continuous recording dendrometers or repeated census on permanent plots. Nevertheless, the use of this technique is limited to short periods of time.

The analysis of tree rings is an interesting exercise which constitutes the base of dendrochronology. Growth curves are obtained plotting tree rings width against time. Adding the values year by year, plots of the variation of DBH over time can be obtained (Figure 8.10). Both types of plots have, in fact, the same information but each of them emphasizes different components of the tree growth.

The growth of a tree is the result of the interaction between internal factors (e.g., genetical background), persistent external factors which act at medium term (e.g., competition with neighbours), and external factors that can change in short periods of time, usually year by year as, for example, the climate components. The pattern of the accumulated curve emphasizes the long-term effects. The genetic background and the effects of crowding, among other factors, lead to different growth patterns that can be described using growth functions that are empirically fitted. The effects induced by climate factors or by changes in trees density due to thinning are best shown by recording the annual growth values.

As can be seen on Figure 8.11, both types of curves are related. The analysis of growth curves reveals some common traits. Looking at the tree rings, the curves show a maximum radial increment during the first years that relies not only on the species but also on the forest life conditions. Over this maximum, tree ring width decreases. The correct interpretation of the ecological mean-ing of such curves becomes more evident if increments of basal area are considered rather than radial increments (Figure 8.12). A lower ring width, in relation to a thicker one, does not necessarily mean that less wood is formed that year. It depends on the rel-

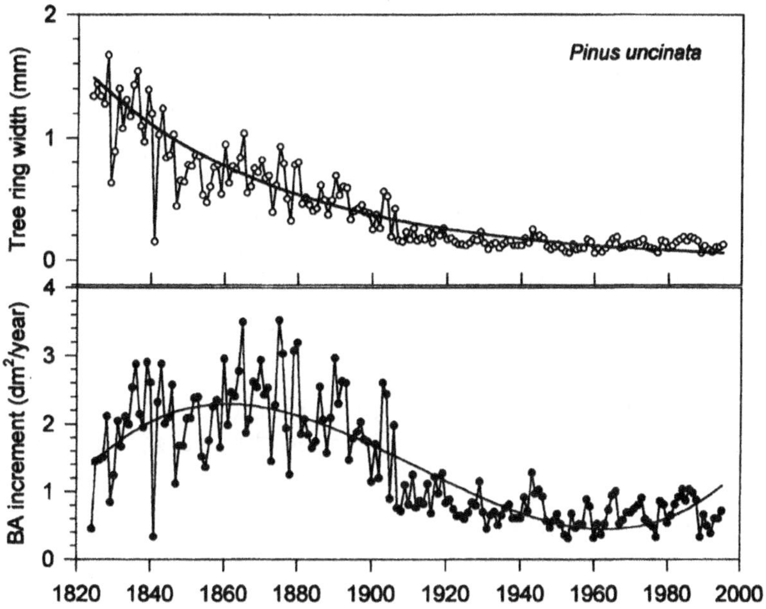

FIGURE 8.12. Tree ring (top) and basal area (bottom) increments show different patterns. Basal area increments are related to the water transport capacity and thus to the leaf area of the tree. In this *Pinus uncinata* from the Pyrenees, the wider ring was formed in 1828, while the maximum basal area increment was reached in 1875.

ative position of each ring in the stem since the same amount of wood will lead to the formation of a thicker ring in small trees.

We have seen that basal area is related with leaf area and with the transport of water in plants. So, from an ecological point of view it is more useful to think in terms of basal area. The lack of correlation between size and age has been pointed out. Growth curves gave the evidence. And this fact has practical consequences.

TABLE 8.7. Mean age and radial increment of *Pinus sylvestris* and *Pinus nigra* as a function of their DBH in the forested area of Solsones (Spain). Slow and fast growing trees are compared

	Age (years)		Radial growth (mm/year)	
DBH (cm)	Slow	Fast	Slow	Fast
Pinus sylvestris				
5-10	22.6	48.9	1.46	0.64
10-15	33.6	60.5	1.68	0.74
15-20	45.6	81.6	1.73	0.78
20-25	52.9	88.1	1.71	0.83
25-30	68.4	142.0	1.63	0.69
Pinus nigra				
5-10	18.1	49.3	2.17	0.71
10-15	23.4	64.8	2.48	0.83
15-20	36.7	65.6	2.37	0.96
20-25	43.8	82.0	2.50	0.97
25-30	52.8	100.0	2.36	1.04
>30	52.1	108.2	2.86	1.06

A traditional forest management practice is based on cutting from time to time the trees that have reached a certain diameter, prefixed by com-mercial reasons. Trees that are below this value are left in the forest until they reach the prefixed diameter value. This practice assumes that all trees from a forest have the same growth rate, statistically speaking, which in most forests is not the case.

In some Spanish forest close to Barcelona, where the age and size of the trees have been analysed, it has been proved that trees with smaller diameter are proportionally older. In addition, their actual growth rate is smaller than those of the bigger trees. 50-year-old *Pinus nigra* trees can have a DBH of 5 to more than 30 cm (Table 8.7), and similar results have been found in other species.

Such management is translated into a negative selection of trees with higher growth rates—those with less time in the forest. This reinforces the trees with slower growth rate and their probability to contribute to the soil seed bank and to the offspring. The practical consequences of this management on future production has not been studied in detail. It is wholly unrealistic and very dangerous to assume any relation between the size of the trees and their age, other than the vague principle that the largest trees in a canopy are likely to be old. If a tree is very young it is likely to be small, but if it is small it may be any age (Harper, 1977).

8.4.2 Tree Density and Tree Distances

When individual trees grow in an uncrowded population there is little or no mortality from com-petition. In a diagram displaying the logarithm of the biomass of an average tree against the logarithm of population density, growth trajectories, therefore, are nearly vertical. But as trees begin to fill up the available space, crowding commences and mortality reduces population density.

As plants grow, the data points representing consecutive growth stages of survivors in a log–log diagram lie along this thinning line from a high population density and small average biomass per plant at lower right, upward and leftward toward the ordinate. The growth trajectories of populations with different starting densities eventually approach a common asymptote, which constitutes the so-called thinning line (White and Harper, 1970; White 1981). It is a line of constraints in that no data points occur above it. For an even-aged plant population in dynamic equilibrium at some crowding density, the power function

$$B = kN^b$$

relates the maximum, average plant biomass to the population density N. Empirically obtained exponents come close to −1.5 for terrestrial plants as different as mosses, ferns, grasses, herbs, and trees. Depending on the kind of plant, the lines elevation (the y-intercept k) may differ by a factor of eight (White, 1981), but the exponent usually remains close to −3/2. This is the well-known −3/2 power law, also called the shelf-thinning or Yoda's law in plant ecology:

$$B = kN^{-3/2}$$

The relationship described by this equation between the average plant biomass and the

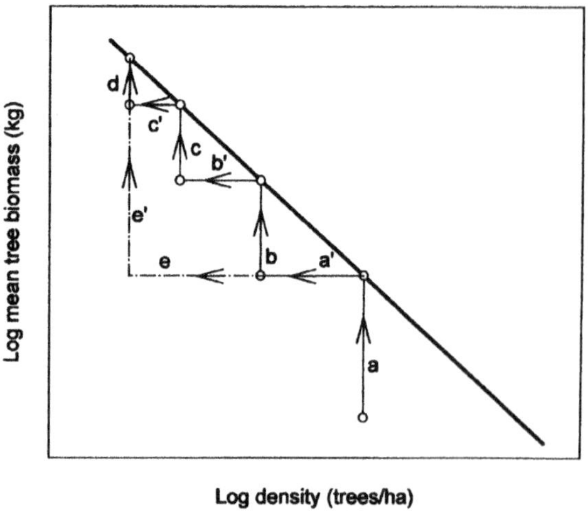

FIGURE 8.13. Generalized diagram showing the relationship between the tree density and the mean biomass per tree in fully stocked stands. The upper line is a constraining line of a slope close to −1.5, the so-called self-thinning line. Natural populations grow from an initial density and mean size, showing vertical trajectories (a). When the thinning line is reached, any further growth of individuals should be compensated by some mortality, and populations follow the line. Human-induced thinning reduces density (a', b', or c') and the subsequent forest growth follows vertical trajectories again (b, c, or d). Management can be done in different ways to go from an initial to a desired situation. In the figure, high-density/small-sized trees are conducted to low-density/big-sized trees by means of three consecutive thinnings (a', b', and c') or by means of only one drastic thinning (e). Such a heavy management technique should be avoided.

TABLE 8.8. Changes of density, aboveground biomass, net biomass increment (*P*), productivity (*P*/*B*) and time of residence of a *Pinus nigra* forest in Catalonia (Spain)

Age (years)	Density (trees/ha)	Aboveground biomass (tm/ha)	*P* (kg/ha/yr)	*P*/*B* (per year)	*B*/*P* (year)
40	2000	46.00	—	—	—
50	1100	57.20	1120	0.02435	41
60	700	68.60	1140	0.01993	50
70	520	74.88	628	0.00915	109
80	400	81.20	632	0.00844	118
90	320	87.68	648	0.00798	125
100	270	88.56	88	0.00100	996

population density, often applies to natural populations, provided that the population is subjected to such severe crowding that competition leads to mortality and self-thinning of the population (Figure 8.13). This is one of the most important principles in plant ecology (White, 1980). It was derived empirically and has subsequently been corroborated by new empirical data even though many empirical studies have produced important deviations from the −3/2 exponent (Weller, 1990).

The causal basis of the thinning rules obviously depends on general, geometric principles of growth and packing. In order to understand the thinning exponents it is necessary to consider the growth mode of individual trees, that is, the way in which trees are shaped and how proportions change with increasing size. The simplest growth model is when plants grow isometrically so that individual plants remain geometrically similar when increasing size. Geometric similarity results when all lengths increase in a constant proportion in relation to a representative length l (usually DBH in trees). From that, it follows that areas increase as l^2 and volumes as l^3. Yoda et al. (1963) derived the −3/2 exponent from empirical data and also gave a geometric explanation of it.

The mortality of the self-thinned trees is a slow process. Before dying, weakened trees can re-main for a long time in the forest. In some slow-growth species or in environments with poor fertility, forests remain visibly unchanged for many years in what can be described as forest stagnation. The mortality rates due to competition are extremely low. In a *Quercus ilex* forest in Prades (Tarragona, Spain) plots cut in the 1940s to obtain charcoal have at present, 50 years later, a density close to 20,000 trees/ha and the mean DBH is only 4 cm. Sukatsthew recognized in 1925 that self-thinning is more intense in high-fertility plots.

8.4.3 The Allen Diagram

Allen (1951) proposed a method to represent the biomass and density changes of a cohort—a set of individuals of a population having approximately the same age—over time. Although originally applied to fisheries, the diagram is general enough to be applied to any type of species and the analysis helps to understand some interesting aspects of exploited populations.

Let us imagine a cohort of which we know the growth and the survival curves. Both curves can be combined plotting survival as a function of the mean individual weight. A diagram expressing changes of biomass over time is obtained. In fact, the coordinates for each point (P_1, Figure 8.14) are tree density and mean weight. It is evident that the product of both values is, at any moment, the total biomass of all the survivors of the cohort. From the analysis of these diagrams a well-known and interesting result is obtained.

Biomass follows a regular pattern increas-

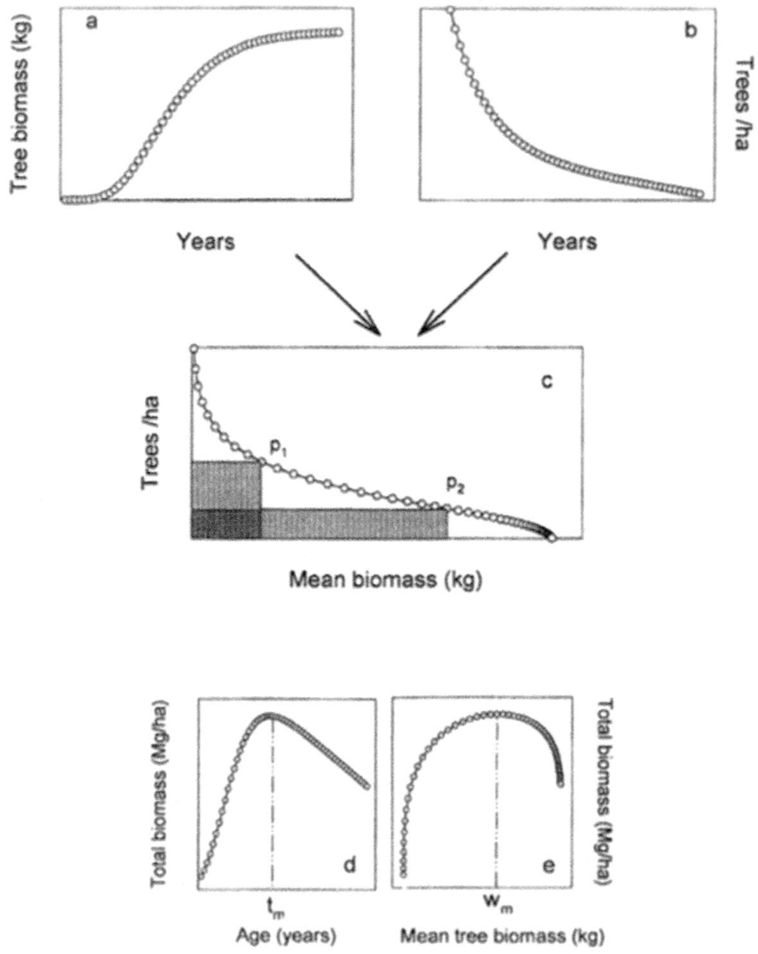

FIGURE 8.14. The Allen's diagram can be drawn combining the growth curve (a) and the survival (b) of a cohort. In such a diagram (c) the number of individuals is plotted against their average biomass. The points of the resulting curve (e.g., P_1, P_2) are an expression of the total biomass of the cohort at any time (the product of number of survivors × mean weight, represented by the rectangles in the figure). The changes of biomass through time are plotted in (d) as a function of the age of the individuals or as a function of their mean weight (e). In both cases total biomass reaches a maximum and starts to decrease. Exploiting the individuals at age t_m or weight w_m the maximum profit will be obtained.

ing at the beginning, reaching a maximum, and then decreasing—or stabilizing for a certain period of time if the trees are very long-lived (Table 8.8). Although trees become bigger their increase in size is not enough to compensate for the decrease in density. It is evident that the maximum biomass of the cohort will be reached at age t_m or when the mean size of the trees will be W_m in Figure 8.14d,e. Of course, technical limitations or economic reasons (as, e.g., a different market value depending on the size) can modify in practice this criterion, but it still remains a good ecological reference when designing management plans.

8.5 Forest Management

8.5.1 The Site Index

The increase in height is considered the best indicator of the capacity of a station for tree development. The increase in diameter depends on the plot density. In a certain plot increased density results in a decrease of tree diameter. The increase in height is less affected by density. In addition the growth in height is very well correlated with the increase in volume when height and volume variations are due to the conditions of the plot. That is the reason why the height is the variable most used to evaluate the quality of a site. The height reached by the dominant trees of a plot at a certain age—usually 25, 50, or 100 years—is the quality index of the site.

The site index is based on the hypothesis that the quality of a stand is similar for all age classes and that the pattern of height growth is the same in all the plots. These two assumptions are not always reached. A growth factor can start to be limiting for a tree at a certain age—or at a certain size. Some authors have demonstrated that height is influenced by the stand density (Jack, 1971). The site index is a purely empirical evaluation that gives no information about the causes that are the origin of the difference in quality among different stands.

8.5.2 Forest Exploitation

Forests have provided humans with materials and products since ancient times. The exploitation of a forest to make a profit from its resources should be based on a knowledge of the biology of the species and on the demographic characteristics of the forest populations. The slow growth of most forests recommends accurate planning to foresee all sort of consequences—hydrological, erosive, effects on regeneration, competition, etc.—that the exploitation of the forest could produce. The forest growth models, most of them taking into account management as a key factor, can help the managers to analyse the medium-term consequences derived from their decisions.

This is true for all kind of forest exploitations. However, in tree plantations a new important factor is added. The time gap between planting and logging, which in the best cases will be 25 or more years, obliges the planner to introduce economic criteria in the analysis on the foreseeable product demand, their market price, and the necessary investments to maintain the plantation in optimal conditions during such long periods of time.

Forest management is carried out in a different way depending on the type of forest. In even-aged forests the extraction of trees for their use can be done by clear-cutting (a final harvesting of all the trees). Afterward, mass regeneration is done either artificially or by germination with seeds from nearby forests or from the seeds of the trees that are cut. An alternative to the clear-cutting consists in leaving a small num-ber of parental trees (usually less than 30 trees per hectare) which will produce the seeds needed for regeneration. In some forests logging is made covering the rotation in several successive thinnings.

Rotation is the number of years necessary from the start of a forest mass till it gives products with the needed degree of maturity. In the uneven-aged forests the exploitation is carried out by felling some selected trees, usually big trees that leave space and resources for the remaining trees. Obviously, before the moment of the final felling, the maintenance of the density in optimal conditions requires some actions. Liberating cuttings in which defective or spare trees are cut may be required. The site quality may deteriorate under rotational harvest, in which case the forest must be fertilized to main-tain its growth.

8.5.3 Forest Planning

From the analysis of the concepts involved in the growth and competition of trees we have seen that it is possible to foresee the growth pattern of a forest. There is no doubt that this analysis is the main tool in forest management. Because of the long-lived species involved in forestry, the consequences of forest management will be evident for decades.

Forest management should be planned in such a way that initial densities, diameters to be reached at any age, successive liberation cuttings, investments, and so on, should be anticipated. Only with planning it is possible to foresee the final products to be obtained as well as the products obtained in the intermediate fellings.

Economic evaluation that takes into

account the necessary investments, the inflation rates, and the expected outputs is essential in forest decision making. Even so, planning cannot avoid the risk of any long-term investment, such as forest fires, an episode of severe drought, or economic fluctuations affecting the markets.

Growth and productivity, although important, are not the only objectives in forest planning. In the economic analysis of forest planning an effort is needed to internalize external values. This point is still a failed subject. It is particularly important in Mediterranean countries where climate, often adverse to forest growth, reduces forest productivity. Nevertheless, the role that forest can play in protecting soils, contributing to the quality of water, to the control of flooding, or as recreational areas, especially close to big cities, is of utmost importance.

The Forest Plan of Andalucía, in Spain, has recently estimated that the reforestation of 1.2 million hectares can help to double the reservoir active life. That will reduce from 20 to 10 pesetas the cost of construction of 1 m^3 of reservoir. It can reduce from 5 to 2.5 pesetas/m^3 the cost of depuration of 400 million m^3 annually used for human consumption. It will also decrease by 40 percent the hydrological works needed as protection against flooding which represent 595 million pesetas per year and contribute to save 443 million pesetas annually in soil restoration and fertilization. The 1.2 million hectares reforested will represent an annual saving of 3437 million pesetas.

8.5.4 Economic Balances of a Forest Exploitation

In addition to the indirect benefits mentioned above, the direct economic value of a plantation relies on the volume of wood produced and its price. The analysis of the economic balance of a forest is done using the production tables like the one in Table 8.9.

In this table, concerning *Picea sitchensis*, trees are classified in three different categories selected according to commercial criteria and

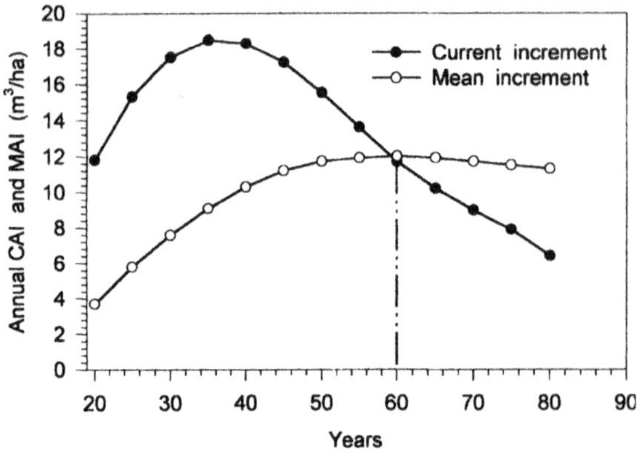

FIGURE 8. 15. The comparison between the current annual incremental (CAI) and the mean annual increment (MAI) can give us an idea of the age of optimal exploitation. If trees are removed at the age at which both curves cross, the benefit, in terms of biomass, will be maximized. Beyond that point, the current production decreases, reducing the benefits that can be obtained. Differences in the economic value of trees of different size may modify the optimal point (data from Williams, 1981, for a *Picea sitchensis* plantation in Great Britain; economic values are recorded in Tables 8.9 and 810.

according to which correspond to three different prices. For each age (first column) the density of trees per hectare, the mean DBH and the basal area of the plantation are recorded. The number of trees per class of quality—or, directly, their volume in m^3—is recorded in the next columns. The number of trees to be felled in each clearing, their DBH, and their volume per class of quality are also recorded. The last three columns represent annuals increments of basal area and the so-called current annual increment (CAI) and mean annual increment (MAI)—e.g., the mean increment obtained during the last period and the mean increment since the start of the plantation. These two variables, CAI and MAI, are used to calculate the moment for optimal exploitation. The point in which these two variables cross (Figure 8.15) is the recommended time for cutting all the trees and beginning to plant again.

If the volume of the trees in the forest or the cut trees in successive clearings is multiplied by their respective prices (Table 8.10), the eco-

TABLE 8.9. Production of *Picea sitchensis* (see Table 8.10). (From Williams, 1981; copyright John Wiley & Sons Ltd. Reproduced with permission.)

	Forest before thinning						Production thinned							Increment				
Age	No. of trees	Mean DBH (cm)	Basal area (BA) (m²/ha)	Vol. (m³) 7 cm	Vol. (m³) 18 cm	Vol. (m³) 24 cm	No. of trees	Max. DBH (cm)	Mean vol. (m³)	Vol. (m³) 7 cm	Vol. (m³) 18 cm	Vol. (m³) 21 cm	Accumulated BA	Accumulated Vol.	BA (m²/ha)	CAI vol. (m³)	MAI vol. (m³)
20	3079	10.2	25.0	64	0	0	930	12.8	0.045	11	0	0	26.8	75	2.34	11.8	3.7
25	2149	12.0	24.2	90	1	0	565	13.5	0.074	41	1	0	38.0	144	2.12	15.3	5.8
30	1584	14.5	26.1	123	10	0	378	14.7	0.111	40	2	0	47.9	228	1.86	17.5	7.6
35	1206	17.3	28.3	144	37	1	250	16.6	0.168	39	3	0	56.6	320	1.62	18.5	9.1
40	956	20.1	30.5	135	83	16	171	18.8	0.245	35	7	0	64.1	413	1.39	18.3	10.3
45	785	22.8	32.1	114	121	46	121	21.0	0.333	29	12	1	70.5	503	1.18	17.2	11.2
50	664	25.3	33.3	93	137	93	79	23.0	0.428	20	16	4	76.0	585	1.00	15.5	11.7
55	585	27.4	34.6	77	136	148	56	24.7	0.521	13	15	6	80.5	657	0.81	13.6	11.9
60	529	29.3	35.7	67	127	201	42	26.3	0.615	9	12	8	84.3	720	0.70	11.7	12.0
65	487	30.9	36.6	58	117	248	32	27.6	0.709	7	9	10	87.5	774	0.59	10.2	11.9
70	455	32.4	37.4	51	107	289	26	28.9	0.793	5	8	10	90.3	821	0.52	9.0	11.7
75	429	33.6	38.1	46	100	323	22	30.0	0.871	4	6	11	92.7	864	0.45	7.9	11.5
80	399	34.7	38.6	42	93	351	—	—	—	—	—	—	94.7	901	0.38	6.4	11.3

TABLE 8.10. Economic balance of the *Picea sitchensis* plantation from Table 8.9. (From Williams, 1981; copyright John Wiley & Sons Ltd. Reproduced with permission.)

	Main trees				Commercial thinning				Total
Years	712 (pts/m³)	1245 (pts/m³)	1695 (pts/m³)	Total	712 (pts/m³)	1245 (pts/m³)	1695 (pts/m³)	Total	value (pts)
20	45,600	0	0	45,600	7,800	0	0	7,800	53,400
25	64,200	1,200	0	65,400	29,250	1,200	0	30,450	95,850
30	87,600	12,450	0	100,050	28,300	2,550	0	31,050	131,100
35	102,600	46,050	1,650	150,300	27,750	3,750	0	31,500	181,800
40	96,150	103,350	27,150	226,650	24,900	8,700	0	33,600	260,250
45	81,150	150,600	78,000	309,750	20,700	15,000	1,650	37,350	347,100
50	66,300	170,550	157,600	394,500	14,250	19,950	6,750	40,950	435,450
55	54,900	169,350	250,800	475,050	10,200	18,750	10,200	39,150	514,200
60	47,700	158,100	340,650	546,450	6,450	15,000	13,500	34,950	581,400

nomic balance of a forest exploitation is obtained. These values give us the income that will be obtained by selling the wood of one hectare of forest at a particular age. For example, the wood of one hectare of a 35-year-old forest will produce, at present, a value of 181,800 pesetas. Furthermore, the comparison of the income in two dates as for example the price of a hectare of a 35-year-old forest, which is 181,800 pesetas, with the price of a hectare of a 40-year-old forest, which is 260,250 pesetas, allow us to estimate the revaluation of the forest that will be 260,250 − 181,800 = 78,440 pesetas in the 5-year period, or 15,688 pesetas per year. That represents an annual rate of increment of:

$$100 \times 15{,}688/[260{,}250 + 181{,}800/2] = 7.1 \text{ percent}$$

Making a series of estimations like the ones described above, it can be seen that the rate of increment decreases at an acceptable level. Strictly considering economic criteria only, the clearing of the plantation should be seriously considered.

8.5.5 Sylvopastoral Systems

Sylvopastoralism is probably one of the oldest practices combining tree and forage production. In some countries the practice is so ancient and traditional that it has a specific name which is used to define the system. The best example of such a tradition is the system managed for centuries in the southwest of the Iberian peninsula, called *dehesa* in Spain or *montados* in Portugal (Campos and Martin, 1986).

Considered as an ecosystem, woody pastures can be defined as a complex structure combining long-term production (timber or fuel-wood) and annual production (forage). Trees have a beneficial effect on organic matter level, nutrient uptake, atmospheric input, and nutrient use efficiency. Nitrogen levels are particularly im-proved when the tree is a legume, e.g., *Acacia caven* in Chile or Bolivia and *Acacia tortilis* in Kenya. Quality and productivity of a sward may be modified. Botanical composition is strongly affected by shading. No common trends have been observed on biomass production. In some cases sward production decreases according to tree density. Relationships between total herbage production and basal area are estimated by linear regression, $P = a + (b \times BA)$, with a ranging from 2300 to 2900 kg/ha and b equalling −68 for *Pinus ponderosa* (Uresk and Severson, 1989), −85 for *Pinus elliotii*, and −102 for *Pinus palustris* (Wolters, 1982).

On the contrary, in Mediterranean and tropical dry areas a certain tree cover enhances the sward production, particularly with Acacia species, increasing from 20 to 95 percent (Etienne, 1996). In the tropics, as well as in the Mediterranean areas, heat stress is one of the main constraints to animal production. The reduction of temperature by shade contributes to better animal reproduction and to better feed conversion. This conversion is the basis of secondary production which on the average is a process that takes place in natural systems with an average efficiency of about 10 percent.

Management of these sylvopastoral systems can help to improve this efficiency in different ways. Figure 8.4 shows the increasing efficiency produced by the wildebeest in the Serengeti. Not just the control of cumulated biomass is important. Animals produce a fertilization effect; trees help to flow nutrients from the deep parts of the soil which are inaccessible to the roots of herbs; the patched areas of sun and shadow under the trees contribute to increase specific diversity; and in Mediterranean areas, affected by a high risk of fires, sylvopastoralism can contribute in a very efficient way to control the understorey biomass and so reduce the risk.

References

Allen, K.R. The Horokiwe Stream. *New Zealand Marine Dep. Fish Bull.* 10:231; 1951.

Bellot, J. Análisis de los flujos de deposición global, transcolación, escorrentía cortical y deposición seca en el encinar mediterráneo de L'Avic (Sierra de Prades, Tarragona). Ph.D. Thesis, University of Alicante; 1989.

Bormann, F.H.; Likens, G.E. *Pattern and Process in a Forested Ecosystem.* New York: Springer-Verlag; 1979.

Campos, P.; Martin, M. *Conservación y Desarrollo de las Dehesas Portuguesa y Española.* Madrid: Ministerio de Agricultura; 1986.

Djema, A. Cuantificación de la biomasa y mineralomasa subterránea de un bosque de *Quercus ilex* L. M.Sc. Thesis, CIHEAM-Mediterranean Agronomic Institute of Zaragoza; 1995.

Etienne, M. Research on temperate and tropical silvopastoral systems: A review. Etienne, M., ed. *Western European Sylvopastoral Systems.* Paris: INRA Editions; 1996:5–19.

Evans, G.N. Evaporation from rice at Griffith, New South Wales. *Agric. Meteorol.* 8:117–127; 1971.

Harper, J.L. *Population Biology of Plants.* New York: Academic Press; 1977.

Jack, W.H. The influence of tree spacing on Sitka sprice growth. *Irish Forestry* 28:13–33; 1971.

Lieth, H. Primary productivity of the major vegetation units of the world. Lieth, H.; Whittaker, R.H., eds. *Ecological Studies,* Vol. 14. Berlin: Springer-Verlag; 1975:203–216.

Lieth, H.; Whittaker, R.H. *Primary Productivity of the Biosphere.* Berlin: Springer-Verlag; 1975.

Lledó, M.J. Compartimentos y flujos biogeoquímicos en una cuenca del encinar del Monte Poblet. Ph.D. Thesis, University of Alicante; 1990.

Margalef, R. *Ecologia.* Barcelona: Omega; 1974.

Martin, H.A.; Specht, R.L. Are messic communities less drought resistant? A study of moisture relationships in dry sclerophyll forest at Inglewood South Australia. *Aust. J. Bot.* 10:106–118; 1962.

McNaughton, S.J. Grassland–herbivore dynamics. Sinclair, A.R.E.; Norton-Griffiths, M., eds. *Serengeti: Dynamics of an Ecosystem.* Chicago: The University of Chicago Press; 1979.

Monsi, M.; Saeki, T. Uber den Lichtfaktor in den pflanzengesellschaften und seine Bedentung for die Stottproduktion. *Jpn. J. Bot.* 14:22–52; 1953.

Penman, H.L. Vegetation and hydrology. *Comm. Bur. Soils Tech. Commun.* 53; 1962.

Piñol, J. Hidrologia i biogeoquimica de conques forestades de les muntanyes de Prades. Ph.D. Thesis, University of Barcelona; 1990.

Rutter, A.J. Studies in the water relation of *Pinus sylvestris* in plantation conditions. *J. Appl. Ecol.* 1:29–44; 1964.

Specht, R.L. Dark Island Heath (Ninety-mile Plain, South Australia): The water relationships in heath vegetation and pastures of the Makin Sand. *Aust. J. Bot.* 5:151–172; 1957.

Specht, R.L. Water use by perennial, evergreen, plant communities of Australia and Papua New Guinea. *Aust. J. Bot.* 20(3); 1972.

Specht, R.L.; Jones, R. A comparison of the water use by heath vegetation at Frankston, Victoria, and Dark Island Soak, South Australia. *Aust. J. Bot.* 19:311–326; 1971.

Sukatschew, W. Einige experimentelle Untersuchungen über den Kampf ums Dasein zwischen Biotypen derselben. *Art. Z. Indukt. Abstamm. u. VererbLehre* 45:54–74; 1925.

Uresk, D.; Severson, K. Understory–overstory relationships in ponderosa pine forests. *J. Range Manage.* 42:203–208; 1989.

Weller, D.E. The self-thinning rule: Dead or unsupported? A reply to Londsdale. *Ecology* 72:747–750; 1991.

White, J. Demographic factors in populations of plants. Solbrig, O.T., ed. *Demography and Evolution in Plant Population.* Oxford: Blackwell; 1980:21–48.

White, J. The allometric interpretation of the self-thinning rule. *J. Theor. Biol.* 89:475–500; 1981.

White, J.; Harper, J.L. Correlated changes in plant size and number in plant populations. *J. Ecol.* 58:467–485; 1970.

Williams, M.R.W. *Decision Making in Forest Management.* New York: John Wiley and Sons; 1981.

Wolters, G. Longleaf and slash pine decreases herbage production and alters herbage composition. *J. Range Manage.* 35:761–763; 1982.

Yoda, K.; Kira, T.; Ogawa, H.; Hozumi, K. Self-thinning in overcrowded pure stands under cultivated and natural conditions. *J. Biol. Osaka Univ.* 14:107–129; 1963.

Further Reading

Gracia, C. La gestión forestal y la conservación de los bosques en el marco de la Unión Europea: La necesidad de una nueva estrategia de los países del Sur de la Unión. Amezaga, I.; Onaindia, M., eds. *Conservación de la Biodiversidad.* Cuadernos de Sección, Ciencias Naturales. Bilbao: Sociedad de Estudios Vascos; 1995:77–92.

Gracia, C.; Sabate, S. La resposta de la vegetació al foc: Els canvis en la estructura de la vegetació. Terradas, J., ed. *Ecologia del Foc.* Barcelona: Edicions Proa; 1996:162–172.

Olson, J.S. Productivity of forest ecosystems. *Proceedings of the Symposium on Productivity of World Ecosystems.* Seattle, WA: National Academy of Sciences; 1975.

Reichle, D.E., ed. *Dynamic Properties of Forest Ecosystems: International Biological Programme.* Cambridge: Cambridge University Press; 1981.

Rodin, L.E.; Bazilevich, N.I.; Rozov, N.N. Productivity of the world's main ecosystems. *Proceedings of the Symposium on Productivity of World Ecosystems.* Seattle, WA: National Academy of Sciences; 1975.

Weller, D.E. Will the real self-thinning rule please stand up? A reply to Osawa and Sugita. *Ecology* 71:1204–1207; 1990.

9

The Agricultural System: Potential Production and Environmental Consequences

Malcolm E. Sumner

9.1 Nature of Agricultural Systems

In the previous chapter, the process of production was presented as a universal property of vegetation, which provides the organic energy transformations that operate the living part of the biosphere. The discussion not only described the physiological actions that lead to the production that humans or other animals can use but also describes the regional patterns of production and then, in the final section, applies this information to forest systems. This is a logical place to begin because forest systems dominate the earth, considering the gradient of the boreal, temperate and tropical forest, with variants such as the Mediterranean, Pacific coast coniferous, and other forest types that make up the world biomes. Grasslands, tundra, and deserts represent other biomes that are unforested and have lower rates of production because of environmental limitations.

Humans have occupied essentially all the biomes for a long time, possibly 300,000 years or more, and have developed adaptive traits appropriate to harvesting some of the natural productivity for human use. We term the people with these cultural traits as hunter-gatherers. Hunter-gatherers depend on natural productivity and move from source to source throughout the year. Their material life is limited as a consequence. However, limitation in size and amount was never a limitation in creativity and imagination. The genius of human artistic expression is seen in hunter-gatherer artifacts and, in those few survivors of this life form, in ritual and ceremony and style of life.

In certain habitats the natural productivity is unusual in being quantitatively large in amount and predictable in time. In this situation, the hunter-gatherer can return again and again to harvest the productivity. Familiarity of place and observation of relationships could lead to simple manipulation of the environment— for example, construction of a shallow channel for water flow in a river flood plain.

Gradually over time, in several places, humans developed the skills to manage plant productivity and to domesticate animal associates to create what we call agriculture. Agriculture emerged in the record available to us about 10,000 years ago. Presumably it had a long prehistory before then as well. In its first form, it represents management of grasses and herbs that produce abundant seeds, which can be harvested, stored, and then converted into food.

The appearance of agriculture is associated with a cultural complex that includes permanent settlement, storage vessels and baskets, grinding implements to convert seed to powder, warriors to defend food stores from hunter-gatherers and animals, and so on. The agricultural transition also had a large impact on human physiology, especially reproduction, health, and creative opportunity. Cohen (1977) draws consequences to human population growth from this revolution in culture. He sees origins of current problems in that long-ago event.

9.1.1 Differences with Natural Systems

Agricultural systems are one of the ecosystems that cover the surface of the planet. As an ecosystem, it contains primary producers (plants), consumers of the primary production, including humans, and organisms that decompose organic material and make it available for recycling. If we define agriculture as human-managed vegetation (including animals), with the exception of forests which were treated in the previous chapter, then about 30 percent of the earth's surface is involved in agriculture. But by agriculture, we include the shifting plots of tropical forest people and the vast maize farms of the American prairies. The fundamental difference that distinguishes agricultural systems, in all their variety, from natural systems is that they require human inputs for their operation. Otherwise they revert gradually to a natural state in which all the processes operate according to ecological principles.

In the Southeast part of the United States, where agriculture land was abandoned in large amounts from 1930 to about 1950, recovery of the landscape is well underway 50 years later. This landscape is pine forest, with a hardwood forest understory. Recovery of a hardwood forest type, which will simulate the original forest of the biome, may occur in 100 to 200 years. To recover the species of plants that may be present in the few remaining uncut stands may require much longer time intervals because these species must migrate to the new stands and their rate of movement is quite slow in many cases.

Grazing lands, mostly rangelands supported by natural rainfall, occupy about 22 percent of the land surface of the world, concentrated in Africa, central Asia, America, and Oceania. Another 23 percent of the land surface that is considered potentially arable is occupied by improved pasture or established forests, while the current arable resources of the world constitute roughly 10 percent of the area.

TABLE 9.1. Relationship between yield increases due to intensification of agronomic systems and fertilizer usage in the United Kingdom from 1911 to 1983 (Wild, 1988)

	1911–13	1981–83
Average crop yields (t/ha)		
Wheat	2.2	6.1
Barley	1.9	4.6
Oats	1.7	4.4
Potatoes	15.0	33.0
Fertilizer usage (000 t)		
N	29	1588
P_2O_5	183	488
K_2O	23	559

Rangelands are usually used extensively for raising meat-producing animals. Provided that they are not overstocked and are appropriately managed using a suitable rotational grazing system, relatively little soil degradation is likely to result. Unfortunately, this is seldom the case, particularly in nomadic systems where increased human population has caused pressure to overgraze the available land with resultant water and wind erosion. On the other hand because of the costs of the inputs required, improved pastures such as those in New Zealand, are usually managed much more effectively, and consequently impact the environment to a lesser degree.

In order to supply the food and fibre need of humans, it has been necessary to intensify production on the arable area which has only doubled in the last century relative to the much higher burgeoning of the population. This strategy has proved to be highly effective in improving productivity, particularly in Europe, the Americas, and parts of Asia, albeit at some cost to the environment. Grain, fibre, and other row crop yields have increased spectacularly since the advent of chemical fertilizers and fossil fuel–driven farm

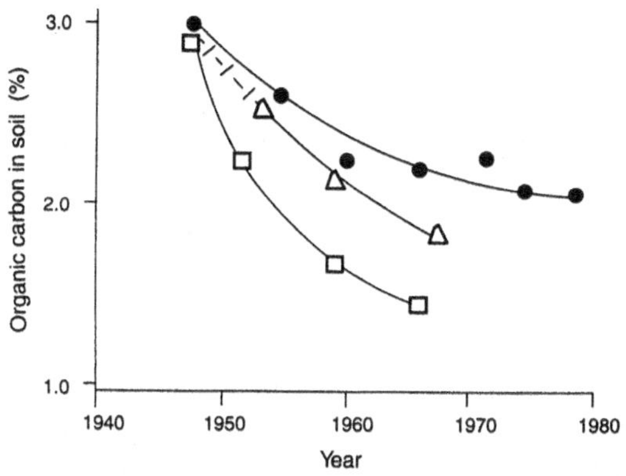

FIGURE 9.1. The effects of different farming systems on the carbon content of Rothamsted soils ploughed out from old grass. (•) Six-course rotation (two roots, three cereals, one year ley); (Δ) three-course rotation (two roots, one cereal); (□)bare fallow (Jenkinson, 1988; reprinted by permission of Addison Wesley Longman Ltd.).

machinery. The rapid increase in fertilizer usage since World War II has been the main driving force for increased yields per unit area (Table 9.1). Recently, further gains have also been made by improved plant resources achieved through plant breeding.

The soil resource has inevitably been degraded by cultivation both from a physical and chemical point of view. The initial degradation stems primarily from the oxidation of organic matter when soil aeration is increased during cultivation. The disappearance of organic matter, which is the primary agent responsible for cementing soil particles together into aggregates, leads to a breakdown of soil structure culminating in poor physical condition of the soil, which on sloping land leads to erosion. With few exceptions, organic matter is lost most rapidly from continuously cultivated systems without crop residue retention (Grace et al., 1992).

From a chemical standpoint, crop removal without fertilizer or manure inputs depletes the soil of nutrients, which is another aspect of soil degradation. Rotational cropping involving a number of row crops and sometimes pasture as well as minimum tillage have been used as means of stemming the loss of organic matter (Figure 9.1) but seldom, if ever, is it possible to maintain the level of organic matter in a cultivated soil at or near that of the pristine soil.

Soil organic matter content can be increased by the use of manures, but in intensive agriculture, their use has been limited by economic constraints. This structural decline is further exacerbated in intensive agriculture by field machinery that causes compaction. In addition, the use of ammoniacal fertilizers and legumes causes further degradation in the form of increased acidity. Fortunately, in arable agriculture, this can readily be neutralized by addition of lime. Irrigation, particularly with low-quality waters, often leads to soil degradation manifested as salinization and sodification. The extent of severe soil degradation caused by agricultural mismanagement has been estimated at 17 percent of the total vegetated land area of the world (Olderman, 1991), but relative to the soil under the virgin cover, all cultivated soils should be considered to have suffered some degree of degradation.

9.1.2 Properties of Agricultural Systems

Because agricultural systems differ from natural systems in requiring human input, we must examine these inputs. But before we do that, it is useful to ask if agricultural systems are fundamentally different from natural systems in their production characteristics. Basically, they do not differ from natural systems. The information in Chapter 8 applies to agricultural systems, too. Productivity involves principally the conversion of solar energy into chemical energy via the process of photosynthesis. The process requires a variety of inputs and if these are limiting then plant production is limited. Plant production feeds animal production, and the dead material is decomposed in the soil, promoting soil structure and releasing the essential elements required for plant growth.

Agriculture differs from natural systems in that humans have had difficulty understanding the processes in natural systems and simulating these in the agricultural system, and humans have also had difficulty in providing all the inputs required for sustainable production. For example, natural systems have developed complex recycling systems that release the elements that compose plant or animal biomass slowly where they can be complexed, stored, and released as the plants require.

These systems of decomposition are extremely complex as one would expect from the complex biochemistry of the tissues that must be broken down into elemental form for uptake. This is the vulnerable point in the natural cycles of essential elements. If the elements leave the living system and enter the physical environment they are potentially lost through the hydrologic cycle and enter streams and the ocean. Tropical forests, for example, growing on very poor substrate have developed almost closed recycling mechanisms that provide the nutrients needed to maintain complex and very large forests (Jordan, 1989). It is these adaptations to the physical environment that have evolved through natural selection and that make natural systems resilient to disturbance and self sustainable in their operation and diversity.

Agricultural systems are designed to yield the maximum possible product for human use, which is a different design objective than is that of the natural system. Maximum production is obtained by several techniques. First, the plant or animal is changed so that that part of primary or secondary production going to the product of human interest is maximized. For example, in wheat the plant breeder has reduced the carbon allocation to root and stem, so that this carbon can be channelled to the seed, which is the part that humans use.

As we learned in the previous chapter, there are limits to primary production, which prevent it passing a maximum point, which is a function of the chemical process itself. In the real world, there are also environmental limits but these are subject to human management. Second, in natural systems the production goes to a large diversity of organisms. Agriculture tries to reduce competition for the product and therefore, uses many devices to reduce or eliminate populations of competitors to humans. Further, the intensification of production requires often the addition of chemical elements and water to provide the optimum growing conditions of the crop.

The result of these inputs is to make the agricultural system unsustainable in a natural sense. It requires human inputs and management to maintain the system at the point where it produces optimal yield. In many instances, traditional agriculture has developed techniques that accomplish this point without large monetary costs. For this reason, it is useful to understand the social factors in agriculture because the traditional system can be useful and should be preserved where appropriate. Further, the system is maintained by monetary inputs to pay for the physical inputs to obtain the highest possible yields. These systems are part of agribusiness and agriculture as a natural process is fundamentally changed into a industrial-like operation that is coupled to urban-industrial needs. The stability of these systems then depends upon the stability of the economic and political systems of world trade, as much as on the environmental systems we are discussing here.

9.1.3 Resources

From the above, it is clear that the resource base for agriculture is the same as that of natural systems. However, in the natural system the organisms have been selected for using those resources in an efficient manner; otherwise, organisms that are more efficient would be selected and would replace the existing flora and fauna. In the agricultural system, we are using a limited set of varieties (the world human population depends upon a few species of crop plants—rice, maize, wheat, barley, millet, potato, and so on) and these are often placed in environments where they are not adapted. In these instances, the farmer must supply the environmental resources required to satisfy the biological and ecological requirements of the plants and animals. Of course, agriculture in the economic sense also requires financial and social inputs to operate as well.

9.2 Limitations to Potential Production

The production process in crops has been modelled for a variety of purposes (e.g., van Keulen and de Wit, 1984), indicating that we understand the key processes sufficiently well to predict performance of the plants if the environmental controls are described. Not only do crops differ greatly in their overall performance but within a crop there is also great variation (Table 9.2).

Note the extreme difference between average yields and maximum observed yields on experiment stations. As agricultural research has advanced, the application of these findings has been applied to the production process with dramatic results (Figure 9.2). This means that there is great variation within the variety, or the species, in its capacity to produce, and that the observed production represents the crop's interaction with a variety of factors that exercise control over production.

These limiting factors are the subject of this section, especially since design of rural areas might allow for alteration of these controls, producing increased yields leading to economic growth. Or, alternatively, control factors can be affected adversely, causing declines in production. Crop ecology studies the role of limiting factors and techniques to alter their impact. Here we will introduce the topic and ask that the rural planner seek further information with a crop ecologist or by researching

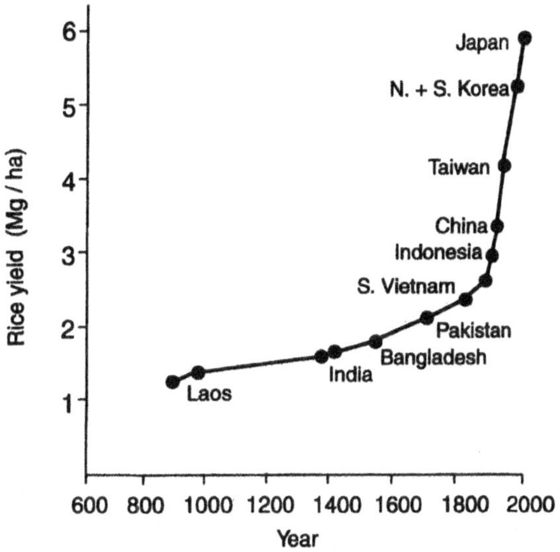

FIGURE 9.2. Average rice yield in Japan from AD 600 to the present. The modern average yields of other nations are located on the yield curve for Japan (Buringh, 1984).

TABLE 9.2. World area harvested, production, and average yield of cereal crops compared to the maximum yields obtained from experiments.

Crop	Area harvested (10^6 ha)	Production (10^6 t)	Mean yield (t/ha)	Maximum yield from experiments (t/ha)
Wheat	266	450	1.9	14.5
Rice in husk	145	386	2.7	26.0
Maize	118	364	3.1	21.2
Barley	94	194	2.1	—
Sorgum	52	69	1.3	21.5
Oats	28	51	1.8	—
Rye	16	32	2.0	—
Millet	55	36	0.7	—
Total	774	1582		

the literature on the subject. Our approach focuses on limiting factors and addresses how they operate in specific situations.

9.2.1 Water

Productivity requires carbon dioxide as a resource that is derived from the atmosphere and enters the leaf through the stomata. When adequate water is available to the plant, it enters the plant through the roots and is transported through the stem to the leaf. The stomata are covered with water and because the atmosphere is not usually saturated with water, water moves from the stomata to the atmosphere by diffusion. As water molecules are departing the leaf, they are replaced by water moving in the transpiration stream from the roots. This process is called transpiration. This assumes an adequate source of water. If the water is inadequate, then the plant closes the stomata, reducing the influx of carbon dioxide and causing a reduction in production. These processes are illustrated in Figure 9.3, which describes the growth process in wheat under optimum growing conditions and under the natural climate when water input varies over the growing season and the plant faces drought conditions after day 60 (van Keulen and de Wit, 1984).

The availability of water is strongly influenced by soil type. Soil is a porous medium in which the particles are arranged in various shapes and sizes so that empty spaces occur between the solids. The spaces, called pores, are filled with air or they may contain water. When water enters the soil through rainfall or irrigation, the pores fill progressively with water, with the smallest pores filling first and the largest last. In the pore, the water is held under suction, and removal of the water requires a counter force. The force needed is inversely related to pore size.

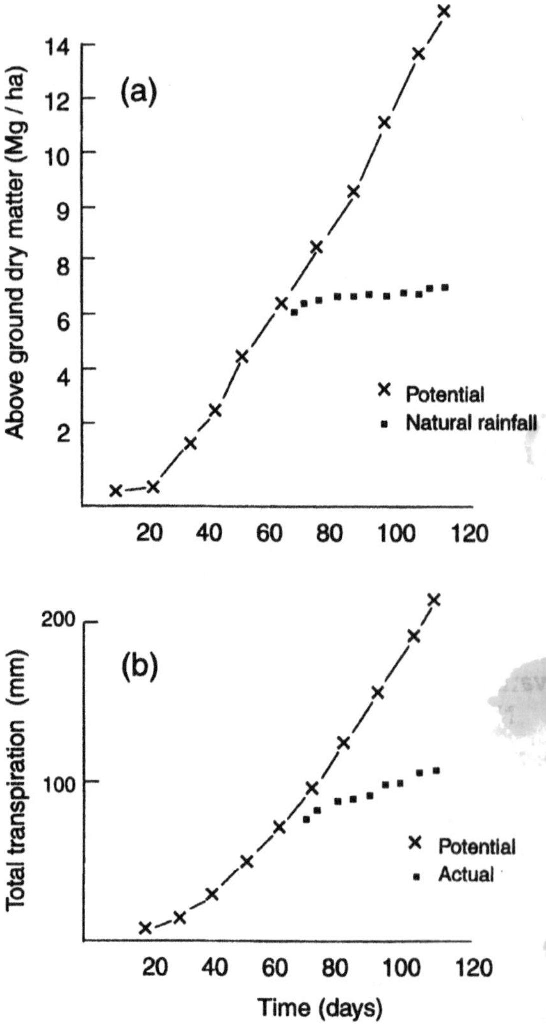

FIGURE 9.3. Time course of total above-ground dry weight for wheat, emerging December 5, under optimum growing conditions and under natural rainfall in Dhaka, Bangladesh: (a) cumulative transpiration for a wheat crop growing under optimum moisture supply; and (b) under natural rainfall (van Keulen and de Wit, 1984).

As indicated in earlier chapters, soil water, if sufficient, can move through soil and enter streams or storage. Alternatively water may move upward as it is evaporated from the soil surface and enters the atmosphere. This process is enhanced if there is a shallow water level or groundwater table. If salts are present in the soil it can lead to salinization. Comparison of yields of wheat in Bangladesh in sandy loam soil and heavy clay soil illustrate the effect of soil texture (Figure 9.4). Clay is characterized by small pores and a heavy texture. Less water can move through the clay soil and less can be extracted. In this example both fields had adequate water at the start but the sandy soil contained sufficient extractable

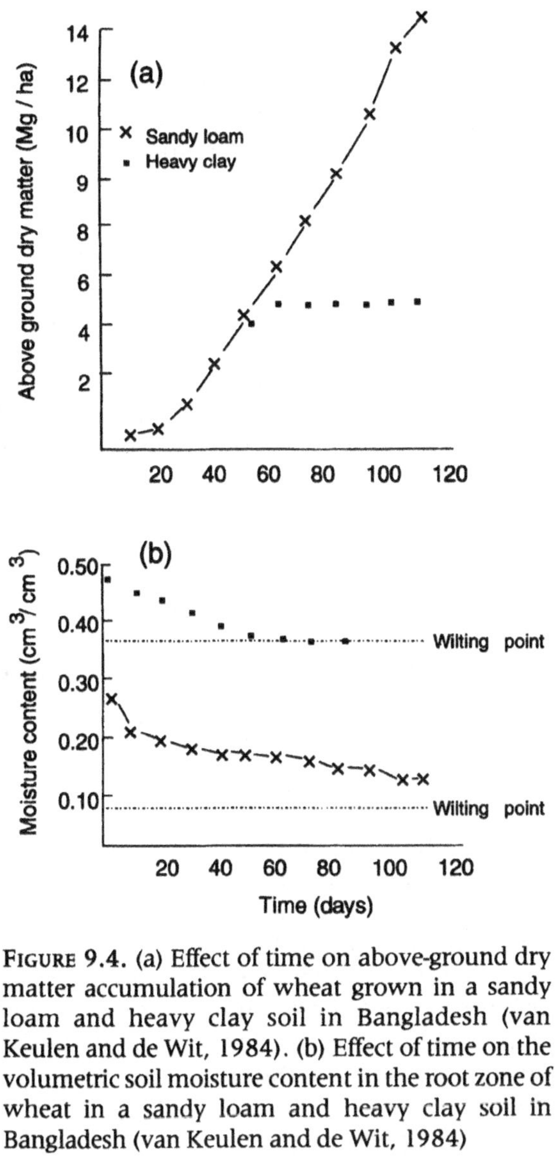

FIGURE 9.4. (a) Effect of time on above-ground dry matter accumulation of wheat grown in a sandy loam and heavy clay soil in Bangladesh (van Keulen and de Wit, 1984). (b) Effect of time on the volumetric soil moisture content in the root zone of wheat in a sandy loam and heavy clay soil in Bangladesh (van Keulen and de Wit, 1984)

water to allow the crop to grow to maturity.

It is this role of water in limiting the crop from achieving optimum production that has motivated the agriculturalist to manage water through irrigation. By irrigation, it is possible to maintain the water supply in the soil at a relatively constant rate so that the plant never experiences wilting conditions. Ancient farmers relied on their intimate relationship to, and steady observation of, their crops. The modern farmer has a variety of mechanical instruments available to support his or her observations.

9.2.2 Supply of Essential Nutrients

Beside the need for carbon dioxide and water, the crop requires other elements to build and maintain its tissues. These elements come from sources in the environment, from the hydrosphere, atmosphere, or lithosphere. If a limiting nutrient is supplied to the crop, the crop responds by increased growth. In this case, we conclude that that element was in short supply in the source compartments of the system. However, the increase in crop production continues only for a period. Eventually the plant no longer responds to added nutrient. If even further element is supplied, it can become toxic and turn from a fertilizer to a pollutant. The point where positive response ends is the point where some other factor becomes limiting. If that factor is added, growth continues to respond. The process continues until the crop is receiving the inputs it requires for optimum growth and we observe maximum performance as described in Table 9.2.

Because the crop harvest removes nutrient capital from the system, there is a continuing need to add nutrients to balance the loss. These patterns are illustrated in a figure of the nitrogen balance of a watershed in Georgia, USA (Figure 9.5). The model shows that fertilizer input was substantial, equalling input from the precipitation and exceeding input by nitrogen fixation. The output of nitrogen was mainly in the harvested grain leaving the watershed as product. In this example, a further property of nitrogen is important. The loss to streams was quite low because the excess fertilizer nitrogen moved in groundwater to the flood plain forest bordering the stream. There, nitrogen was taken by trees with deep root systems. Further, as the flood plain forest was covered by water after heavy rain, the supply of oxygen to soil bacteria was reduced, favouring the activity of denitrifying forms. These bacteria use nitrate as a substrate and convert it into nitrogen gas, which enters the atmosphere. The consequences of these processes in the flood plain is that very little nitrogen added to the fields enters the streams,

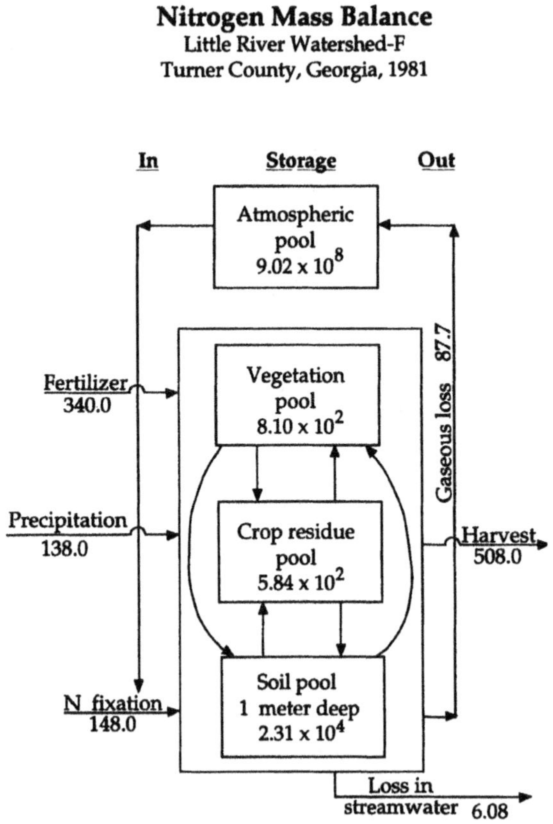

FIGURE 9.5. An example of a conceptual model that describes the estimated amounts of total N in major storage compartments and transfers among compartments for a watershed (partially based on data from Kessler and Meentemeyer, 1989).

thus retaining stream purity. Here is a further example of the need to think at several scales at the same moment, because a positive effect at one scale may produce a negative effect at another scale. Rural planning seeks a win–win situation where positive events are seen at all relevant scales.

Each of the micronutrients acts in a similar fashion; however, each has its own peculiar dynamics. For example, phosphorus binds with soil colloids so that it is unavailable to plants while being present in relatively large amounts. The immobilization of phosphorus is due to the fact that many of the reactions involving phosphorus are sorption reactions on soil mineral surfaces resulting in very low equilibrium P concentrations in the soil solution.

9.2.3 Conclusions

These relationships of the crop with limiting environmental factors depend on the nature of the crop species, the distribution of biomass within the plant, the economic situation that supports subsidy, and the presence of adequate sources of the limiting factors. These relationships create a complex system of interactions, which are amenable to agricultural research and predictive system modelling. The great advances in crop production seen in the last 100 years, with its concomitant supply of higher-quality food to a growing human population, is the consequence of this research programme.

9.3 Management of Agricultural Systems

9.3.1 Water Management

9.3.1.1 Rainfed Systems

Under natural rainfall, water inputs cannot be managed except to ensure maximum acceptance of the rain at the soil surface. This can be achieved by ensuring that the soil surface is protected by a mulch layer or standing crop. This protective soil cover intercepts the energy of the raindrops, preventing the formation of a seal at the surface (Sumner and Stewart, 1992).

Increasing water entry into the soil not only renders more water available to the crop but also reduces erosion due to excess runoff. In areas where rainfall is limiting to crop production, yield gains can be achieved by reducing water loss due to evaporation and here again a mulch is of great benefit (Troeh et al., 1991). In addition, the use of drought-resistant crops can also increase the use efficiency of the water available in the root zone. In heavy soils with low permeability, both surface and internal drainage systems may be necessary to ensure that the soil does not become waterlogged during periods of high rainfall (Troeh et al., 1991).

9.3.1.2 Irrigated Systems

Water can be managed and controlled much more efficiently under irrigated than rainfed conditions. Irrigation scheduling can be used to ensure a continued supply of water to the crop at all growth stages in proportion to the needs of the crop. This is usually based on estimates of evapotranspiration from evaporation pans corrected for crop characteristics from which the amount of water required to replenish the water-holding capacity of the soil can be calculated. Alternatively, soil sensors such as tensiometers or other probes that measure moisture content can be used to continuously monitor soil water with irrigation being scheduled when a certain deficit of moisture is reached (Miller and Donahue, 1995).

Probably the easiest part of managing irrigated land is that of supplying the requisite amount of water. Unfortunately in many irrigation schemes, too much attention has been focused on this aspect without adequate attention being paid to the potential effects of the applied water on the soil. All waters, both surface and subsurface, contain salts which remain in the soil because evapotranspiration removes only pure water. This build-up of salts increases the osmotic potential of the soil causing the crop to suffer drought stress even when the soil is wet. To prevent such salt accumulation, excess irrigation water must be occasionally applied to leach the salts out of the root zone. This, in turn, requires a freely drained subsoil or the installation of drains to remove excess water and salts (Singer and Munns, 1996). In turn, such saline drainage water poses a threat to the water bodies into which it is discharged.

In fact the USA National Research Council (1989) concluded that: "Irrigation...over time cannot avoid causing an adverse off-site effect. This must be acknowledged: it can be minimized, internalized, or rejected, but it cannot be ignored." Furthermore, irrigation with waters that are sodic (i.e., containing high levels of sodium and bicarbonate and low levels of calcium and magnesium) can lead to sodification, which impairs the physical condition of the soil. As a result, such sodic soils become poor rooting media (low air space, permeability, and high soil strength). Clearly, in such cases, prevention is better than a cure that can be achieved by treating the water or soil with gypsum before irrigation is commenced to ensure that the water quality is suitable for the particular soil (Sumner and Naidu, 1997).

9.3.2 Soil Management

9.3.2.1 Fertilization

Application of fertilizers is usually required for the production of high-yield crops on most soils. The quantities of nutrients required to be applied in the form of fertilizers and soil amendments can be accurately estimated from soil tests calibrated for given regions (Brown, 1987). This is particularly true for phosphorus, potassium, and lime but, in the case of nitrogen, soil tests are often unsatisfactory because it is difficult to assay the quantities of nitrogen that will be mineralized from organic sources by means of a rapid soil test. Consequently, nitrogenous fertilizers are often overapplied, leading to potential water pollution. Use of organic sources of nutrients such as animal manures or sewage sludges makes the evaluation of available nutrients even more complicated and the biggest offsite nutrient effects are often from these sources (Sharpley et al., 1987). This occurs despite the fact that organic materials act as slow-release fertilizers.

9.3.2.2 Weed Control

In the past, weeds were controlled by cultivation but the advent of herbicides has revolutionized this facet of crop production. Herbicides are now available which specifically target particular spectra of weed populations and are available in both preemergence and postemergence varieties. The new herbicides are generally environmentally more benign than the older materials, with some such as glyphosate being nontoxic to animals. As a general rule, if herbicide applications are made according to label specifications, they are both environmentally friendly and very efficient at controlling weeds.

9.3.2.3 *Tillage and Soil Conservation*

The best way of conserving soil is to disturb it as little as possible during the cultivation of crops. There is no doubt that conventional tillage operations such as ploughing, disking, and rototilling exacerbate soil erosion. The question should always be asked: Is it necessary to till this particular soil? If the answer is no, one should resort to modern conservation tillage methods such as minimum or no-tillage techniques in which seeds and fertilizer are directly drilled and knifed into the soil without prior tillage. These techniques have been shown to cause little or no soil loss and therefore are being widely adopted as strategies to improve the sustainability of production.

9.4 Environmental Impacts of Agricultural Systems

9.4.1 Soil Degradation

9.4.1.1 *Acidification*

Although soil acidification is a natural process, intensive agriculture accelerates the rate by interruption and modification of N and C cycles, and by the use of N fertilizers (ammoniacal-N) and legumes (amino-N) to enhance the N nutrition of crops and pastures. Both forms are readily oxidized during nitrification to form nitric acid (HNO_3) with the extent of acidification depending on the amount of nitrate (NO_3) leached below the root zone. In general amino-N (urea and legumes) produces about half the acidity produced by ammoniacal-N (ammonium sulphate) (Helyar and Porter, 1989). Interruption of the C cycle takes place by the removal of crop and animal materials in which the alkaline organic salts of Ca, Mg and K are carried off the land leading to its acidification. The long-term consequences of acidification arising from the continued use of ammoniacal-N fertilizers such as loss of cation exchange capacity (CEC) and basic cations, increase in labile (toxic) Al, and crop yield decline are illustrated in Table 9.3. In many situations, Mn becomes sufficiently soluble to become toxic also.

9.4.1.2 *Salinization*

Irrigation of soils with water containing salts leads to the build-up of salinity in the absence of adequate drainage and excess water for salt removal. Given that all natural waters contain salts, salinization will occur whenever insufficient water (irrigation plus rainfall) is applied to leach the accumulated salts below the root zone. In some cases, natural drainage is sufficient but often the installation of drainage systems is necessary. Over 10 million hectares of irrigated land are abandoned annually as a result of salinization (Szabolcs, 1989). The increased level of salts in the soil leads to crop water stress and reduced productivity as a result of the osmotic competition for water. In addition, clearing of natural vegetation (forest) can result in a rise in saline groundwater under rainfed conditions as has frequently been observed in Australia and North America, leading to severe salinization of the root zone and crop failure (Sharma and Williamson, 1984; Halvorson, 1990).

9.4.1.3 *Sodification*

A sodic soil is one containing elevated levels of exchangeable Na. Sodic soils develop as a

TABLE 9.3. Effect of continued application of ammoniacal N fertilizers for 15 years on soil chemical properties and yield of sorghum (Sumner, 1991).

Soil pH	CEC at field pH*	Extractable* Al	Extractable* H	Σ Basic cations*	Sorghum yield (t/ha)
4.7	2.72	1.88	0.16	0.61	0.16
5.1	2.89	1.07	0.09	1.49	2.00
5.7	3.10	0.35	0.14	2.64	3.60
6.2	4.62	0.00	0.08	4.66	4.20

*$cmol_c/kg$.

result of the removal of salts from a saline soil by leaching with irrigation water and also by subjecting a nonsaline soil to irrigation water containing elevated levels of $NaHCO_3$ and low concentrations of Ca and Mg.

The extent to which sodic soils exhibit poor physical condition depends on the concentration of salts in the soil solution. As the level of exchangeable Na in a soil increases, ever-increasing electrolyte concentrations are required to maintain the clay and organic matter in the soil in a flocculated condition (Sumner, 1993). In the absence of adequate concentrations of salt, clay, and organic matter disperse, become mobile, and tend to block the pore space, resulting in drastically reduced permeability. Sodic soils that do not have adequate salt present to maintain the clay in a flocculated condition are often waterlogged during wet periods.

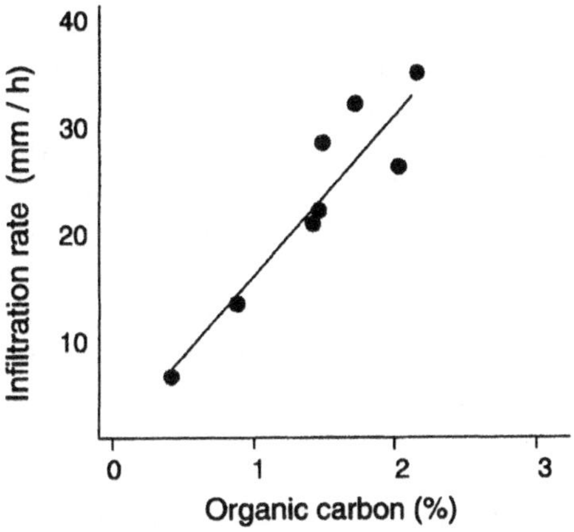

FIGURE 9.6. Effect of organic carbon content on the infiltration rate of Australian red brown earths (partially based on data from Grierson et al., 1972).

9.4.1.4 Erosion

The loss of organic matter under cultivation results in the rupture of organic–inorganic bonds responsible for stable soil aggregation allowing clay to become dispersible under certain circumstances. Clay dispersion increases as a result of inputs of energy (impacting raindrops), elevated levels of Na, and low concentrations of salt in the soil solution (Sumner, 1992, 1993). When clay disperses at the soil surface usually under the influence of impacting raindrops, the mobile clay blocks pores forming a less pervious seal which drastically reduces water entry into the soil. The relationship between organic C content and infiltration rate is illustrated in Figure 9.6 which clearly indicates the importance of maintaining the level of organic matter as high as possible. When soil dries, a surface crust often forms which can limit seedling emergence and the exchange of gas between soil and atmosphere. The reduction in infiltration rate results in increased runoff which then has the potential to cause erosion. The extent of the resulting erosion depends on soil characteristics, slope, rainfall intensity, and extent of soil cover (Sumner, 1992). Because tillage disturbs and exposes soil to raindrop impact, some increased level of erosion usually results. This, in turn, impacts water quality in terms of turbidity, sediment, and nutrient loads.

9.4.1.5 Compaction

Soil compaction, which is usually brought about by in-field wheeled vehicles and trampling by livestock, increases soil bulk density, changes pore size distribution toward finer pores, and reduces total porosity (Swinford and Boevey, 1984). Consequently, compaction results in poorer aeration and aggregate stability which limits root proliferation and, ultimately, crop yield. In general, soils falling in the sandy loam to loamy sand textures are most prone to compaction because of the enhanced packing ability of the nearly equal proportions of sand, silt, and clay.

9.4.1.6 Contamination

Soil pollution can occur due to inappropriate agrochemical application, deliberate disposal of contaminated liquid and solid wastes, and inadvertent addition of contaminants in amendments and fertilizers. However, although gross contamination seldom occurs in agricultural systems, land application of wastes such as sewage sludge can add unwanted nonnutrients (Cd, Hg, Pb, As). There is evidence that these metals can adversely impact the activities of soil biota (Witter, 1993). Because rock phosphates often contain high levels of Cd and Ra, soils treated with single superphosphate for continued periods have resulted in unacceptable levels of Cd in the crop, particularly in the presence of high levels of Cl in the soil which promotes Cd uptake (McLaughlin et al., 1996).

9.4.2 Water Contamination

9.4.2.1 Surface Water

Protection of surface and groundwaters free from pollution by sediments, nutrients, and pesticides is essential to assure a desirable environment. However, in the past, agriculture has contributed to the nonpoint pollution of these water resources.

Relative to reduced or no-till, conventional tillage contributes greatly to the suspended sediment and total N and total P loads in surface waters. On the other hand, concentrations of NO_3 and dissolved P (DP) are higher in water leaving no-till land, because fertilizers have to be applied on the soil surface in such systems (Laflen and Tabatabai, 1984).

Soil vegetative cover merely interdicts sediment and sorbed pollutants as it moves in surface runoff but has little effect on the dissolved constituents. Pesticide behaviour is similar to that of P, depending largely on whether the material occurs in solution or adsorbed to soil surfaces. Thus, in pasture and cropped systems, the concentrations of NO_3 and DP in runoff water can be similar, but total concentrations are higher in the latter.

The presence of livestock on a pasture increases the loss of NO_3 and DP in runoff due to the high concentrations in urine patches (Schepers and Francis, 1982). Fertilizers and manures, when applied on the surface as in pastures, make substantial contributions to surface water contamination with NO_3 and DP, often exceeding drinking water standards and accelerating eutrophication.

In fact, it is doubtful whether N levels in runoff from highly productive agricultural land can be maintained below recommended levels. On the other hand, DP can be controlled to a greater extent by incorporating fertilizers and manures into the soil mechanically, allowing time for the P to react with soil-fixing surfaces.

On 20 Southern Plains watersheds in the United States, Sharpley et al. (1987) found the impact of fertilizers on surface water quality to be mainly due to DP with some contribution from particulate P. The P concentration in runoff is determined by the extent to which DP is able to interact with the sorption sites during overland flow. Crop residue retention on the soil has little effect in reducing DP concentrations in runoff water. The greatest threat to water quality is likely to occur immediately after fertilizer or manure application, and the effect decays rapidly in subsequent runoff events.

9.4.2.2 Groundwater

Pollution of groundwater with NO_3 is of growing concern worldwide. In the United States, over 50 percent of community and rural wells are contaminated, mainly from fertilizers and manures applied to agricultural land, with 2 percent of the wells exceeding acceptable drinking water standards. Because NO_3 is not sorbed by most soils, it readily leaches to groundwater.

Conservation tillage and fallow systems tend to increase NO_3 leaching. In the former, more water enters the profile and is available for leaching while, in the latter, the absence of a crop reduces evapotranspiration (Baker et al., 1987; Juergens-Gschwind, 1989).

The greatest losses of NO_3 usually occur between crops when roots are largely absent in the soil. Winter cover crops with prolific root systems can be used to intercept part of the NO_3 before it leaves the root zone with nonlegumes being more effective than legumes (Meisinger et al., 1991). However, the N stored in the cover crop will be mineralized in the spring as a slow-release fertilizer requiring synchronization of N crop requirements with the supply. This is usually very difficult to achieve. Although legumes can be used as an alternative to fertilizer and usually act as slow-release sources of N, substantial losses can occur in rotations involving legumes when the legume is killed (Owens, 1990).

On the other hand, perennial deep-rooted legumes and grasses can be efficient scavengers for deep NO_3 left over from pervious crops or released under feedlots (Owens et al., 1994). Usually leaching losses from ungrazed low-input pastures can be low, but substantial losses can occur under heavily fertilized systems with NO_3 levels exceeding the drinking water standard (Roberts, 1987). Introduction of grazing animals greatly increases losses due to the high concentrations of N in urine and dung patches (Ball et al., 1979).

Because cropping systems require substantial inputs of N to achieve optimum production, concentrations of NO_3 in leaching waters can be considerable (Weil et al., 1990). Even at economic optimum rates of N application, NO_3–N concentrations in drainage water have been found to be excessive (Jemison and Fox, 1994).

Under horticultural crops, NO_3 leaching is about double that of agronomic crops because the soil is often without cover, is intensively cultivated, and large amounts of residues are returned (Juergens-Gschwind, 1989). In all cases, the quantities reaching groundwater are largely determined by the excess of rainfall over evapotranspiration. The long-term (30 years) effects of row cropping on groundwater quality in the Big Spring Basin (Iowa, USA) (Figure 9.7) show that by 1985, NO_3 concentrations had increased tenfold over the value in 1930 and closely followed the increase in fertilizer and manure inputs. The marked decrease in fertilizer inputs during the payment-in-kind (PIK) programme during which farmers were subsidized to take land out of production is clearly reflected in the reduced NO_3 concentration in the water a few years later highlighting the delayed sensitivity of water quality to fertilizer inputs. Excess NO_3 can take a long time to appear in subjacent aquifers, and conversely, this means that contaminated groundwater could take even longer to be reclaimed.

Although an intensely grazed legume pasture should be an ideal agroecosystem (low input, organic matter accumulates, N-cycle should prevent leaching), this has not proven

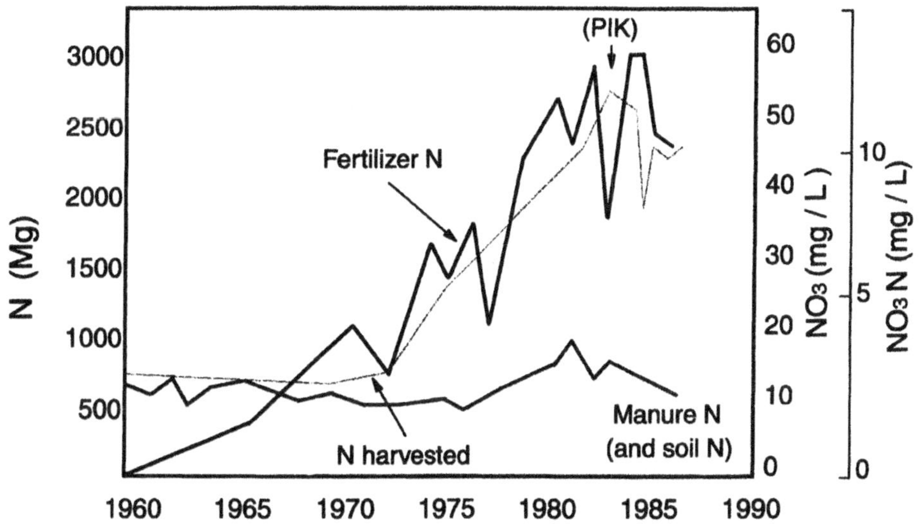

FIGURE 9.7. Changes in groundwater nitrate (NO_3) concentration resulting from increases in manure, and fertilizer usage in the Big Spring Basin in Iowa, USA (partially based on data from Hallberg and Keeney, 1993).

to be the case. More than fivefold increases in NO_3 have been found below grazed than ungrazed pastures (Baber and Wilson, 1972; Ryden et al., 1984). High NO_3 concentrations have been found under legume-based pastures in Australia (Scott and Dillon, 1989). Fertilization of nonlegume-grazed pastures always results in NO_3 leakage more or less in proportion to inputs, with concentrations frequently exceeding the drinking water standard (Roberts, 1987).

Under irrigated systems, the risk of NO_3 leaching is greatly increased. Together with the higher rates of N application this poses a serious problem. Again the level of NO_3 leached is directly related to the amount of water that passes through the profile (Hergert, 1986). Although the selection of water-efficient irrigation systems may reduce NO_3 leaching, reducing losses to zero appears to be impossible particularly on sandy soils. In drier areas where irrigation is the main source of water for the crop, salinity in drainage water is often a problem. For example on the Colorado River, 37 percent of the salt in the lower reaches (Los Angeles) stems from irrigated agriculture (Holburt, 1984). In addition, certain elements (Se, B, Mo) become a threat to wildlife where inland reservoirs have been used to store and evaporate drainage waters from irrigated areas.

9.4.3 Food Contamination

Because of the impurities contained in rock phosphates from which all P fertilizers are manufactured, certain nonessential elements (mainly Cd) have been inadvertently applied to soils, particularly in single superphosphate. This condition has been especially acute in Australia where the phosphate source has been Oceania with high Cd contents. Uptake of Cd by crops is promoted by high levels of Cl in the soil which is frequently the case in Australia (McLaughlin et al., 1994).

9.5 Management of Agricultural Systems for Sustainability and Minimal Environmental Impact

9.5.1 Pasture and Rangeland

Generally, rangelands receive much lower levels of management than improved pastures. In New Zealand, rangeland on steep terrain has

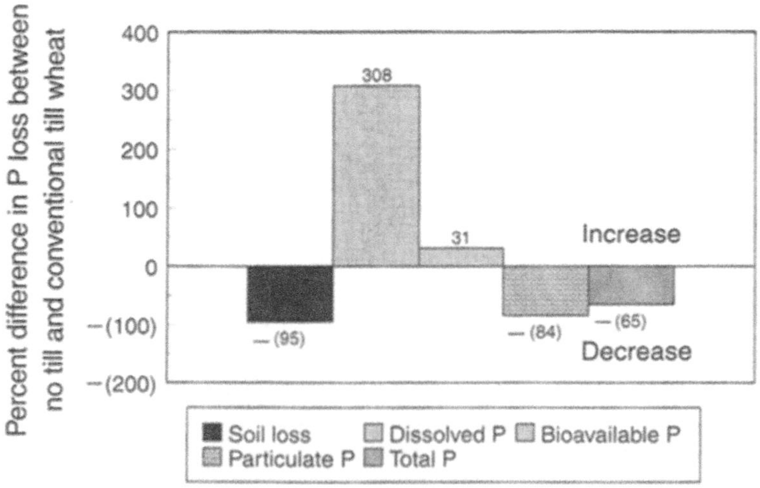

FIGURE 9.8. Differential effect of no-till wheat in the Southern Plains on soil loss and categories of P in runoff water (partially based on data from Sharpley et al., 1994).

been improved by flying on superphosphate at relatively low rates which increases both the quality and quantity of herbage with minimal threat to the environment. Fertilization of rangeland in Africa has usually resulted in severe degradation in botanical composition of the sward and is seldom practised. Because overstocking under rangeland conditions increases erosion losses, stocking rates should closely match the biomass available to the animals. Rotational grazing schemes with strategically placed watering points can minimize rangeland degradation with its attendant problems.

Pastures are usually improved by either ploughing out the original grassland and establishing stands of exotic species or by reinforcing the native vegetation with other species such as legumes. In both cases inputs of fertilizer and lime may be required to achieve adequate production. Wherever possible, most of the fertilizer and manure inputs should be incorporated before planting to reduce potential losses in surface runoff. Losses become inevitable with time as all future nutrient inputs must be placed on the surface.

Because pasture land is seldom ploughed, the incorporation of lime to neutralize the acidity produced by the legumes in the stand is possible only occasionally. Thus, to maintain productivity, pasture reestablishment with appropriate amendments becomes necessary from time to time. Application of manure on frozen soil should be avoided at all costs as this is readily transported during a thaw. In addition, because overwintering stock on pasture drastically increases runoff and soil losses, this practice should be avoided whenever possible.

Potential nutrient losses both in surface runoff and by leaching can be reduced by timeous applications matched to the growth rate of the crop. Nitrification inhibitors and slow-release N sources can be used to slow down nitrification, but this merely delays the inevitable (Owens et al., 1992). Where manures such as poultry litter are surface applied on pasture, consideration should be given to pretreating the manure with chemicals (alum) to reduce the level of soluble P. Manipulation of the animals' dietary intake of nutrients can be used to lower the concentrations in the excreta. Composting of animal manures before application can reduce the potential for leaching and may be one of the most effective ways of reducing the levels of labile N and P. Because on sandy and organic soils leaching of P particularly under intensive grazing can be a problem, the use of alternative P sources such as rock phosphate can result in marked reductions in the levels of P entering water bodies. However, because animals produce highly soluble sources of nutrients in their excreta, it is unrealistic to believe that losses (particularly N) in runoff and drainage to depth can be reduced to zero while maintaining a profitable grazing enterprise.

On average, erosion losses are much lower from pasture and rangeland than from cropped soils.

9.5.2 Row Cropping

Erosion on cultivated lands can be reduced in a number of ways, including the use of conservation tillage, return of residues, contour strips of grass, and grassed waterways. Although these practices limit sediment and particulate nutrient transport, the reverse is true for soluble nutrients (Figure 9.8). In this case, no-till decreases soil and particulate P loss but increases dissolved and bioavailable P. To minimize the latter categories, fertilizer inputs should be ploughed in prior to planting or knifed into the soil under conservation tillage to maximize the time available for the soluble nutrients to be rendered less soluble in the soil. To reduce leaching losses of NO_3 and P, fertilizer applications should be made at times when the crop is actively taking up nutrients and in proportion to the needs of the crop. Soil testing for NO_3 prior to planting can effectively predict N response and the amount of additional N that needs to be applied to produce optimum yields of crops following in the rotation (Bundy et al., 1992). This is an important strategy in reducing NO_3 leaching. The use of rapid tissue assays can improve the accuracy of determining the amount of N

TABLE 9.4. Best management practices (BMP) for controlling soil (sediment), nitrate, and adsorbed nutrients (e.g. phosphate) movement to water supplies*

BMP	Sediment	Soluble nutrients	Adsorbed nutrients	N loss to water
Conservation tillage	■	□	■	◆
Contour/terracing	■	◇	■	◆
Strip cropping	◇	◇	◇	□
Filter strip	◇	□	◇	□
Cover crop	◇	◇	◇	◇
Crop rotation	■	◇	■	□
Nutrient management	□	■	□	■
Irrigation management	◇	◇	◇	◇
Pasture management	◇	□	◇	□
Manure management	■	■	■	■
Grassed waterway	◇	□	◇	□
Critical area planting	■	◇	◇	□
Riperian forest	◇	□	◇	□
Streambank/waterbody protection	■	□	■	□
Composting	□	■	■	■

*□ = no control to low effectiveness; ◇ = low to medium effectiveness; ■ = medium to high effectiveness; ◆ = may increase loading.

required for topdressing (Edmisten et al., 1992). In addition, fertilizer recommendations need to give credit for the N and P applied in manures which is not always done. Field borders, filter strips, and riparian zones can serve as filters for subsurface flow of NO_3 (Cooper et al., 1995) but to be sustainable, periodic removal of biomass would be necessary to prevent build-up adjacent to streams.

Newly developed soil specific crop management technology (precision farming) has the potential to be not only environmentally friendly, but also economically advantageous in terms of savings in fertilizer and pesticides (Marks and Ward, 1993). The fact that young, well-educated, and environmentally concerned farmers are more likely to apply appropriate rates of fertilizers and pesticides, highlights the importance of education which should address the linkages between N management and deteriorating water quality, and also how recommendations are made to increase farmer confidence in applying re- duced levels of N. Nevertheless, it should be realized that arable agriculture is the major culprit and no matter how carefully the recommendations are made and executed, a "window" for NO_3 to enter groundwater between tillage and crop establishment will always exist.

Where irrigation is practised, care must be taken to ensure that a minimum of excess water is applied to reduce leaching of soluble nutrients. Reducing NO_3 losses to zero under irrigation on sandy soils is impossible. Again, N applications prior to sowing should be minimal with the bulk of the fertilizer N being applied when the crop has developed a sizeable root system to maximize uptake. Rigid irrigation scheduling will be needed to make sure that excess water does not leave the root zone. However when soils are or have become saline, additional quantities of water for leaching must be applied to remove the salts.

Keeney (1989) has made the plea for more agroecosystem-oriented concepts to be incorporated into best management practices (BMP) to reduce NO_3 and DP leaching. In developing these BMPs, attention should be focused on the following:

(i) account accurately for N and P contributions from manures, soil organic matter, crop residues, and stored inorganic N and P in making fertilizer recommendations;

(ii) tailor make fertilizer applications to the developing needs of the crop, using split applications, slow-release sources, and/or nitrification inhibitors;

(iii) use deep-rooted cover crops to scavenge for NO_3 before it enters groundwater;

(iv) use more N-efficient crops;

(v) schedule irrigation as accurately as possible, taking rainfall into account;

(vi) control and intercept runoff from dairy barns and feedlots;

(vii) recycle drainage waters through forest, grassland, or wetlands to reduce N and P levels;

(viii) focus on any sensitive areas such as sandy soils; and

(ix) adopt a systems approach to solve problems peculiar to a given region.

A summary guide to BMPs for water quality protection is presented in Table 9.4.

9.6 General Discussion

Overall, agriculture has been highly successful in keeping pace with the increasing demands for food and fibre required to meet the needs for the population explosion. Continued increases in production per unit area can be expected, but ultimately a ceiling will be reached. The decision will then have to be taken to consider the cultivation of more marginal soils.

Where food production has increased to the greatest extent since World War II, the adverse effects of agriculture on soil, water, and food quality have been the most noticeable. Nutrient and pesticide inputs, as well as intense cultivation, have significantly affected the environment in terms of water contamination with sediment, nutrients, and pesticides particularly in the developed areas of the world (North America, Europe, and parts of Asia). On the other hand, in developing countries, inputs are generally low, and often the soils are less resilient to intensive cultivation resulting in soil erosion, compaction, and soil structural degradation. To some extent this is also true of high-input systems, particularly with respect to compaction by machinery.

Only a portion of the nutrient inputs made to improve productivity are exported in the crop, with the remainder either accumulating in the soil, being lost by erosion or leaching, or volatilized to the atmosphere. Irrigation makes it more difficult to control transport of nutrients, pesticides, and sediment into surface or groundwaters. Because of the inherent leakiness of agriculture, some balance must be struck between the benefits of an assured high-quality food supply and some off-site impacts of the materials lost from the land.

We have the knowledge of the management practices required to minimize the negative impacts of agriculture on the environment. However, adoption of these measures has often been slow because this would drastically alter the economics of production. Therefore, the challenge for the future is to develop sustainable production systems where economic incentives are introduced to protect the environment and our natural resources (soil and water) while continuing to guarantee the food and fibre needs of the world.

References

Baber, H.T.; Wilson, A.T. Nitrate pollution of groundwater in the Waikato region. *J. N. Z. Inst. Chem.* 56:179–183; 1972.

Baker J.L.; Logan, T.J.; Davidson, J.M.; Overcash, M.R. Summary and conclusions. In: Logan, T.J.; Davidson, J.M.; Baker, J.L.; Overcash, M.R., eds. *Effects of Conservation Tillage on Groundwater Quality: Nitrates and Pesticides.* Chelsea, MI: Lewis Publishers; 1987:277–282.

Ball, P.R.; Keeney, D.R.; Theobald, P.W.; Nes, P. Nitrogen balance in urine affected areas of a New Zealand pasture. *Agron. J.* 71:309–314; 1979.

Brown, J.R. *Soil Testing: Sampling, Correlation, Calibration, and Interpretation.* Madison, WI: Soil Science Society of America; 1987.

Bundy L.G.; Schmitt, M.A.; Randall, G.W. Predicting N fertilizer needs for corn in humid regions: Advances in the upper Midwest. Bock, B.R.; Kelley, K.R., eds. *Predicting N Fertilizer Needs for Corn in Humid Regions.* National Fertilizer and Environmental Research Center Bulletin. Y-226. Muscle Shoals, AL: Fertilizer and Environmental Research Center; 1992:73–89.

Buringh, P. The capacity of the world land area to produce agricultural products. Golley, F.B., ed. *Workshop Agroecology*; 1984 January; Mediterranean Agronomic Institute of Zaragoza, Zaragoza, Spain. Options Méditerranéennes, Series Studies 1984-1; Zaragoza: CIHEAM; 1984:15–32.

Cohen, M.N. *The Food Crisis in Prehistory.* New Haven, CT: Yale University Press; 1977.

Cooper, J.R.; Gilliam, J.W.; Daniels, R.B.;

Robarge, W.P. Riperian areas as filters for agricultural sediment. *Soil Sci. Soc. Am. J.* 51:416–20; 1995.

Edmisten, K.L.; Wood, C.W.; Reeves, D.W.; Tracy, P.W. Determination of cotton nitrogen status with a hand-held chlorophyll meter in Alabama and Missouri. Herber, D.J.; Richter, D.A., eds. Proceedings of the Beltwide Cotton Conferences; Vol. 3. *Cotton Soil Management and Plant Nutrition Conference*; 1992 January 6–10; Nashville, TN. Memphis, TN: National Cotton Council of America; 1992:1101–1102.

Grace, P.J.; Ladd, J.N.; Sjemstad, J.O. The effect of management practices on soil organic matter dynamics. Pankhurst, C.E.; Doube, B.M.; Gupta, V.V.S.R.; Grace, P.R. eds. *Soil Biota: Management in Sustainable Farming Systems.* Melbourne: CSIRO Publications; 1992: 162–171.

Grierson, I.F.; Jijne, J.W.; Greenland, D.J. Changes in some physical properties of red-brown earths of different texture with increasing content of organic matter. *Expt. Rec.* 6:16–22; 1972.

Hallberg, G.R.; Keeney, D.R. Nitrate. Alley W.M., ed. *Regional Groundwater Quality.* New York: Van Nostrand Reinhold; 1993:297–322.

Halvorson, A.D. Management of dryland seeps. Tanji, K.K., ed. *Agricultural Salinity Assessment* and Management. New York: American Society of Civil Engineers; 1990:372–392.

Helyar, K.R.; Porter, W.M. Soil acidification, its measurement, and the processes involved. Robson, A.D., ed. *Soil Acidity and Plant Growth.* Sydney: Academic Press; 1989:61–101.

Hergert, G.W. Nitrate leaching through sandy soil as affected by sprinkler irrigation management. *J. Environ. Qual.* 15:272–278; 1986.

Holburt, M.B. Colorado River salinity: The user's perspective. French, R.H., ed. *Salinity in Watercourses and Reservoirs.* Boston, MA: Butterworth Publishers; 1984:13–22.

Jemison, J.M.; Fox, R.H. Nitrate leaching from nitrogen-fertilized and manured corn measured with zero-tension pan lysimeters. *J. Environ. Qual.* 23:337–343; 1994.

Jenkinson, D.S. Soil organic matter and its dynamics. Wild, A., ed. *Russell's Soil Conditions and Plant Growth.* New York: John Wiley and Sons; 1988:564–607.

Jordan, C., ed. *An Amazonian Rain Forest.* Man and Biosphere Series; Vol. 2. Paris: UNESCO; 1989.

Juergens-Gschwind, S. Groundwater nitrates in other developed countries (Europe): Relationships to land use patterns. Follett, R.F., ed. *Nitrogen Management and Groundwater Protection.* Amsterdam: Elsevier; 1989:75–138.

Keeney, D.R. Sources of nitrate to groundwater. Follett, R.F., ed. *Nitrogen Management and Ground Water Protection.* Amsterdam: Elsevier; 1989:75–138.

Kessler, B.T.; Meentemeyer, V.A regional analysis of total nitrogen in an agricultural landscape. *Landscape Ecol.* 2(3):151–163; 1989.

Laflen, J.M.; Tabatabai, M.A. Nitrogen and phosphorus losses from corn–soybean rotations as affected by tillage practices. *Trans. Am. Soc. Ag. Eng.* 27:58–63; 1984.

Marks, R.S.; Ward, J.R. Nutrient and pesticide threats to water quality. Robert P.C.; Rust, R.H.; Larson, W.E., eds. *Proceedings of Soil-Specific Crop Management Conference.* 1992 April 14–16; Minneapolis, MN. Madison, WI: American Society of Agronomy; 1993:293–299.

McLaughlin, M.J.; Tiller, K.G.; Beech, T.A.; Smart, M.K. Soil salinity causes elevated cadmium concentrations in field-grown potato tubers. *J. Environ. Qual.* 34:1013–1018; 1994.

McLaughlin, M.J.; Tiller, K.G.; Naidu, R.; Stevens, D.P. The behavior and environmental impact of contaminants in fertilizers. *Aust. J. Soil Res.* 34:1–54; 1996.

Meisinger, J.J.; Hargrove, W.L.; Mikkelsen, R.L.; Williams, J.R.; Benson, V.W. Effect of cover crops on groundwater quality. Hargrove, W.L., ed. *Cover Crops for Clean Water.* Ankeny, IA: Soil and Water Conservation Society of America; 1991:57–68.

Miller, R.W.; Donahue, R.L. *Soils in Our Environment.* Englewood Cliffs, NJ: Prentice-Hall; 1995.

Olderman, L.R. *World Map of the Status of Human-Induced Soil Degradation.* Wageningen: UN Environment Programme and International Soil Reference and Information Center; 1991.

Owens, L.B. Nitrate–nitrogen concentrations in percolate from lysimeters planted to a legume–grass mixture. *J. Environ. Qual.* 19: 132–135; 1990.

Owens, L.B.; Edwards, W.M.; Van Keuren, R.W. Nitrate levels in shallow groundwater under pastures receiving ammonium nitrate or slow-release nitrogen fertilizer. *J. Environ. Qual.* 21: 607–613; 1992.

Owens, L.B.; Edwards, W.M.; Van Keuren, R.W. Groundwater nitrate levels under fertilizer grass and grass–legume pastures. *J. Environ. Qual.* 23: 752–758; 1994.

Roberts, G. Nitrogen inputs and outputs in a small agricultural catchment in the eastern part of the United Kingdom. *Soil Use Man.* 3:148–154; 1987.

Ryden, J.C.; Ball, P.R.; Garwood, E.A. Nitrate leaching from grassland. *Nature* 311:50–53; 1984.

Schepers, J.S.; Francis, D.D. Chemical water quality of runoff from grazing land in Nebraska: Influence of grazing livestock. *J. Environ. Qual.* 11:351–354; 1982.

Scott R.I.; Dillon, P.J. Reconnaissance of soil water nitrate and related microorganisms near Mount Gambier, South Australia. Summer Student Project Report 20. Glen Osmond, SA, Australia: Australian Centre for Groundwater Studies; 1989.

Sharma, M.L.; Williamson, D.R. Secondary salinization of water resources in Southern Australia. French, R.H., ed. *Salinity in Watercourses and Reservoirs.* Boston, MA: Butterworth Publishers; 1984:571–582.

Sharpley, A.N.; Chapra, S.C.; Wedephol, R.; Sims, J.T.; Daniel, T.C.; Reddy, K.R. Managing agricultural phosphorus for protection of surface waters: Issues and options. *J. Environ. Qual.* 23:437–451; 1994.

Sharpley, A.N.; Smith, S.J.; Naney, J.W. Environmental impact of agricultural nitrogen and phosphorus use. *J. Agric. Food Chem.* 35:812–817; 1987.

Singer, M.J.; Munns, D.N. *Soils: An Introduction.* Englewood Cliffs, NJ: Prentice-Hall; 1996.

Sumner, M.E. Soil acidity control under the impact of industrial society. Bolt, G.H., ed. *Interactions at the Soil Colloid–Soil Solution Interface.* Dordrecht: Kluwer Academic Publishers; 1991: 517–541.

Sumner, M.E. Gypsum and acid soils: The world scene. *Adv. Agron.* 51:1–32; 1992.

Sumner, M.E. Sodic soils: New perspectives. *Aust. J. Soil Res.* 31:683–750; 1993.

Sumner, M.E.; Naidu, R. *Sodic Soils: Distribution, Properties, Management, and Environmental Consequences.* New York: Oxford University Press; 1997.

Sumner, M.E.; Stewart, B.A. *Soil Crusting: Chemical and Physical Processes.* Boca Raton, FL: Lewis Publishers; 1992.

Swinford, J.M.; Boevey, T.M.C. The effects of soil compaction due to infield transport on ratoon cane yields and soil physical characteristics. *Proc. S. Afr. Sugar Tech. Assoc.* 58:198–203; 1984.

Szabolcs, I. *Salt-Affected Soils.* Boca Raton, FL: CRC Press, Inc.; 1989.

Troeh, F.R.; Hobbs, J.A.; Donahue, R.L. *Soil and Water Conservation.* Englewood Cliffs, NJ: Prentice-Hall; 1991.

USA National Research Council. *Irrigation-Induced Water Quality Problems.* Washington, DC: National Academy Press; 1989.

Van Keulen, H.; de Wit, C.T. To what extent can agricultural production be extended? Golley, F.B., ed. *Workshop Agroecology*; 1984 January; Mediterranean Agronomic Institute of Zaragoza, Zaragoza, Spain. Options Méditerranéennes, Series Studies 1984-1; Zaragoza: CIHEAM; 1984:55–76.

Weil, R.R.; Weismiller, R.A.; Turner, R.S. Nitrate contamination of groundwater under irrigated Coastal Plain soils. *J. Environ. Qual.* 19: 441–448; 1990.

Wild, A. *Russell's Soil Conditions and Plant Growth.* New York: John Wiley and Sons; 1988.

Witter, E. *Heavy Metal Concentrations in Agricultural Soils Critical to Microorganisms.* National Swedish Environmental Protection Board Rapport 4079. Solna: Maturvardsverket; 1993.

Part IV
The Human Use of Rural Space

10

Human Organization in Rural Areas

Luisa M. Frutos

10.1 The Rural Areas Concept, Its Corresponding Features, and the Role of the Agriculture

Rural areas have traditionally been identified with agrarian zones, the primary reasons being that rural societies have until recently been mostly agrarian and also the perception that these spaces and the way of life there was obviously different from urban life and thus visually recognizable as the "country" (Clout, 1972).

Many ruralists have attempted to give a more precise definition: (i) by juxtaposition with urban spaces (Cherry, 1976; Lassey, 1977; Johnston et al., 1987); (ii) by using qualitative or morphological elements (Gou rou, 1973; Dollfus, 1976); (iii) by considering the use of the predominating soils, the dynamic, the intensity of exploitation, and productivity (Wibberley, 1972); and (iv) by placing emphasis on characteristics of population such as size and morphology of the nuclei, or on demographic features, including the production orientation of the inhabitants or the social structure (Kayser, 1990).

Even so, studies of rural areas by the OECD (1990) and the EU Commission in its report on the future of the rural world (MAPA, 1992a,b) state that these areas consist of open systems, which are multifunctional, humanized, and dynamic, although the combination of diverse elements offers an enormous typology at the world level. This can clearly be seen in the typology prepared by the Délegation pour l'Amenagement du Territoire et l'Action Régionale (DATAR) for France and in the proposal made by the Institut National de la Statistique et des Etudes Economiques (INSEE), both sources of data collected by Kayser (1990). In these studies, no less than five distinct classes (*deep rural, semi-agricultural rural, rural zones in mutation, static rural,* and *intermediate rural*) and fifteen types were established according to the dominant em ployment of the population and the dynamism of its economy.

However, regardless of definition, the rural zone is important from a strictly geographical or spatial point of view, as it covers from 80–90 percent of the *oikuméne,* that is, the land surface of the planet that is habitable and accessible to humanity. It is also important from a social standpoint when one considers that, despite the rapid increase in the urban population at the expense of the rural, the population statistically defined as rural makes up 50 percent of the total population of the planet. Also when absolute values are used one finds that the global total of inhabitants of rural zones is higher than 2000 million and that the United

Nations predict this figure to reach approximately 2750 million at the end of the millennium, with a notable increase in absolute figures, especially in Third World countries (Bonnamour, 1993). One should also bear in mind that rural and urban spaces are mutually dependent, although cities are the centres for decision making about economic policy and agrarian markets. It is impossible to understand the functioning of any territorial or global economy without considering agrarian areas and what they produce.

10.1.1 The Components of Rural Spaces

The organization of rural spaces has been the task of the human societies that inhabit them since Neolithic times. Their social and judicial characteristics, and the techniques they use, are a decisive influence on the way they are organized, thus creating a landscape marked by population distribution, structure, and morphology of agrarian exploitation and the uses of the land. These all come together to act as an expression of the different activities of the inhabitants. These human groups, which in turn are subject to a changing dynamic, also make decisions based on the conditions of the environment in which they live and which is their means of livelihood. Their activities affect the environment in many different ways, either maintaining an ecological balance or acting as a debilitating factor on the ecosystem. Whatever they do, their activities always lead to environmental changes, and the results are a consequence of the factors already mentioned. In short, the components of rural areas, which can be seen in Figure 10.1, combine in many ways to create a rich variety of territorial structures.

In the following pages, after some conceptual notes, we shall deal with the human aspect of the organization of rural spaces—in other words with the population, the essential agent, its system of settlements, and socio-juridical structures that regulate soil use.

10.1.2 The Role of Agriculture in the Organization of Rural Areas

If one considers those zones of extensive productive use and low population density, and the territories included in the census as rural, the sum total of agricultural and forestal activity is extremely important at an international level. The *FAO Trade and Production Yearbooks* noted that in 1993, out of a total of 13,422 million hectares of the planet's surface, nearly 8450 were occupied by cultivation, livestock use, and forest, which makes up 65 percent of the total. Of this area, 1450 million is given over to cultivation, making it 10.7 percent of the rural space. Pastoral land and forests should also be included as they also have a part to play in the livelihood and lifestyle of the occupants. Also, although there are areas which have not been exploited because of certain characteristics or by choice of the communities occupying them, they could also be included in the total as Agriculturally Usable Land (AUL). It is true that the extension of rural land space, including cultivated land, is not homogenous across the continents or countries and regions. However, the proportions in comparison to urban and unproductive zones are always high (Table 10.1).

It is certainly true that the population working in agriculture has steadily been decreasing during the course of this century, and that it continues to do so (Table 10.2) especially in developed countries, where it constitutes a relatively unimportant segment of the active population. However, despite this, at a worldwide level more than a third of the population are actively occupied in agriculture, and in some countries this ratio reaches more than half of the working inhabitants. Along with the drop in population there is a corresponding decrease in the contribution that agriculture makes to macroeconomic output (Table 10.2) because of the lower value given to primary products and the rapid increase and increasing complexity of industry and services.

However, at the beginning of the decade,

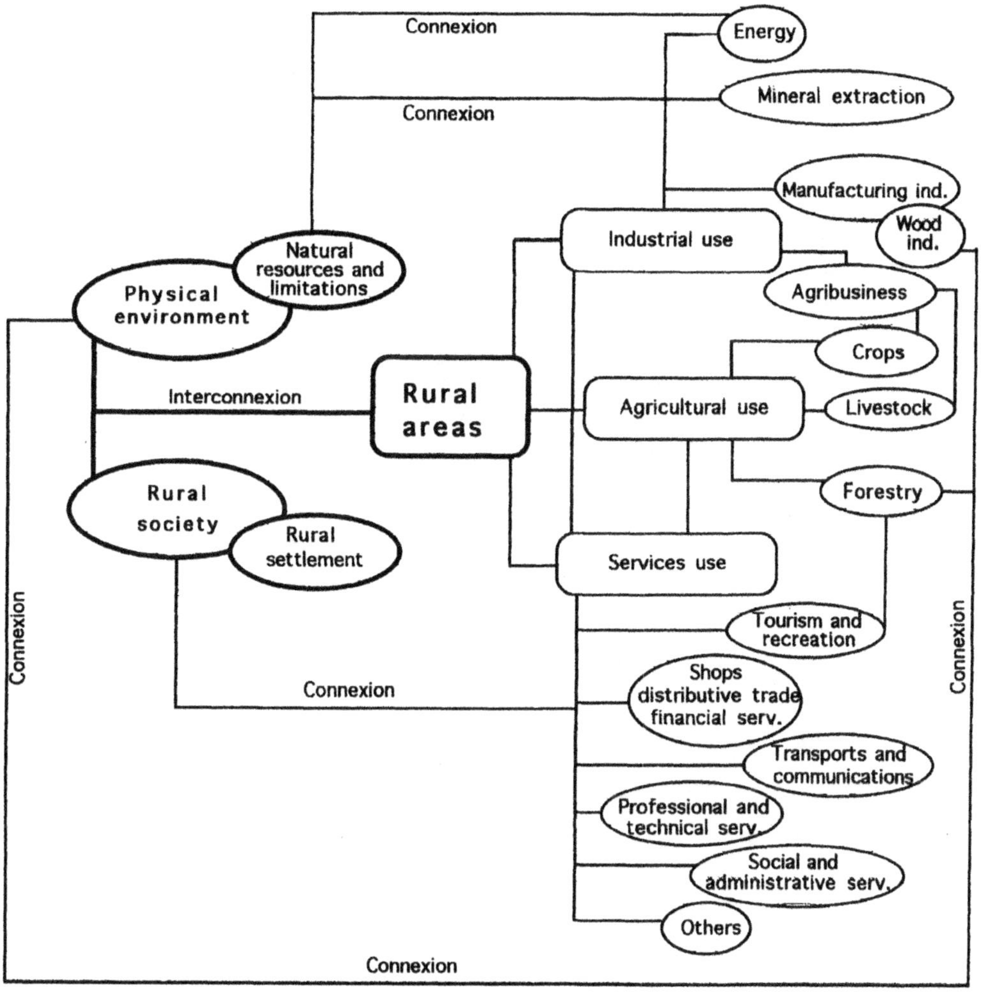

FIGURE 10.1. The function of the rural areas.

TABLE 10.1. The land use in the world (Source: FAO, 1993a)

Territory	Arable lands, pastures, and forest (000 ha)	% of total lands	Arable lands (000 ha)	% of total lands
Africa	1,759,315	58.0	182,545	6.0
North and Central America	1,345,440	58.0	271,341	11.8
South America	1,430,223	80.3	113,093	6.3
Asia	1,794,331	65.0	459,164	16.6
Europe (all countries)	375,361	77.0	136,632	28.0
Europe (EU)	195,380	82.0	81,921	34.7
Oceania	639,095	74.9	52,756	6.0
Ex-USSR*	1,355,420	60.5	228,920	9.5

*Data of the year 1991.

TABLE 10.2. The percentage of the rural population (RP), active population in primary sector (excluding fisheries) (AP) and the contribution of agriculture to Gross National Product (GNP) in selected countries in two different recent periods (1975–80 and 1990–92) (Source: UN, 1990; Instituto Geografico de Agostini, 1983–1985; El Pais, 1983–1995; MAPA, 1983–1985–1988)

Country	RP 75–80	RP 90–92	AP 75–80	AP 90–92	GNP 75–80	GNP 1992
Algeria	57.0	55.3	47.0	22.3	8.0	15.0
Egypt	55.0	51.2	50.0	30.0	36.0	18.0
Nigeria	74.5	64.8	51.0	63.7	26.0	37.0
Tanzania	86.7	67.5	80.0	79.8	52.0	61.0
Argentina	15.5	17.8	12.4	9.7	12.0	6.0
Brazil	32.5	23.1	36.0	22.6	10.0	11.0
Cuba	31.0	27.2	22.0	18.1	23.0*	15.0
Mexico	34.8	27.4	34.0	28.1	9.0	8.0
Peru	41.7	29.8	36.0	35.1	13.0	12.0
USA	26.3	27.4	2.0	2.1	3.0	2.0
China	70.4	73.6	58.0	65.2	33.0*	27.0
India	76.7	72.0	62.0	61.7	36.0	32.0
Israel	13.1	10.1	6.4	3.8	7.0	9.0
Japan	23.8	23.1	9.7	5.2	5.0	2.0
Nepal	96.6	90.4	92.0	91.0	62.0*	52.0
Turkey	56.1	51.6	51.0	41.4	27.0	15.0
France	24.0	25.9	7.9	4.2	5.0	3.0
Italy	41.7	35.8	14.4	5.2	8.0	3.0
Holland	11.6	11.0	4.0	3.2	6.0	4.0
Portugal	75.6	66.4	25.0	14.9	13.0	5.0
Spain	44.5	21.6	16.0	9.1	9.0	5.0
UK	17.0	7.5	1.9	1.8	2.0	2.0
Austria	45.9	42.3	18.3	4.3	5.0	3.0
Poland	40.5	38.0	29.0	16.5	16.0	7.0
Sweden	17.3	16.0	4.6	3.9	4.0	2.0
Switzerland	43.3	43.4	4.8	3.0-	—	4.0
Australia	14.0	14.5	5.4	4.5	5.0	3.0

*Data of the year 1981.

despite a relative loss of importance in the previous years, foodstuffs and raw materials of vegetable or animal origin made up approximately 10 percent of worldwide production. This figure would be even higher if agro-industrial products were included. It should also be mentioned that in some developing countries the economy revolves around agrarian or forest products, although these countries have to import foodstuffs. In developed countries the exportation and importation of products of this type are of considerable importance (Table 10.3).

In certain countries the importance of the exportation of a small number of agricultural or forest products is such that the country depends almost entirely on the international markets. Such is the case of Paraguay, with exports of 82 percent based on cotton, meat, wood, oils, and tobacco, or Guatemala, where, like other Central American countries, almost half of its exports are bananas, sugar, cafe, cotton, and meat.

The same can be said for many African countries such as Ethiopia with 83 percent of its output from coffee, skins, leather, oil-producing seeds, and meat, or the Ivory Coast, the exports of which consist of coffee, cocoa, palm oil, and wood, and which make up 63 percent of the total. All these products mentioned were produced in rural areas.

It is clear that the agrarian sector has a large part to play in the organization of rural space, and, directly or indirectly, has been incorporated into the international economic system. Thus, it is not surprising that the socioeco-

TABLE 10.3. The agrarian sector in the export trade of selected countries in 1993* (Source: FAO, 1993a,b)

	Agricultural and livestock products		Forestry products	
Country	Import	Export	Import	Export
France	10.0	15.1	2.7	1.8
Greece	12.2	29.0	2.7	1.1
Holland	12.7	21.6	3.5	1.8
Spain	11.8	16.4	3.7	1.9
Algeria	30.9	0.9	4.6	—
Nigeria	13.0	2.5	1.0	0.1
Argentina	5.9	51.0	1.7	0.9
Brazil	11.5	25.0	0.8	4
Mexico	11.6	11.8	1.9	0.5
USA	4.7	10.2	2.7	2.7
India	5.8	14.6	1.4	0.1
Japan	13.0	0.4	5.4	0.4
Malaysia	5.8	10.2	1.0	7.8
Pakistan	17.0	12.5	1.4	—
Turkey	7.7	23.6	0.9	0.3
Australia	4.4	25.9	3.0	1.1
New Zealand	7.3	46.5	2.1	12.2

*Data of the import and export agricultural products in percentage of the total foreign trade of the countries.

nomic imbalances that affect rural areas are to a great extent caused by the difficulties in adapting traditional agrarian thinking to present day production methods and modern business concepts.

10.1.3 Organization, Management, Planning and Development in the Countryside

Throughout this century the global evolution of the world economy and population has had especially significant repercussions for rural zones, where traditional activities and values have entered a stage of crisis. The populations of these zones are in the process of emigrating to urban areas and nontraditional activities, with a notable impact on the environment.

This has caused profound changes in rural society and its relationship with the space that it inhabits. Thus, the results brought about by this "voluntary organization" are not always the most effective ones. They may be the cause of imbalances in the society itself, which loses its quality of life and is impoverished, and may also affect the relationship with the potential resources in the area, which are either under- or overexploited. The consequence is that intervention becomes necessary, which introduce the concepts of "development" and "planning." Planning therefore seeks a point of equilibrium between economic and social development of the society and the conservation of the environment, which is the basic resource and supply centre of material support. In short, a sustained and harmonious development is called for.

In the first third of this century some highly interesting antecedents of this type of intervention can be found, such as the work by Stamp (1928) on the use and abuse of the soil, or the subsidized activities for the control of natural resources, especially water, such as the *Plan Integral de la Cuenca Hidrográfica del Ebro* in Spain, the planning of the Tennessee Valley, or the Italian subsidy program (Barbero, 1961; Labasse, 1966; Frutos, 1995). Other worthwhile agrarian reforms have been carried out where property and land exploitation are concerned.

In the last few decades, and a long way behind urban planning procedures (López Ontiveros, 1983), rural planning has been considered from two basic viewpoints. The first of these, with a clear technical and economic focus, can be summed up as orientation versus development. This viewpoint has in turn gone through two stages. The first stage was specifically agricultural and took the form of the so-called "Green Revolution." This was based on the idea that the reason for the underdevelopment of certain rural areas was, primarily, out-of-date technology. Thus, these areas needed to be part of the rapid advances in biochemistry—high-yield hybrid seeds (VAR), agrochemistry, and irrigation—which meant the application of government and FAO programmes and subsidies.

Although results were favourable in the short term, there was also a heavy dependency on technology, producing an internal socioeconomic imbalance (Arnón, 1987; Molinero, 1990; Becker, 1995) as a result of the application of the same scheme worldwide, as noted by Pelissier and Sautter (1984). The other stage, principally applicable to more dynamic rural areas, was based on the concept of integral development, implying that there was a need for an integral and multifunctional development influenced by exogenous forces. The best example of the latter is the EU's LEADER programme.

The more recent orientation in planning of the rural environment has come about as a result of the concern with the environment, which has come under pressure from demographic expansion, urbanization, and "touristification" of the countryside and the management of the environment with ever more heavy-handed techniques. Thus, it was decided to find a balance that would result in a sustainable development. The recuperation of more extensive agrarian regions was recommended, and in some cases agro-ecological programmes were initiated and protected natural areas were set up. These initiatives were not always well received by the inhabitants of the regions. At the same time, there was a convergence between a pluriactive rural environment and the aforementioned protectionist measures, this being termed integrated rural development.

Before initiating any kind of activity that might affect the rural environment, it is essential to take into account the society that inhabits the area, because it is the primary and most important agent. It is necessary, therefore, to acquire an understanding of the bases of this society.

10.2 The Rural Population and Rural Settlement

The population is the organizing agent of territory. The settlement is a primary symbol of occupation, the fruit of the social organization of the community, an expression of the techniques that are at its disposal and the type of exploitation and land use. In the following section we will consider the essential characteristics of rural communities, their demographic features, and the nature of their settlements.

10.2.1 Rural Society and Its Demographic Characteristics

All experts agree that there are four features that can be used to define the rural population: (i) relatively low population density; (ii) agglomeration in small communities; (iii) a constant loss of population caused by emigration (a process that has accelerated during the second half of this century); and (iv) the profound changes experienced in the last few decades in its social and demographic structure and in its mentality.

Low population density is characteristic of a community that requires large areas of land for the production system that supports it. However, we have previously mentioned the relative character of these relationships. It is essential to establish the relationship between the density of the rural area and the urban zones of the same region or country. This relationship is obvious when we contrast areas with more than 500 inhabitants per km^2 (in the agrarian zone of the Nile delta, for example) with other areas where the densities are five inhabitants per km^2, found in desert and subdesert regions, and also in temperate zones such as central Spain.

The identification of rural settlements with dispersed buildings and small nuclei is a consequence of a statistical and administrative concept of classification. It is necessary to use these criteria prudently. Certain authors have called attention to these criteria (Casas Torres, 1982; Pahl, 1968) and to the problems they cause when one tries to establish the proportions between rural and urban populations in different areas, and also when trying to calculate the development within a country when the official censual criteria are different from country to country. This means that one starts with relative concepts; however, the data are sufficiently valid for comparative purposes. It should be added that it is normal for the smallest nuclei to be more isolated and to lack more services. Thus, the relationship can be visualized as a question of urban hierarchy, with rural settlements occupying the lowest rank.

Changes in social structures and in mentality are strongly related to the increasing urbanization and the improvement of communications, as well as the crisis of the traditional system, the globalization of the economy, and the mobility of rural population.

The continuous loss of members of the community because of emigration is today, in the opinion of many ruralists, one of the most characteristic constants of the rural population. This combines, like the positive and negative of a photograph, with the steady increase of the urban population. This movement from country to city is certainly not a new phenomenon. It began to grow in the first period of the Industrial Revolution in the developed countries, as Ravenstein noted in 1885, and accelerated after the end of the World War II at practically a worldwide level. There are numerous examples to bear this out, as in the data presented by Clout (1972) or in the studies carried out for the Regional Conference on Mediterranean Countries about rural depopulation in 1986 (Faus and Higueras, 1986).

Cases were presented of several countries: Spain (very important after 1960), France, East Germany, Poland, Finland, Brazil, South Africa, Zambia, and Kenya. Even during a brief time period (Table 10.2) the relative drop in population could be clearly seen.

In developed countries the evolution is negative in absolute terms, where a slow process of depopulation occurs, hardly slowed by a halting return movement to the country, which has been called counter-urbanization. However, in those areas where the population continues to rise in absolute terms, this growth was caused by a positive biological balance that has not been sufficiently affected by emigration, although the tendency toward growth has been halted to some extent. In each case the planner should study the demographic contingencies and analyse the causes and effects.

The consequences of this migratory current are of great importance. There may be notable differences between developing and developed countries and also between regions in the same country. In developed countries (e.g., Australia, France, Denmark, Sweden) the emigration has profound repercussions on a rural area's biological and social structure, and also has a certain impact on the innovative abilities of the community. The more dynamic and enterprising individuals of working age tend to be the ones to leave the region, thus reducing the potential work force. This loss not only affects the work force active in agriculture but also reduces the quantity of individuals preparing for professions outside the agricultural sphere, or those who are working in the commercial or social service sector. The indicators of demographic structure in Table 10.4 show the different impacts that population loss have on different countries. What stands out is the marked tendency toward maturity or old age in the developed countries, with a Veyret–Vernier index higher than 0.4 (which demarks an aging from a

TABLE 10.4. Demographic structure indexes of some countries; data of 1988* (Source: UN, 1990)

Country	Total *RP*	*E*	*R*	*F*	*S*	*D*
Zaire	18,743,673	0.33	0.60	0.54	93.4	149.50
Tunisia	3,289,980	0.12	1.45	0.53	103.9	90.70
Colombia	9,124,379	0.13	0.71	0.44	113.4	53.67
Malaysia	6,871,739	0.11	0.66	0.42	99.3	83.14
Australia	2,083,360	0.31	0.35	0.24	112.1	57.17
Portugal	6,914,465	0.45	0.34	0.26	94.8	62.25
Spain	13,143,528	0.52	0.30	0.26	99.0	64.02
France	14,473,380	0.45	0.29	0.17	100.0	53.20
Denmark	826,897	0.64	0.30	0.18	113.0	53.19
Sweden	1,406,963	1.10	0.22	0.19	112.0	63.01

**RP*: Rural population; *E*: Elderly rate, Veyret–Vernier ageing index = (*PR* > 65 years)/(*PR* < 15 years), the threshold is 0.4; *R*: Replacement of generations rate = (*PR* 15–19 years)/(*PR* 40–64 years); *F*: Fertility rate = (*PR* 0–4 years)/(woman 20–24 years); *S*: Sex-ratio = 100 (male/female); *D*: Dependency rate = (< 15 years + > 64 years)/(*PR* 15–64 years)

TABLE 10.5. Demographic structure indexes of some Aragonese (Spain) areas* (Source: Instituto Aragones de Estadistica, 1991)

Rural area	*E*	*R*	*F*	*S*	*D*
Sobrarbe (Pyrenees)	1.92	1.15	0.28	114.97	61.79
Prepirineo (Pyrenees)	3.31	0.97	0.16	111.93	60.05
Monegros (Ebro plain)	1.55	1.08	0.27	105.22	60.75
Tierra de Belchite (Ebro)	3.20	0.83	0.19	103.82	61.44
Maestrazgo (Iberian System)	1.93	0.97	0.33	109.88	59.60
Daroca (Iberian System)	2.73	0.72	0.22	104.62	60.25

**E*: Elderly ratio (Veyret–Vernier ageing index); *R*: Replacement of generations rate; *F*: Fertility rate; *S*: Sex-ratio; *D*: Dependence rate.

young population), except Australia, and also a reduced level of fertility and replacement rates. Another feature is that of high indices of masculinity and dependency, as in the extreme case of Sweden.

In the developing countries (e.g., Zaire, Tunisia, Colombia, Malaysia) the situation is somewhat different, in that the high birth rate counteracts the aging level, which is normally below a threshold of 0.4. Thus, the age pyramid still presents a relatively youthful picture, although signs of the emigration of individuals younger than 30 years of age can already be discerned. The elderly rate is still at a relatively high level, although replacement of the generations is beginning to fall, while sex ratios are similar to worldwide and urban levels. The dependency rate shows similar values to those of developed countries under the same variety of situations.

Also worthy of consideration are those countries that have reached a reasonable economic level but have yet to attain the level of the richest countries. In these cases demographic behaviour is mixed, showing elements from the two groups previously mentioned. Such is the case in the so-called peripheral countries of the European Union. Portugal and Spain are two examples, as can be seen in Table 10.4. Finally one should mention the regional and local levels, where rural emigration has created an enormous demographic hole. The region of Aragón (Spain) is an example of this, the indicators of which are similar to developed countries but without reaching the same economic level. All this can be seen in Table 10.5, where some of the most markedly rural of the regions have been analysed. Many of these regions are in the Pyrenees or in the Iberian System.

Another effect of selective emigration is that of disarticulation and loss of cohesion in rural society, caused by the loss of a sizeable section of the intermediate generations, especially women, young people, and children. This effect can be even graver in tightly knit communities because of the difficulty in their accepting changes and their possible rejection of new residents that might regenerate the area, as Stevenhagen (1976) remarked. Obviously the family and social structure in each community is unique and this should be borne in mind.

The features we have considered are so characteristic of the populations of rural areas

TABLE 10.6. Index of rurality: Variables list

1. Population density (per acre or km^2)
2. Population change (% intercensus)
3. Population over age 65 (% total population)
4. Population men age 15–45 (% total population)
5. Population women age 15–45 (% total population)
6. Occupancy rate (% population at 1.5 per room)
7. Occupancy rate (households/dwelling)
8. Household amenities (%)
9. Occupational structure (% in socioeconomic groups)
10. Commuting out pattern*
11. In-migration (% population resident for less than 5 years)
12. Out-migration (% population moved out in last year)
13. In/out migration balance (%)
14. Distance from urban centre of 50,000 inhabitants**
15. Distance from urban centre of 100,000 inhabitants**
16. Distance from urban centre of 200,000 inhabitants**

*Residents in employment (%), working outside the rural district.

**The nearest urban centre.

that they have been used as a starting point for studies by many ruralists, such as Clocke (1977), for the creation of an Index of Rurality. This was applied to England and Wales, and the principal components were listed. Sixteen variables were found, the majority of which are demographic and are related through a multiple lineal regression (Table 10.6).

It is clear that Clocke's method is valid essentially for developed countries, although it could be adapted for others. However, the rural–urban gradient characteristic of developed countries does not function in the same way in developing ones. There, the discontinuity between both areas is more marked, and other methods should be used, such as Nelson's specialization index, which considers sectorial activity as a feature of great importance (Nelson, 1955).

10.2.2 Settlement Systems and Their Place in Rural Space

If it is essential to gain an idea of demographic behaviour in order to understand organization in rural areas to support planning, then the settlement system is the most immediate result of the human occupation of a territory. It affects not only the countryside, of which it is an essential component, but also the structuring of the production methods and the social relationships. The occupants of these settlements are a high proportion of the world population total, to the extent that at the end of the 1960s it was estimated that there were some 14 million rural settlements occupied by more than 2000 million inhabitants. Thus, two out of every three individuals at that time lived in villages, hamlets, or dispersed dwellings (Doxiadis, 1968, quoted by Pacione, 1984).

The basic unit of these *settlement systems* is the *rural house*. The house may be isolated or be a part of small groups which are ordered to a greater or lesser extent. They may also form larger-sized nuclei which show at a reduced scale some of the features of urbanism. During the first half of this century the rural house with its double function of dwelling place and site of production was the object of study by geographers and anthropologists. They looked for differentiations in ways of life, the countrysides, and the social importance of the farms. They also studied the connections between the inhabitants and the environment, the cultural or ethnic inheritance deduced from the size and layout of the settlements, the distribution of out buildings, the materials used in construction and the ornamentation, including its symbolic value.

The changes that have occurred in the rural world, new research methods, and a marked shift to an economic view in the social sciences relegated this field of research to a second place position. Now, however, a renewed humanistic interest and a more modern focus have again stimulated study of the traditional rural house and its value as an item of cultural heritage. This has awakened the interest of other professionals (e.g., urban and rural architects, sociologists, economists), who have made in-depth studies of the subject. They have also protested against the distortions produced by the process of standardization of the new suburbs and the rural dwelling itself. Thus, the rural house is now considered as a valuable resource, from a social and touristic perspective. This is the case not only in the developed countries, but also in the world as a whole.

The key to settlement systems is the distribution of houses in a concentrated or dispersed, ordered or random fashion. The implications of the diverse systems are important not only because they establish differences between cultural landscapes, but also because they presuppose certain demands on the society and the administration. These demands take the shape of the need for appropriate communication networks, different conceptions of the services that should be available to the settlement, and other problems with their possible solutions.

Thus, it is understandable why this subject inspired so much interest among researchers, especially geographers, who sought explanations for the diversity of settlement systems where cultural, physical, and historical factors were concerned. They attempted to systematize and map the resulting typologies by means of simple mathematical formula, which related the number of buildings or homes with the number of inhabitants and the land surface occupied (Demangeon, 1920, 1928, 1939; Bernard, 1931; Colas, 1945). All of these methods pose problems: for example, different data can be combined to give indices of equal value, and the scale of references is imprecise. Some authors have pointed out these problems (Tricart, 1956, Bonnamour, 1970; Suarez Japón, 1982).

In the 1960s new methods were introduced that concentrated on the use of locational analysis to classify settlement systems, relating the dispersed or concentrated character of the settlement to the society's behaviour and type of diffusion in the area. Christaller's theories of the search for regularity or variations of Bylund's models (Haggett, 1976; Pacione, 1984) were the bases for this new method.

Also, the variables under consideration were formalized, the most common method utilized being the Clark-Evans model, also known as the *nearest neighbour* model. This is based on the theory that the human group expands into a territory from a primary settlement. The expansion takes shape depending on the existence of resources, ease of expansion, and so on, and leads to a concentrated occupation with all the settlements in close proximity, or an ordered or random dispersion expressed as:

$$R = \frac{\overline{D}_0}{\overline{D}_a}$$

where:

$\overline{D}_a = 1/(2\sqrt{n/s})$
$\overline{D}_0$ = average real distance between the nearest nucleii neighbours.

The range of values of R, which vary from 0 to 2.15, and how it is related to the settlement system, can be seen in Figure 10.2. It must be pointed out, however, that this type of index of locational character also has limitations. It only concerns the dispersion of settlement points in space; it fails to reflect the relative importance of those settlements, which constitutes a particularly important aspect of any organization on the space based on given degrees of connectivity, access to the resources, and services. However, some of the above-mentioned methods can be combined to produce a more realistic description.

Any system of settlement has its dynamic characteristic, because the society that creates it modifies it and shapes it in line with its own evolution. It is important to consider two different stages in any analysis that is carried out because any change in the period can be noted, and from then on trends can be discovered or simulations established for a programmed intervention.

Studies carried out on population changes show a tendency for the population to be concentrated in fewer nuclei, beginning with the abandonment of scattered habitats, those nuclei that are very small, with few services and poor connections to the general communications network, as noted by Clawson (1966). This process is particularly noticeable in long-established populations, as in Europe. A case in point is the Province of Teruel (Aragón, Spain), the greater part of which is situated in the ranges of the Iberian System (Frutos et al., 1995), a zone that has undergone one of the most significant processes of depopulation in Spain. This example is supported by other studies on Spanish and European regions. They have been called automatic trends (Clout, 1972; Pacione, 1984) since they, in essence, are adjustments to the new socioeconomic organization of extensive areas.

It must be added that the transformation of the system of settlements is not only a result of depopulation but is also consistent with new settlements, whether caused by official programming or because of spontaneous population shifts. The latter could be linked with the occupation of land in *pioneering fronts*, nowadays infrequent although it is worth mentioning the case of Israeli *kibbutzim* or Brazil's rainforest migrants, or with the migratory flow from the most unproductive areas to potentially better ones. This occurs in some underdeveloped countries with an imbalance in economy and resources.

This is the case with the so-called *young towns* in Peru, precarious settlements generally on the fringes of the city where the settlers take possession of a small space and fence it to acquire squatters' rights and establish residence. The developed countries are also experiencing a repopulation of rural areas. Not only is this a case of individuals returning to the villages—what Kayser (1990) has called *the renaissance of rural areas*—but also whole family groups are occupying depopulated and abandoned nuclei, and these groups represent the true essence of those who have come to be called neorurals.

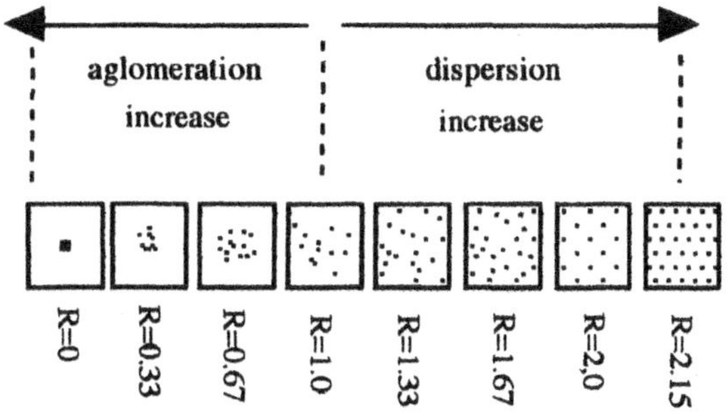

FIGURE 10.2. Clark–Evans index. *R* values.

Where official programming of settlement networks is concerned, this is obviously governed by territorial planning and development incentive measures. These may depend on various factors and always produce an obvious impact, as Rawson (1981) points out in the case of the rural areas of Britain, where this process began early; or as noted in the studies of Newman (1964, quoted by Clout, 1972) in connection with Ireland, France, and the Low Countries, Martínez Cordón (1988) in Aragón (Spain), or Lassailly-Jacob (1984) in the case of the Ivory Coast.

Making decisions in this area is not something to be taken lightly as there is much weighing of the social, economic, and political advantages or disadvantages of a dispersed or concentrated habitat, and these are taken into account by the different administrations before deciding which model to adopt. Thus, Pacione (1984) mentioned four possible types of settlement considered by the *Norfolk County Council* in 1976. These were defined as: (i) concentrating growth in the largest settlement; (ii) dispersal-1, allowing a standard rate of growth in all villages; (iii) dispersal-2, making use of spare capacity in services taking account of physical constraints; and (iv) dispersal-3, developing villages of 500–1000 population. Each one of these implied a different cost to the community in services and variations in social benefits.

In the same line, Clocke (1979) pointed out two trends in the planning of the rural population. The first trend leant toward the concentration of settlements, which usually produced a reinforcement of the local authority (administration and services), implying social and economic benefits. The second trend represented the alternatives that dispersion offered in connection with functional interdependence and maximum access to services for inhabitants of dispersed settlements.

However, settlement planning may also be responsible for a change in use of a rural area, such as the draining of marshes (the Italian *boniffica*) or the installation of irrigation. Different solutions have been adopted when weighing up the advantages or disadvantages of creating a particular type of settlement. Thus, while in Tunisia it was decided to create a zone with a network of detached houses, although close to each other, in areas colonized by the Spanish, with irrigation installed, a series of villages with new buildings was set up. These villages were small (no more than 600 inhabitants), were close to each other and the municipal capital, and the colonists were those who were to work the land. The case studies were based on the Badajoz Programme (Extremadura) and the Bardenas–Upper Aragón Programme (Aragón) (de los Ríos, 1966; Rodríguez Cancho, 1984).

In addition, the increase in urbanization and the existence of a society with more leisure time has meant the springing up in rural areas of small nuclei of holiday homes which have modified the countryside in numerous areas and have introduced a new element to affect the countryside and traditional inhabitants of the rural environment, as in the synthesis by Clout (1972) and Pacione (1984).

10.3 Landscape and Agricultural Systems: The Basis of Human Land Organization

It has already been mentioned how large a part agriculture plays in the human organization of rural space, as the source of foodstuffs and raw materials. But it is also important to consider the part that other social, political, and legal factors play when decisions are made by the farm-workers and the influence these factors have on the diversity of uses of rural space. Thus, not only should the population, with its dynamic, socioeconomic structure and settlement system, be studied, but also the forms of land appropriation and transmission. These are the basis of a greater or lesser fragmentation or parcelling and the legal system that regulates land exploitation and the management of agrarian transactions.

The analysis of agrarian structures, be it of exploitation or property, make up one of the factors to be considered in any rural project. Imbalances in land usage, a feature that is characteristic in developing countries and others with social problems, have brought about reforms and agrarian revolutions. These have been mentioned in this chapter, and the characteristics of exploitation make up one of the variables used in multiple typologies and organization models of rural spaces. This is not only because of the economic connotations involved, but also land possession has long been, and continues to be, a mark of social status. Thus, it is obligatory for any planner to be cognizant of these factors as a deep understanding of the situation is necessary for it to be improved, and, at the same time, any change is likely to have unforeseeable socioeconomic repercussions.

10.3.1 The Sources: Agricultural Statistics and the Census

Access to information that would lead to a deeper understanding of the forms of land appropriation and exploitation is one of the serious problems at a general level. Information is not always available, and the data that are available are at times unreliable. The systematic registering of properties, with their boundaries clearly defined and the type of ownership described, in which it is stated that the owner exploits the land him/herself or leases it out to third parties, make up a complex quantity of data. This is brought together by land registries and agrarian censuses and is to be found in many statistical sources. These registries are fairly complete, relatively accessible, and are acceptable as source material in developed countries. In developing countries it is a different matter. There, the data are more irregular. In many countries of the Third World there are no official publications on the subject and any knowledge of the matter is acquired by means of incomplete surveys carried out by the country's government or by international institutions such as the FAO.

From their inception, these registries did not have a purely informative purpose. They were also fiscal in nature. There exists the possibility of confusion between the terms register and census, despite both words having relatively recent origin in terms of meaning. The register is a compilation of the farms in a given region for tax purposes, while the census is a statistical survey for obtaining information. Though the first registers and censuses are of ancient origin, between the end of the nineteenth and the first half of the tentieth century the land registers of the western countries were instituted and they are now studied with other intentions in mind, the analysis of which enables one to understand some of the present-day problems of property and exploitation structure.

Present-day agrarian censuses with systematized information constitute the most accessible source of information and are thus the most frequently used, not only in research work but also in planning and development projects, for the application of agrarian policy. But when using these sources one should also consider their reliability and the possible problems resulting from the criteria used in their elaboration. The most common method of information gathering is that of the survey, which the farmer must by law to fill in.

In some countries these types of statistics were initiated in the last century, but it is more frequent to find documents established more recently (Ilbery, 1985). These statistics include information not only about the types of exploitation but also types of ownership, soil uses, cultivations and livestock, what type of work is done by whom according to age, sex, and the degree of mechanization. Thus it is possible to discern the *spatial variation in farming over wide areas and to study the evolution of these changes* (Clark et al., 1983) although problems have also arisen for researchers, which have been commented on by many ruralists (Ilbery, 1985; MAPA, 1987; Kayser, 1990).

The subject matter of conventional censuses includes the exploitation structure, the total of land registered, the useful agrarian surface (UAS) and the soil use, the classification of exploitations according to the number of plots and the technical and economic orientation (TEO), the different types of land tenure (direct ownership, rent, sharecropping and others), the management, and the work. Information about livestock refers to he number of heads of different species and livestock units (LU). Finally, technical data is given about irrigation and the land area that it covers and machinery used, either owned or rented.

However, this apparent plenitude of information quickly loses its value when one attempts to use it for planning at the local level; this is one its principal limitations. A large amount of this information cannot be applied to the smaller units of administration such as the municipalities. Other problems also exist: the survey methods do not always give optimum results, it has been shown that false answers and a large number of omissions occur. This obviously makes the results less reliable.

Another important limitation is the change of criteria in the formulation of the different censuses, which can be seen in the census sections. This makes comparison of concepts difficult and it is also hard to obtain a consistent evolution throughout the period registered. A final problem is that of the long time period between censuses; although revisions of the censuses are made from time to time, diffusion of the material is less widespread.

Many of the criticisms made, which could equally be applied to the censuses of other countries, are applicable to the new publication, according to the TEO. Perhaps some of the problems could be mentioned here. Firstly spatial desegregation is minimal, the published information is only at the national and regional level. Secondly, the researcher does not have a detailed knowledge of the data base (the selection or sample of exploited lands, the calculation of which has been used to calculate the gross standard margin), which is important given the heterogeneous nature of the physical conditions and output of countries like Spain. It should also be added that peculiar slants have been detected in the principal technical orientation in both cultivation and livestock. This is caused by the pluriactive nature of Spanish exploitation and tends to falsify or invalidate the data to some extent.

All that has been mentioned above makes it clear that censuses are an obligatory source of information but that they should be used with caution. Verification, checking, and comparison by means of field work are necessary before coming to any definitive conclusions. This is even more important when planning for the development of a zone. Of all the other data of this type in national or international statistics, be they yearly general or agrarian reports, the most notable for their continuity are those of the EU and those published by the FAO (Kostrowicki, 1980).

10.3.2 Land Property Classification

Land appropriation is the first activity of any rural society that organizes its territory and this appropriation is subject to constant adjustments and changes, as is society itself. There exists a wide range of ownership types, each one bearing a certain socioeconomic status. The fundamental variables that should be

TABLE 10.7. Agrarian property typology (based on López Ontiveros, 1983, and Molinero, 1990)

Collective property	Communal	Tribal, patriarchal Of township Of town council	
	State	Socialist	Kibbutz, moshav chitufi, and ovdim (Israel) Collective farm (people farm)(Cuba) Self-management farm (Algeria) Others
		Capitalist	Public property land Of state inheritance
	Private legal entity	Agricultural co-operatives Other companies or associations For purchase of communal land (resident associations); undivided	
Individual property	Large estate Big property Middle property Small property Small farm		

known about property systems are size and the public or private, individual or collective, nature of the judicial process.

This is not merely an academic exercise for the simple reason that the balance or imbalance in the division of goods and power and their resulting effects on the landscape depend on these factors. Land control also signifies its adaptation to the uses that the human group impose on it by means of ploughing, clearing, or destruction of the indigenous vegetation.

Ploughing and land control are so closely united that many documents testify to the giving of virgin lands which were owned by a tribe, a government, or member of the aristocracy, which are then ploughed and cultivated. This is as valid for Medieval Europe as it is for the colonization of America. Also, demographic pressures in a territory lead to the ploughing of lands that were generally once common property. Technical advances also enable societies to plough new areas with different techniques (George, 1975) and cultivate them.

When looking at the varieties of property ownership from a juridical point of view, they can be divided into two broad groups: the collective and individual, which in their turn contain a varied typology. This is what Kostrowicki (1980) means when he refers to *common and social property*.

In an attempt to systematize this typology, Molinero (1990) and López Ontiveros (1993) elaborated a table that took elements from other studies. I introduced some modifications as a result of the recent changes in Russia and Central/Eastern Europe and used the numerous studies of tenure as a basis (Table 10.7).

It is clear that this plan is something of a simplification of a very complex phenomenon. A more expanded treatment of the subject, concerning countries and socioeconomic systems, is in the work of Meynier (1958), Rodríguez Castro (1979), Kostrowicki (1980), Golte (1987) and Cebrían (1993). State property constitutes the base of agriculture in socialized systems and forms the patrimony of public utility in the remaining, where it includes Natural Protected Spaces. Collective property may be a survival of more or less modified traditional systems, as is the case of the Mexican *ejido* or the *communal* or *municipal lands* in Spain, or under new forms, as the cooperatives, which involves a large number of farmers.

The diversity of types of land ownership gives, as a result, differences in the degree of concentration of lands in owner's hands with socioeconomic consequences. That degree can be quantified by means of Gini's concentration index, which can be obtained either from the computation of the concentration area of

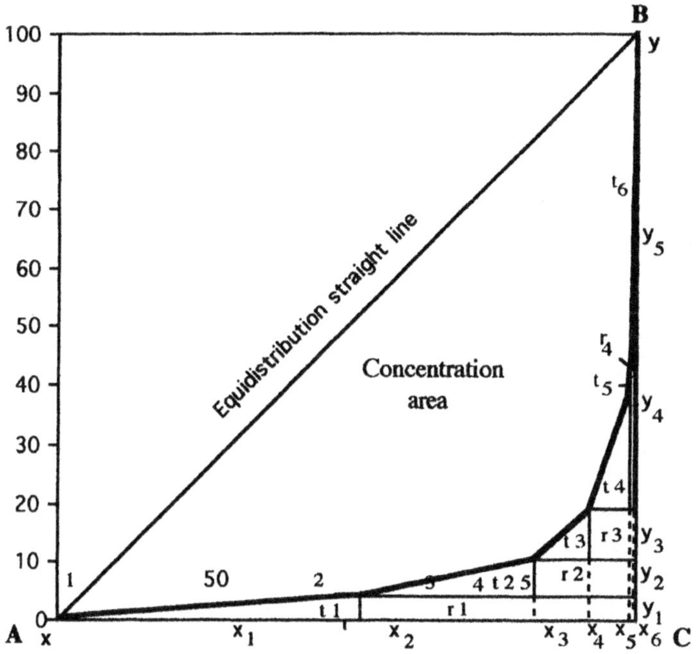

Official land data

classes	x % number properties	y % hectare for class	Properties accumulate %	Hectares accumulate %
<1	53.3 = x_1	4.2 = y_1	52.3	4.2
1-5	30.2 = x_2	6.3 = y_2	82.5	10.5
5-10	9.2 = x_3	8.4 = y_3	91.7	18.9
10-50	6.7 = x_4	18.8 = y_4	98.4	37.4
50-100	0.8 = x_5	8.7 = y_5	99.2	46.4
>100	0.8 = x_6	53.6 = y_6	100.0	100.0

$$\text{Index} = \frac{2.\ \text{Concentration area}}{\text{Square area}} = 0.8589 \ (\text{high concentration})$$

$$\text{Square area} = 100.100\ ; \qquad \text{Concentration area} = \left(\frac{100.100}{2}\right) - (\Sigma\, t_i + \Sigma\, c_i)$$

FIGURE 10.3. Lorentz's curve to calculate the concentration area and Gini index of land property in Spain. The axes x and y represent the accumulate percentage of properties and the accumulated percentage of total land (ha), respectively. Symbols: t_i = triangles; r_i = rectangles. (Source: Official land data in Saenz Lorite, 1988.)

Lorentz's curve or directly from the formula:

$$G = \frac{1}{2}\sum_{i=1}^{n}\sum_{j=1}^{n} |x_i - y_j - x_j - y_i|$$

In the Lorentz's curve (Figure 10.3) the concentration is expressed by a nondimensional value (Gini index) which ranges between 0 and 1, from less to more uneven distribution. The concentration area is the difference between the area of the triangle ABC minus the area under the curve, which is the sum of the areas of the triangles t_i and the rectangles r_i.

This concentration is very significant in ex-colonial countries (Table 10.8), as is borne

TABLE 10.8. Gini index: Land property concentration* (Source: Based on Dourojeanni, 1982)

Index values	Latin America (%)	Other nonindustrial countries (%)	Industrial countries (%)
>0.80	70.5	14.3	12.5
0.70–0.79	29.4	19.0	25.0
0.60–0.69	0.0	33.0	25.0
0.50–0.59	0.0	19.0	18.7
<0.49	0.0	14.3	18.7

*Total number countries sample: Latin-America 17, other non-industrial countries 21, industrial countries 16.

out in some Latin American countries (Dourojeanni, 1982). This situation also exists in some European countries, such as Spain or Italy, with Gini indices that are higher than 0.75. These examples tend to show the dichotomy of a high number of farm workers with less than 5 ha in comparison to a low percentage of large estate owners, with properties of more than 1000 ha who control between 35 and 95 percent of the land. A slight reduction of these figures has been registered, but, nonetheless, it is not difficult to understand why many agrarian reforms in the Third World have concentrated on land redistribution.

10.3.3 Land Tenure and Farm Structure

10.3.3.1 Land Tenure

Despite the importance supposedly given to land ownership as a social basis of the rural environment, the agricultural enterprise does not identify with this system because, by utilizing its rights, many owners lease a part or all of their property to third parties, who exploit the land subject to regulations established by written law or by custom. Using the available data on the relationship between the head exploiter and the land it is possible to discern three dominant types: (i) direct ownership, which constitutes an important segment in some countries, making up to 80 percent of the total; (ii) tenancy, a system that implies the payment of a fixed sum to the owner; and (iii) sharecropping, a large percentage of which is the historical residue of the relationship between owners and tenants.

At present its peculiarity in the western countries is the payment of a variable amount which is proportional to the size of the harvest. The most common type of payment is in kind or in cash, according to the agreement established between both parties (Kostrowicki, 1980). Other types, such as the colony, the regime of common lands, or trusteeship are, relatively speaking, of little importance, although in some zones they may grow to be more significant.

Indirect systems imply a division between capital (land) and work (the farmer) and vary in the intensity of their effects depending on the country. There are some ruralists defending or against of one or the other land-tenure systems (Fromont, 1961). What is without doubt true is that these different forms of direct and indirect land ownership have social and economic repercussions, and there is evidence to support both points of view. Thus the planner should bear in mind the specific situation in each case and analyse the types of land ownership in the area of study.

The types of land ownership are subject to the same dynamics as the other factors in the environment and rural society. There has been a notable evolution throughout this century, which has increasingly limited the sharecropping system. This system is a remnant of outdated situations but in some tenanted areas has begun to grow again as a result of the marked growth of rural emigration.

10.3.3.2 The Farm Structure

From what has been stated, it can be deduced that agricultural exploitation is the result of a legal system of ownership and tenancy whose structure makes up the basis of the agricultural production system, of indisputable social and economic importance. Thus, it is essential for the planner to carry out a study of the situation in the area in question, analyse the results to detect any problems and possible solutions, arrive at a diagnosis, and, if necessary, propose any actions to modify the structure.

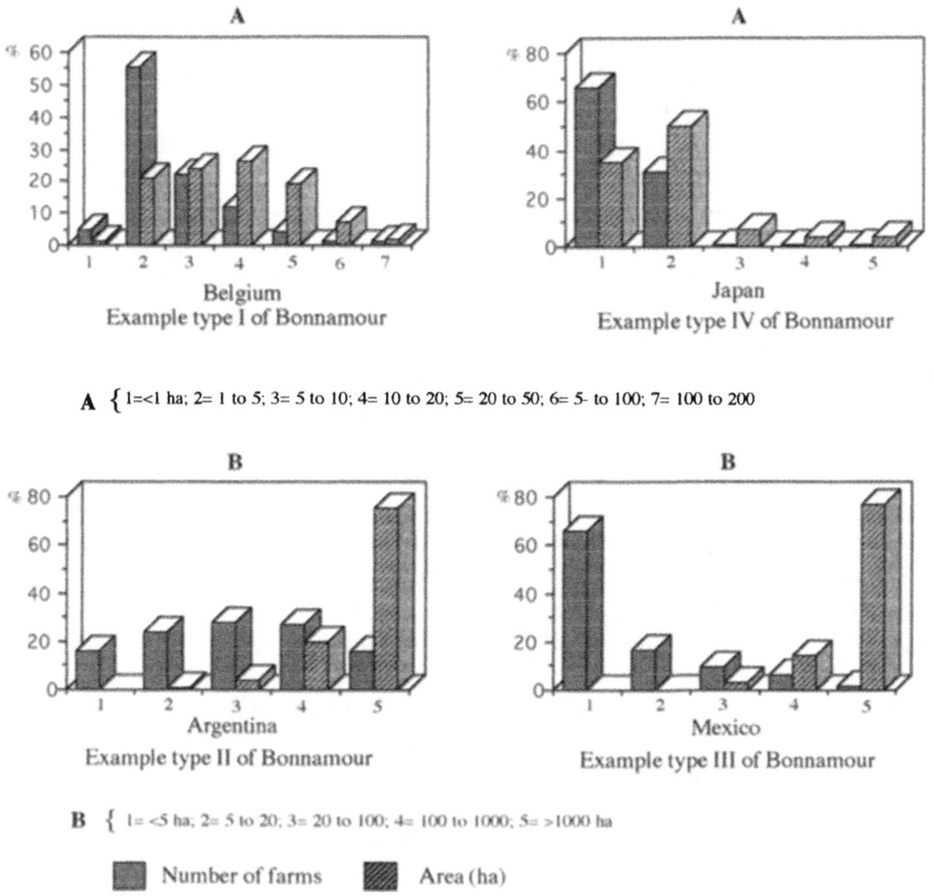

FIGURE 10.4. Examples of farm typology. (Source: A, FAO data in Bonnamour, 1970; B, CEPAL data in Méndez and Molinero, 1991.)

The typical concept of land exploitation takes into account, as with any piece of real estate, the physical area. Comparisons, both past and present, can be made by taking the average area as the best approximation. However, if the dispersion index of this value is unknown, a serious misreading could result.

Frequency histograms express in more detail, and in relative values, the number of exploitations of each class and the accumulated area they occupy. Using this method Bonnamour (1970) chose four reference models, with various subtypes in each, according to the resultant distribution of the land and interpreting them in terms of the socioeconomic relationships reflected. These models make it possible to compare the frequency curves of the structure of the exploitations of any agrarian system analysed and to interpret the socioeconomic data as indicated by each type or subtype as the starting point for a suitable structural policy.

This policy may be by means of a better distribution of income deriving from confiscation or redistribution of land. It may be by orienting those excessively small land holding systems towards more economically viable sizes by stimulating the formation of cooperatives or other forms of association by programmes that help to reform the structure in question, for example, the creation of Land Banks to which small holders would have access by means of buying or renting, or by contributing to early retirement schemes of the chief farmers, whose land might then be added to other exploitations. In these models, all types of land are considered, but they can also apply exclusively to exploitable agricultural land or tilled land.

The four groups that Bonnamour (1970) pres-ents, set out graphically in Figure 10.4, will now be explained. Group I, whose data correspond to European countries, is typically composed of small and medium-sized farms, with an absence of large land areas, but with a fairly even land distribution that permits their classification in the previously mentioned subtypes. The socioeconomic interpretation of this first group is straightforward: income distributed fairly evenly between these family exploitations whose main problem is their smallness and whose efficacy depends on intensification, on the correct use of adequate technology and on marketing know-how.

In Group II, European and American countries, a less even distribution of land can be noted. This unevenness becomes more pronounced in Group III, where the predominance of large landowners is increasingly marked. With respect to the last two groups mentioned above, the socioeconomic problems of the agrarian system derive from the growing dichotomy between the large number of very small exploitations which account for a small amount of land (scarcely enough for a family to survive on), and a small number of large holdings which account for the greater part of the land under cultivation and furnish permanent or temporary employment for landless day-labourers.

It is clear that this structure entails an unfair distribution of income, which will make itself felt in discontent and social problems. By way of righting these imbalances, the administration's response, often forced by the poorest peasant labourers, is agrarian reform, which will be discussed briefly below.

Group IV is somewhat exceptional, referring as it does to areas of almost absolute control of small and medium-sized farms. However, while in some cases this means an even distribution of land (Japan), in others it refers to a structure of a colonial and even primitive nature, collective lands with a very high number of small farms controlling very small areas; and then there are a few—sometimes only one or two—of medium to large size that occupy an extensive area.

The problems associated with this fourth group have the odd point in common, but are, in essence, of a different nature. In the first case, where small exploitations are predominant, it is necessary to promote cooperative systems and other systems previously mentioned. In the second case it is evident that profound agrarian reform is the only possible measure for the improvement of the smallholders' predicament.

However, this model of analysis is only valid as training, since agricultural ventures of equal size may have different economic dimensions, as shown in Figure 10.5 with data from the EU. Thus it must be understood

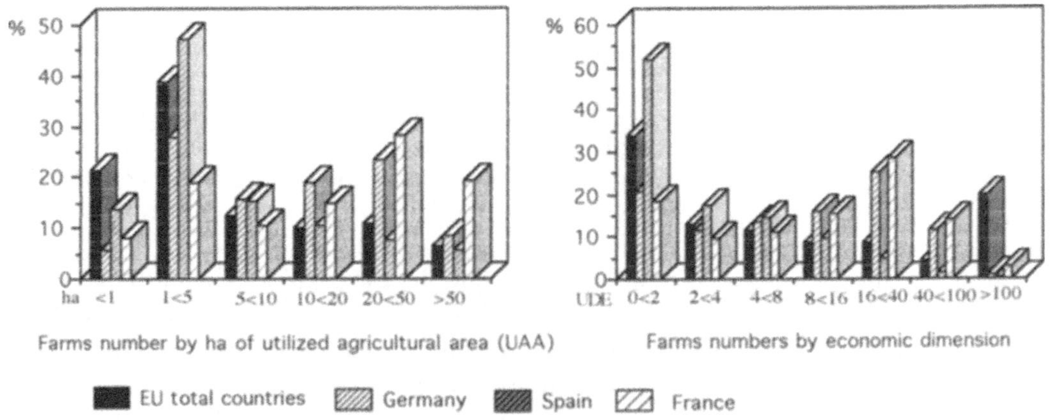

Figure 10.5. Distribution of farms by size groups in the EU and other countries. (Source: Eurostat, 1995.)

that the area of a property is only a variable in consideration of its viability, because a minimum size is necessary to be able to achieve an economy of scale (Morgan and Munton, 1975). Other variables to be considered, all of them interrelated, are: natural resources, the intensity of use and techniques employed, the uses the land is put to (crops only, mixed farming, etc.), and the market value of the produce—whether the market is free or regulated by a particular agricultural policy, as was mentioned when discussing property. Therefore it is not possible to establish a general scale of reference that would provide a minimum threshold or an optimum size. Mansholt (quoted by Clout, 1972) made this clear in his report on European Community agriculture. A concrete analysis of the zone being studied would be necessary before any conclusion could be drawn.

The limited breakdown of existing information, as was stated when analysing sources, enables us more easily to reach macrospacial, macrosocial, or macroeconomic conclusions. But if we want to approach the study on a local scale, it is necessary to resort to microanalysis, implying extensive fieldwork. As regards the member countries of the EU, consultation of the Agrarian Accounting Network may well be of great use when interpreting the viability of the farms in specific areas. However this network does not exist in every country, so a direct survey is necessary.

The farms structure is not static, but is subject to the same dynamism as society (Bowler, 1983), as numerous local and regional studies have shown, collated very efficiently by Molinero (1990), Méndez and Molinero (1991), and others. In European countries there has been reduction in the number of exploitations as a result of the rural exodus, with the corresponding increase in the average size and a tendency towards concentration in medium-sized to large farms, although perhaps not so rapidly as expected (Bonnamour, 1970; Calmes, 1981). In Japan, on the other hand, demographic pressure brought about intense fragmentation after the war, while in the United States, although with different characteristics between the states, there has been an increase in large farms as a result of the setting up of capitalist ventures, veritable agribusiness. In Latin America, economic crises and inflation have ruined many small and medium-sized farmers, thus reinforcing the power of the big farms, although they may have lost some of the characteristics of the traditional *latifundium*.

In short, although one might have expected a rationalization of agricultural ventures, especially after recent structural reforms in many countries, this has generally not come about. In addition, there has been a significant proletarianization of the peasant class, both those who have small holdings, eking out their existence in the more dynamic countries where agriculture is only one of their occupations (OECD, 1978; Etxezarreta, 1985), as well as day labourers, often without a steady job, and whose number is growing in underdeveloped countries, where the large ventures hold sway.

10.3.3.3 Parcelling Out Land

The morphology of land division is the clearest visual expression of an agricultural system in the countryside. A parcel of land is defined as that extension of land within one single boundary; in other words, surrounded by lands, buildings, or water not belonging to that exploitation, and thus is considered to be the technical unit of that exploitation. It may be fenced or not, but its boundaries are sufficiently clear when the land parcels are cultivated.

In countries with a long agricultural history, land parcelling is like a palimpsest, resulting from the superimposing of different stages of ploughing and conquest of new land, whether adapted to the topography, stages of demographic pressure, introduction of certain techniques, and changes of use of the land or not. In the same way, it is a reflection of the exercise of the right of transfer, regulated according to type of ownership and the hereditary system of an area, which has split the countryside into thousands of land parcels in areas where there is traditionally the right of division.

In contrast, in the large, sparsely inhabited spaces of the so-called *new world countries*, whose land was dedicated to mixed farming

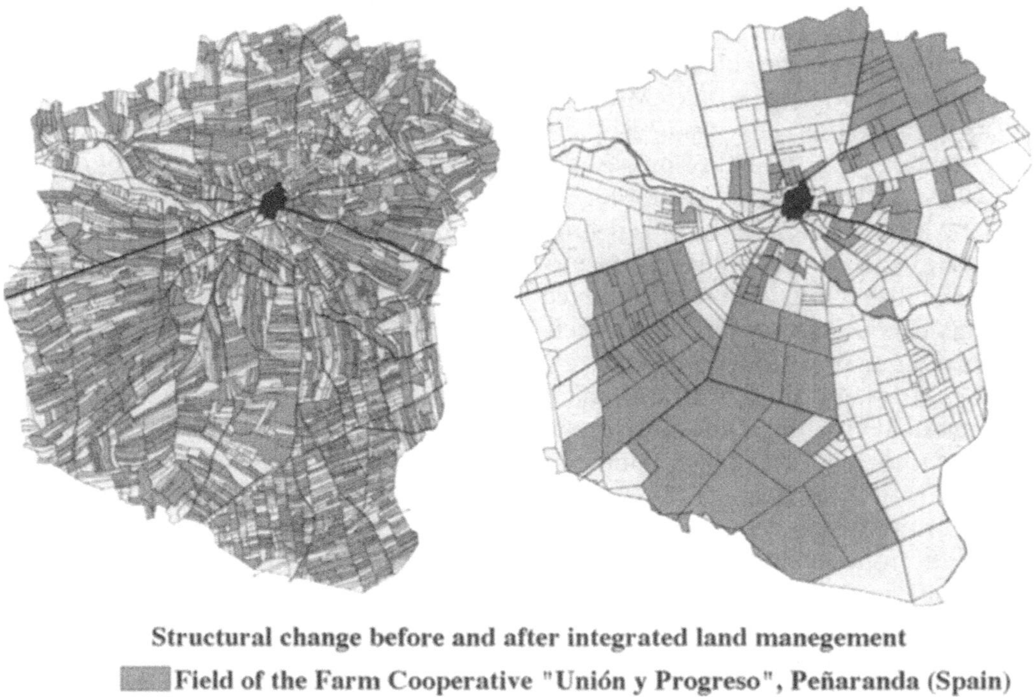

FIGURE 10.6. Change in the field structure. (Peñaranda, Spain). (Source: IRYDA, 1973).

from the nineteenth century onward, the land parcels are extensive and uniform, such as the ranches of the United States or the haciendas of Argentina, and are even unrecognizable in the countryside, such as the vast Australian pasturelands. A similar land-parcelling morphology can be seen in the Soviet structure originating from the 1917 revolution.

As a result of a long process, many agricultural ventures are formed by various small land parcels, often dispersed over an extensive area. The number and size of the parcels making up an exploitation and to a lesser degree the existence or otherwise of fencing, represent a complementary factor of agricultural structures. Many ruralists have highlighted the socioeconomic and even environmental impacts of these structures (Chisholm, 1972; Pacione, 1984), emphasizing the difficulty in rationalizing a farm divided into small scattered fragments, the physical energy expended in the farmers' efforts to work them, the environmental impacts that have resulted from the elimination of hedging to enable modern machinery to gain access to the plot, or the installation of electric fencing to control stock with less labour.

The problems presented by an excessively fragmented and dispersed land-parcelling system have forced many European countries to undertake technical reform oriented toward concentrating existing land parcels into larger ones, so as to create compact units more in line with present-day agricultural practice (Clout, 1972; Bowler, 1983; Pacione, 1984).

France, where in 1940 it was estimated that 14 million ha were in need of concentration, was a pioneer in this effort, affecting half the territory in the 1960s. The initiative in applying for concentration is left to the owners, although the Government encourages these requests and later acts on them. In Spain, where the system is the same, concentration was begun in the 1950s and has now been completed in extensive areas in the interior (Figure 10.6), although there is still much to be done (Arias Abellán, 1983; de la Riva, 1990; Alarios, 1991; Frutos, 1994). This process is also under way in Italy and Germany, as one more factor in planning.

10.3.4 Agricultural Reforms

Social and economic stagnation in conjunction with the structure of farm ownership have been the catalyst for many governments to initiate various types of agricultural reform. At the same time, it is a crucial problem that planners should bear in mind because any rural development program, even when proposed as sustainable for the benefit of the environment, has its base in a reorganization of agrarian structures according to their characteristics and imbalances; aspects which have already been considered.

For this reason it is helpful to know more about some of the changes or reforms that have been brought about in different countries so as to give greater orientation to the same reforms in the future, and to correct any defects. These reforms have been oriented toward structural modification in order to achieve greater social equality and greater economic efficiency. At times these reforms have been successful, but at other times they have failed.

Land reform can be broadly defined as "all changes relating to agriculture which have been introduced in an evolutionary or revolutionary manner by public administration or private institutions" (Kostrowicki, 1980). More precisely, George (1970) defines it as "l'ensemble des opérations visant à transformer les rapports fonciers, donc économiques et sociaux dans un Etat, une région, une communauté agraire, par la libération des contraintes sociales devant permettre une amélioration quantitative et qualitative de la production agricole. Au sens le plus étroit du terme, elle consiste dans la redistribution de la propriété foncière, par le partage des 'latifundia' ou la modification des modes de tenure."

In its most explicit sense, Gutelman (1981) defines it as a social product, resulting from a measure that means expropriation of land from one social class and the transfer of this land to another class, whether compensation is paid or not.

In any case, an agricultural reform is not something to be taken lightly, because of its many environmental, economic, social, and political implications. It is a measure with diverse degrees and focuses. From mere morphological transformation (as in the case of the previously mentioned land parcelling) to complete revolutions that imply a radical change in the system of ownership and exploitation of the land. From partial reforms of a technical nature which bring about the transformation of land by means of irrigation or fertilization, but with social repercussions if we include the distribution of land and settlement in colonies, such as the Spanish, Italian, or Tunisian settlement schemes, to reforms which exclusively affect the structure of land ownership.

We cannot go into this topic in detail, but it has been addressed by various authors, such as Le Coz (1975), Molinero (1990), and Méndez and Molinero (1991), as well as others who have realized analyses of particular areas.

References

Alarios, M. *Significado Espacial y Socioeconómico de la Concentración Parcelaria en Castilla-León.* MAPA, Serie Estudios. Madrid: MAPA, Secretaría General Técnica; 1991.

Arias Abellán, J. *La Política de Concentración Parcelaria: Un Ejemplo Andaluz.* Granada: Instituto de Desarrollo Regional, Universidad de Granada; 1983.

Arnón, I. *La Modernización de la Agricultura en Países en Vías de Desarrollo.* México, DF: Limusa; 1987.

Barbero, G. *Land Reform in Italy. Achievement and Perspectives.* FAO Agricultural Studies No. 53. Rome: FAO; 1961.

Becker, B.K. Agricultura de subsistencia, nueva agricultura y hambre en América Latina. Una visión a partir de Brasil. López Ontiveros, A.; Naranjo Ramírez, J., eds. *Propiedad, Actividad Agraria y Medio Ambiente en España y en América Latina; Vol. II Ponencias.* Proceedings of VII Coloquio de Geografía Rural, Cuarta Ponencia, Serie Estudios de Geografía No. 9. 1994 March 22–25; University of Córdoba, Spain. Córdoba: AGE/University of Córdoba; 1995:149–169.

Bernard, J. Une formule pour la cartographie de l'habitat rural avec application au Département de L'Yonne. *Proceedings of International Congress;* Vol. III. Paris: UGI, CRCI G; 1931:108–117.

Bonnamour, J. *Structures Agraires.* Paris: CDU; 1970.

Bonnamour, J. *Géographie Rurale. Position et Méth-*

ode. Recherches en Géographie. Paris: Masson; 1993.

Bowler, I. Structural change in agriculture. Pacione, M., coordinator. *Progress in Rural Geography*. London: Croom Helm; 1983:46–73.

Calmes, R. L'évolution des structures d'exploitation dans les pays de la CEE. *Ann. Géogr. XC* (500):401–427; 1981.

Casas Torres, J.M. *Población, Desarrollo y Calidad de Vida*. Madrid: Rialp; 1982.

Cebrían, A. *Las Regiones del Sistema Mundo*. Murcia: Universidad de Murcia, Secretariado de Publicaciones; 1993.

Cherry, G. *Rural Planning Problems*. London: Leonard Hill; 1976.

Chisholm, M. *Rural Settlement and Land Use*. London: Hutchinson; 1972.

Clark, G.; Knowles, D. J.; Phillips, H. L. The accuracy of agricultural census. *Geography* 65: 115–120; 1983.

Clawson, M. Factors and forces affecting the optimum future rural settlement pattern in the United States. *Econ. Geogr*. 42:283–293; 1966.

Clocke, P.J. An index of rurality for England and Wales. *Regional Studies* 11:31–46; 1977.

Clocke, P.J. *Key Settlements in Rural Areas*. London: Methuen; 1979.

Clout, H.D. *Rural Geography: An Introductory Survey*. Oxford: Pergamon Press; 1972.

Colas, R. Répartition de l'habitat. *Bull. de l'Association Géographes Français* 1:51–56; 1945.

De la Riva, J. La política de concentración parcelaria en Aragón. *Estudios Agrosociales* 151:191–236; 1990.

De los Ríos, F. *Colonización de las Bardenas, Cinco Villas, Somontano y Monegros*. Cuadernos de Aragón. Zaragoza: Diputación Provincial de Zaragoza, Institución "Fernando El Católico." 1:181–230; 1966.

Demangeon, A. L'habitat rurale en France, essai de classification. *Ann. Géogr. XXIX*:352–375; 1920.

Demangeon, A. *La Géographie de l'Habitat Rural*. Report of the Commission on Types of Rural Settlement. Newtown, Montgomeryshire: International Geographical Union; 1928.

Demangeon, A. Une carte de l'Habitat. *Ann. Géogr. XLII*(42):225–232; 1939.

Dollfus, O. *L'Espace Géographique*. Collect. Que Sais-Je? Paris: PUF; 1976.

Dourojeanni, M.J. *Recursos Naturales y Desarrollo en América Latina y el Caribe*. Lima: Universidad de Lima; 1982.

El País. *Anuario Estadístico*. Madrid: Ediciones El País; 1983–1995.

Etxezarreta, M. *La Agricultura Insuficiente*. Madrid: MAPA; 1985.

EUROSTAT. *Estadísticas Básicas de la UE. Estadísticas Generales, Serie A*. 32nd ed. Luxembourg: Statistical Office of the European Communities, 1995.

FAO. *FAO Yearbook: Production*. Rome: FAO; 1993a.

FAO. *FAO Yearbook: Trade*. Rome: FAO; 1993b.

Faus, M.C.; Higueras, A., eds. *Rural Depopulation*. Regional Conference on Mediterranean Countries. Proceedings of Sessions of Commission on Population Geography; 1986 August 25–30; International Geographers Union (IGU), Barcelona. Zaragoza: Department of Geography, University of Zaragoza; 1986.

Fromont, P. *Economía Agrícola*. Madrid: Aguilar; 1961.

Frutos, L.M. Los espacios rurales. Bosque, J.; Vilá, J., coordinator. *Geografía de España; Vol. 5*. Barcelona: Ed. Oceano/Instituto Gallach; 1994:856–931.

Frutos, L.M. Las confederaciones sindicales hidrográficas (1926–1931). Gil Olcina, A.; Morales Gil, A., eds. *Planificación Hidráulica en España*. Alicante: Fundación CAM; 1995:181–255.

Frutos, L.M.; Solans, M.; Chueca, M.C. Cambios en el Sistema de asentamientos rurales: La provincia de Teruel (Aragón, España). *A Península Ibérica: Un Espaço em Mutação*; Vol. I. Proceedings of the VI Coloquio Ibérico de Geografía; 1992 September 14–17; Porto, Portugal. Porto: Publicações da Universidade do Porto; 1995: 284–290.

George, P. *Dictionnaire de la Géographie*. Paris: PUF; 1970.

George, P. *Geografía Rural*. Barcelona: Ariel; 1975.

Golte, J. *La Racionalidad de la Organización Andina*. Lima: Instituto de Estudios Peruanos (IEP); 1987.

Gourou, P. Pour *Une Géographie Humaine*. Paris: Flammarion; 1973.

Gutelman, M. *Estructura y Reformas Agrarias. Los Problemas Agrarios y Los Métodos Para Su Estudio*. Barcelona: Fontamara; 1981.

Haggett, P. *Análisis Locacional en la Geografía Humana*. Barcelona: Gustavo Gili; 1976.

Ilbery, B.W. *Agricultural Geography: A Social and Economic Analysis*. Oxford: Oxford University Press; 1985.

Instituto Aragonés de Estadística. *Información estadística de Aragón. Censo de Población de Aragón*. Zaragoza: Instituto Aragonés de Estadística; 1991.

Instituto Geografico de Agostini. *Calendario Atlante*

de Agostini. Novara: Instituto Geografico de Agostini; 1983–1985.

IRYDA. *Fines y Actividades.* Madrid: Ministerio de Agricultura; 1973.

Johnston, R.J.; Derek, G.; Smith, D.M. *Diccionario de Geografía Humana.* Madrid: Alianza Editorial; 1987.

Kayser, B. *La Renaissance Rurale: Sociologie des Campagnes du Monde Occidental.* Paris: Colin; 1990.

Kostrowicki, J. *Geografía dell'Agricoltura.* Firenze: Franco Angeli; 1980.

Labasse, J. *L'Organisation de L'espace: Eléments de Géographie Volontaire.* Paris: Hermann; 1966.

Lassailly-Jacob, V. La charge de population en question: Colonisation planifiée des rives du lac Kosson en Côte d'Ivoire. ORSTROM, ed. *Le Developpement Rural en Question.* Col. Mémoires. Paris: CNRS; 1984:403–416.

Lassey, W.R. *Planning in Rural Environments.* New York: McGraw-Hill; 1977.

Le Coz, J. *Las Reformas Agrarias.* Barcelona: Ariel; 1975.

López Ontiveros, A. Ordenación del territorio y espacio rural en ambitos regionales. *Proceedings of the VII Coloquio de Geografía;* Vol. II; 1981 September 29–October 2; Pamplona, Spain. Pamplona/Salamanca: AGE; 1983:487–493.

López Ontiveros, A. La actividad agraria. Bielza, V., coordinator. *Geografía General; Vol. II. Geografía Humana.* Madrid: Taurus Universitaria; 1993: 158–210.

MAPA. *Anuario de Estadística Agraria.* Madrid: MAPA, Secretaría General Técnica; 1983, 1985, 1988.

MAPA. *Estructura y Regímenes de Tenencia de la Tierra en España.* MAPA, Serie Estudios. Madrid: MAPA, Secretaría General Técnica; 1987.

MAPA. *El Futuro del Mundo Rural.* Documentos de la CEE. Madrid: MAPA; 1992a.

MAPA. *La Nueva Política Agraria Común.* Madrid: MAPA; 1992b.

Martínez Cordón, A. Los poblados de colonización en el sistema de riegos Bardenas-Alto Aragón. *Geographicalia* 25 (2ª época):145–155; 1988.

Méndez, R.; Molinero, F. *Espacios y Sociedades. Introducción a la Geografía Regional del Mundo.* Barcelona: Ariel Geografía; 1991.

Meynier, A. *Les Paysages Agraires.* Paris: CAC; 1958.

Molinero, F. *Los Espacios Rurales.* Barcelona: Ariel Geografía; 1990.

Morgan, W.B.; Munton, R.J C. *Geografía Agrícola.* Barcelona: Omega; 1975.

Nelson, H.J. A service classification of American cities. *Econ. Geogr.* 31:189–210; 1955.

OECD. *L'Agriculture à Temps Partiel Dans les Pays de l'OECE.* Paris: OECD; 1978.

OECD. *Nuevas Tendencias en Política Rural.* Madrid: MOPU, Instituto del Territorio y Urbanismo; 1990.

Pacione, M. *Rural Geography.* New York: Harper and Row; 1984.

Pahl, R.E. The rural/urban continuum. Pahl, R.E., ed. *Readings in Urban Sociology.* Oxford: Pergamon Press; 1968:299–329.

Pelissier, P.; Sautter, G. Preface. ORSTROM, ed. *Le Developpement Rural en Question*: Col. Mémoires. Paris: CNRS; 1984:IX–XIV.

Rawson, J. The impact of rural settlement policies of concentration and dispersal in rural Britain. Paper presented to the PTCR Summer Annual Meeting. Warwick: University of Warwick; 1981.

Rodríguez Cancho, M. *Análisis Geográfico del Regadío en Extremadura.* Cáceres: Caja de Ahorros de Badajoz; 1984.

Rodríguez Castro, I. *El Ejido Colectivo Como Empresa.* Investigaciones Sociales No. 5. México, DF: Instituto Nacional Indigenista; 1979.

Saenz Lorite, M. Geografía agraria. Puyol, R., ed. *Geografía de España;* Vol. 7. Madrid: Síntesis; 1988.

Stamp, L.D. *The Land of Britain: Its Use and Misuse.* London: Longmans; 1928 (new edition, 1962).

Stevenhagen, R. *Las Clases Sociales en las Sociedades Agrarias.* Madrid: Siglo XXI; 1976.

Suarez Japón, J.M. *El Hábitat Rural en la Sierra de Cádiz: Un Ensayo de Geografía del Poblamiento.* Cádiz: Excma. Diputación Provincial; 1982.

Tricart, J. *L'Habitat Rural. Cours de Géographie Humaine.* Paris: CDU; 1956.

UN. *United Nations Demographic Yearbook: Data of 1970–80 and 1988.* New York: Department of International and Social Affairs, Statistical Office; 1990.

Wibberley, G.P. Conflicts in the countryside. *Town and Country Planning* 40:259–265; 1972.

11

Rural Development Engineering: Labourers of Development

José Luis Salinas

11.1 Introduction

The diversification of rural areas has resulted in a transformation of policies and assignments of public and private training development agents. The following pages are the result of personal experience as a trainer in the CIHEAM Institutes, which is a good observatory from which to detect the transformation of Mediterranean regions through the concerns of students of many nationalities.

In this chapter, the focus is on the question of the priority duties of agricultural extension and rural development agents, with the following objectives: (i) allowing governing bodies to better negotiate with the state; (ii) enhancing the value of products and services; and (iii) opening themselves up to markets, to better energize professional organizations and the relational fabric of civil society. On the other hand, the problem of accompanying technical innovation through an endogenous process of learning is also considered. Finally, we consider the question of integrating information and training which should allow rural societies to manage their own development through self-focused, integrative, participatory processes.

The conditions of agricultural production, the role of the state, the system of population, and so on, as well as the aspiration and demands of the society with regard to rural areas, are quite varied in the Mediterranean regions. Indeed, these regions are heterogeneous, heavily populated or practically deserted, perched on the coastline or withdrawn in the hinterlands, and endowed with more or less socioeconomic decentralization. However, they present a common element in being confronted with the challenge of sustainable rural development focused on three key areas: (i) contributing to the food supply; (ii) obtaining a standard of living and revenues for peasant families; and (iii) managing natural resources and the ecological balance.

This challenge is, in part, taken up by a diversified group of professional development agents: researchers, trainers, extension agents, rural organizers, technical advisors, those in charge of studies, and heads of projects. These experts are available in geographic areas, sector-based programmes, or technical or specialized support groups in production techniques, in economic management, and in social organizations. The transformation of rural societies today and the evolution of rural policies or programmes of structural development have led to the need to explore other ways, and new duties have emerged around which those devoted to the difficult profession of rural and agricultural development have to specialize.

This professional transformation is based on a threefold problem: (i) rural development and its regionalized socioeconomic planning;

(ii) the management of socio-technical systems of the renovated "green revolution"; and (iii) the apprenticeship in social innovation—that is, new synergies and participatory approaches.

11.2 Rural Development

Rural is no longer a residual concept, nor a populist aspiration. Since the 1980s, the concept has been modified by three phenomena.

The observation of a "dis-agriculturalization," that is, (i) the decline of agricultural work, (ii) the decrease of the relative weight of agriculture in the GNP, and (iii) the diversification of rural economies. This process has sometimes been accompanied by the diminishing, even the disappearance, of many kinds of agriculture (i.e., rural exodus) because modern technology has brought about the regrouping of modern farms and the replacement of labour-intensive activities by capital-intensive activities. Some rural families have resisted thanks to part-time agricultural work and the discovery of the possibility of creating an adaptable production pattern based on the family, getting around the crisis by the seasonal or sector-based mobility of its active members. They put a relative self-subsistence strategy into practice, a short circuit of markets, or a hidden economy (Arkleton Research, 1990).

Between extreme positions, a new paradigm of territorial and regional development has been diffused, which places primary importance on endogenous resources and local initiatives (Friedmann and Weaver, 1979). Its ambition is to realize regional planning based on policies of decentralization and the principle of subsidiarity in which the state delegates the role of reactivating production structures and human resources to the economic actors and local leaders.

The sociopolitical reevaluation of the interest of rural areas is dictated by four reasons. First, by the principles of equity and interregional solidarity. Second, by recognizing that rural areas contain a threefold wealth: human, natural, and cultural treasures that are part of the world's heritage (i.e., sites and environments worthy of being preserved). Third, by the desire to slow migratory movement of rural population toward already heavily congested urban areas. Finally, by energizing rural areas in a strategy of regulating crises brought on by technological changes and the internationalization of markets (Vázquez Barquero, 1993).

The importance of sustainable rural development was: (i) developed in the rural conferences in Rio in 1992 on Environment and Development, in Copenhagen in 1995 on Social Development, in Cairo in 1995 on Population and Development, in Istanbul in 1996 on Habitation and, especially, at the world summit on Agriculture and Food held at the seat of the FAO in Rome in 1996; (ii) recommended directly by international institutions such as the United Nations Development Programme (UNDP), the OECD, and the World Trade Organization, and by the Council of Europe in its charter on rural areas; and (iii) financially supported by multi- or bilateral institutions of cooperation such as the World Bank, the International Monetary Fund for Innovation and Development, and so on. Also, the European Commission's "green book" presents its main ideas on rural development and the content of a variety of programmes that focus on this issue, such as the FEDER, FEOGA-O, and LEADER II initiatives.

Who then can say that rural development is not a current topic? These ideas are being discussed vigorously but the protagonists (designers, politicians, etc.) involved in the rural renaissance (Kayser, 1990), combine them in different ways depending on the characteristics of the country and its position in the process of internationalization.

First, a new social role is attributed to farmers, which makes a distinction between agriculture for commercial production (borne by a relatively small number of farmers entrepreneurs resolutely professionalized in agribusiness) and a type of agriculture that maintains the structures and jobs that have the more social role of maintaining family revenues and ecological balance. As for the latter, many endeavour to find new activities that can be integrated into farming so as to render it socioeconomically viable through diversification (Alexandratos, 1995).

Secondly, a territorial paradigm is used to develop the whole strategy of overall development in a given area. This is planning in the shape of an hourglass: on the one hand, the region is treated as a "programming space" and policies are elaborated that coordinate resources and initiatives. On the other hand, more or less homogeneous rural areas (the comarca, the countryside) are areas for the "redistribution of skills" where diverse economic and social actors work and coordinate on the pretext of a rediscovered sociocultural identity, with a willingness to live and work in the country (Dascon, 1984). This type of development seeks to resolutely favour the spirit of enterprise and the endogenous mobilization of actors. Human factors are reconsidered, and the economic and social actors take their place as partners of the administration and the state in order to create rural jobs (OECD, 1995).

Finally, ecological concerns increasingly influence action strategies. That is to say, the environmental context of agricultural projects, industrial farms, the habitat location, and the use of nonrenewable energy and resources are identified and assessed. How can these developments be made compatible with the economy of natural resources and the environment (Faucheux and Noël, 1995)? Rural development, from this perspective, is neither exclusively economic nor anthropocentric; it must also respect the rules of tenable coevolution of life and careful management of our ecological home (Morin, 1993).

11.3 The "Green Revolution"

The World Summit on Food (FAO, 1996) proposed a renewal of the green revolution, among other resolutions, in order to confront increasing demographic growth and reduction of available agricultural land. These dual processes continually challenge us to increase agricultural productivity (not only production) and to give to the poorest regions the means of escaping from food insecurity.

The green revolution, initiated in the 1950s in industrialized countries, consisted of a worldwide technological effort oriented toward the introduction of improved varieties of animals and plants, irrigation of land, mechanization, and use of fertilizers and pesticides for the principal crops. It was accompanied by investments in institutional infrastructures (access to credit, land ownership, and professional organizations) and by research and diffusion/extension programmes (Hayami and Ruttan, 1985). These efforts are still being felt today, but the results were not always beneficial to the disadvantaged areas and the poorest populations (FAO, 1991).

The ultimate goal of the green revolution was, of course, social: improve the living conditions of the rural population as well as provide urban centres with foodstuffs for mass consumption. The means to reach these goals were primarily economic: increase farm revenues through a combined set of agricultural policies, through influence over the markets, and through management advice. The proposed solutions were always technical (based on the processes of research and development, the extension and supervision and regular provision of inputs), that is to say, they focused on how production and productivity per hectare, per livestock unit, and per work unit could be increased.

The proposed reformulation of the green revolution (FAO, 1996) addresses two main concerns: (i) that of large-scale conservation of sensitive natural areas, according to the recommendation of the action programme of the Rio declaration of 1992; and (ii) that of attacking food insecurity from which almost 750 million world inhabitants suffer (UNDP, 1996).

Two types of scenario are possible (McCalla, 1994). The first supposes the production of large food surpluses in industrialized countries that have a production potential, technical capacities, and commercial infrastructures capable of guaranteeing sustained growth of the supply in world commerce. But this scenario raises the question: Will poor countries from the south be able to make up the food deficit from their meagre exports of goods and services? Is it reasonable to imagine that governments would accept such a food dependency from the outside? How can new possibilities of revenue and work be generated to guarantee better food provision coming from imports for the poor in cities and countrysides?

The second scenario consists of multiplying research and extension activities in order to promote food-crop production in southern countries. This means a significant improvement of advice and training services for peasants so that they can take advantage of research results and technological progress adapted to the solution of their problems: sensitive, marginal land; unstable land conditions; and drought.

This process requires the reformulation of current technological packages that are generally beyond the reach of small and medium-sized farms as well as being below par as far as their problems and economic level are concerned. This also requires the adaptation of management techniques oriented toward family farms that minimize risks more than maximize profits. Finally, this requires facilitating regular access to production factors (credit and provisions) and finding openings on the market at equitable prices (Torado, 1994).

There may be multiple intermediate solutions between these extreme scenarios, but whatever the choice, it is difficult to imagine renewing the green revolution without supplementing agricultural development by sustainable rural development. Even if biotechnology can bring about substantial advances for management and production systems, it is hard to imagine technical innovation that would not be accompanied by social innovation and new types of organization and learning.

11.4 Social Innovation: New Synergy and Participatory Approaches

In general, rural areas, especially disadvantaged areas, suffer from several structural weaknesses: (i) declining and aging population; (ii) the exodus of young people and highly skilled workers; (iii) threatened and low-paying jobs; (iv) isolation and distance from centres of decision-making, information, research/development, and training; and (v) a lack of associative fabric and community spirit. In this case, how can the critical mass—essential to the promotion of innovative activities—be attained, and how can the desire be inspired in rural communities so that they work for and manage their own development? Should everything be expected to come from the state? the market? civil society?:

(i) *The market.* If development of the economy of a small rural region from its own resources is sought, these resources will come mainly from agricultural and forestry production and from tourism. Given that agricultural, forestry, and tourist economies are strongly integrated into specialized channels of a transnational dimension, and that added value is concentrated toward the lower end of the channel (where, moreover, decision-making centres are located), how can one imagine that these resources may be used for local development as a matter of priority?

The interdependence all along these channels has replaced the solidarity within an area. The mechanisms of international markets cannot, by themselves, launch and guarantee full employment in disadvantaged rural areas. They are generally attracted toward richer areas, more successful sectors or basins of the most rapid economic growth, according to the law of comparative advantages.

(ii) *The state.* What can be expected of the state? Can the state overcome the split among rural people who have succeeded (and because of this, have taken a place in the economic activity and in the job market), the subproletariat receiving welfare, and the social outcasts who are condemned to live marginally? The providential state has exhausted its resources and is being replaced progressively by another kind of state that will play the role of mediator of social change and promoter of new contractual relationships between economic and social actors. Can this development be considered a reaction to the crisis that ideologies are going through, from a collectivist utopia to liberalism?

(iii) *Civil society*. What conditions does society offer to ensure sustainable development in rural areas? From a historical perspective, it has adopted diverse nonmerchant and nonstate modes of production of goods and services by relying on different types of relationships: family bonds, neighbourhoods, solidarity, and so on. This type of solution has been put forth by so-called alternative or autonomous communities, but up to now has been considered as being marginal, not able to create by itself the necessary conditions to maintain the process of sustainable development.

11.4.1 Learning New Synergies

In response to these limits a variety of approaches are presently being tried, either through programmes of the European Union (LEADER) or in programmes of bilateral or international cooperation that put participatory approaches into practice by combining three socioeconomic supports: (i) the basic philosophy of a market economy—risk, economic efficiency, and innovation; (ii) principles that are the basis of state action—social justice, mediation, and agreements made with local governments; and (iii) phenomena proper to civil society—cooperation, solidarity and membership.

This "cocktail" goes beyond the traditional division of tasks between the market and the state; that is, between the economic and the social. It gives a political dimension to development programmes insofar as each action of development carries within it the germ of a social plan. For this reason, rural development may generate crises, from which spring the necessity for negotiation strategies and learning programmes.

Participatory approaches have as one of their goals bringing to light local knowledge and reinforcing the capacities of rural communities for the analysis of potentialities and constraints, for planning, and for the realization and assessment of development activities. A multitude of new approaches to participatory planning have appeared: MARP (Accelerated Method of Participatory Research), PAR (Participatory Action Research), DELTA (Development, Education, and Leadership Teams in Action), and so on. All these approaches (Schönhuth and Kievelitz, 1994) are a reaction toward the methods and practice of rural development that have been traditionally conceived without the people who were to be affected. These approaches have sometimes been for them but rarely with them and their participation. This has been especially true of programmes focusing on rural women (UNDP, 1996).

Participatory approaches have followed two main paths: one geared to research development (e.g., Farmer Participatory Research), and the other concerning rural extension and community development. Many NGOs have understood participatory approaches to be a means by which rural groups can reinforce the control over their lives and better negotiate with the state (Norton and Alwang, 1995). Their methodology is rooted in the liberation pedagogy advocated by Paolo Freire and by the movements of popular education, and is expressed by the phrase "see, think, act, and assess." Other experiences and programmes (e.g., UNDP/FENU) go further toward a participatory eco-development (Lazarev et al., 1993); that is, toward a new kind of engineering of rural development.

Indeed, rural development is at a crossroads of ideological, political and socioeconomic perspectives that take place schematically between four categories of participating partners:

(i) *Resource provider partners*. They include physical persons and organizations that obtain money, material, and energy (inside and outside the region or country). States, as well as agencies like the World Bank, are working on this aspect of rural development in international co-operation or in urban–countryside coordination. These partners have legitimate concerns of knowing if the sums granted will be used correctly and efficiently. This is especially true if these funds are credits that must be reimbursed directly or indirectly from the investments. Starting from the motto "he who pays, commands," the partners sometimes fall into the trap of wanting to impose models of rationality and procedures that do not fit the logic and expectations of rural areas and collectivities.

(ii) *Decision and legitimacy provider partners*. In the case of regional development, the authority legitimates projects and programmes and sets the framework for their realization. In the case of bilateral or multilateral cooperation with the exterior, this political authority plays the key role (even if it is local) for the official use of resources. In the crises of public finances and the drying up of sources of financing, this authority easily accepts programmes and projects paid in cash so as to balance its monetary budget. At the regional level, the argument for investments capable of producing jobs or professional taxes is determining, without considering the well-foundedness of the programmes at the medium or long term.

(iii) *Service provider partners*. These are development professionals, national or international experts, civil servants, or members of NGOs, from universities and research departments, who are in charge of providing services for a rural clientele. Researchers, experts, technicians, and professionals in the field are included in this category. These men and women are generally devoted and competent, they know where to find the resources, where the political legitimization is, where technical information is, and how to tie these three subsystems to the needs of the field. Administrative processes, expertise and assessment (*ex ante*) in the formulation of projects, responses to call for proposals, and dealing with red tape take up a substantial percentage of the energy and motivation and willingness of these partners. Their concern for organization and service to the rural community sometimes remains blunted by bureaucracy or by their personal strategies.

(iv) *Beneficiary client partners*. These are the end-users of services and the receivers of rural development programmes. They are rarely organized and well-informed. They serve as an excuse for setting up a whole complex of projects, but are often disappointed, forgotten, marginalized, and are used more as a pretext—a captive client—than as a partner. However, the success of sustainable development is to be found here, in their capacity to negotiate, propose, manage, and give meaning and power to their projects, to become "actors" (Crozier and Friedberg, 1977).

11.4.2 Crossroads of Learning and Support Services (CLSS)

In the traditional management of rural development, the flow of information, resources, and power generally follows a descending hierarchical path from the resource provider to decision and legitimacy providers who order projects and services from service providers so that the latter can carry them out for beneficiary clients.

Innovative management of participatory development suggests another model of social communication. Clearly reversing the relationship of power and doing nothing else would not only be revolutionary, but useless and demagogic. However, seeking out strategies to allow people to go beyond their present situation toward new crossroads of participatory management that serve as a basis and fertile ground for rural projects, through mutual learning and negotiation, is not utopian. In this model, it is necessary to accompany technical innovation with the organization of social institutions if we are to accomplish our desire to create opportunities for self-learning of rural actors and communities (Gaulejac-Bonetti, 1989).

These aforementioned crossroads of negotiation provide personalized and close relationships, materialized in a team representing the four subsystems of partners in the rural development programme. They are the instruments of management, training, and information. They will be efficient if they have clear mandates that define their action and the means to achieve it; that is to say: (i) the confidence of resource providers (budget, material, etc.); (ii) professional expertise of service providers (emphasizing information resources, and technical references); and (iii) the support of the population of beneficiary clients of which they are the delegate core (in representation and participation).

Is the development agent the builder of this learning enterprise, the labourer on the fertile ground of this negotiation, the mediator in this crossroad of multiple functions? Is it a question of an isolated or a networked team?

The knowledge of new possibilities opened up by research and their immediate technical application has become the main source of innovation and economic growth for our societies (Freeman and Perez, 1986). Genetic engineering, bio-technologies, computer sciences, and so on, will change the possibilities that certain farmers and rural entrepreneurs have to produce, as well as their capacity to decide how and what to produce. Here there is a great ethical responsibility for intellectual institutions of research and development, who will have an important role to play as centres of innovation in the preparation and management of a changing world (Sebillotte, 1993). Where will these ethical discussions be held? How will societies of diverse cultures and purposes find common ethical grounds?

And without close collaboration between grass-roots people—the training agents—how can the operational technical itineraries be adopted? How can solutions be found where problems come up? How can adequate answers to questions that have not yet been formulated be given? How can technical itineraries for people who do not want to go or wanting to go (sometimes rightly) in the opposite direction be laid down?

Learning, as Piaget (1967) suggests, is not knowing how to repeat the knowledge of one's teacher; rather; it is knowing how to create one's own knowledge, one's own solutions to problems. Expressed in another way, field instructors are needed to apply a pedagogy adapted to the social creativity of each region, to embody technical innovation in social organization. A new research–education administration of rural and agricultural development is emerging. It is centred more on the creativity of socioeconomic actors than on the appearance of new structures of intervention dictated by the state and by scientific rationality conveyed as an ideology or as an instrument of domination.

11.5 Conclusion

Rural communities today are the object, and at the same time the potential actor, of development: the actor has the capacity to create and to undertake change. In this context, duties of social creativity, mediation, and interphase appear, around which development agents must specialize. These priority duties can be summarized in four categories:

(i) Technical advice and elaboration and diffusion of information on different farming systems will always be a primary duty of extension agents serving as a bridge between research institutes, centres of experimentation, and systems of real production.

(ii) The research development function, the hinge between the specialized technicians, business advisors, and groups of innovative farmers is crucial. This function concerns as much technical references for the types of products as the processes and means of production. Clubs of technical innovation using expert methods can generate more or less sophisticated and effective consulting activities in which the computer sciences allow immediate use of advice or diagnostics. They also allow a distinction between public service and private service, and, for this reason, the ability to diversify the means of financing.

(iii) Development engineering, the subject of which is to bring out and supervise appropriate development programmes for, or with, rural communities, professional organizations, associations for the management of irrigated perimeters, and so on; to establish the diagnosis of the area's potentialities; and to help farmers formalize their projects and fit them into regional development.

(iv) The consolidation of communication and training networks in order to have at one's disposal content and pedagogical materials responding to in-service training needs of farmers, rural women, young people, and other members of the rural community.

Advice for rural enterprises, the study of production systems from a vertical point of view, and the analysis of the development of markets in order to support professional grouping or farms that start new types of production, new activities, or want to better enhance their products will be included in these materials. This function always involves a capacity for synthesis that integrates technical, economic and legal elements as well as human and social criteria.

The desire to live and work in the country means mobilizing the resources to produce goods and services for a market. Proposing these main foci of orientation and creativity, the breeding ground of enterprises and support centres for revitalized rural small and medium-sized companies through the integration of young people, is perhaps the key piece for tomorrow's rural development. That is where the development agent and the rural population will let themselves be educated for the future by anticipating it.

The changes that have taken place over the past few years in the very conception of rural development have led us to infer a certain number of new functions for development agents and those of private or public extension services.

Mediterranean regions are teeming with trials, experiments, and approaches all going in this direction. It is too early to conclude that these efforts have led to success and that certain strategies are to be favoured. In this field, as elsewhere, action research remains wide open.

References

Alexandratos, N. *Agriculture Mondiale: Vers l'An 2010*. Rome: FAO; 1995.

Arkleton Research. *Cambio Rural en Europa*. Madrid: MAPA, Serie Estudios; 1990.

Crozier, M; Friedberg, G. *L'Acteur et le Système. Les Contraintes de l'Action Collective*. Paris: ed. du Seuil; 1977.

Dascon, F. *Des Contrats de Pays aux Chartes Intercommunales. Revue POUR (Janvier–Février)*. Paris: Privat; 1984.

FAO. *Tercer Informe Sobre los Progresos Realizados en la Aplicación del Programa de Acción de la Conferencia Mundial Sobre Reforma Agraria y Desarrollo Rural*. Rome: FAO; 1991.

FAO. *Sommet Mondial de l'Alimentation*. Documents d'information technique. Rome: FAO; 1996. 2 vol.

Faucheux, S.; Noël, J.F. *Economie des Ressources Naturelles et de l'Environnement*. Paris: Colin; 1995.

Freeman, Ch.; Perez, C. *The Diffusion of Technical Innovations and Charge of Techno-Economic Paradigm*. Brighton: University of Sussex, SPRU; 1986.

Friedmann, J.; Weaver, C. *Territory and Function*. London: Edward Arnold; 1979.

Gaulejac-Bonetti, M. *L'ingénierie Sociale*. Paris: Syros; 1989.

Hayami, Y.; Ruttan, V.W. *Agricultural Development and International Perspective*. Baltimore, MD: Johns Hopkins University Press; 1985.

Kayser, B. *La Renaissance Rurale*. Paris: Colin; 1990.

Lazarev, G.; Kalberhatten, G.; Michel, B. *Vers un Eco-Développement Participatif*. Paris: Harmattan; 1993.

McCalla, A.F. *Agriculture and Food Needs to 2025: Why We Should Be Concerned*. Washington, DC: Consultative Group on International Agricultural Research; 1994.

Morin, E. *Terre-Patrie*. Paris: ed. du Seuil; 1993.

Norton, G. W.; Alwang, J. *Economía del Desarrollo Agrario*. Madrid: Mundi Prensa; 1995.

OECD. *Créer des Emplois pour le Développement Rural*. Paris: OECD; 1995.

Piaget, J. *La Psychologie de l'Inteligence*. Paris: Armand Colin; 1967.

Schönhuth, M.; Kievelitz, V. *Les Méthodes de Recherches et de Planification dans la Coopération au Développement: Diagnostic Rapide en Milieu Rural, Diagnostic Participatif–Une Introduction Commentée*. Esschborn: GTZ; 1994.

Sebillotte, M. *Avenir de l'Agriculture et Futur de l'INRA*. Rapport à M. le Président de l'INRA. Paris: INRA; 1993.

Torado, N.P. *Economic Development*. New York: Longman; 1994.

UNDP. *Rapport Mondial sur le Développement Humain*. Paris: Economica; 1996.

Vázquez Barquero, A. *Política Económica Local*. 2nd ed. Madrid: Piramida; 1993.

Part V

Rural Economics and Economics of Natural Resources

12

The Farm and Rural Community As Economic Systems

Felisa Ceña

12.1 The Nature of Agriculture

All human activity takes place within a context shaped by certain relationships between society and the biophysical environment, or the rest of nature. This interaction is of particular importance in the case of agriculture, an activity that "artificializes" the environment through the management of natural resources. Development necessarily entails the transformation of these relationships. The interaction between man and nature has evolved gradually over time, and above all in response to the degree of development attained. This, in turn, has given rise to different forms of agriculture, ranging from traditional forms (based essentially on animal or live energy) to intensive or modern forms (involving high fossil-fuel consumption), by way of today's ecological or low-input forms (using environment-friendly technology). These and other, intermediate forms of agriculture correspond to different models and aims of development throughout history, each of which assigns to agriculture a different role. This chapter will provide a broad context for rural planning from an economic perspective. The planner needs to understand and apply this perspective in order to suggest viable and sustainable alternatives.

12.1.1 The Role of Agriculture in Development

As a sector, agriculture has generally been subordinated to the demands of the rest of the economy. Its role has thus always varied as a function of the dominant concept of development in different countries and in different decades. Two main schools of thought may be discerned within the overall approach that identifies development with growth (1950s–1960s): (i) those who hold industry to be the only sector capable of stimulating development, and thus accord to agriculture a very minor role; and (ii) those for whom agriculture plays a fundamental part in the development process.

For the first group, the mission of the agricultural sector should be restricted to providing manpower to the industrial sector, which drives growth. Proponents of this view argue that investment in industry generally leads to greater growth and even, via industrialization, to the development of agriculture itself. Moreover, industrialization also creates an institutional and socioeconomic climate that is more conducive to the spread of innovations. The transformation of agricultural societies into industrialized societies thus lies at the heart of the development process.

The dual-model strategy, based on the

Lewis (1954) model, aims at optimal allocation of resources; this involves transfer of manpower from agriculture, a backward sector in which marginal productivity is very low or nonexistent due to surplus resources (unlimited supply of labour at subsistence wages), to the industrial sector, where productivity is far greater. Moreover, low agricultural wages and prices allow the creation of a surplus in industry, through low real wages; and reinvestment of the surplus by industrialists increases urban employment and GNP.

Exclusive concern with the removal of the agricultural labour surplus encourages underestimation both of agriculture's potential to co-operate in development and of the difficulties faced by the agricultural sector in adapting to the new situation. This approach, propounded by a number of economists, is based on several arguments, as follows:

(i) *Empirical observations showing a decline in the contribution of agriculture to the economy during economic growth.* One reason for the decline in the contribution of agriculture to GDP is the low elasticity of supply/demand for agricultural products, whose causes are explained by Engel's law and by Malassis (1973), who argues that demand for agricultural produce comes increasingly not from the end-user but from agricultural and food industries. It has been shown that, during industrialization, demand for agriculture grows more slowly than food demand. These two effects combine to trigger a strong downward trend of relative agricultural income, because the demand for agricultural produce is a central component of agricultural income, which increases less rapidly than national income.

(ii) *The historical decline in terms of trade.* Analysing the historical evolution of international trade, in which terms of trade were unfavourable for raw materials, in other words the exports of developing countries, structuralists advocated the Import-Substituting Industrialization (ISI) strategy, i.e., substitution of industrial imports; short-term investment in agriculture was therefore not considered advisable (Prebisch, 1963; Furtado, 1968).

(iii) *A lack of linkage between agriculture and the rest of the economy.* This argument is based on Hirschman's (1958) requirement for a sector to be able to drive development. The linkages thesis maintained in his book The Strategy of Economic Development states that investment in certain sectors stimulates investment in other sectors with which these maintain output/input relationships. Agriculture, according to Hirschmann, is not one of these sectors. Because investment in agriculture would not generate the same effects as investment in other "industrializing" industries, it is not of the slightest interest.

For the second group, who defends the value of agriculture in development, the enormous importance of industry in the development process does not in itself signify that agriculture has no part to play. There is, in fact, little agreement regarding the relative importance of industry and agriculture in the development process. Some economists would even defend a reversed order of importance. Bairoch (1974), in his study of France and Great Britain, argues as follows: economic growth began with the development of agriculture, which stimulated that of industry. Economic development in both countries was triggered by increased agricultural productivity and increased demand by agriculture for industrial inputs.

Other authors point to even more specific contributions. According to Kutznets (1972), agriculture can contribute to development in a number of ways: (i) in terms of output, through its contribution to GDP; (ii) in market terms, through sales to and purchases from national and foreign sectors; and (iii) in terms of factors of production, by providing labour and capital to the industrial sector.

The extent of such contributions, however, is governed by the degree of industrialization attained but also, and more particularly, by certain specific features of each country: food, agricultural employment for the rural population, manpower for the industrial sector, production of industrial inputs (fibres, dyes, aromatic and medicinal plants), financing of nascent industry, foreign currency for the import of technology and other inputs, the market for industrial goods, and ecological and territorial balance. By demanding industrial products, agriculture creates linkages with industry, and this has far-reaching implications for the design of development strategies.

The 1960s saw a shift away from industrial fundamentalism (Ruttan and Hayami, 1971) toward growing emphasis on the significance of increased agricultural production and productivity for the development process as a whole. This shift in approach stemmed from observation of the real situation in countries where a rapidly increasing demand for food could no longer be met by an agriculture which, because of the strategy of uncompromising industrialization dominant in the 1950s, had failed to make any technological progress. Instead of remaining a passive partner in the development process, agriculture was now assigned a primary role in the initial thrust of development, as well as contributing to the subsequent maintenance of a high growth rate. Johnston and Mellor (1961), following Rostow (1960), distinguish three historical stages in the evolution of agriculture:

(i) *The prerequisites of agricultural development*. This stage is characterized by the institutional and behavioural changes essential for increased output: improvement of land infrastructure, access to consumer goods markets, information on available technology, and a changed attitude and enhanced receptivity of farmers to progress.

(ii) *Increased efficiency of the production process through the spread of labour-intensive and capital-saving innovations.* This is the form taken by development when agriculture is the predominant activity and when the amount of capital required for expansion of the industrial sector is small. The demand for agricultural products increases with increased population and per capita income.

(iii) *Growth of agricultural production, characterized by capital-intensive and labour-saving technology—in other words, the industrialization of agriculture*. The contribution of agricultural production and agricultural employment to the economy as a whole undergoes a rapid decrease.

Incorporation of the *basic-needs* approach—according to which the main aim of development should be to meet the basic needs of the poorest population: food, health care, education, and housing—into the new concept of development during the 1970s, highlighted the importance of agriculture. After two decades of development, a new consensus was reached: agricultural development is essential to, if not a prerequisite of, industrialization and general economic growth.

The changing orientation of economic development entailed a more commanding role for agriculture in development programmes. The reasons may be summarized as follows: (i) in developing countries, most of the population lives in rural areas; (ii) low agricultural productivity is considered the main cause of poverty; (iii) food prices are the major determinant of the purchasing power of both rural and urban poor; and (iv) urban industry has proved unable to absorb the labour force resulting from rapid population growth.

The structural adjustment plans introduced in the 1980s exerted a profound effect on agriculture in developing countries, by removing government support for the sector. But at the same time, due to the new development strategy based on increased exports (liberalization of external trade and production) and to widespread support for what was termed "adjustment with a human face," agriculture resumed a key role as the driving force for growth, reflecting the importance of the primary aim of rural development: to feed the population at adequate levels. By the end of the 1980s, the access of developing countries to self-sufficiency in food had become a priority objective in development models (Mellor, 1988; Cornia, 1993).

12.1.2 Models of Agricultural Development

Over recent decades, various models of agricultural development have been proposed for developing countries. The most representative of these are mentioned below.

12.1.2.1 The Technology Transfer Model

This assumes that farmers in underdeveloped countries can increase productivity through more efficient allocation of resources and by adopting technology from developed countries, and more particularly the technology developed by the United States for its own agriculture, that is, mainly mechanization.

Agricultural technology was duly transferred, and the United States model of agricultural extension was promoted. Transfer of technology, supplemented by what was termed "community development" took place in the Cold War climate of the 1950s, a period in which Western aid programmes to Third World countries were based on the assumption of nonrevolutionary rural change in these countries. This model of development assumes that farmers in contact with development specialists can express their needs, and together they can plan and implement self-help programmes in order to promote development in the rural environment.

12.1.2.2 The So-called "Green Revolution" Model

The emphasis placed by Schultz's study (1964) on agricultural research and extension, together with certain events taking place during the 1950s,[1] led to a new strategy for growth of the agricultural sector, based on increased per hectare yield achieved through the use of high-yield seeds which demanded a high level of inputs such as water and fertilizer. By this means, small farms gained access to technological (biological and chemical) progress in which inputs were wholly divisible and adaptable to the size of the farm, a feature not offered by the mechanical progress envisioned in the previous model.

But initial enthusiasm for the Green Revolution waned as time passed and data gradually became available regarding its effects. Although its value as a means of increasing a country's grain production was not in question, empirical data regarding distribution of the profits that might derive from that increase proved somewhat disappointing. Nevertheless, in the 1980s, high-yield varieties accounted for almost a third of the area sown to wheat and rice in the Third World.

12.1.2.3 Models of Institutional Reform, or "Agrarian Reform"

In addressing domestic obstacles to development in Latin American countries, the Economic Commission for Latin America and the Caribbean (ECLAC) was led to view agrarian reform as essential to development. This type of reform goes beyond simple changes in land ownership and tenure regimes, encouraging increased production via technical aid and appropriate pricing and credit policies (Prebisch, 1963). This type of strategy was to receive the support of the Alliance for Progress and FAO.

Other authors (Dorner and Kanelt, 1971; Gutelman, 1971; Barraclough, 1973; Berry, 1975) advocated the economic need for agrarian reform based on experience showing that economies of scale in agriculture were smaller than expected and that increases in per-hectare productivity and manpower employment on small family holdings were greater than on large farms. Yet although in the developing countries there was evidently scope, in terms of economic efficiency, for agrarian reform as an instrument of rural development, this approach failed to canvass the necessary political support, and during the 1960s was

[1]These events included the following: (i) the population boom in the Third World countries, which generated strong pressure on world food supplies (countries in Africa, Latin America, and Asia, which in the 1930s were net grain exporters, by the 1960s had become net importers); (ii) the failure of two successive monsoons in the Indian–Pakistani subcontinent, and the consequent disastrous effect on crops; and (iii) changes in United States agricultural policy, which replaced surplus-derived food aid to friendly underdeveloped counreies by the supply of fertilizers. It was now felt to be essential that poor countries improve agricultural output in order to avoid large-scale social unrest.

gradually discarded (Bromley, 1984; de Janvry, 1984).

12.1.2.4 The "Induced Innovation" Model

The model put forward by Ruttan and Hayami in the 1960s (Ruttan and Hayami, 1971) marked a significant step forward, in that it considered technological progress as an endogenous variable in the development process, whereas previous models of agricultural development had considered it to be exogenous.

It was felt that the price ratio between factors of production might steer researchers in the direction of a more efficient technology. In other words, each country should choose the technology best fitted to its specific availabilities and characteristics, rather than simply imitating the technology developed or used in other countries under different circumstances. Innovation, moreover, should be applied not only to technology but also to public and private institutions.

12.1.2.5 The "New Green Revolution" Model

This model is based on the following reasoning: the Green Revolution of the 1960s had shown that technological progress could be introduced into agriculture, and that it allowed the creation of a substantial important marketable surplus, an increase in agricultural earnings, and an increased demand for rural population due to increased employment. Therefore, a new technology based on the use of drought- and disease-resistant varieties would ensure technological progress in both subsistence and marketable production, thus increasing overall agricultural output.

This new model assigns to governments an important role both in financing and implementing rural infrastructure, as well as in agricultural research and the supply of means of production and education. This stands in marked contrast to the strategy of the first Green Revolution, which saw the market as the only mechanism for economic and social change. But both approaches coincide in highlighting the need to replace institutional reforms and direct measures to improve income or the means of production in rural areas by technological progress and the expansion of agriculture (Griffin, 1989). This strategy for technological change in agriculture is considered capable of contributing to the increased exports required to pay for capital-intensive goods (fertilizers, pesticides, fossil fuels) and thus create indirect employment in other sectors.

In the industrialized countries, although the model followed has involved enhancing agricultural productivity through a more intensive use of off-farm inputs of all types, the appearance of surpluses and new requirements regarding conservation of the environment and natural resources, are leading to a shift away from the intensive model in favour of a more ecological form of agriculture characterized by the use, or the more extensive use, of environment-friendly technologies.

Yet the increasing complexity of the food and agriculture system, the shift of food decision making from production to distribution, new farm policies, and world-level competition between producing areas are leading to a major crisis in the agricultural sector. In food-glutted societies, the sector has found no alternative raison d'être, and agricultural development models characterized by productivity and the professionalization of farmers have become obsolete. Alternatives such as low-input models account for only a tiny part of the whole, and there is growing evidence of agricultural decline.

12.2 From Agricultural to Rural System

In developed countries, rural economies have ceased to be exclusively agricultural. The relative contribution of agricultural product to gross domestic product has diminished, and new strategies adopted by farmers seeking to adapt to the changing socioeconomic circumstances forced upon them by the global economy have given rise to a steady decline in employment in the sector. Labour-saving technology and alternative employment in other sectors have led to a loss of importance of agriculture in rural areas. The multiple economic activities of farm owners, and the diversification of earnings of agricultural households through the employment of family members in other sectors, or through state allowances such as rural unemployment benefit or retirement pensions, lead in many cases to a decline in agricultural activity. This decline is thus prompted by the transfer of added value to other sectors such as the services sector, from which the farmer contracts agricultural labour.

The waning importance of agriculture as an economic activity has far-reaching consequences, not only for the economy as a whole but more particularly for the rural community. The loss of jobs in agriculture drives out the working population which will seek employment elsewhere, so that agricultural workers—for centuries the majority of the working population—will become a minority.

The move from a food-shortage society to a food-glutted society has prompted a distinction between the concepts of agriculture and food supply. This new dichotomy, coupled with a growing distinction between agricultural produce and food, has deprived farmers of any clear appreciation of their role in society. For the first time, farmers are being encouraged—through extensification policies—to reduce output, without the support of price guarantees. Direct aid for nonproduction is turning them into subsidy managers rather than managers of natural and human resources or mechanical and biochemical capital. They are thus deprived of the almost sacred status as food producers that they hitherto enjoyed in society, a status which ensured society's willingness to grant them aid either directly or via pricing policies. That favourable attitude on the part of society is essential if the flow of subsidies—still enjoyed in industrialized countries—is to be maintained.

The need to retain population in rural areas, thus avoiding the growth of marginal belts of unemployment in urban areas; the need to hold together a community which can no longer be held together by agriculture as an economic activity; and the need to maintain the ecological balance are giving rise to a shift away from agriculture in favour of other rural area activities that might meet those needs. The loss of agriculture's economic importance has thus led to major changes in rural communities in industrialized countries.

12.2.1 Transformation of the Rural Community

The changes taking place in agriculture have had the most profound effect on the rural economy. The transformation of the agricultural sector, particularly since the 1950s, has been so radical that it might be said to have totally disrupted rural life. Although other factors have also influenced this transformation, there is little doubt that changes in the rural community have coincided with changes in agriculture itself. The process of change can be broken down into two broad stages: the first marked by the transition from traditional to modern agriculture, and the second by the incipient shift from modern to what might be termed postmodern or sustainable agriculture (Ceña, 1992).

Yet the transformation has not been limited only to the agricultural sector; it extends to the whole of the rural community which, especially in Western European countries, has undergone profound changes over the last forty years. Among the most striking of these are: (i) demographic changes resulting mainly from the mass rural exodus of the 1960s and, in some areas, the trend toward counterurbanization (i.e., the influx of the urban population into rural areas) from the 1970s onward; (ii) economic changes arising from the decline of agriculture and urban society's new view of the rural community, which has given rise to a diversification of its activities and functions; and (iii) institutional changes resulting from political decentralization and the supranationalization of agricultural policy.

There are two major reasons for these changes: (i) the decline in agriculture; and (ii) the intensive urbanization.

The decline in agriculture, in terms both of the population employed and agriculture's contribution to GDP, has had radical consequences for all rural areas. By far the most important of these have been the drastic reduction of job opportunities for the rural population and the reduced access of this population to public and social services.

The origin of these problems must be sought in the concentration/diffusion model of industrialization implanted in the 1950s and 1960s, characterized essentially by urbanization and technological progress. These characteristics governed both the form of agriculture to be developed and, effectively, the socioeconomic changes affecting the rural community. The need for increased agricultural productivity, in order to ensure a food supply that would satisfy the growing urban demand, was met by capitalization of agriculture, via mechanical (labour-saving) capital and chemical/biological (land-saving) capital, and by an agricultural policy of support prices and protection against potential outside competition.

Moreover, this policy aimed to match agricultural income to that of other sectors of the economy (Ceña, 1992). The results are well documented: surpluses of protected produce and a growing disparity between regional incomes and between farmers. Many farmers, in areas where production was not protected, or on small farms, were forced to totally abandon agricultural activity (migration) or reduce it to a part-time activity.

Moreover, the type of technological progress introduced (mechanization) led to the driving out of thousands of agricultural workers, who sought employment in the cities. All this prompted an unprecedented rural exodus, which in turn led to major demographic and social changes. This bore out the truth of Myrdal's Principle of Circular and Accumulative Causation (1968): the decline in agricultural employment leads to selective migration, which prompts a fall in demand for public services and goods at local level, which in turn triggers a further reduction of rural employment, and thus a reduction in rural income, thus intensifying the drift of the population toward the cities, and hence the devitalization of rural society.

Intensive urbanization during the early decades of development, resulting from the concentration of economic activity characteristic of the model of industrialization implemented in EC countries, has also been one of the causes of mass depopulation of large rural areas. According to Funnell (1988), urbanization is still taking place in a small number of large cities, although within the EC this process differs as a function of the degree of industrialization of the country concerned. Thus, while urbanization appears to have stabilized in the more industrialized countries, in the South the largest cities continue to expand relatively rapidly, following—albeit at a slower rate—the traditional model of rural–urban migration (Commission of the European Union, 1991).

The relative importance of decreases in transport costs and the location of productive activities is governed increasingly by the location both of markets and of the concentrations of economic activity associated with large metropolises. As the urban population grows, so does the local economy, and in this way the trend toward concentration is reinforced. This weakens the ability of peripheral areas, which are generally rural, to attract new economic activity.

The changes taking place in the 1980s were dictated by the location and specific characteristics of the rural areas involved. The Commission of the European Union itself (1988) classified rural areas into: (i) areas close to large conurbations in the centre and north of the community and in coastal regions; (ii) Mediterranean rural areas; and (iii) mountainous and underdeveloped areas.

In the first group, the most important problems at present stem from competition for land use, the undermining of the ecological balance, indiscriminate building of housing, and seasonal overpopulation due to tourism. In Mediterranean rural areas, the essential problem continues to be rural decay, the loss of economic activity and thus a decline in the standard of living of the rural community. Finally, in mountainous and underdeveloped areas constant depopulation and abandonment of land endangers not only the survival of the small population remaining but also the maintenance of the ecological balance due to soil erosion, forest fires, and desertification; all these are in effect problems of rural land management.

The rural community in industrialized countries today has to tread a delicate path between the different functions demanded of it: it must produce food and raw materials within the framework of a sustainable agriculture, while at the same time affording an essential space for ecological balance and for leisure and recreational activities.

In all rural areas, the development of mass media and telecommunications, and the advent of new forms of transportation (particularly the car) have alleviated the isolation of earlier ages. At the same time, the social and economic life of the rural community has lost much of its uniformity, due to the introduction of new economic activities and the diversification of sources of income for farmers and farm workers.

The presence of the mass media and the appearance both of new businesses and new residents, on a temporary or permanent basis, expose the rural community to new ideas and influences. Incoming businesses introduce new ways of working, of organizing work schedules (holidays, regular pay, etc.). The influx of new residents brings with it new professions and different tastes, and therefore changed preferences and new demand for goods and services. In those areas where this incoming population is of significant size (areas generally less than 100 km from a large city, coastal areas, rural areas of tourist interest) this helps to create a greater diversity of commercial and leisure activities, as well as improving public services (education, health care, transport, etc.). All this enhances rural life in these areas, providing a wider range of social and cultural opportunities.

The centres of decision making, regarding both the future of agriculture (new CAP) and other activities in the rural community, are no longer located even in nearby cities, but at an international level (Commission of the European Union, World Trade Organization, International Conferences on the Environment, multinational corporations, etc.). The rural community has become peripheral within the framework of the growing internationalization of all national economies.

12.2.2 The New Economic Space

All these changes, and others not outlined here, show the closely interdependent nature of relations between the rural community and the rest of the economy in general, and the urban community in particular. Although this interdependence is found at various levels, its most significant expression is in economic relations, where it is evident in the commercial flow of agricultural and manufactured goods, the financial flow and the flow of natural and human resources.

Urban areas decide, through market forces, on the allocation of the different resources associated with the rural environment, be they natural (for leisure activities, second homes, tourism, etc.), human (migration of the population to areas where jobs are concentrated) or financial (switch of savings towards urban areas). Yet this allocation, as we shall see later, is far from ideal.

The creation of multinational economic areas (EU, NAFTA, MERCOSUR) places farming and other rural activities within an international network of commercial and political relations. The rural economic space not only spreads to include other nonagricultural activities, but also transcends the boundaries of regional and national economies. Agricultural produce is subject to international trade terms; rural tourism involves foreigners as well as nationals; multinational companies set up in rural areas; city-dwellers choose rural areas as places of residence and for leisure pursuits, and so on. The rural community is thus increasingly integrated into the urban environment, and as a result its economy undergoes increasing diversification. This new situation is being encouraged by agricultural, environmental, and rural development policies.

Consequently, however remote they may be, rural areas are by no means independent either of nearby urban centres or of large cities. Yet although this interdependence is widely recognized, there is still a tendency, both in conceptualization and in policymaking, to treat the rural economy as though it existed independently of national and international economic activity. In 1988, for example, the OECD reported that: "The rural economy has integrated into the national and international economy;...the processes of diversification and internationalization have created a new interdependence between all the sectors of the rural economy. In response, member states increasingly treat the rural economy as a complete economy." It would appear ill-advised, however, to treat the economic activity of the rural community as a separate entity from national and international economic activity. The need for an international perspective is constantly growing, because the agricultural sector, which is still so important in many rural areas, is wholly immersed in an increasingly large-scale international interdependence.

From all the foregoing, it may be concluded that a new approach is required if the problems affecting the rural community, of which agriculture is an integral part, are to be solved. Such an approach must take into account the specific features of that community and its importance for the well-being of society as a whole. Moreover, it is essential to review the passive role assigned to the rural community by the global development models of the past, grounded on the traditional assumption that its contribution to national economic growth was of secondary importance.

12.3 Farming Systems

The changes taking place in agriculture have not only been apparent at macroeconomic level (reduction of agricultural product, decline of its contribution to GDP). The structure of the agricultural sector in general has also been transformed: production systems have been modified—in some countries more than in others—in terms of both their relationship with other sectors of the economy and the internal links between their various components.

In the past, agricultural systems were designed either to enhance the productivity of land, labour, and manmade capital, or to ensure survival. In both cases, farmers were managers of a set of scarce resources. Two broad agricultural systems have focused the interest of researchers: traditional and modern agriculture.

The term "traditional agricultural system" applies equally to peasant smallholdings in developing countries and to the large extensive-crop estates found in some of these countries. Both meet the requirements implicit in the definition of the term, that is, use of biological energy, nonreproducible capital and low labour productivity.

During the early years of development, the farmer in developing countries was lacking any economic rationale. His decisions were held to be dictated more by tradition or conservatism than by economic incentives. Prevailing economic theory therefore was inapplicable to the study of peasant agricultural systems. Other interpretations of peasant behaviour may be found in Chayanov's work (Chayanov, 1966).

In the 1960s, Schultz's (1964) book *Transforming Traditional Agriculture* argued that such assertions had no grounds in reality, and that allocation of resources by peasant farmers was in fact efficient. According to Schultz, the peasant agricultural system is based on an equilibrium achieved generation after generation using the same factors of production. In other words, it would be extremely difficult to increase output by reallocation of available resources. The problem, therefore, is not cultural but simply economic: the marginal productivity of labour is so low that there is little point in working more; the very low marginal productivity of capital equally renders saving pointless. In studies carried out in Guatemala and India, Schultz actually found that the rate of yield of factors of production in traditional agriculture, including land, is only 3 or 4 percent.

Schultz's thesis met with harsh criticism, directed against both his methodology and his "efficient but poor" theory. But even so, it may be concluded that the behaviour of farmers in traditional agriculture is not aimed at maximizing profit, nor even at maximizing expected profit, but rather at minimizing the likelihood of a below-subsistence output, and at ensuring survival by mitigating the consequences of a poor harvest (primitive social security; extended family and broad-based intergroup, interclan, headman–client relationships). The peasant's fear of risk is closely linked to survival and to the absence of insurance of the sort available to farmers in developed countries. Subsistence strategy involves diversification of crops, even on the same land. (For survival algorithms see Pyle and Turnovsky, 1970; Roumasset, 1976.)

Schultz's contribution was significant to the extent that it challenged classical theories regarding the causes of low productivity in traditional agricultural systems and the inability of such systems to become dynamic, high-productivity sectors. In questioning the crucial role attributed by prevailing theory to physical capital in the development of the sector, and in highlighting the need for human resources in general, and for improved education in particular, Schultz paved the way to a better understanding of peasant-farmer behaviour in developing countries, even though that understanding was achieved largely through criticism of his theories.

In the developed world, traditional agricultural systems were turned into modern systems mainly through the capitalization of farms and their integration into the global food and agriculture system. The dramatic decline of the rural population (examined earlier), higher agricultural wages, and the abandonment of land through migration led the remaining farmers to capitalize their operations by means of labour-saving technology.

The internal competition to which farmers have been subjected and the implementation of certain agricultural policies designed to increase output have created farming systems based on intensive use of off-farm inputs. Farm management increasingly resembles the management of a capitalist business rather than a family farm of the type which has prevailed in most countries. Globalization of the food and agriculture complex and the falling level of support through agricultural policy have left farmers in developed countries open to competition from those of developing countries. The resulting systems are geared toward selling rather than producing, giving rise to a radical transformation of modern agricultural systems.

In industrialized regions such as the EU, agricultural models have largely been dictated by the size and agrological characteristics of the farms involved. While large farms have tended to specialize in extensive monoculture, medium-sized and small farms with good soil, water, and climate conditions have favoured input-intensive operations (e.g., fruit and vegetables, livestock). Some small farms, and farms whose economic results do not justify year-round employment of manpower or sufficient revenue, have opted for part-time agriculture. Plural activity may remain wholly within the agricultural sector (in the case of larger or more profitable farms), or may involve other sectors of the economy (construction, tourism, etc.). The importance of part-time agriculture varies from country to country, but in many cases accounts for over 50 percent of all farmers (Arnalte, 1992).

Growing integration into national and world markets, coupled with off-farm employment alternatives, has radically transformed the predominant concept of family farming, of the family as a farm production and management unit and of the family economy. Farmers' families are abandoning common objectives, and the strategies of each family member vis-à-vis the farm itself are imposed by considerations wholly independent of those of other members (Baptista, 1995). The transfer of knowledge has also changed. Due to the speed and increasing complexity of technological advances, and to the integration of farms and farming families into a market-orientated system, knowledge is no longer handed down within the family, but rather provided outside the family in the form of officially recognized professional training. The farming family is increasingly becoming a unit for the management of available resources rather than a labour unit (Arnalte, 1992). The family farm model that for years prevailed in Western Europe is becoming progressively more abstract and difficult to control; at the same time, the new model resulting from the amalgamation of family farms is acquiring greater significance.

The negative effect of input-intensive systems and of monoculture on the environment is gradually giving way to a more extensive and mixed farming that offers greater sustainability from the standpoint of ecological balance. The introduction of afforestation on farms traditionally devoted to agricultural crops is one aspect of the new agro-environmental objectives imposed, yet again, by society on the agricultural sector. New forms of production are aimed at achieving: (i) lower fossil-fuel consumption; (ii) reduced use of chemical pesticides; (iii) reduction of soil erosion; (iv) reduced leaching of nitrates into water; (v) reduced pollution of soil water; and (vi) efficient use of water for irrigation.

The environmental demands on agriculture do more than just transform cultivation techniques used in production systems. They also give rise to a transfer of rights regarding the use of resources (land, water, etc.) from farmers to society as a whole, which in turn obliges farmers to conserve resources. This obligation may take the form of legislation penalizing certain forms of fieldwork, or may be reflected in the characteristics demanded of agricultural produce by the consumer or by processing plants (themselves bound by quality and safety regulations). In countries where water is a scarce resource (e.g., countries in the Mediterranean basin), the farmers' historical water rights are being questioned, and even withdrawn, in response to new social demands.

Food-glutted societies that pursue agricultural policies aimed at extensification often question the use of water for irrigation. However, while in general terms this view may be reasonable, it is less justified in specific cases where irrigation, rather than simply increasing per-hectare yield, is the only procedure actually ensuring production. This is the case in regions with very hot climates, which every year face long periods of total drought, and in which a shortage of water would endanger the obtaining of any harvest at all. This does not of course signify that the water used for irrigation should not be managed more efficiently, both in terms of saving and in terms of the type of crop irrigated. New agricultural systems have to manage natural resources with a view to their total value, and not just their useful value. Unless the value of their conservation (option or existential value) is taken into account, allocation will not be efficient.

12.4 Agricultural Support and International Markets

Since the economic crisis of the 1930s, the agricultural sector has been supported by all types of government, either because of its strategic status (food security), or else to stimulate or otherwise modify output or to maintain farmers' earnings in the face of falling prices when supply outstrips demand. Sometimes the productivity or competition demanded of agriculture by the remainder of the economy can only be achieved through major structural transformation of the agricultural sector as a whole. Such changes are not always prompted by market indicators, signalling the need for reallocation of inadequate or non-existent resources; it is then that governments must intervene to facilitate the appropriate structural changes.

Aid to agriculture furnishes it with greater resources than would be made available to other sectors. Agricultural policies have also affected methods of production and farm size.

12.4.1 Government Intervention in Agriculture

Agriculture is treated very differently in industrialized and developing countries. While in industrialized countries governments generally intervene to push domestic agricultural prices above those of the world market, in developing countries intervention tends to be aimed at bringing prices below international levels.

In developing countries, where agriculture is generally the largest economic sector, agricultural pricing policies have become the main instrument for influencing the economy. Policies aimed at transforming farm structure have been important only during specific sociopolitical periods, when governments have launched agrarian reforms whose results, in terms of the structural transformation of the agricultural sector, have for various reasons failed to prove significant (de Janvry, 1984).

The most common aims of agricultural policies in these countries have been: (i) to increase output, reduce import needs, and thus attain self-sufficiency; (ii) to stabilize farm prices or reduce retail prices in cities; (iii) to generate income for the government through taxation of the agricultural sector; (iv) to increase foreign currency reserves by boosting exports; and (v) to provide raw material for the food processing and textile industries at below free-market prices.

The effects of such policies have been widely documented in studies of development strategies since the 1960s (Little et al., 1970; Balassa, 1971; Krueger, 1978, 1983; Krueger et al., 1981). Their specific effect on agriculture is evident in the fact that agricultural pricing policies implemented in most developing countries discriminated against agriculture from 1960 to 1984 (Valdés and Schiff, 1992).

The policies benefitted: (i) governments, which increased their revenue; (ii) urban consumers, who were able to buy food cheaper than on the free market; and (iii) industry, which acquired raw materials and other inputs at lower prices. The losers were the farmers. Subsidies to cover input costs barely offset the fall in revenue for the sector as a result of intervention in product markets. The main objective of pricing policies would thus seem to have been the removal of resources from agriculture. According to Mundlak (1985): "policies which tax agriculture reduce investment in agriculture, increase outmigration, and reduce the implementation of new techniques." All this leads to a lower growth rate.

In most industrialized countries the aims of government intervention in the agricultural sector differ from those pursued in developing countries, in that the former seek to protect agriculture. Protection of agriculture in developed countries increased from the 1950s onward, and by the mid-1970s had accelerated substantially.

Intervention seeks to ensure both efficiency and equity and, since the beginning of the 1990s, to attain sustainability. Attempts may sometimes also be made to facilitate structural transformation in order to achieve those aims. In the member states of the EU, for example, initial agricultural policy sought to increase the final agricultural product (through increased sector productivity) and match the income of farmers to that of other social groups, assure the consumer of sufficient food at affordable prices and stabilize domestic markets. The instruments used were pricing and market regulation policies, designed to maintain domestic prices above world prices through state purchase of agricultural produce at support prices, and protection against third-country competition by means of import tariffs.

Measures of this nature had several major effects:

(i) the intensification of agriculture through subsidized chemical and mechanical inputs;

(ii) growth of the agricultural sector borrowing requirement and farm specialization, leading to the creation of veritable production basins;

(iii) a dramatic increase in agricultural productivity (i.e., supply outstripped domestic demand, giving rise to substantial surpluses);

(iv) increased competition between regions and between individual farms, leading to disparities in earnings levels.

EU subsidy policies and U.S. export enhancement measures prompted a further fall in world prices, already declining in real terms, and there were major fluctuations in the 1970s and early 1980s. EU member states improved their self-sufficiency in grains and livestock products, while developing countries progressively increased grain imports.

Low and falling agricultural prices and the low elasticity of food supply/demand may partly account for the fall of international food prices in real terms, and seasonal factors may account for fluctuations around these trends. Even so, the rate of decrease and the increasing variation in prices from the early 1960s onward, together with a change in the models of food trade between developed and developing countries, suggest that government intervention has caused increased distortion (Tyers and Anderson, 1992). Growing international chaos on world food markets began to prompt distortions of the economy, which eventually led to negotiations for agricultural policy reform. As a result, between 1986 and 1993 a number of proposals were made in the course of the GATT Uruguay Round.

12.4.2 Agricultural Policy Reform

The negotiations taking place within the framework of the Uruguay Round, together with the need to reduce the financial costs entailed by price guarantee policies, led OECD countries in the late 1980s to launch a reform of traditional agricultural policies. The aims of this reform were: (i) to make farms more sensitive to market indicators, in order to reduce distortion and thus reduce support for increased output; and (ii) to effect structural adjustments in cases where a direct relationship exists between aid received and structural transformation of farms.

One of the major instruments of this reform is the system of direct payments to farmers. The key aim of this policy is to clearly define the group of farmers who will benefit from aid, in order to exert the most direct influence possible on the variable selected by policymakers. Moreover, the eligibility requirements to be met by benefiting farmers are defined.

In implementing farm income support schemes of this sort, it is essential to distinguish between farm income and the farmers' standard of living. There are two major reasons to highlight this distinction: standard of living depends on consumer prices, and the income of many farming families is also derived from other non-agricultural activities. Consequently, measures of this type must be clearly targeted toward eligible farmers. Adequate implementation of direct aid depends on the availability of full statistical data regarding all sources of income of farmers; otherwise, resources are in danger of being transferred from the rest of society to farmers whose standard of living is higher than that of the average taxpayer (as happens in the EU). The development of a reliable database and in-depth follow-up, though entailing considerable operating costs, allows better management of public funds and greater efficiency in the achievement of objectives such as income support for the most needy farmers and reduced pesticide use.

For the first time, as a result of the Uruguay Round, agricultural policies have become the object of international agreement; the negotiations have laid the foundations for national agricultural policies for the next decade. Henceforth, national policies will be bound by GATT agreements. The main aim of the Uruguay Round, in agricultural questions, was to attain a greater liberalization of agricultural markets, and to bring under GATT rules and discipline all measures affecting import market access and export competition. The final agreement on agriculture takes as its long-term objective: "...to establish an equitable, market-orientated system of farm trade, [and to] launch a process of reform by means of the negotiation of agreements on aid and protection, and by means of the establishment of strengthened GATT rules and discipline and more efficient functioning." This long-term objective consists of "forecasting a substantial progressive reduction of aid and protection for agriculture, sustained throughout the agreed period, as a result of which restrictions and distortions in world agricultural and livestock markets can be corrected and prevented."

The progressive reduction of protectionist measures favours and stimulates competition between developed and developing countries, by opening domestic markets to outside influence. In the short term, no great impact is expected either from changed tariff policies (because these represent only a partial liberalization) or as a result of the sought-after reduction of farm subsidies, which is small and spread over several years. Distortions will therefore continue to exist on world commodity markets. According to FAO, farm prices are expected to show a moderate rise (between 5 and 10 percent for products from temperate climates, and less for products from tropical climates).

Food-importing countries will face an increase in world prices, and those that depend on food aid may see aid reduced, as surpluses in donating countries fall following the decline in hitherto supported domestic prices. Reduction of export subsidies and increased world prices will raise the import prices paid by developing countries.

Effects on price stability remain unclear. On the one hand, the elimination of all nontariff duties on imports will expose all countries to the vagaries of the world market, which will in turn affect the stability of their own markets. On the other hand, the general trend toward liberalization may lead to a fall in government management of intervention stocks, which might prompt a generalized downturn in overall stocks, thus giving rise to wide price variations. At the world level, this could trigger—indeed has already triggered—a fall in stocks.

12.5 Agriculture and Environmentally Sustainable Agriculture

Agricultural development is today no longer seen as simply a matter of increased productivity; the political agenda also includes considerations such as efficient land use, efficient management of natural resources and protection of the environment. These aspects are vital to sustainable development, and therefore have to be taken into account in designing agricultural development programmes.

The term "sustainability" was first used in 1980 by the International Union for the Conservation of Nature and Natural Resources (IUCN, 1980) in its "World Strategy." It is, however, a difficult term to define, in that it embraces a range of aspects, from ethical to biological, from economic to social and anthropological.

The demand for sustainability has been interpreted in various ways by authors and international organizations. The definition approved by the FAO Council in 1988 runs as follows: "Sustainable development is the handling and conservation of natural resources and the orientation of technological and institutional change so as to ensure the continuous satisfaction of human needs for present and future generations. This sustainable agriculture (in the agriculture, forestry, and fishing sectors) conserves water and plant and animal genetic resources, does not degrade the environment, and is technically appropriate, economically viable, and socially acceptable" (FAO, 1989). Other definitions may be found in Pezzey (1992) and Ruttan (1991).

As a theoretical aim, sustainability is challenged by virtually no one. The problems arise with its implementation. When including the sustainability of agricultural production as an objective of development, it is essential to understand and recognize trade-offs with other objectives such as maximization of yield or better distribution of profits.

Indeed, one problem commonly encountered in the analysis of sustainability is precisely the comparison with economic efficiency. But a key distinction must be drawn between individual efficiency (from the standpoint of private users of the production system) and social efficiency (the way in which production systems affect the allocation of resources in society as a whole). Individual efficiency is not always a guarantee of social efficiency, due for example to the negative external effects of agricultural systems on third parties.

In agriculture, one striking example of the market's failure to achieve an efficient allocation of resources is the existence of external factors or public goods which affect the environment, that is, when farmers produce public (environmental) goods or positive external (landscape) factors whose market value reflects only the valuation of the private buyer, leading to underestimation of the real social value.

Similarly, certain types of intensive agriculture exert negative effects on the environment (contamination of soil and water, soil erosion, exhaustion of agricultural and woodland resources, etc.) whose costs can no longer be borne by society and are not reflected in market prices. The cost of degradation of a natural resource, or user cost, is measured in terms of the future nonavailability of that resource, that is, its option value (intergenerational equity). As a result, market operations lead to production levels which are far from optimal for society as a whole.

But the relationship between agriculture and the environment has to be viewed within the larger framework of the network of relationships linking population, resources, and environment, whose main overall aim is to feed a growing population over the forthcoming decades, under the constraints posed by limited resources and a vulnerable environment. Agriculture, in short, must guarantee that the interaction between population and natural resources will ensure their availability for present and future generations. It must therefore conserve and maintain natural capital and biodiversity as well as man-made and human capital.

In developing countries, degradation of natural capital stems generally not from input-intensive agriculture or industry, but from excessive pressure exerted by the population on available resources in order to achieve food and energy survival. Structural adjustment programmes introduced at the start of the 1980s, forced the economies of many developing countries faced with considerable foreign debt, to concentrate on the immediate problems of servicing that debt rather than on long-term problems such as sustainable management of natural resources. Moreover, these resources may be being exploited in order to obtain the foreign currency required to service the debt. Thus, items more properly viewed as capital—in this case natural wealth—are treated as revenue. This leads to loss of potential for future development, thus making it impossible to achieve the object of intergenerational equity of sustainable development (Ceña, 1994).

Agriculture in developed countries is a sector particularly affected by this new concept of sustainable global development. Indeed, the sustainability taken for granted by society as a whole is translated into new demands on the rural community: (i) food that is healthy in terms of both composition and processing methods, as well as being environmentally friendly, thus requiring major changes in agriculture; (ii) availability of recreational and leisure pursuits; and (iii) conservation of the landscape and maintenance of the ecological balance. In short, there is a demand for sustainable agriculture, that is, agriculture closely linked to ecological and environmental systems, but also to economic systems, and for a rural development that affects both rural and urban communities.

According to Dankelman and Davison (1988), the requirements of sustainable agriculture in developing countries are as follows: (i) equal access of farmers to fertile land, to credit, and to agricultural information; (ii) maintenance and support for an independent agriculture controlled by male and female farmers; (iii) development of methods of farming, food-processing, and storage that facilitate the employment of women; (iv) high degree of crop diversification to maintain flexible harvesting models; (v) conservation of soil fertility by recycling of organic matter, in order to prevent dependence on imported nutrients; and (vi) appropriate use of water and fossil fuels.

In developed countries, farmers' aims for sustainable agriculture may be somewhat different: (i) to maintain and improve natural resources; (ii) to protect the environment; (iii) to ensure profits; (iv) to conserve energy; (v) to increase productivity; (vi) to improve food quality and safety; and (vii) to create a more viable infrastructure for farms and rural areas.

If we are to have an ecologically and economically healthy world, agricultural systems and environmental systems must be integrated onto sustainable systems. The problem of agricultural/environmental integration has been neatly presented by Young (1991): "To what extent do macro-agricultural policies take into account their effects on environmental quality, and to what extent do environmental policies take into account their impact on agricultural output, income and prices?"

The problem has to be defined in these terms if the close relationship between the two fields, and the potential for mutual reinforcement between agricultural and environmental decision-makers is to be fully reflected. This is an essential prerequisite for sustainable development. Of course, this does not mean that agricultural policy needs to be obliged to pursue purely environmental aims, nor that environmental policy is to be subordinated to agricultural needs. Integration of policies through appropriate tradeoffs between the interests of the two sectors can lead to economically and environmentally rational, efficient and sustainable development. Greater integration of policies, it is felt, will enhance the sustainability of agricultural policies.

In order to maintain a given population level in rural areas, it is essential to aim at diversification of activity, because the agricultural sector cannot by itself generate sufficient employment or the required added value. This may lead to conflicts with certain environmental objectives, because certain productive activities, including agriculture, have negative effects on the environment. Similarly, measures aimed at nature conservation (national parks and natural reserves) may at times prevent the development of productive activities, giving rise to social conflict.

The strategy currently being implemented involves the integration of sectorial policies (transport, energy, agriculture, etc.) and environmental policies, as well as the creation of special-interest groups representing all the social agents (consumers, technicians, experts, unions, and professional classes) and different administrative bodies at local, regional, national, and even supranational level, to analyse environmental problems and draft rural development policies. Cooperation, complementarity, and partnership are the principles underlying the drafting and implementation of these development policies, in accordance with the agreements reached at the Rio Summit, which focus attention on "consensus-building" as a key aspect of sustainable agriculture, given that it is crucial to go beyond theoretical agreement in policymaking to practical agreement in the implementation of policies (Desai, 1994).

References

Arnalte, E. Agriculturas del sur de Europa y reforma de la PAC. *Rev. Econ.* 12; 1992.

Bairoch, P. *Révolution Industriel et Sous-Développement.* 4th ed. Paris: Mouton; 1974.

Balassa, B. *Development Strategies in Semi-Industrial Economics.* Baltimore, MD: The Johns Hopkins University Press; 1971.

Balogh, T.S. *La Economía de la Pobreza.* Barcelona: Vicens-Vives; 1972.

Baptista, F.O. Politiques agricoles, strategies paysannes et développement rural. Mimeo. Tunis: IRMC; 1995.

Barraclough, S. *Agrarian Structure in Latin America.* Lexington, MA: Lexington Books; 1973.

Berry, R.A. Special problems of policy making in a technological heterogeneous agriculture: Colombia. Reynolds, L.G., ed. *Agriculture in Development Theory.* New Haven, CT: Yale University Press; 1975:253–297.

Bromley, B.W. The role of agrarian reform in economic development: Comment. Eicher, C.K., Staatz, J.M., eds. *Agricultural Development in the Third World.* Baltimore, MD: The Johns Hopkins University Press; 1984:275–278.

Ceña, F. Transformaciones del mundo rural y políticas agrarias. *Rev. Estud. Agro-soc.* 162:11–35; 1992.

Ceña, F. Planteamientos económicos del desarrollo rural: Una perspectiva histórica. *Rev. Estud. Agro-soc.* 169:11–53; 1994.

Commission of the European Union. Document 501; 1988.

Commission of the European Union. *The Regions in the 1990s.* Brussels: Commission of the European Union; 1991.

Cornia, G.A. Is adjustment conducive to long-term development? The case of Africa in the 1980s. Vagi, G., ed. *From the Debt Crisis to Sustainable Development: Changing Perspectives on North–South Relationship.* New York: St. Martin's Press; 1993:165–206.

Chayanov, A.V. *The Theory of Peasant Economy.* Homewood, IL: Irwin Inc.; 1966.

Dankelman, I.; Davison, J. *Women and Environment in the Third World.* London: Earth-scan Publications, Ltd.; 1988.

De Janvry, A. The political economy of rural development in Latin America: An interpretation.

Eicher, C.K.; Staatz, J.M., eds. *Agricultural Development in the Third World.* Baltimore, MD: The Johns Hopkins University Press; 1984: 263–275.

Desai, N. The Rio Earth Summit: A year beyond. Serageldin, I.; Steer, A., eds. *Valuing the Environment: Environmentally Sustainable Development.* Proceedings Series No. 2. Washington DC: World Bank; 1994:10–13.

Dorner, P.; Kanelt, D. The economic case for land reform: Employment, income distribution, and productivity. Dorner, P., ed. *Land Reform in Latin America.* Madison, WI: Land Tenure Center, University of Wisconsin; 1971:41–56.

Food and Agriculture Organization of the United Nations (FAO). *El Estado Mundial de la Agricultura y la Alimentación. Part III.* Roma: FAO; 1989.

Funnell, D.C. Urban–rural linkages: Research themes and directions. *Geograf. Annal.* 70B(2): 267–274; 1988.

Furtado, C. *Economía y Política del Desarrollo Económico.* Mexico: Siglo XXI; 1968.

Griffin, K. *Stratégies de Développement.* Paris: OCDE; 1989.

Gutelman, M. *Structures et Réformes Agraires.* Paris: Maspéro; 1971.

Hirschman, A.O. *The Strategy of Economic Development.* New Haven, CT: Yale University Press; 1958.

International Union for the Conservation of Nature and Natural Resources (IUCN). *World Conservation Strategy: Living Resource Conservation for Sustainable Development.* Gland, Switzerland: IUCN-UNEP-WWF; 1980.

Johnston, B.F.; Mellor, J.W. The role of agriculture in economic development. *Amer. Econ. Rev.* 51(4):566–93; 1961.

Krueger, A.O. *Foreign Trade Regimes and Economic Development: Liberalisation Attempts and Consequences.* Lexington, MA: Ballinger Press for the National Bureau for Economics Research; 1978.

Krueger, A.O. *Trade and Employment in Developing Countries; Vol. 3. Synthesis.* Chicago, IL: University of Chicago Press; 1983.

Krueger, A.O.; Lary, H.B.; Akrasanee, N. *Trade and Employment in Developing Countries; Vol. 1. Individual Studies.* Chicago, IL: University of Chicago Press; 1981.

Kutznets, S. *Croissance et Structures Economiques.* Paris: Calman Levy; 1972.

Lewis, A.W. *Economic Development with Unlimited Supplies of Labour.* The Manchester School of Economic and Social Studies XXII(2):139–191; 1954.

Little, M.D.; Scitovsky, T.; Scott, M.F.G. *Industry and Trade in Some Developing Countries.* London: Oxford University Press; 1970.

Malassis, L. *Economie Agro-alimentaire; Vol. I. Economie de la Consommation et de la Production Agro-Alimentaire.* Paris: Cujas; 1973.

Mellor, J.W. Food policy, food aid and structural adjustment programs. The context of agricultural development. *Food Policy* 13(1):10–17; 1988.

Mundlak, Y. *The Aggregate Agricultural Supply.* Working Paper 8511. Rehvot, Israel: Centre for Agricultural Economic Research; 1985.

Myrdal, G. *Asian Drama. An Inquiry into the Poverty of Nations.* New York: Pantheon; 1968.

Pezzey, J. *Sustainable Development Concepts. An Economic Analysis.* World Bank Environment Paper No. 2. Washington, DC: World Bank; 1992.

Prebisch, R. *Hacia una Dinámica del Desarrollo Latino Americano.* México, DF: Fondo de Cultura Económica; 1963.

Pyle, D.; Turnovsky, S.J. Safety-first and expected utility maximization in mean-standard deviation portfolio analysis. *Rev. Econ. Statist.* 23; 1970.

Rostow, W.W. *The Stages of Economic Growth: A Non-Communist Manifesto.* Cambridge, MA: Cambridge University Press; 1960.

Roumasset, J.A. *Rice and Risk: Decision Making Among Low-Income Farmers.* Amsterdam: North-Holland; 1976.

Ruttan, W. *Sustainable Growth in Agricultural Production: Poetry, Policy, and Science.* Department of Agricultural and Applied Economics. Staff Paper P-91. St. Paul, MN: University of Minnesota; 1991.

Ruttan, W.; Hayami, Y. *Agricultural Development: An International Perspective.* Baltimore, MD: The Johns Hopkins University Press; 1971.

Schultz, T.W. *Transforming Traditional Agriculture.* New Haven, CT: Yale University Press; 1964.

Tyers, R.; Anderson, K. *Disarray in Worldfood Markets: A Quantitative Assessment.* New York: Cambridge University Press; 1992.

Valdés, A.; Schiff, M. *The Political Economy of Agricultural Pricing Policy. Vol. 4: World Bank.* Baltimore, MD: The Johns Hopkins University Press; 1992.

Young, M.D., ed. *Towards Sustainable Agricultural Development.* London: Belhaven Press; 1991.

13

Introduction to the Economics of Resource Use in Planning

Caroline Saunders and Martin Whitby

13.1 Introduction

Natural resources are a pervasive part of any economy and they pose particular problems for any society trying to plan their use. By natural resources we mean all resources that are produced by very long, slow geological processes which are not easily reproduced by human activity. Major natural resources in- clude land, water, and the earth's atmosphere. They are natural resources in the sense that they cannot be reproduced and that their management has profound effects on the welfare of whole populations. Because of their special characteristics they are particularly difficult to allocate to appropriate uses through unfettered markets, and planning and policy systems have evolved in developed countries in an attempt to improve their use. In this chapter we review the economic underpinning of resource management, as practised through the planning system.

The story begins with an explanation of how markets work and a review of the problems that arise where they do not. Renewable and nonrenewable natural resources are contrasted, and then we consider the central problem encountered in planning, namely, the allocation of land between competing uses. The text then moves on to consider market failure where property rights are not adequately specified, especially in the case of public goods and what are known as externalities. The final section then seeks to bring these arguments together in their application to cost–benefit analysis. First, though, we review some of the ingenious methods now available for valuing goods not traded through markets, and then we present a worked example of how this information is put together to indicate the value to society of alternative courses of action.

13.2 Basic Economic Principles

Economics is the study of how to allocate scarce resources among infinite wants, the objective being to maximize the well being of an individual, a group, or a society. To achieve this implies efficiency of resource use, which depends, firstly, upon the optimum use of scarce resources. Thus: "The economy is producing efficiently when it cannot produce more of one good without producing less of something else" (Samuelson and Nordhaus, 1995). Secondly, the consumption of resources is by those who value them the most. When these two conditions have been met we have maximized social welfare and have optimum efficiency. often referred to as Pareto efficiency, where "No one can be made better off without making someone worse off."

Economics focuses particularly on the allocation of scarce resources. A good or service is said to be scarce if, at zero price, the amount demanded exceeds the amount produced or supplied. If goods or services are scarce then choices have to be made about their consumption. Economics is concerned with how these choices are made, that is, how producers choose to use scarce resources in the production of goods and services and distribute these among consumers.

The existence of scarce resources and infinite wants raises the necessity of choice, which implies that there is a cost associated with any decision. Economics measures this cost in terms of the next best alternative available. This concept is central to economics and is called opportunity cost—the cost of an activity defined in terms of the next best opportunity foregone.

There are a number of economic systems, and these range from command or socialist economies where resource allocation is controlled by the state, to pure market economies. In reality most economies are mixed, that is, they exhibit features of both extremes mentioned above. In this chapter we concentrate upon the working of the market economy and how this operates in relation to the allocation of natural resource or environmental goods.

A market economy works on the basis that the selfish behaviour of individuals can lead to the most efficient allocation of resources. This theory was developed by Adam Smith (1776), who stated in *The Wealth of Nations* that: "It is not from the benevolence of the butcher, the brewer or the baker that we expect our dinner but from their regard to their own interest" (Smith, 1776).

By trying to maximize our own benefits, selling for as high a price, and buying for as low a price as possible, the market ensures that the goods and services produced are those having the greatest value to society and that they go to those who value them the most. This therefore achieves the criteria of Pareto efficiency.

Markets bring together buyers and sellers. Sellers supply goods and services, and how much they are willing to supply depends on a variety of factors, such as price, the price of similar goods, technology, and costs of production. Buyers demand goods and services, and how much they demand likewise depends on a variety of factors, including price, price of related goods, income, and tastes. The common factor influencing both buyers and sellers is price, and it is price that determines what is actually bought and sold.

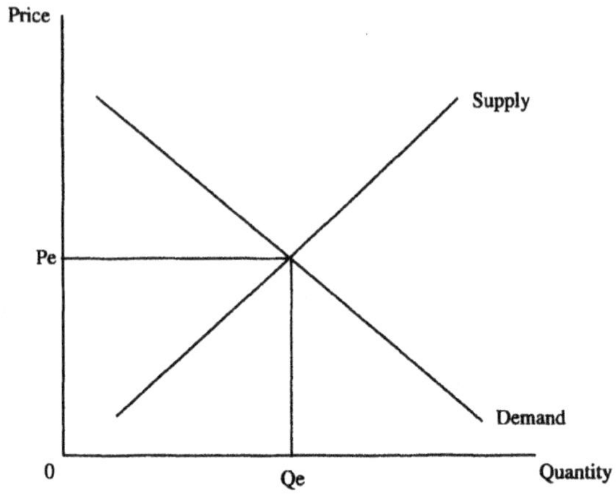

FIGURE 13.1. Market diagram.

This relationship is illustrated by Figure 13.1, which shows how the quantities demanded and supplied are assumed to change with respect to price. In the case of demand, the higher the price the less will be demanded and as price drops demand will increase; therefore the demand curve illustrated in Figure 13.1 slopes downward. In the case of supply, producers are assumed to be more willing to produce a good or service as the price rises, so the supply curve slopes upward, as illustrated in Figure 13.1. Other factors affecting both demand and supply will also cause the demand or supply curves to shift. In the case of demand, if, for example, the price of a substitute good fell, this would reduce the price of the good under analysis at each and every price on the demand curve, causing the demand curve to shift to the left, as consumers moved their expenditure to buying the substitute.

As stated above, other factors affecting supply include the price of *jointly produced* goods and the costs of production. If the price of a complementary or jointly produced good rises, then this will increase both its supply and the supply of the good in question and both supply curves will shift to the right. The equilibrium price will fall and the quantity both sold and purchased will rise.

The market will clear, that is, it will find an equilibrium with no tendency to change, where demand precisely equals supply. It is this that determines what is actually bought and sold at what price. In Figure 13.1, this is at price *Pe* and quantity *Qe*.

But markets will only achieve efficient allocation of resources under a certain set of assumptions, including: (i) perfect knowledge of both sellers and buyers; (ii) no restrictions of entry and exit to and from the market; (iii) homogeneous products; (iv) many buyers and sellers so that no trader can influence the market price by ceasing to trade; and (v) fully defined property rights. Under these circumstances, all costs and benefits of an activity are passed through appropriate markets. If any of these conditions cannot be met, then we have market failure and markets may not achieve an efficient allocation of resources.

The existence of market failure is one of the principal arguments for government intervention. Some causes of market failure can be solved through minimal intervention or regulation. For example, the requirement for perfect information can be met by the government encouraging the establishment of market information systems. In fact most markets could not operate without an institutional structure to provide information and enable traders to make contact. Other forms of market failure may require government regulation, such as monopoly legislation, to ensure that no individual firm, or collaborating firms, have excess power to influence prices or prevent entry of other firms into the market. However, it is market failure due to ill-defined property rights that is most frequently seen, with environmental goods and natural resources often requiring direct intervention, as outlined in more detail below.

Other problems also exist with the market allocation of resources especially where society may consider that the distribution of income created by the market is not acceptable. So societies relying on the market structure to allocate resources have established ways of redistributing income by collecting taxes and redistributing the proceeds through welfare systems.

13.3 Application of Economics to Natural Resources

Natural resource or environmental economics is a branch of economics that has developed due to their special characteristics. This is because property rights are frequently less easy to define in the case of natural resources or environmental goods. Other important distinguishing features of natural resources or environmental goods include them having a fixed stock or stock that changes at a natural or biological rate. This feature means that time is of central importance in the allocation of natural resource economics.

Economists generally use discounting as a technique for evaluating the allocation of resources over time. Discounting is based upon the presumption that given a choice between \$1 today or \$1 in a year's time individuals would choose to receive it now. To entice someone to give up \$1 today in return for a \$1 in a year they would have to given some compensation. How much that compensation might be depends on individual preferences reflecting perhaps the alternatives for which the \$1 could be used now and in the future. At the level of the whole economy the alternative is usually considered as placing the money in the bank and obtaining the market rate of interest on it. So the choice is between putting \$1 in the bank or investing in a project.

This is equivalent to conserving a natural resource. If it is in the bank then the return after one year would be the original \$1 augmented by the rate of interest (r) so the investment project must return at least that to persuade someone to invest. Therefore:

$$FV = PV(1 + r)$$

where:

FV = future value
PV = present value
r = rate of interest

or:

$$PV = FV/(1 + r)$$

For example, if the present value is \$100 and the market rate of interest is 10 percent, then the future value would be:

$$FV = 100\,(1 + 0.1) = \$110$$

This formula can be used to calculate the outcome for any number of years in future, for example:

After 1 year = 100 (1 + 0.1) = 110
After 2 years = 110 (1 + 0.01) = 121

Summarize

$$FV = PV(1 + r) + PV(1 + r)^2 + \cdots$$
$$FV = PV(1 + r)^t$$

So

$$PV = FV/(1 + r)^t$$

If present value is greater than the cost of the investment project, then it is considered to be profitable. This formula can be applied to the allocation of natural resources.

Natural resources differ from others in that there is a fixed stock of resource that changes at a rate outside human influence. We can distinguish three main types of natural resources: non-renewable, renewable, and land. In all there is a stock of resource, but in the former this is static; in the second the stock changes at a biologically determined rate; and in the case of land the stock is fixed, but it is the service flows that are of interest. These will be dealt with in turn.

13.3.1 Nonrenewable Natural Resources

The supply of nonrenewable natural resources is fixed. Although the extraction rates may increase due to technological advances we will assume here that the amount is static. The decision, then, is how much resource to extract now and how much to leave for future time periods. The future demand for a natural resource is not necessarily different from other investments. Thus if the return from a stock of natural resources is expected to be greater in the future than from other alternatives—here assumed to be the market rate of interest—then there will be an incentive to preserve the stock. If the expected return is lower, there is no incentive to preserve the resource and depletion will occur. The optimum allocation over time occurs when the price of the resource is expected to rise at a rate equal to the rate of interest, that is known as the Hotelling principle or time path of price.

Thus, for optimal depletion of resources over time, the present value of the resource must be equal over all time periods. If this is not true, then increased benefit could be obtained by transferring some of the resource depletion between time periods. As it is the rate of interest that distinguishes the value between periods, the price of the resource rises

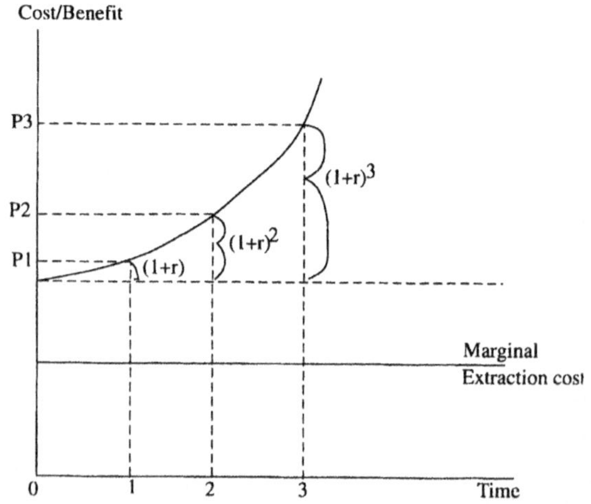

FIGURE 13.2. Time path of price.

at the rate of interest giving the time path of price in Figure 13.2.

Thus the price of the resource will rise steadily over time with smaller and smaller amounts of the resource depleted. Depletion of the resource is therefore gradual, which allows switching (substitution) to the use of other resources as the price rises. For example, at the turn of the century whale oil was a material used commonly for lighting, fuel, and other general household uses. As whale oil became more expensive relative to other oils its use declined and now it is only used for specialist uses in very small amounts such as space exploration where no substitute has yet been found.

The best-known example of resource scarcity was predicted by Malthus, writing at the end of the eighteenth century. He predicted that, whereas food production might grow arithmetically over time, population tended to expand geometrically. The result he predicted was that, if population growth could not be constrained by what he called "moral restraint" (e.g., later marriage) then it could only be limited by vice (in which he included practices such as infanticide and conventional birth control) or pestilence. Not surprisingly, in a religious world, this view was hotly contested and widely challenged. It is still argued that Malthus was wrong in that he omitted to recognize the power of new technologies which allow increasingly effective use of natural resources and hence provide for an expanding population. Despite such arguments, there remain really difficult development situations where Malthusian checks to population growth are important despite the immense expansion in world food production since his time.

The Malthusian experience is important in that it emphasizes the dynamic nature of economic processes. In particular, the examples above show that new technology is constantly changing the basis of production, lowering costs, and increasing output. While this process continues, the fear of the Malthusian spectre is in retreat. Consequently, although the evidence of many resource prices has shown them to be falling in real terms, this does not invalidate the market theory that began this chapter. Falling prices in the long term are due not only to the continual discovery of new stocks but also to new and lower cost methods of extraction. Each time a new method of extraction is discovered the cost curve shifts to the right, and costs for all levels of extraction are lower. These trends should not be used to encourage complacency in resource allocation. Societies will still benefit from a wise use of resources, and their economic and social goals are more likely to be met if they allocate their resources efficiently and equitably.

13.3.2 Renewable Resources

To illustrate the allocation of renewable resources the Schaefer model is used (Schaefer, 1957). This model represents the way biological populations grow in relation to their size, and although the following example relates to fish, it can be applied equally to many other renewable resources. The Schaefer model, as illustrated in Figure 13.3, describes the average relationship between growth of a fish population and its size, and fluctuations in growth rate are averaged out to give long-term trend.

Figure 13.3 shows that from $S1$ to S^* the rate of growth increases as fish stock does, but from S^* to S the rate of growth declines. There-

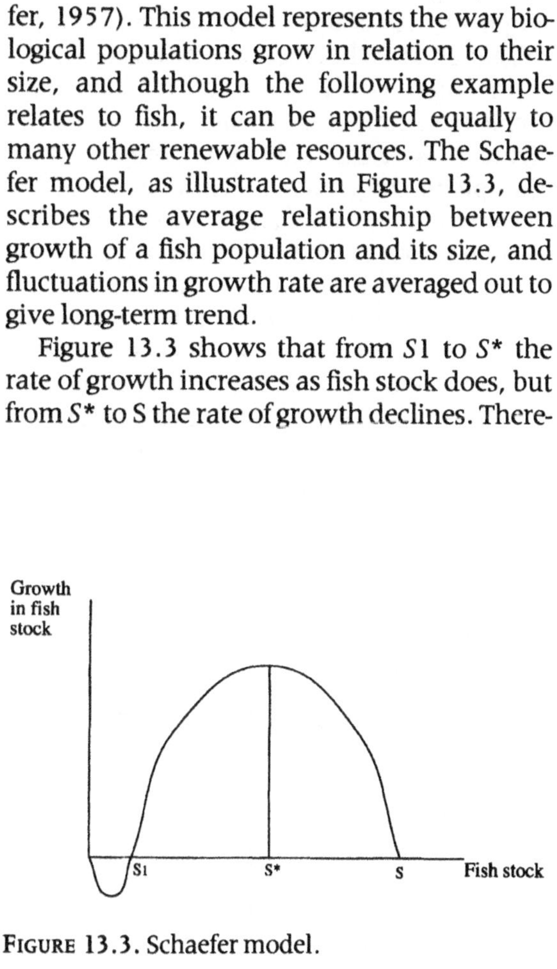

FIGURE 13.3. Schaefer model.

fore, *S*1 is the critical minimum stock where if stocks fall below this level the species becomes extinct. The value of *S* is the carrying capacity of the environment which would be the level of fish stock if there was no fishing, and the value of *S** is the maximum growth rate.

Fish could be harvested, or this resource could be depleted. So, as long as the extraction is equal to the growth rate, the level of stock remains the same: the catch is then said to be sustainable. The value of *S** is thus known as the level of population to give the maximum sustainable yield (MSY), that is, the level of population that gives maximum growth rate. If more than the MSY was harvested in the short run by removing stock, then this would not be sustainable.

Note that the MSY is not the economic optimal level of catch because it does not account for costs and benefits. To calculate this, for simplicity, it is assumed that the price of fish is constant and as the stock falls it becomes harder and more costly to harvest fish. The economic optimum level of catch is illustrated in Figure 13.4, which converts Figure 13.3 by multiplying the growth in fish stocks, *F*(*x*), by the price of fish, *Px*, to obtain the total revenue obtained from extracting fish at each level of stock. Therefore the growth rate becomes a sustainable revenue function and represents the revenue that can be obtained sustainably from this resource. Imposing on the diagram the total costs of extracting the resource represented by the total cost curve, this increases as fish stocks decline, indicating that the lower is the stock of fish the harder it is to extract them.

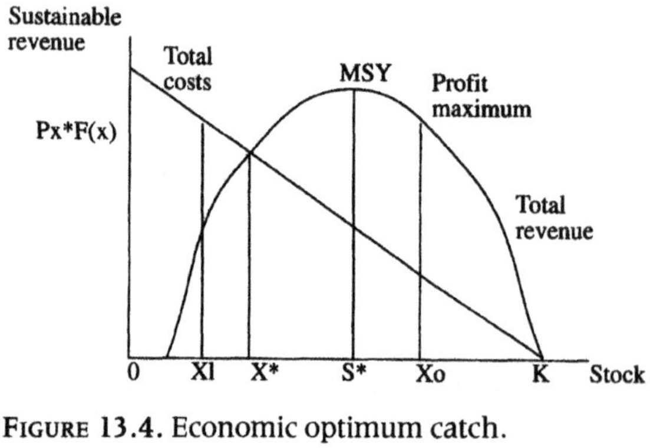

FIGURE 13.4. Economic optimum catch.

The economic optimal harvest is where the difference between revenue and costs are at a maximum and therefore profits (defined as revenue minus cost) are maximized. In Figure 13.4 this is at *Xo*. Profits will still be positive at *S**, at the maximum sustainable yield, but they will not be as great as at *Xo*. Therefore harvesting at the economic optimum leads to greater stock than harvesting at maximum sustainable yield. However this is a static solution, which implies that there is no opportunity cost of leaving stock in the sea (i.e., the discount rate is taken as zero).

In a dynamic solution the discount rate is assumed to be positive, and therefore there is an opportunity cost of leaving fish in the sea. This implies that there will be a greater extraction of fish than the level implied above, but it is likely to be close to or below the level of stock required for maximum sustainable yield. Finally, there is the extreme case where the growth rate of the fish is lower than the discount rate and the cost of extracting the last fish is low enough for it still to be profitable. In this case, extinction will occur.

Take the example of a whale population growing at an annual growth rate of 5 percent with a prevailing discount rate of 10 percent. If the full current value of the whales is £1000 million, then a choice of harvesting them all now for a value of £1000 million, which at the rate of interest of 10 percent would be worth £1100 million in a year's time. Alternatively, if the whales were allowed to grow and were then harvested after a year, they would only be worth £1050 million. However, because of the rising importance of the cost of extraction, extinction is rarely "efficient."

In practice, the problem of overdepletion of fisheries is not related to biological growth rates or even to a recommended economic harvest rate, but rather to property rights. In a closed fishery where the property rights are well defined, efficient allocation is not generally a problem. Nevertheless, the problem does arise with open access fisheries, in particular ocean fisheries, where there are not well-defined property rights over fish stocks. In such a case potential fishermen will be attracted into the industry by the profits. This will reduce the stock at first until the sustainable catch is reduced to where no profits are made, at X^*, and there is no incentive for new entrants.

This in itself does not lead to any problems regarding exploitation versus conservation of stocks. It is still a sustainable harvest. But the point where X^* is may be close to the critical zone and this may endanger species. This is more likely to occur than is at first apparent. As firms see profits and enter the industry, catches at first are likely to be greater than later, because stocks and the growth rate will still be around the level of MSY. Fishing effort will eventually deplete the stock and may well overshoot X^* and end up at $X1$, a loss-making position that may also endanger the species. The catch is no longer profitable, but because fishing requires a large element of fixed costs in the purchase of boats and equipment it is difficult for fishermen to leave the industry, and therefore there is a tendency for fishing effort to increase beyond sustainable levels.

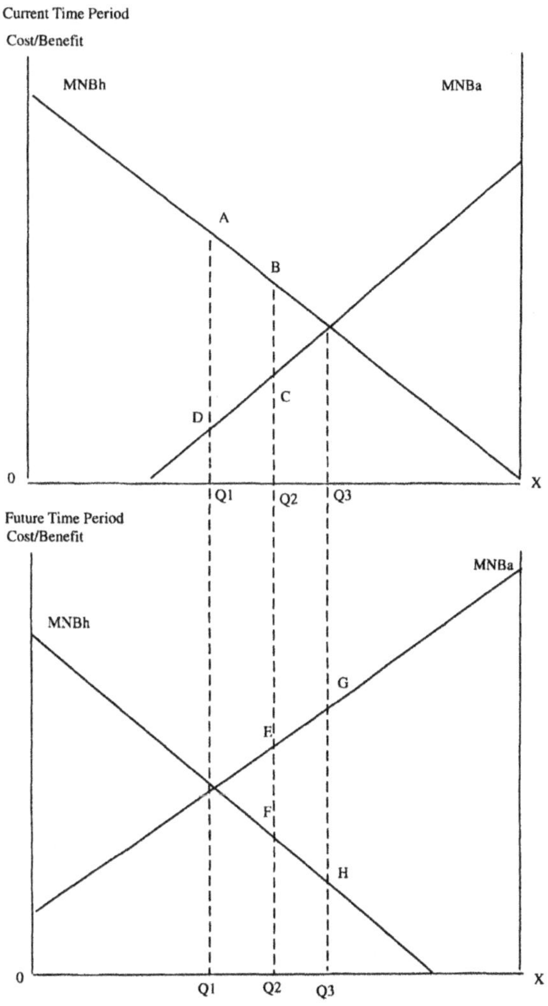

FIGURE 13.5. The allocation of land between competing uses (Source: McInerney, 1976).

13.3.3 Allocation of Land

Land itself is a stock resource, but we rarely demand land as a stock but as the service flows it provides. It differs from other resources in that the current consumption of service flows does not generally affect the resource stock, or even necessarily future service flows. Also what service flows are not utilized in any time period are lost forever. The problem of the allocation of land is its distribution among various uses which is particularly important if land use is irreversible over a normal planning horizon. Thus the use of land now must consider future demand for land use. This is illustrated by Figure 13.5.

The illustration shows the allocation of land between two uses, agriculture and housing, in both the current time period and in some future time period. The diagrams show the respective marginal net benefits of housing (MNBh) and agriculture (MNBa) in relation to the amount of land in each uses. The horizontal axis shows the total stock of land—all land is allocated to housing on the left-hand side of the graph at 0, and all land to agriculture on the right-hand side at X—with different combinations between these extremes. The optimum allocation of land between these two uses in the current time period is where the marginal net benefits are equal, that is, at $Q3$, giving $0Q3$ in housing and $XQ3$ in agricultural land use.

The future expected demand for land in housing and agriculture is represented in the lower diagram, which shows an expected increase in demand for agricultural land, and the optimum allocation being $0Q1$ in housing and $XQ1$ in agriculture. Clearly, the demand for land use in the two time periods differs, and therefore some compromise is needed.

The optimum allocation between the two time periods is where the cost to future generation of losing land to housing, illustrated by the area *EFGH* equals the gain to the current generation of having more land in housing, represented by the area *ABCD*. So the optimum allocation would be 0*Q*2 land for housing and *XQ*2 for agriculture (McInerney, 1976). This model allows identification of time over the two periods.

13.4 Market Failure, Property Rights and Natural Resource Allocation

The allocation of natural resources outlined above usually assumed the perfect operation of the market as discussed in Section 13.2. This section examines allocation of natural resources or environmental goods and services when there is market failure. In the case of natural resources this may arise due to lack of perfect information. In particular, lack of information leads to uncertainty with respect to environmental factors such as the weather or scientific uncertainty, such as the impact of the depletion of the ozone layer. But such reasons are not often sufficient to invalidate the operation of the market entirely, although they may generate a requirement for some regulation.

Where property rights are incompletely defined and assigned, market failure is most likely to require direct intervention. Property rights constitute ownership (ownership being the exclusive right to use, legally defined) and are an essential prerequisite for a market to function—for who would pay for something if they could not use it? To be effective they must also be transferable.

The fundamental base of trade is not the physical movement of goods but the transfer of property rights, for example, land buying and selling does not mean movement of land, but the actual transfer of rights of ownership. Even purchasing a good is the transfer of right to use that good. Stealing does give physical possession of a good but not the right to use it.

To be effective, property rights must be enforceable and require the support of some kind of penalty system: unenforceable rights are not rights at all. The penalty for violating rights must be such that it exceeds the benefit from the violation. For example, under pollution control policies the probability of getting caught, and the level of fine if caught, is too often less than the benefit of polluting.

In economic language, property rights must be nonattenuated, that is, they must: (i) be completely specified—perfect information must be available to all, including the restrictions on use and penalties for violation; (ii) transmit all rewards and penalties resulting from the use of a right to the owner or users; (iii) be transferable—so rights can move to the highest value use; and (iv) be completely enforceable.

This process is not costless. Specification, transfer and enforcement of rights require the use of scarce resources. The value of these resources is called the *transactions cost*. It may be borne by either the private individual, who protects the right to use his or her own property, or by the state. Each different specification of property rights gives rise to a different market solution and a different incidence of transactions cost. Thus, prices are a function of property rights, for example, if property rights for farmers to develop land were changed, this would influence the market for land for development and hence its price and the amount offered for sale.

It is necessary to reduce conflict between owners and nonowners of property rights. So rights accompanying ownership have to be specified. How these are specified will depend on legal, cultural, and economic systems and practices.

Perfect specification and assignment of property rights is generally prohibitively costly, so they are rarely perfectly determined. Owners frequently choose not to exercise all their rights: the benefit of enforcing a right does not exceed the cost of doing so. Where this occurs, rights are particularly likely to be appropriated by others wishing to use them. According to Bazel (1989), rights that are not used are likely to return to the public domain, from which they may be appropriated by other potential users. Many government policies that regulate the use of resources have a profound influence on the assignment of property rights and are likely to be controversial amongst existing owners. This is particularly the case with planning laws and policies that determine the allowable use of resources.

As economies develop and change, rights are constantly being redefined and reassigned as new rights are created in response to new economic forces. For example, Demsetz (1967) cites the development of heavier pelts in Labrador, which had a lower value until the European market became available and the price rose again. Before then the animals were a common property resource, but afterward they became private, thus showing that the right was not worth establishing until the value of the resource rose.

Conflicts arise due to the imperfect definition of property rights. Changes in the value of a resource for which the rights of ownership are imperfectly developed can cause conflict. They also can cause major changes in the distribution of wealth and income. The associated cost of transactions adds to the undesirable side effects. Dual- or multiownership makes this further complicated.

13.4.1 Property Rights and Market Failure

Nonexclusive property rights or attenuation of property rights lead to market failure, which, in turn, tends to lead to: (i) underprovision of goods and services; (ii) overexploitation of a resource; and (iii) underinvestment in management, conservation, and productive capacity of a resource.

For private goods, efficient allocation is assured by the selfish behaviour of individuals, because the benefits or costs of a private good are exclusive to the user and rival in nature. That is, the benefits or costs accrue to the owner of the rights only. When it is not possible to exclude others from consumption of a good or service or there is nonrivalry in consumption (i.e., the availability of the good is not depleted by the addition of further consumers), then there is market failure due to the existence of public goods.

Thus, we can identify different types of goods: (i) rival, exclusive goods—i.e., pure private goods; (ii) rival, nonexclusive goods—e.g., ocean fisheries; (iii) nonrival, exclusive goods—e.g., noncongested roads; and (iv) nonrival, nonexclusive goods—i.e., pure public goods.

In fact, these categories are rarely definitive. Rather, they represent a continuum along which the degree of publicness varies, as shown in Figure 13.6. Absolutely pure public or private goods are comparatively rare. For example, with wine, although one person may be excluded from consuming the wine in another person's glass, they may not be able to avoid any displeasure resulting from associated drunken behaviour. Similarly, although many lakes and beaches are so large as to be effectively nonrival, many are small enough for some people to find the pleasure they obtain from them diminish due to the presence of other users.

13.4.2 The Allocation of Public Goods

There is no reason to suppose that the production of public goods is any different from private goods, both being subject to the same constraints. It is in consumption that the differences occur. In the case of private goods total consumption is a summation of the individual demands of all buyers. So the amount of good produced equals the sum of each consumer's demand. In the case of pure public goods consumption by an individual does not affect that of another. Therefore, the total demand for this good is a summation of each individual's willingness to pay for the good. This leads to a problem of provision of public goods.

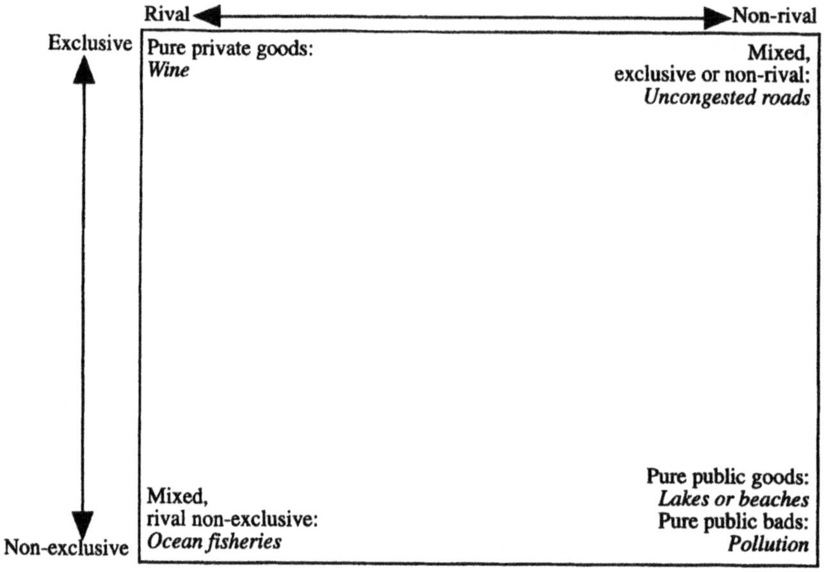

FIGURE 13.6. A classification of public and private goods.

The allocation of public goods is illustrated in Figure 13.7. This assumes we have two consumers, *a* and *b*, whose demand for the good is represented by *Da* and *Db*. In the case of a private good, the total demand curve would be calculated by summing *Da* and *Db* horizontally so demand given price of P1 would be $A + B = C$. In the case of a public good, because there is nonrivalry in consumption, the demand for a good is summed vertically because both *a* and *b* can consume the good consecutively. So total demand for quantity Q is $W + Y = Z$.

If the benefits for all individuals who value the public good could be added together, then it might well be possible to provide the good via a market. Because there is no incentive for any one individual to reveal their willingness to pay as they expect that the good will be provided in any case, then there is a case for government intervention. This is known as the free rider problem.

Even if it was possible to charge users directly for the public good this may still underestimate its value and therefore provision, as it would only account for the value of the public good for users and ignores nonuse values. Such nonuse values represent the value placed on a good by those who do not benefit directly from it. Nonuse values may occur for the following reasons: (i) existence—valuing the good because of it being there (e.g., Antarctica); (ii) bequest—valuing good for future generations (e.g., the ozone layer); and (iii) option—valuing goods for the option of using them in the future. These can only be obtained by government intervention and may have to be paid for out of general taxation.

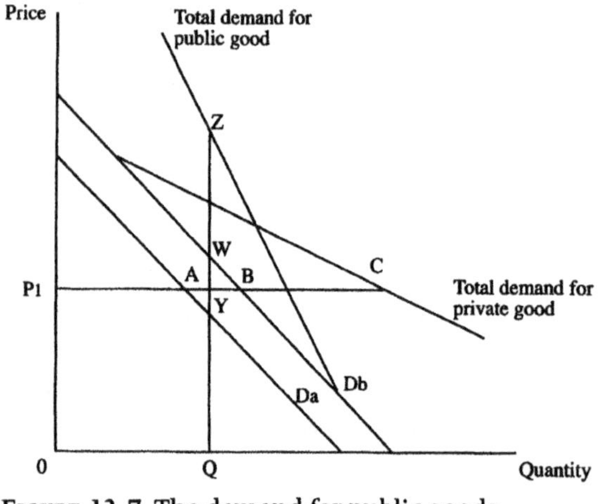

FIGURE 13.7. The demand for public goods.

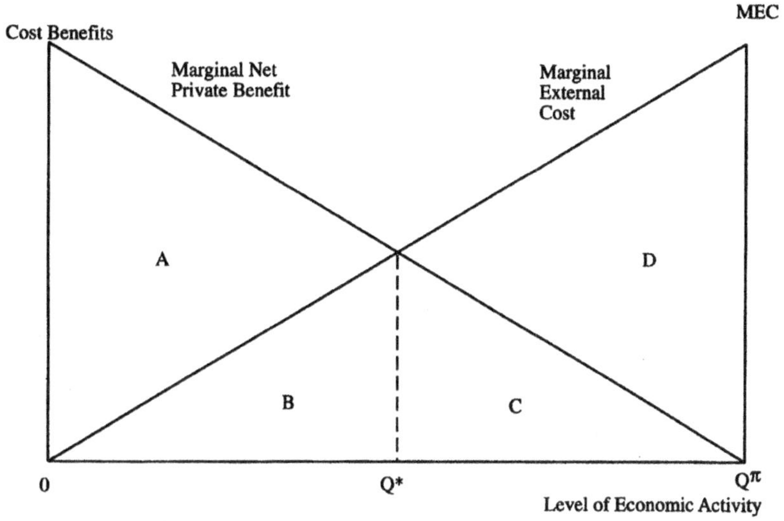

FIGURE 13.8. Optimal production with existence of an externality.

13.4.3 Externalities

A form of public goods of special interest in natural resource or environmental economics is what is known as externalities, that is, byproducts of production or consumption processes that have unintended impacts, either positive (such as a pretty landscape produced as a by-product of agriculture) or negative (such as pollution). Goods or services generating positive externalities tend to be underprovided, as the positive benefit to society has not been included, so there is a divergence between private and social benefits. Alternatively, goods or services generating negative externalities are overprovided because the negative impact of the externality on society has not been considered, and therefore private and social costs diverge.

Thus, there is the concept of optimal externality in economics, which is illustrated below using negative externalities as an example. (The theory can be equally applied to positive externalities, but in reverse.) Most economic activities involve some form of externality, and it is rare that there are no externalities. The most common situation is between zero and the level of externality which would exist, given the operation of an unrestricted market.

The optimal level of externality is illustrated in Figure 13.8. The figure has the level of economic activity, along the horizontal axis, from zero (or where the external effect starts) to the maximum level of economic activity if the external effect is ignored Q^{π} As economic activity increases, the marginal net private benefits, that is, the profit of increasing output, decline. At the same time, as the economic activity increases, it is assumed that the marginal external cost (MEC) rises, that is, the additional cost of the externality.

The optimal level of externality is where the marginal net private benefits (MNPB) equal MEC, at Q^* in Figure 13.8. At low levels of economic activity, below Q^*, an increase in output will lead to a large net benefit compared with the relatively small external cost, yielding a net increase in welfare to society. At levels of economic activity beyond Q^*, the MNPB from increasing output is less than the MEC incurred, and society thus loses welfare.

The diagram shows:

Q^* = optimal level of economic activity
B = optimal level of externality
$A + B$ = optimal level of net private benefits
A = optimal level of net social benefits
$C + D$ = nonoptimal externality
C = net private benefits socially unwarranted
Q^{π} = level of economic activity that generates maximum private benefits

The fact that a negative externality exists means that external costs are present which are not accounted for by the market. So, to achieve a social optimum, some form of government intervention seems to be necessary. Coase (1960) argued that the market could achieve a social optimum given the specification of property rights and communication between polluter and sufferer. For example, if a sufferer has the right to no pollution, then it would pay the polluter to compensate the sufferer for pollution so long as the benefit to the polluter exceeds the cost to the sufferer. There will therefore be a tendency to move toward the social optimum. Conversely, if the polluter has the initial right to pollute, then it will benefit sufferers to compensate polluters to reduce output or pollution, again toward the optimum level.

If this is the case, then there is no need for the government to regulate the market to reduce the externality as the market will achieve the optimum. However, Coase's theorem is based on a number of limiting assumptions such as perfect information, possibilities for bargaining, no initiation costs in negotiation, and perfect operation of markets. In practice, many of these conditions do not apply, so government intervention in the market is generally sought. This usually takes the form of standards in the case of pollution control or subsidies in case of positive external benefits—the latter seen most clearly with incentive payments for countryside stewardship.

It is in the case of public goods and externalities that government intervention is frequently required to overcome market failure. This can be through a number of mechanisms including regulation altering property rights and agreements with or without compensation. Compensation can be paid in a number of ways via tax incentives, permits, fixed direct payments, or individually negotiated payments.

Hence the need for government intervention to improve the allocation of resources and the provision of goods where property rights are not well defined. Economists have developed a technique, cost–benefit analysis, to evaluate the provision of goods by the government or public sector.

13.5 Cost–Benefit Analysis

Cost or benefit analysis simply compares costs and benefits with and without the change. There are three classes of cost or benefit analysis: (i) financial, which just includes cash flows, as an accountant would in the private sector; (ii) public exchequer cash flows, which are of particular interest to policymakers; and (iii) social or economic cost–benefit analysis, which attempts to include all the impacts on society of the change. It is the latter that is of interest to economists and will be developed here.

Social cost–benefit analysis raises two problems: (i) valuing goods where there is market failure; and (ii) the market value does not represent the opportunity cost of an activity to society. This can be overcome in two ways. Firstly, in the case of goods that are marketed, but where the market is not perfect, the opportunity cost of the good must be estimated. This involves shadow pricing the good or service, which can be done by referring to markets that have fewer imperfections and therefore represent more closely the resource/opportunity cost of production. An example of this could be using world prices of agricultural commodities instead of the protected prices at which they are produced in many countries. Secondly, in the case where there is no market for the good, techniques of nonmarket valuation are used. There are three main alternative techniques developed in the nonmarket evaluation literature: contingent valuation, travel cost, and hedonic pricing, as discussed below.

13.5.1 The Contingent Valuation Method (CVM)

CVM was originally proposed by Davies (1963) to value environmental benefits. Since their introduction by Davies (1963), contingent valuation methods (CVMs) have been used in studies of recreation and the environment to elicit values for nonmarket goods. Goods such as clean air, landscape, and water quality are not traded in a competitive market, and therefore command no market price.

Contingent valuation depends on the definition of a hypothetical, or contingent, market in which some usually nonpriced good or commodity can be thought of as being bought or sold. The approach relies on responses to questions in which individuals are asked the maximum amount which they would be willing to pay, or the minimum amount of compensation they would be willing to accept, for some change in the current level of a specified good. Although no money usually changes hands, this process involves some form of payment vehicle: a mechanism by which money could in principle change hands.

The environmental applications to which CVMs have been applied are many and varied, and range from valuations of widescale effects such as visible air pollution (Rowe et al., 1980), water quality (Desvouges et al., 1983), and nature conservation (Willis, 1990), to assessments of the benefits gained from specific activities such as hunting (Boyle et al., 1985) or skiing (McCollum et al., 1990), and the benefits of specific recreational locations such as wilderness areas in the United States (Walsh et al., 1984). While much of this work has been carried out in the United States, Europe is now adding significantly to the body of current research with important studies being implemented in Europe.

CVMs differ from other nonmarket valuation methods, such as those based on hedonic pricing or travel cost, in that they measure a preference expressed directly by respondents, rather than relying on a physical linkage to a private (marketed) good (such as travel to a site or the location of a house) through which respondents' observed behaviour is used to reveal a preference.

In addition, the range of applications for CVMs is far wider than those for the other methods. Thus, while the travel-cost method and hedonic pricing are restricted to the valuation of specific environmental goods, CVMs can be used to value both specific goods and a range of more general goods. Furthermore, they are not restricted to those individuals who use or live in a particular site or area, and can be extended to take in the preferences of nonusers, or to examine the potential effects of some hypothetical change in a good which might arise in the future (Willis and Garrod, 1991).

Contingent valuation works by directly questioning a sample of consumers about their willingness to pay (WTP) or willingness to accept compensation (WTA), for a change in environmental goods or services in a hypothetical market. The total value of an environmental effect is obtained by vertically summing WTP or WTA, as indicated by Figure 13.7. The total values can be separated into use value and non-use values.

13.5.1.1 Application of Contingent Valuation

The application of contingent valuation is by following these stages:

(i) *Establish and describe a hypothetical market*. This stage includes determining the good or service to be valued. This may entail breaking down the good or service into its attributes, which can then be valued. For example, a CVM study to estimate the WTP/WTA of design externalities connected with the Barcelona ring road broke down these externalities into extra entry/exit points, covering of two-thirds of the road, public facilities and green open spaces, and provision of 1000 parking spaces (Riera, 1993).

Crucial to CVM study is the presentation and level of information provided that describes the change in the good or service to be valued. Information may be presented verbally, in text form and/or pictures and models, for example, the Barcelona study used a mix of the above including models. At this stage the bid vehicle to be used in the study is also determined, that is, the method of payment that the study uses to obtain WTP/WTA. Typical methods of payment employed in CVM studies are income taxes, entrance fees, permits, sales taxes, higher utility (electric, gas, water) charges, and payments to charitable trusts. The choice of payment vehicle of instrument in a CVM study plays a crucial role in the development of the hypothetical contingent market.

(ii) *Administrator survey.* This can be done using a number of methods, each of these with their own benefits and costs. Face-to-face interviews are expensive in terms of both time and money, but generally give the greatest quality of response. Mail surveys may involve a damaging degree of nonresponses, and telephone surveys may lead to an unrepresentative sample being taken. Mitchell and Carson (1989) conclude that the current consensus of opinion among researchers using these methods favours no particular survey instruments for all types of questions. Indeed, the context of each particular survey should play the greatest role in the choice of survey instrument.

(iii) *Calculate average bid.* The results of the questionnaire are then analysed and the average bid values calculated. Frequently the potential determinants of WTP or WTA bids are examined such as respondents' income, age, and education, which are extremely useful to consider both in determining how representative the sample is and in aggregating the results.

The average bids are aggregated to the total population affected by the good or service and the total WTP/WTA estimated. Thus, the appropriate total population has to be determined as well as the method of aggregation and what variables to consider if the sample needs adjusting to be representative of the total population. In the Barcelona study the average WTP for design externalities was just over £200 in 1990 values. This was aggregated to obtain total benefits by the total population of Nou Barris, the main beneficiaries of the project. This gave a total benefit of £17 million compared to a cost estimate of £14 million of the charges.

13.5.1.2 Problems with Contingent Valuations

There are a number of well-documented problems with CVM, mainly due to the biased estimates (i.e., systematic under- or overstating of WTP/WTA). These biases include:

(i) *Strategic bias.* Respondents may underestimate their WTP/WTA if they think this will be used to actually "price" the good or service (either throughout taxes or some other means), this is known as the "free riderî problem. On the other hand, respondents may actually overstate WTP/WTA if they expect to benefit from a change yet do not expect to pay. In general, good survey design may help to overcome most of these problems.

(ii) *Design bias.* How contingent the valuation study is designed can influence the results. This includes the way information is represented to respondents and the bid vehicle used. In general this bias is lower where there is a relatively high level of knowledge of the good or service to be valued. This was the case with the Barcelona study, with respondents having prior knowledge of the design externalities.

(iii) *Mental account bias (part–whole bias).* This occurs where respondents find it difficult to separate and isolate a specific case from overall considerations. This will tend to make people overestimate value of resource. Thus, when, for instance, respondents are asked to value a specific environmental good, they may make that valuation on the basis of a much larger number of goods in that environmental class. This practice of allocating utility to groups of commodities rather than to individual commodities should reduce this bias and obtain a more realistic estimate of value. For example, in the Barcelona study (Riera, 1993), the design externalities were deliberately broken down into component parts to try and reduce the bias.

13.5.2 Travel Cost Method (TCM)

The second method of eliciting nonmarket values of goods is travel cost. This places a value on environmental goods by estimating a demand curve by assuming benefits are equal to the costs of consuming environmental goods. These costs are typically: (i) travel costs; (ii) entrance fees; and (iii) other costs.

Travel costs are divided into: (i) distance costs—the cost of getting to the site, so visitors are generally surveyed to ask for means of transport, petrol costs, type of car, as well as their origin; and (ii) time costs—the value of time spent travelling to the site; it is usually estimated as the opportunity cost of leisure time measured in terms of income foregone.

Entrance fees are often not charged, but should be recorded where they exist. Other costs are, for example, capital equipment to enjoy the leisure site, such as clothing, fishing equipment, boats, and so on.

TCM estimates the impact of the following variables on the number of visits to a specific site, as outlined below:

$Vij = V(Cij, Popi, Pij, Si, Qj, Feej)$

where:

Vij = number of visits
Cij = cost from i to j
$Popi$ = population of origin
Pij = measure of visits to alternative sites
Si = socioeconomic data (e.g., income, age)
Qi = quality of site (only used if more than one site)
$Feej$ = entrance fee
i = visitor
j = sites

There are two ways of estimating number of visits either on an individual basis or grouping them into zones. Zoning means categorizing visitor into predetermined zones around the site. If the latter is used, then travel costs can be obtained by estimating an average cost from zone to site. Moreover, zoning allows an estimate of visits per origin zone population, which allows predictions of visitor numbers possible.

The travel cost method gives estimates of total benefits of visiting the site, but excludes nonuse values. There are also other problems with the technique including how to apportion cost among multipurpose trips, how to value costs of non-daytrippers (i.e., residential holidaymakers), and how to value time or whether it should be valued at all.

13.5.3 Hedonic Pricing

The third technique is hedonic pricing, which seeks to find a proxy value for the nonmarketed good by using a marketed good. This is typically between house prices and some environmental good, for example, air quality and houses. Generally an econometric relationship between the price of houses and other factors affecting price, including the environmental good, is determined. So, for example:

$Ph = P(Si, Nj, Qk)$

where:

Ph = price of house
Si = site characteristics, such as number of rooms, gardens, garage
Nj = neighbourhood characteristics, such as schools, shops, transport facilities
Qk = environmental characteristics, such as noise, air quality, view

The implicit price of the environmental good or service is therefore obtained by discovering how much extra consumers would be willing to pay for a house with or without the environmental characteristic.

The problem with hedonic pricing is that it relies upon the availability of an extensive database of house price sales, which is not always available, especially given the low proportion of houses that are marketed each year. The results are vulnerable to omitted variables, which explain fluctuations in house prices and can therefore bias the results. It also assumes that owners have full knowledge of all the environmental variables that affect the value of the house. As well as assuming that the quantity of environmental good is measured by changes in location of individuals, this may not be the case because there are other methods of altering the impact of the externality, such as double-glazing.

13.6 A Case Study

To exemplify the above techniques we present a short case study of a development project with important developmental, environmental, and agricultural effects. This is a hypothetical case, one that demonstrates how these techniques of evaluation may be used. It relates to a small region, covering 20,000 hectares of land, in the Mediterranean basin where depopulation has been a serious problem and where it is proposed to establish a tourist village with associated attractions.

The present situation is that the current population of the area is 150 persons, with an age structure biased toward the elderly and with a preponderance of women. The levels of employment and income are extremely low. Depopulation over the last two decades has reduced this economy to one quarter of its previous size, leaving a very poor prognosis. If there is no economic development, the entire population is expected to disappear in the next decade.

Most of the land is uncultivated natural grassland and is extensively grazed by 2000 sheep. Small amounts of cheese and meat are produced for sale, and the whole agricultural activity employs five people, all over 50 years old. Dead sheep provide food for raptors, of which there is a significant population, and ornithologists visit the area to study these and other birds during the spring, summer, and early autumn. Mountain pastures produce orchids, which are a particular attraction to visitors in early summer. Fungi are collected by the local population, mainly for home consumption. This tradition has begun to be eroded by visitors to the area competing with local gatherers.

The substantial winter snow has allowed the establishment of ski-tourism on a small scale. This provides an important source of employment and income. There is substantial unused residential capacity in the region that could be used in summer as well if "green tourism," based on the ecological interest of the area, could be developed.

The basis of green tourism would be the educational provision of ecological holidays in the ski lodges. These could provide for up to 100 bed-nights throughout the summer season, attracting the urban school children and others interested. It is assumed that the season will last for six months and that, on average, 40 percent occupancy will be achieved. Student guides would be required for the enterprise, but some of the food and all heating fuel would be provided from local sources. As the project develops local sources of these inputs will expand. It is also possible to develop the local fungi gathering using a system of safety advice and locally sold licences to ensure that revenue remains in the district.

A manager for the enterprise is available in that the school nearby is about to close and the schoolteacher is looking for employment. His local knowledge and managerial skills are adequate for the task, but it would be desirable that he is given some training. The ski lodge already has an accommodation management structure which will be expanded to cope with year-round working. A possible expansion path for the enterprise that will be explored would be into conference provision. This lucrative trade will provide extra funds for further development.

The proposal is to develop the extension of the ski enterprise over a three-year period, after which it will have paid for itself and a further stage of expansion may be considered. Grants are available from the EU and from central government to cover capital works and some operating costs. The conversion of the present seasonal ski enterprise to a year-round operation will provide substantial local employment benefits. Given the high rate of local unemployment, these are important. Also highly significant are the opportunities offered to visitors to enjoy the local wildlife. This opportunity is highly valued, but there is no market mechanism that would allow people to indicate their valuation of this option, except insofar as it is reflected in the demand for overnight accommodation.

There is also a potential group of day-visitors who might come to the area from the various towns within 100 km. These people may spend nothing locally and cannot be excluded from participating in the wildlife benefits in the region. The valuation such people place on the opportunity to visit can only be assessed by one of the evaluation methods listed above. As there are currently few visitors to the region, apart from those coming to ski, it will be necessary to conduct a contingent valuation study amongst visitors to the region. This will estimate the value of nonmarketed benefits generated by the project.

A simple financial analysis of this project is set out in Table 13.1. This shows only annual costs and revenues, assuming that no capital investment is needed for the project. The first block of information presents the costs of the project, consisting of employees, food fuel, and services. The manager's vehicle has been expressed as an annual cost, including repairs and depreciation as well as fuel.

The flow of benefits is more diverse, including the payments of eco-tourists, exchequer savings from reduced Social Security payments, and extra tax payments due to the income generated by the project. These amount to the *financial* benefits of the project which sum to less than the total costs recorded. On purely financial grounds the project would therefore be rejected. However, if we include the fact that there will be some increased day-visiting to the area and value the visits using contingent valuation, then there is a substantial increase in the recorded benefits without any addition to costs.

From this simple analysis the conclusion is therefore that if a narrow financial view of costs and benefits is taken, then the project would be rejected, whereas it would be implemented on wider economic considerations, including an evaluation of the public good benefits acquired by day visitors. In this analysis the project has been treated in annual terms only, and no allowance has been made for changing values over time. This has been a defensible way of carrying out the analysis because all relevant factors can be presented as annual flows. Where this cannot be done a more complex analysis is needed. To exemplify these possibilities there are two ways in which the analysis could be further developed: first, to allow for changing relative values over time, and, second, to consider the

TABLE 13.1. Annual costs and benefits of eco-tourist development project

	Daily rate ECU	Annual rate ECU
Annual costs		
Labour		
- Four domestic workers, six months at 50 ECU/day	200	38,600
- Manager, chef and domestic supervisor		100,000
Food, 40% occupancy, 6 months at 10 ECU/day	400	77,200
Heating, firewood at 1 ECU/person/day	40	7,720
Other services (water, electricity, etc.)	40	7,720
Vehicle for manager		5,000
Manager's training		2,000
Total costs		256,240
Benefits		
Revenue from eco-tourists	20	154,400
Exchequer savings		
- Reduced Social Security payments	50	9,650
- Extra income tax paid, 10% of salary bill		13,860
Value of fungi gathered		30,000
Grant for manager's training		2,000
Total financial benefits		207,910
Net financial benefits		−28,330
Contingent valuation of benefits to 50,000 day visitors p.a.	0.85	42,500
Total economic benefits		242,080
Net economic benefits		14,170

possibility of discontinuous, or capital, investments which pay off over a period longer than one year.

If this project was implemented and relative values of costs and benefits did not change over time, then there would be no need to extend the analysis. However, in practice such changes do occur and should be considered in the interest of precision. What is envisaged here is that the value of some aspects of the project will change faster over time than others. This could arise because of sudden permanent changes in input prices, with other prices remaining unchanged, as happened in the OPEC "oil shocks" of the 1970s. (In those particular cases, the changes proved to be rather more temporary than they seemed at first.

While the sudden surge in oil prices did bring about some sharp changes in other economic values, the price of oil quickly returned to the same relative position in most countries.)

Where the prices of some goods do diverge from others over time this should be allowed for, especially when a project pays off over a long period. Such changes may occur either through the demand side of the market, where they may follow from income growth, generating changes in expenditure patterns for goods where the income elasticity of demand is more or less than unity, or, as in the OPEC case, where producers manipulate supplies in order to increase prices.

If the demand for a good is income elastic, then a rise in consumers income will be followed by a rise in the demand for it. This means either that more will be consumed or that its price will rise, or both. It is widely believed that environmental public goods are income elastic in demand, which means that rising incomes will increase their value or the amount consumed.

As supply for most environmental goods changes only slowly, it is a short step to argue that rising incomes will lead to increasing unit values for these goods. For example if their income elasticity is 2.0 and incomes are rising at 2 percent per annum, then the value of these goods will grow faster than 4 percent per annum. It is not justified to apply such elasticities to a population over a long period because they are not, themselves, fixed for all time.

As consumers become wealthier, their pattern of preferences changes and their income elasticities for various goods may reflect this change. It is also difficult to predict the path of income growth over long periods. Such problems may be dealt with by making rather cautious assumptions about future changes both in income and in income elasticities.

In order to apply such arguments to the estimates in Table 13.1 we need income elasticities of demand for the different types of cost or benefit considered. Detailed estimates are not available, but we can make assumptions on the basis of known similar types of good. For example, the fungi-gathering operation produces a saleable output which is a normal good in the countries where it is consumed. This would suggest that it is probably income inelastic.

However, there is strong consumer interest in goods such as this which are harvested from the "wild." Maybe there is a case for arguing that it has a higher elasticity than most foods—perhaps even as high as one. By way of contrast, the ecotourism benefits and those from day visits could be argued to be more highly income elastic. Such goods may have an elasticity as high as two or even three. In the absence of specific estimates and given the problems of fluctuating income growth, it is probably reasonable to assume that the elasticity for these goods is two and that incomes will grow at 1 or 2 percent per annum in the next ten years. That would imply annual values of benefits and costs as displayed in Table 13.2.

The inelastic share of benefits includes exchequer savings and the value of fungi gathered. The manager's training is a one-off event that may therefore be ignored. The annual value of costs is constant, as is the value of the inelastic benefits produced from the project. However, the annual value of benefits grows steeply due to income growth over time, more than doubling at low income growth rates and increasing by a factor of six when the income growth rate doubles.

The presentation in Table 13.2 is unsatisfactory in that it includes too much information to be appreciated and further analysis is required to condense this to criteria that will allow comparison of this project with others and make the results easily comprehensible.

This is done by discounting, as described earlier. The process reduces the future stream of benefits and costs to a single number, the net present value (NPV), reflecting the value of that project to its owner at present. The discount rate is a critical variable in determining project choice and Table 13.3 demonstrates this by applying a number of different discount rates to the flows in Table 13.2.

These calculations show that the project produces a positive social value ranging from 0.5 million to nearly 3 million ECU, over ten years, depending on the discount rate applied

TABLE 13.2. Annual value of shares of benefits and costs over ten years, not discounted (ECU)

Year	Annual value of costs	Annual value of benefits: Inelastic share Elasticity = 1 Income growth = 1	Elastic share Elasticity = 2 Income growth = 1	Elastic share Elasticity = 2 Income growth = 2
0	236,240	53,510	196,900	196,900
1	236,240	53,510	200,838	204,855
2	236,240	53,510	208,952	221,741
3	236,240	53,510	221,741	249,717
4	236,240	53,510	240,020	292,583
5	236,240	53,510	265,001	356,657
6	236,240	53,510	298,435	452,327
7	236,240	53,510	342,808	596,836
8	236,240	53,510	401,654	819,329
9	236,240	53,510	480,013	1,170,203

and the rate of growth of the value of income elastic benefits. This range in values is not particularly wide given that the range of discount rates used varies from zero to 10 percent per annum. That range was chosen simply to show the impact of the discount rate. If, as is usual, the rate was externally chosen and known, then the reader has only to decide which income growth situation is more credible of the two displayed. That must be a matter for judgement based on knowledge of the economy in which the project is set.

The fact that all conceivable outcomes are positive means that the project would be acceptable, at all discount rates chosen, because it produces a positive NPV. The size of this NPV is an important indicator of project scale and could be taken as an indicator of the potential of this project to support some investment as well as undertaking the development proposed. Or, to put it another way, if the project was redesigned to include some investment which paid off over, say, ten years, that investment could be quite substantial.

The size of investment the project could carry is indicated by the NPV calculated. For example, at a discount rate of 6 percent the project produces an NPV between 0.7 and 1.8 million ECU. In principle, it could therefore be capable of supporting an investment costing that much without becoming economically unattractive. The analyst would have to decide which growth scenarios was most likely to occur, whether it would produce 0.7 million or 2.8 million or a sum in between, and that would indicate the smallest size of investment that could be made. Indeed, further expenditure would be justified up to the limit of the flow of benefits from the investment. In this sense the example only takes us part of the way into the design of the project: it should be seen as the opening review of project possibilities, rather than the definitive analysis of the situation.

Another aspect of the project that may be important is the distributional impact of benefits and costs. If the costs of the project are borne exclusively by the poor and the benefits enjoyed only by the rich, it might well be less acceptable politically than if they were paid for

TABLE 13.3. Project net present values (NPV): high and low growth scenarios (ECU)

NPV	Discount rate i (%): 0	4	6	8	10
Low growth	1,029,062	789,021	695,547	626,835	547,491
High growth	2,733,848	2,054,617	1,792,342	1,598,797	1,380,324

by the rich and enjoyed by the poor. The only simple basis for making such an assessment is to consider the individual elements in Table 13.1. The obvious point to note is that several of the items recorded in Table 13.1 are paid for through markets, whereas some are not.

The workers' salaries are transmitted through the labour market: the other costs are similarly dealt with. The benefits to eco-tourists and consumers of fungi are also distributed through markets, whereas the other items—e.g., reduced Social Security payments, extra taxes paid—are impacts which benefit other tax payers, who have to pay less tax because of the existence of the project. Meanwhile the people who enjoy day visits to the area but do not pay for them are gaining a benefit from the public exchequer at the expense of the taxpayer and from local providers of jointly supplied public goods. The complexity of these flows and counterflows for even a simple project like this shows why distributional implication are often ignored in project appraisal. Nevertheless they should be considered in a serious appraisal. At least they should be drawn to the attention of decision-makers, to whom they may be an important criterion for project selection.

Finally, any CBA should conclude with a critical appraisal of the study itself, particularly noting the issues that have not been fully covered. This might be done in the form of an environmental impact assessment (EIA), if one has not preceded the CBA, in which case it would focus on the way in which the local ecosystem and the wider environment might be affected by the project. It should include all untreated aspects of the project such as, for example, its impact on global warming. That would have had to consider how much extra car mileage resulted from the evolution of eco-tourism and day-visiting resulting from the project, as well as the energy used in accommodating visitors. The report should be fully explicit on these questions and should be written in the expectation that it will be subject to professional scrutiny and, possibly, challenge. By these means the quality of information on which public decisions are based will be improved.

References

Bazel, Y. *Economic Analysis of Property Rights*. Cambridge: Cambridge University Press; 1989.

Boyle, K.J.; Bishop, R.C.; Welsh, M.P. Starting point bias in contingent valuation bidding games. *Land Econ*. 61:188–194; 1985.

Coase, R. The problem of social cost. *J. Law Econ.* 3:1–44; 1960.

Davies, R.K. Recreational planning as an economic problem. *Nat. Resour. J.* 3:239–249; 1963.

Demsetz, H. Towards a theory of property rights. *Am. Econ. Rev.* 5(2):347–359; 1967.

Desvouges, W.S.; Smith, V.K.; McGivney, M.P. *Comparison of Alternative Approaches for Estimating Recreation and Related Benefits for Water Quality Improvements*. Washington, DC: U.S. Environmental Protection Agency; 1983.

McCollum, D.W.; Alphonse, H.G.; Peterson, G.H. *The Net Economic Value of Day Use Cross-Country Skiing in Colorado and Comparisons with Sites in Vermont*. U.S. Forest Service, Rocky Mountain Forest and Range Experiment Station; 1990.

McInerney, J. The simple analytics of natural resource economics. *J. Agric. Econ.* 27:31–51; 1976.

Mitchell, R.; Carson, R. *Using Surveys to Value Public Goods: The Contingent Valuation Method, Resources for the Future*. Washington, DC: Resources for the Future; 1989.

Riera, P. *The Social Benefits of Infrastrutures: The Case of the Barcelona's Ring Roads*. Madrid: Civitas; 1993.

Rowe, R.D.; d'Arge, R.C.; Brookshire, D.S. An experiment on the economic value of visibility. *J. Environ. Econ. Manage*. 7:1–19; 1980.

Samuelson, P.A.; Nordhaus, W.D. *Economics*. 15th ed. New York: McGraw-Hill; 1995.

Schaefer, M.B. Some aspects of the dynamics population important to the management of the marine fisheries. *Bull. Inter-Am. Tropi. Tuna Comm.* 1:25–56; 1957.

Smith, A. *The Wealth of Nations*. Glasgow; 1776.

Walsh, R.; Loois, J.; Gilman, R. Valuing option. Existence and bequest demands for wilderness. *Land Econ.* 60(1):14–29; 1984.

Willis, K.G. Valuing nonmarket wildlife commodites: An evaluation and comparison of benefits. *Appl. Econ.* 22:13–30; 1990.

Willis, K.G.; Garrod, G. Landscape values: A contingent valuation approach and case study of the Yorkshire Dales National Park. ESRC Countryside Change Intitative. Working Paper 21; 1991.

14

Public Policies, Markets, and Externalities

François Bonnieux and Hervé Guyomard

14.1 Introduction

The absence of externalities is one of the conditions required for competitive markets to achieve efficient resource allocations. Externalities occur in situations where the activities of one economic agent affect or spill over into the technology, consumption set, or preferences of another agent. Externalities, which may be positive (in that case, they have beneficial external effects) or negative (in that case, they have adverse external effects), are one example of market failures. They mainly arise from the fact that environmental goods and services often have no market. The environment is considered as a free resource, which is zero priced and therefore overused. When interaction between private agents does not result in an efficient resource allocation, government intervention is required in order to internalize external effects. Nevertheless, even in the absence of market failure, government intervention may also be justified mainly for equity or national industry protection reasons. This chapter addresses only the problem of externalities and environmental policy strategies, emphasizing those relevant for agriculture activities and resources.

The chapter is organized as follows. The first section gives an overview of the range of instruments that may be proposed in order to solve the problem of externalities. We first present a simplified model which allows us to introduce the concept of negative externalities and the two basic instruments, that is, a tax on polluting emissions or an emission standard, generally used to solve the efficiency problem resulting from these externalities. The relative merits of these two basic instruments, and more generally of control-and-command policies and market-based approaches, are then discussed in connection with the political objectives assigned to environmental policies.

The second section addresses the problem of agricultural policies and environmental degradation. The Common Agricultural Policy (CAP) example is used to illustrate the discussion. The pre-reform CAP allows us to analyse the consequences of the price support mechanism, while the May 1992 CAP reform allows us to consider the potential effects on environment of alternative support instruments, that is, mainly a "decoupled" direct income support combined with a set-aside programme.

The third section is practically oriented as it concentrates on policy responses to major environmental problems due to agricultural activities and policies in the European Union (EU). The various case studies presented in this section have been selected to illustrate some of the generally applicable policies in order to solve the problem of nonpoint source pollution in agriculture and promote the potential positive effects of agriculture on the environment.

The concluding section gives an insight into the foreseen evolution of policy design. As far as sustainable development is concerned, a shift in approach toward the integration of environmental and agricultural policies in a more unified and integrated framework can be expected.

14.2 Environmental Policy Design

The background is introduced with a very simple model that shows the need for public intervention if pollution occurs. Because economic theories offer two basic approaches to the problem of pollution—a tax-subsidy approach or a regulatory approach—the second subsection presents a more thorough discussion of price and quantity control to correct market failures. The last subsection is devoted to a brief presentation of the various environmental policy instruments in connection with a range of criteria that play a major role in government decision making.

14.2.1 An Introduction to Negative Externalities and Their Control

By relying on the pioneering contribution of Sidgwick (1887), Pigou (1920) emphasized the need for public intervention to correct market failures stemming from externalities. The essence of his argument is as follows. When the supplier of one good does not bear all of the costs of producing the commodity, his marginal private cost (MPC) is less than his marginal social cost (MSC), and he produces more than is socially optimal. The difference between MPC and MSC at each level of output is the marginal cost of the externality (MCE), and the corresponding misallocation of resources, that is, the market failure, results in a loss of welfare (Zilberman and Mara, 1993).

In the regulation of negative externalities there is a tradeoff between the damage that people suffer from pollution and the resources required to reduce polluting emissions. The simple model depicted in Figure 14.1 may be used to illustrate this tradeoff.

Let us consider a single farm (the emitter) which applies inputs to the soil that are carried in runoff into a river. The emissions have a negative impact on the activity of an aquacultural

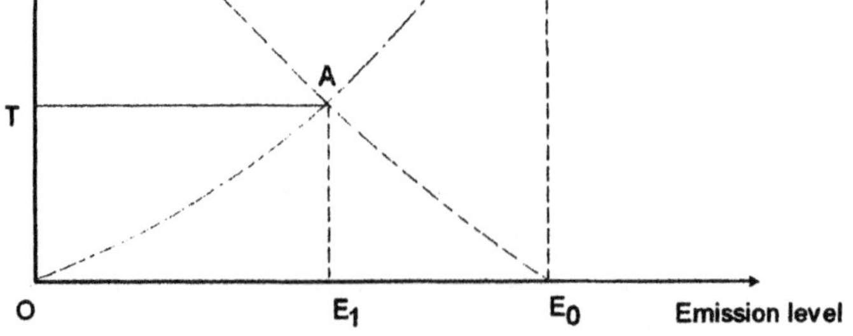

FIGURE 14.1. Efficient pollution control: Pigouvian tax on emissions versus emission standard.

farm (the receptor) located downstream, and this impact is expressed in monetary terms using a damage function. The emitter can reduce the amount of effluents discharged into the river by abating pollution (or controlling pollution). Abatement includes a wide array of means for reducing emissions: diminution in output, changes in production technology, input switching, recycling of residuals, treatment, abandonment of land, and so on. In all cases, abatement involves the use of resources and expenditures which will affect production costs for the emitter.

Marginal abatement cost (MAC) and marginal damage (MD) functions are shown in Figure 14.1. The emitter marginal abatement cost corresponds to the sum of all costs for achieving a one-unit decrease in emission levels or, alternatively, the costs saved if emissions are increased by one unit. The corresponding curve starts at the uncontrolled emission level E_0 and is upward-sloping to the left. The receptor marginal damage function corresponds to the sum of all the damage that stems from a unit change in emissions. The corresponding curve starts at zero and is upward-sloping to the right because marginal damage is assumed to increase with emission levels.

In the absence of government intervention, the emitter has no incentive to take into account the negative environmental externalities arising from his production because damage and associated costs are incurred only by the receptor. In other words, the farmerís production decisions are solely based on his private interest. Accordingly, he will decide to produce an output quantity Q_0 that corresponds to uncontrolled emission levels E_0. This private choice is suboptimal from the point of view of society because potential social costs are completely ignored. Starting from point E_0, Figure 14.1 clearly shows that a decrease in emission levels leads to an increase in abatement costs which is more than offset by the induced decrease in damage costs. Marginal damage is greater than marginal abatement costs until point A where MD and MAC curves intersect. Accordingly, point A corresponds to emission levels E_1 which are optimum for the society. The optimum level of pollution for the society is thus always less than that which is optimal from the private perspective of the profit-maximizing farmer. As shown in Figure 14.1, it is not zero but strictly positive. When emission levels are E_0, total damage costs are OE_0B, and total abatement costs are zero. When emission levels are reduced to optimum levels for the society, that is, E_1, total damage costs are limited to area OE_1A. Total abatement costs are no longer zero, but E_0E_1A. Therefore, there is a net social gain corresponding to area E_0AB.

To summarize, the socially optimal level of emissions corresponds to the point where marginal damage and marginal abatement costs are equal. Higher emissions expose society to greater costs stemming from increased environmental damage, and lower emissions require society to pay for the costs in the form of resources devoted to abatement activities. The gap between the private optimum E_0 and the social optimum E_1 illustrates the failure of private decisions and market mechanisms to maximize society's welfare in the presence of external effects when there is no government intervention.

The basic instrument used to correct the market failure due to an external effect is to levy a Pigouvian tax on producers to internalize the negative social costs resulting from "excessive" production. Consider the simple case of an emission tax where the same tax is levied on each unit of emissions, regardless of the level of total emissions. The equilibrium quantity of emissions is determined by the intersection of the tax line (i.e., T) and the supply curve for abatement (i.e., MAC). It clearly appears that the optimal tax T_1 corresponds to marginal damage evaluated at social optimum point A. This optimal tax gives the emitter the incentive to reduce his emissions from E_0 to E_1: beyond this point E_1, the emitter will prefer to pay the tax than to continue to reduce emissions. Under a Pigouvian tax fixed at optimal level T_1, the total costs borne by the emitter are abatement costs (i.e., area E_0E_1A) and tax costs (i.e., area OE_1AT).

The use of a Pigouvian tax forces the polluter to pay for the activities needed to reduce emissions until the socially optimal level is reached, as well as for the residual emissions. This corresponds to the implementation of the polluter-pays principle (PPP), which acknowledges that society holds the right to use the environment and, hence, that firms must purchase this right in order to emit.

An alternative approach is to regulate directly the quantity of pollution that is permitted. A rapid inspection of Figure 14.1 shows that the enforcement of an emission standard equal to E_1 also achieves the social optimum A. But, in that case, the costs borne by the emitter are restricted to abatement costs because he does not pay for the use of the environment. Both approaches, price control (here a Pigouvian tax on emissions) and quantity control (here an emission standard) are equivalent in terms of social optimality because the same level of emissions is obtained. Nevertheless, they differ in terms of wealth redistribution between pollution emitters and pollution receptors.

14.2.2 Price Versus Quantity Control, or Market-Based Versus Command-and-Control Strategies

Economic theories offer two basic strategies to control emissions of a particular substance, which are either to implement a Pigouvian tax on emissions or to enforce the equivalent standard. The implementation of these strategies requires governmental intervention. Public regulators usually rely on a specific body such as an environmental agency that holds the right to collect the environmental taxes or to enforce the standards. Under the tax strategy, the environmental agency places a negative price on emissions which results in a decentralized reduction in emission levels. Under the regulation strategy, the environmental agency sets the amount of emissions that are permitted without penalty. In the simple case of one emitter only, both strategies are optimal and lead to the same amount of abatement for the same resource cost. This is no longer the case with several polluting firms which have different marginal abatement costs. The essence of the argument is depicted in Figure 14.2.

Figure 14.2 refers to the case of two polluting farms, 1 and 2, respectively, with different marginal abatement cost functions, MAC_1 and MAC2, respectively. The two curves start at E_0 on the horizontal axis and are upward-sloping to the left, but MAC_2 is less steep than MAC_1. Let us now assume that the environmental agency sets an emission tax at level T. In response to this uniform tax, firm 1 will reduce its emission level until E_1 but firm 2

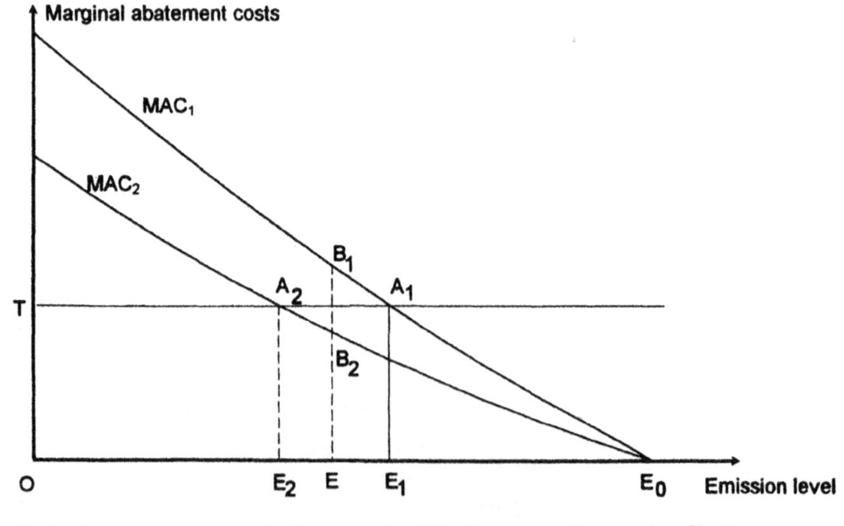

FIGURE 14.2. Emission tax versus emission standard in the case of several polluting sources.

will reduce its emissions to a greater extent, that is, until E_2. Total abatement costs are measured by the sum $E_0E_1A_1$ (borne by producer 1) + $E_0E_2A_2$ (borne by producer 2) for a total emission level of $E_1 + E_2$. It appears that the firm with the lower marginal abatement cost will proportionally reduce its emission level more.

The same reduction in total emissions may be achieved by using a uniform standard set at level $E = (E_1 + E_2)/2$. In that case, farms 1 and 2 will produce an output quantity corresponding to the common standard level of emissions, that is, E. Total abatement costs are now equal to the sum E_0EB_1 (borne by producer 1) + E_0EB_2 (borne by producer 2). This second strategy will result in a situation where marginal abatement costs of producers 1 and 2 differ at the optimum.

While the industry offers the same total amount of abatement under the tax and the standard, the total resource costs differ. Because $E_1A_1B_1E$ is greater than $EB_2A_2E_2$, the total resource cost is lower under the tax strategy than under the regulation approach. This result is due to the fact that the tax encourages the more efficient supplier of abatement, that is, producer 1 with the lower marginal abatement cost, to accept a larger share of the abating while the less efficient supplier of abatement, i.e., producer 2 with the higher marginal abatement cost, bears a smaller part of the abating. The tax strategy is thus more efficient than the standard approach because it allows resource abatement costs to be minimized.

Figure 14.2 illustrates one of the strongest arguments in favour of taxation in controlling multiple sources of emissions in a way that satisfies the equimarginal principle. If the same tax rate is applied to different sources with different marginal abatement cost functions, each source will reduce its emissions until its marginal abatement cost equals the tax. Marginal abatement costs are then equalized across all the sources, resulting in minimum overall abatement costs. At this stage, it is important to note that this outcome is achieved even if the environmental agency does not know with certainty the marginal abatement cost functions of the different sources of emissions. On the contrary, the problem of information on abatement cost functions rules out a strategy based on the enforcement of individually tailored standards, such as E_1 for firm 1 and E_2 for firm 2. Furthermore, the political feasibility of this individual approach is questionable.

At this stage, the three following remarks are in order. First, one possibility to achieve the lowest resource cost solution through quantity control is to introduce a free market for discharge permits. The crucial decision is then related to the total number of discharge permits (emission credits) to be put into circulation, the permit price being the outcome of the market working. A second very strong argument in favour of the tax strategy is that it provides a continuing incentive for improved abatement performance (via the adoption of less costly abatement technologies, for example) while there is no incentive for abatement beyond that required by the standard under the regulatory approach. Finally, one easily understands why polluting firms prefer the regulation strategy rather than taxes. In the second case, they pay an additional cost corresponding to the tax on unabated emissions.

14.2.3 An Overview of Environmental Policies

14.2.3.1 Environmental Policy Instruments

Environmental policy instruments used by the public authority to reduce the level of environmental damage resulting from production and/or consumption activities may be classified as relying primarily on regulation instruments or economic instruments (Baumol and Oates, 1988). Table 14.1 follows this classification and thus first considers command-and-control approaches and second market-based strategies. Because many instruments are on the borderline between direct regulation and economic approach, these two categories may overlap in some cases.

Under a command-and-control environment regulatory strategy, the public authority

TABLE 14.1. A synthesis presentation of environmental policy instruments

Command-and-control strategies
Standard of different kinds
Emission and effluent standards
Ambient standards
Process standards
Product standards
General and specific regulations
Land use control
Prohibitions
Mandatory permits
Eco-labelling
Environmental audits
Market-based strategies
Taxes (and charges)
Emission charges
Product charges
User charges
(Tax differentiation)
Subsidies
Abatement subsidies
Grants
Soft loans
Tax allowances
Market creation
Transferable discharge permits
Tradable quotas
Definition of property rights
Liability laces
Covenants
Disposit–refunds
Performance bonds
Research and education
Research and development
Information
Moral suasion

simply mandates the behaviour in law in order to achieve a socially desirable objective (Field, 1994). This is a centralized approach of the externality problem, which mainly consists in relying on standards and regulations of various types. Emission standards correspond to never-exceed levels directly applied to emissions coming from pollution sources, whereas ambient standards correspond to never-exceed levels for some substances in the ambient environment. The standard approach may also be used to specify the characteristics of technologies (process standards) and commodities (product standards). General and specific regulations include a large number of approaches. For example, land use control is a powerful tool to ban polluting activities from certain locations as well as to protect wildlife habitats and landscape beauty. Prohibitions and mandatory permits are widely used to deal with dangerous substances, which are selected on the basis of various criteria such as toxicity, persistence, or bioaccumulation. Approaches aimed at giving public authorities, firms and consumers the relevant information about the environmental effects of certain activities may also be considered as command-and-control policies. Among those, eco-labelling and environmental audits are expected to play an increasing role in the future.

Economists have long favoured the idea of incorporating economic instruments in environmental policy in order to internalize the negative externalities of production and consumption activities on the environment. Taxes and charges are very often considered as synonyms. In practice, they differ in the way collected revenues are allocated: tax revenues are simply a resource for the public authority while charge revenues are specifically used for financing environment protection measures. According to OECD (1990), emission charges may be defined as charges on the discharge of pollutants (noise included) into air, water, or soil based on quantity or quality. Product charges are levied on substances that are harmful to the environment, when used in production processes, consumed or disposed of. User charges may also be used to internalize the social costs resulting from negative externalities.

The tax or charge approach attempts to restore efficiency by manipulating, directly or indirectly, the price of polluting goods. This approach follows the PPP and implies that private agents pay for some of the costs associated with the production of negative externalities. In the same way, subsidies also aim to modify the cost–benefit ratio of polluting activities in order to achieve a reduction in residuals or emissions. As opposed to tax and charge schemes, they obviously cannot be regarded as an application of the PPP.

In very general terms, market creation strategies use government authority to establish the total quantity of emissions, but use markets to allocate emissions among the various potential polluters (Randall, 1987). Like regulation strategies, they are based on rationing. They nevertheless differ from command-and-control approaches through their much greater flexibility. In effect, like tax and charge schemes, they (theoretically and under certain assumptions, perfect competition in particular) make it possible to achieve the lowest resource cost outcome and encourage innovation in pollution abatement. Market creation strategies include two broad categories: (i) transferable discharge permits (i.e., emission trading); and (ii) tradable quotas (i.e., tradable credits).

Instruments such as liability insurance, covenants, deposit-refund, and performance bonds are based on a clear definition of property rights, which states that the polluter is liable for negative externalities. These instruments are market-oriented since their aim is to apply the PPP. The last category of instruments presented in Table 14.1 is heterogeneous. It includes flexible instruments that try to bring about voluntary behavioural changes. Moral persuasion involves some political and public pressure on polluters, whereas information and extension are based on a softer process of influence.

14.2.3.2 Environmental Policy Evaluation Criteria and Empirical Problems in Externality Analysis

As far as policy evaluation is concerned, a number of issues, summarized in Table 14.2, have to be considered. Efficiency covers several criteria. Environmental effectiveness refers to the capability to reach a predetermined environmental target. Economic efficiency refers to the question of resource costs needed to achieve a given level of abatement. Administrative costs include several components, that is, transaction costs, control costs, monitoring costs, and enforcing costs. Encouraging the development and market penetration of advanced pollution abatement technology is also clearly an important environmental policy objective. The issue of equity (i.e., how total benefits and costs of a given policy are distributed among various income groups) is a central one in evaluating every policy. Finally, the macroeconomic effects of environmental policies, mainly effects on competitiveness, must be considered.

TABLE 14.2. Main criteria for evaluating (environmental) policies

Efficiency
Environmental effectiveness
Economic efficiency
Administrative costs
Incentives for technical development
Equity (or fairness)
Effects on income distribution
Macro-economic effects
International competitiveness

In the EU and other industrialized countries, policy instruments are, in practice, being used in combination to deal with negative externality control. Standards are still very popular, mainly because they are simple to implement as they clearly set the specified target. However, their use raises several important issues. Some of the questions to be addressed include: (i) the "level" of the public authority in charge of setting the standard; (ii) the list of substances to be covered by the regulation; and (iii) the problem of monitoring and control. Other points are also to be made in relation with compliance itself. If there is no penalty for failure to comply, it is unreasonable to expect compliance. The equilibrium level of compliance clearly depends on the system of penalties. Even if the setting of such a system is not conceptually difficult, experiences have showed that government agencies have experienced political pressures militating against such effective penalties (Randall, 1987). Firms seem to prefer the command-and-control approach because they assume that they have more influence on regulation instruments, via negotiations, than on market-based policies. Furthermore, Buchanan and Tullock (1975) argue that standards are imposed more often than taxes because producers of the externalities can earn rents with standards.

Even if the scene has traditionally been

dominated by command-and-control strategies, there is now a great deal of interest and growing experience in the use of market-based instruments. Following Rendleman et al. (1995), four advantages of market-based strategies may be mentioned:

(i) they are more likely to achieve a more efficient allocation of environmental protection than a regulation policy;
(ii) they provide a clear incentive for research into and development of less polluting activities;
(iii) they mean less government interference and greater flexibility, that is, less administrative costs; and
(iv) they can provide a source of government revenues.

Theory suggests that emission charges or taxes create the right incentive to abate pollution. In practice, however, they are very often set at a low level so that they cannot have any steering effect. If emission charge strategies are implemented unilaterally, they are likely to handicap domestic industry on international markets. In that respect, product charges present a clear advantage compared with emission taxation if exports are not charged. With respect to command-and-control approaches, charges represent an additional cost for polluters and therefore can result in an increase in commodity prices. In addition, they can have significant distribution effects. Because environmental charges operate, in many cases, like taxes on consumption, they can have a more important negative effect on low-income households than on high-income households. Nevertheless, it is important to note that most environmental charge schemes actually used in OECD countries do not show significant effects on income distribution.

14.3 Agricultural Policy and Environmental Quality

The main objectives of agricultural policies in industrialized countries are to support farm income and ensure a stable and reasonably priced food supply. These objectives can be achieved by using a large array of instruments. Accordingly, while there is a considerable degree of similarity between agricultural policy objectives throughout the developed world, there are striking differences in the level of support to the agricultural industry and the way it is provided (Commission of the European Union, 1994). For a long time, the most common approach was market price support. As an example, the CAP has traditionally provided the bulk of its assistance to EU agriculture in this form. Nevertheless, in recent years, the proportion of transfers ascribable to other support methods has been rising fast. In the EU, the 1992 MacSharry reform has clearly changed the direction of the CAP in shifting, but only partially, the burden of support from the consumers to the taxpayers via the support price cuts and the introduction of compensatory payments. Every income support mechanism has unique, secondary, and unintentional effects on environmental quality.

The farm programmes were not originally developed to affect environmental quality. Even if farm programmes are not the only factor affecting agricultural management decisions, it is now well accepted that their impact on land use, output supply and input demand levels, goods prices, producer incomes and fixed or quasi-fixed netput returns, and so on, is significant. The effects of farm programmes on the agricultural sector and on the economy as a whole are continually being analysed. There is, in particular, a recent literature on the effects of farm programmes, and more generally of agricultural activities, on environment. This rising interest stems from the fact that agriculture is often a significant source of environmental pollution and degradation.

Whatever agricultural policy instruments are used, the consequences of agricultural support mechanisms pose no particular problem, at least from an environmental point of view, as long as production of the supported commodities does not involve environmental threats. This is no longer the case when agricultural policies increase soil degradation or encourage overuse of agricultural chemicals which are detrimental to the environment or, in other words, when agricultural policies encourage the negative environmental effects that modern agriculture tends to create (Mahé and Rainelli, 1987; Reichelderfer, 1990). Environmental degradation can obviously occur in the absence of agriculture support if commodity prices do not fully reflect the negative effects of agricultural production. In that context, free trade is no panacea for environmental degradation because world prices do not fully reflect environmental costs (Anderson, 1994). Finally, it is important to recognize that modern agriculture also provides positive externalities (e.g., habitat protection, scenic landscape, etc.). The problem is that agricultural policies generally tend not only to an oversupply of negative externalities, but also to an undersupply of positive externalities.

14.3.1 Commodity Price Support and Environmental Degradation: The Case of the Pre-Reform CAP

Agricultural policies can affect environmental quality by influencing management decisions that result in two sorts of fundamental processes, that is, changes at the extensive margin of production—the amount of land used in agricultural production—and changes in the intensity of land use—the amount of nonland inputs used per hectare (Kuch and Reichelderfer, 1992).

In the EU, it is now widely recognized that pre-reform CAP instruments (i.e., mainly the price support mechanism) have resulted in "excessive" use of fertilizers and pesticides detrimental to the environment. The pre-reform CAP has thus probably contributed to a socially inefficient use of chemicals. They have been used to achieve higher yields and maximize government programme payments on agricultural lands, but also to expand production onto marginal lands or to support crops in marginal areas. In the absence of specific government intervention, farmers had no incentives to take into account the potential negative environmental externalities arising from excessive use of chemicals. The resulting potential costs were completely ignored, farmers' decisions being based on private interests solely. In the context of the 1960s, land scarcity resulted in increasing both the price of land and the intensification level (i.e., amount of chemicals per unit of land). The price support mechanism boasted these two movements, and it was only a second-best policy for supporting farm labour income because a significant part of the support was capitalized in land prices. This policy also led to negative environmental effects at the extensive margin because production shifted onto marginal lands and ecologically sensitive and vulnerable zones were converted to agricultural production.

14.3.2 Alternative Instruments of Assistance to Agriculture and Environmental Degradation: The Case of the 1992 CAP Reform

The pre-reform CAP, and in particular high and stable output–input price ratios, provided strong incentives for European producers to both increase the intensity of land use and expand intensive production techniques onto marginal lands. By reducing output–input price ratios, the 1992 CAP reform should induce farmers to reduce variable input use at the intensive margin. Nevertheless, three factors may offset the positive effect of the reform on land use intensity. First, the output institutional price cut does not necessarily mean an equivalent decrease in market price. Second, a downward adjustment of chemical prices may partly offset a part of the output price cut. Third, and perhaps most importantly in the long run, price elasticities of yields and input use may be low around the pre-reform "equilibrium" point. As a result, the effects of price changes are likely to be limited. Some empirical studies suggest that price elasticities are rather low in the short run when primary factors (i.e., labour, land, and capital) are fixed. They are much higher, in absolute value, in the long run when primary factors may adjust. Per-hectare compensatory payments of the CAP cannot be considered as fully decoupled, mainly because they remain tied to the obligation of producing certain crops, that is, cereals and oilseeds (CO crops). It is likely that the area in CO crops should decrease without these compensatory payments. In other words, the definition of a fixed eligible base area for CO crops and the obligation of producing these CO crops to receive direct aids imply that land and labour should not substantially adjust in response to the policy instrument changes introduced in May 1992. Relevant price elasticities are thus short-run or restricted elasticities. They are probably low. Accordingly, the effects of support price cuts on yields and variable input use at the intensive margin should be limited, even in the long run.

On the extensive margin of production, a support price cut should induce farmers to convert some low-quality land, previously cultivated with intensive techniques, to more extensive techniques requiring lower amounts of chemicals. But again, the three factors mentioned above, that is, the imperfect transmission of institutional price changes to market prices, the downward adjustment of market prices of chemical inputs, as well as the quasi-fixity of primary factors and the subsequent low values of price elasticities, are expected to offset the potential positive effect of the reform at the extensive margin.

The previous discussion suggests that the most important policy parameter which could affect the level of demand for fertilizers and crop protection chemicals is the reduction in the arable cropping area due to the set-aside requirement. At this stage, it is thus particularly important to highlight the rationale of the set-aside programme and, more generally, the rationale of the new instrumentation of the CAP. The logic of the reform is clear and may be described as follows. Support prices for grains were substantially cut down, from a current average buying-in price of 155 green ECU per tonne in 1992–1993 to align on a target price of 110 green ECU per tonne and an intervention price of 100 green ECU per tonne in the 1995–1996 marketing year. The objective of this support price cut was to induce an increase in the domestic demand and especially in its feed component thanks to an improved price competitiveness of interior cereals with respect to imported feed ingredients (i.e., protein cakes and substitutes of grains). The support price cut, which is compensated by "fixed" direct payments per hectare, should induce producers to both reduce variable input use at the intensive margin and adopt less intensive production techniques, which should reduce growth rates in yields in the medium term. In the short term, an improved balance for grains is obtained by a set-aside programme.

The decision to decrease EU support prices for grains should provide substantial gains, essentially in terms of lower user prices for both final consumption and derived demand. The closer the EU prices are to world prices, the greater these gains are. One of the main limits of the 1992 CAP reform is that it fixes the minimum institutional price at a higher level than the world price, which alters the price competitiveness of European grains with respect to third countries and implies the continuous use of subsidies to export to foreign markets. Furthermore, the first years of CAP reform application have clearly showed that market prices may be substantially higher than these minimum institutional prices. If the EU does not fulfil its GATT commitment in terms of subsidized export volume reduction, the easiest way to operate would obviously be to increase the set-aside rate, which means that the negative effects of this regulatory instrument would be aggravated, that is, increased costs, decreased competitiveness on the international scene, and fewer incentives to farming more extensively.

An alternative way to operate would be to set the domestic price at the world price level, which would allow the EU to export without restitutions (GATT commitments would no longer apply in that case), to increase its competitiveness not only on world markets but also on the domestic scene with respect to substitutes, and to abandon the requirement for land to be set aside. Removing this requirement would allow farmers to operate at minimum total cost and would induce them to use less intensive farming. In the context of the current instrumentation of the CAP, a first step in that direction would be to introduce a double market for grains in which land actually set aside would be cultivated with more extensive techniques (i.e., with low per-hectare chemical input levels) and would be remunerated at world price levels.

The new CAP mechanisms introduced in May 1992 were primarily designed and implemented to solve the problems of surplus production for grains, not the ill-defined environmental problems that modern farming may create. One weak feature of the reform is the lack of integration between the environmental stewardship role of agriculture on the one hand, and the food production and trade roles of agricultural activities on the other hand. The analysis developed above assumes implicitly that agricultural pollution problems in the EU, such as nitrate contamination of groundwater supplies, eutrophication of surface water supplies, as well as marine waters and smell, arise mainly from the increased use of chemical inputs and animal waste disposal. Given the link between agricultural pollution and the increasingly intensive agricultural production, the solution advocated by economists, and also by CAP reform promoters, in order to reduce agricultural pollution is to return to lower-intensity farming by reducing the use of agricultural chemicals and by shifting to less intensive land uses.

The CAP reform should, at least theoretically, induce a movement toward lower intensive farming. Nevertheless, many factors may offset this positive trend: The imperfect transmission of institutional price changes to market prices, the downward adjustment of market prices of chemicals, the quasi-fixity of primary factors and low values of price elasticities, the current set-aside programme which implies that it is more profitable to set aside low quality land first, and so on. With few exceptions, the use of chemicals at high levels will remain profitable. Our current system of agriculture has developed because of these chemicals and is likely to remain very dependent on them. Because most of the pollution costs are external to agricultural production, society cannot expect the agricultural problem to be reduced drastically without direct intervention, even if the CAP reform is, but only indirectly, a first step in the right direction.

In the specific case of nitrates, pollution is derived from nitrogen contained not only in commercial fertilizers, but also in livestock manure. The CAP reform clearly does not provide the political framework for solving the livestock manure excess problem. Livestock manure may be viewed as a more problematic source of nitrates than commercial fertilizers—first, because farmers generally treat manure as waste and not as a productive input as fertilizers, and, second, because the reform has very limited effects only on animal production intensity.

14.4 Integrating Agricultural and Environmental Objectives and Policies: Three Case Studies

Three case studies are presented. The first two case studies address pollution due to farming. The first illustration deals with the nitrate issue in the EU and shows how policymakers try to reduce the level of environmental damage resulting from intensive agricultural activity by using simultaneously command-and-control instruments and market-based strategies. The second illustration deals with soil contamination. The problem here is related to the potential accumulation of pollutants in the soil. The proposed strategy is based on liability laws with the enforcement of an insurance mechanism in order to cover the high costs of cleanup. At this stage, it is important to underline that nonpoint source pollution is much more difficult to regulate than point source pollution mainly because individual source pollution cannot be monitored easily. Accordingly, environmental policies in the agricultural sector must generally be based on observable proxies for pollution loadings, and not on emissions.

The third case study considers some of the beneficial effects of agriculture. As farming in marginal areas, particularly in high uplands and wetlands, is increasingly less and less viable, a process of land abandonment is occurring in many "sensitive" areas of the EU which has important ecological and social negative consequences. The case study presents the EU policy for promoting wetland protection.

14.4.1 Water Quality and Nitrate Policy

Concerns over the eutrophication of fresh waters and coastal waters and the consequent disruption of the ecosystem have brought attention to the excessive levels of nitrates in surface waters. Nitrate pollution of surface waters reduces its use for drinking, fishing and recreation purposes. The eutrophication process also favours the growth of algae which disturbs the oxygen balance and finally results in fish death. Because groundwater is generally of better quality than surface water, it has a higher value and its contamination by nitrates may be viewed as a more serious problem. Groundwater resources are substantially different from surface water resources in terms of their regenerative capacity. Contaminated aquifers are likely to be lost for use for a long time because of the very slow rate at which they are replenished. This should lead to careful management of groundwater resources because they are the main source of drinking water: in Europe, for example, it accounts for about 75 percent of drinking water supply.

According to OECD (1989), most industrialized countries cite agriculture as a widespread source of water pollution through nitrates from inorganic fertilizers and animal waste. In particular, the development of intensive livestock rearing has increased the availability of animal manure, and its concentration in specific areas has created high spots of nitrate contamination. As emissions can be widely controlled using appropriate techniques when fertilizing and feeding, many environmental policy schemes aim at the adoption of new farming practices, for example, more efficient application of fertilizers by more accurate timing or rates of application, better use of animal wastes, and so on. Most schemes rely on command-and-control strategies. Up to now, there has been limited experience only with incentive-based strategies.

14.4.1.1 Command-and-Control Strategies

In the EU, the directive on the quality of water intended for human consumption (Directive 75/440) sets standards for drinking water in the form of maximum acceptable concentrations and guideline concentrations. A maximum admissible level for nitrates of 50 mg/l is thus legally enforceable throughout the Union. Even if this general regulation is not directly oriented toward the control of nitrate emissions from agriculture, it may lead to specific limitations of farming activities.

Because nitrates from nonpoint agricultural sources form the main cause of the nitrate problem, a specific legislation, that is, the nitrate directive (Directive 91/676), has been issued in 1991. The main purpose of this directive is to restrict the leaching of nitrates in "vulnerable" zones. Crudely calculated, this results in a situation where manure (including that from grazing livestock) may not be applied in amounts that exceed 170 kg nitrogen per hectare. An exemption up to 210 kg nitrogen per hectare can be granted until 1999. In order to achieve this objective, member states must present action programmes for vulnerable zones. These programmes should be consistent with the prescription contained in the new code of good agricultural practices, and their implementation should be no later than December 1999. A programme would include the definition of periods where the application of certain types of fertilizers is prohibited and quantified criteria for their application. Monitoring the effectiveness of action programmes and submitting reports on the results are required. Moreover, financial incentives in order to encourage methods that respect the environment can be included in the programmes.

By banning or regulating farming in recharge areas, risks of contamination by nitrates can be lowered. In combination with the general legislation based on uniform standards, banning can be effective in water protection by implementing further restrictions with regard to local variations in water resources and vulnerability to pollution. Water authorities in many countries rely heavily on zoning as they design water catchment protection areas. Protection policy is usually expressed in a document reviewing activities that are acceptable within different zones around well heads (OECD, 1989). Activities include disposal of waste, general development, manufacture and storage, mineral and energy exploration and exploitation, agriculture, and application of sewage sludge. Activities are ranked, some being prohibited and some being restricted (e.g., agriculture) with a maximum limit for nitrogen application. Most schemes incorporate three or four zones defined with respect to the vulnerability of the resource to pollution and the various degrees of protection required. Moreover, in some countries, there is specific legislation on classified installations for environmental protection. This specific legislation covers intensive rearing activities.

14.4.1.2 Incentive-Based Strategies

For point sources, water effluent charges occur in several countries (e.g., Belgium, France, Germany, Italy, the Netherlands, and the United Kingdom). They cover a number of substances, including nitrogen (Kageson, 1993). Charge systems generate revenues that are used to finance certain environmentally related expenditures such as treatment plants. They can be considered to have been successful in controlling emissions.

Some EU countries have decided to put a levy on emissions from livestock farming. In 1987, the Netherlands introduced a charge on excess production of manure. A similar charge was recently introduced in Belgium by the Flemish authorities. In 1991, the French government decided to gradually extend the charge system implemented in other economic sectors to rearing installations. All the required information in calculating charges is provided by an environmental audit that includes: (i) a detailed description of polluting emission flows with an emphasis on building sitting, storage facilities for manure, runoff, leaching, and infiltration; and (ii) an assessment of agricultural practices in order to establish a nitrogen balance at both the plot and farm levels. This audit initiates a negotiation process between the eligible farmer and the water agency with a view to reaching an agreement whose prescriptions deal with the improvement of farm buildings and the promotion of better farming practices. Investments will be subsidized, and there are financial and technical commitments involving a possible reduction in the charge amount.

Emission taxation is much more difficult for diffuse emissions. A charge on the primary commodity may then be a second-best solution. Product charges are levied on substances that are harmful to the environment when used in production processes. They can be applied to substances that: (i) are used in large quantities and are easy to identify; (ii) have possible substitutes; and (iii) have an elastic price demand function. Few countries (i.e., Austria, Finland, Norway, and Sweden) have implemented a charge system on mineral fertilizers (Kageson, 1993). The leading argument against fertilizer taxation is that the demand is price inelastic. As a result, tax rates should be very high to achieve a significant reduction in fertilizer consumption, which is not politically feasible. Nevertheless, this argument is essentially valid in the short run because there is empirical evidence that the long-run demand for mineral fertilizers is price elastic (Bonnieux et al., 1995).

The process of taxation is depicted in Figure 14.3 where the price and quantity of fertilizers are denoted by P and Q, respectively. Without any regulation, the market equilibrium is M_0 and the use of fertilizers is Q_0. In order to achieve a decrease in the use of fertilizers, the environmental agency implements a charge equal to the difference $P_1 - P_0$. The market equilibrium shifts from M_0 to M_1 leading to a reduction in fertilizer use equal to $Q_0 - Q_1$. Farmers' surplus loss corresponds to area A + B. Area A is the net social loss due to the reduction in output induced by the taxation scheme, and area B is the revenue of the tax. Because farmers have to pay a higher price for

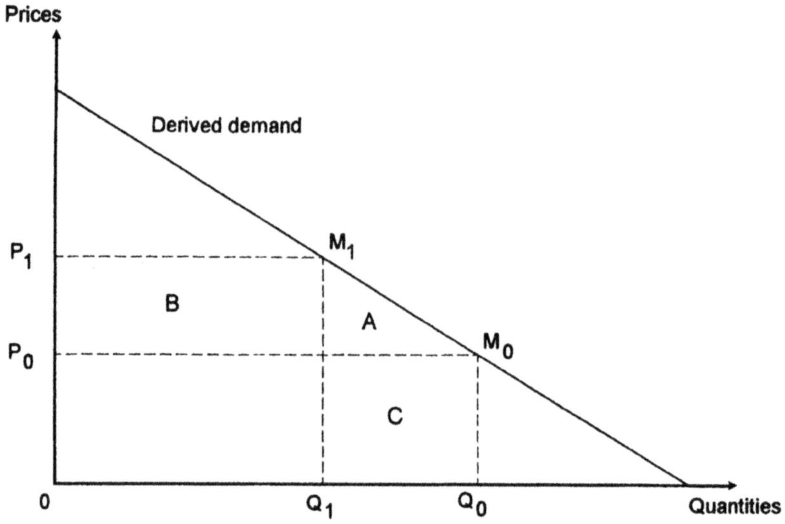

FIGURE 14.3. Demand for mineral fertilizers.

their actual use of nitrogen, their organizations are likely to strongly lobby against such a policy. Accordingly, it is often proposed to neutralize the income transfer from agriculture to society (i.e., area *B*) by reimbursing the tax revenue.

This compensation scheme would not affect farmers' incentives to decrease nitrogen use provided that the tax revenue is paid back in the form of lump-sum payments. A simple compensation scheme that respects this condition is to reimburse the tax revenue at a flat rate per hectare. The crucial factor is then average fertilizer use because this compensation scheme results in an income redistribution from farmers over the average to farmers under the average. The burden for the fertilizer industry is given by area *C*. Total social loss is therefore (area *A*) + (area *C*). This sum has then to be compared with the social benefits resulting from a reduction in the polluting emissions.

A system of tradable quotas could also achieve the desired reduction in fertilizer use. If the quota market is competitive, a price equal to the tax amount (i.e., $P_1 - P_0$) would emerge in an equilibrium. Quotas could initially be allocated on the basis of a uniform reduction in fertilizer use based on consumption in the previous years.

14.4.2 Farming and Soil Contamination

Soil contamination results from the accumulation of flows of pollutants that become bound to the soil. Constituents of chemicals, manure, and sewage sludge, such as phosphates and heavy metals, can accumulate in the soil. One of the main sources of cadmium emissions is related to fertilizers—copper and zinc, which are added to feeding stuffs, are found in pig manure, etc. Agricultural pollution by pesticides occurs both as a result of handling and storage at specific sites on the farm as well as widespread application in crop fields. Application of fertilizers and pesticides may result in transport and transformation from the farm to other sites. This dynamic process is typically induced through soil erosion, surface runoff, and leaching into groundwater.

In the case of soil contamination, public policy focuses on two main issues, that is, restoration of contaminated soils and prevention of future contamination.

14.4.2.1 Restoration Issues

The economic costs of a one-time hazardous substance release are of two types: (i) the costs associated with foregone or lost service flows

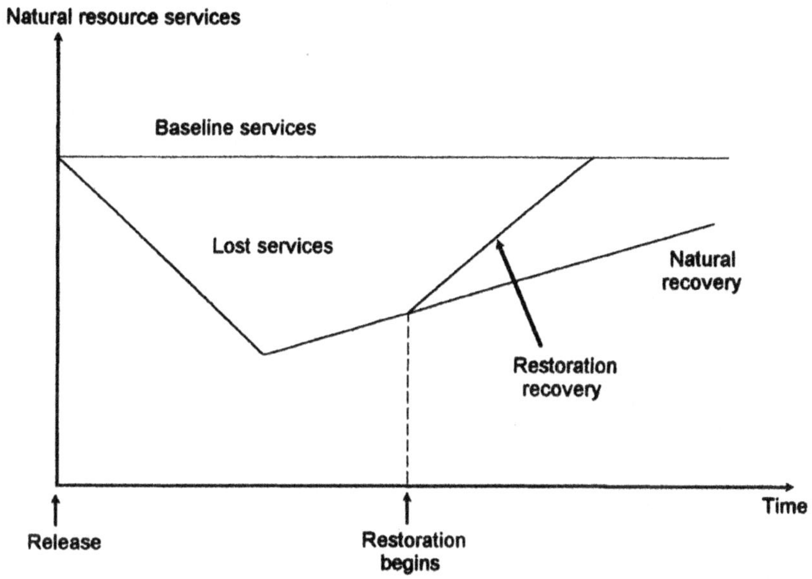

FIGURE 14.4. Loss of natural resource services after a release and their recovery rate.

from the damaged natural asset; and (ii) the costs associated with restoration or replacement of the resource. The tradeoff between restoration costs and reduction of lost service flows determines the efficient level of restoration. An important issue in any quantification of damage to the environment is the resource recoverability as measured in reference to baseline level services, that is, the conditions existing before the release. It corresponds to the amount of time needed for an injured resource to recover and reach its initial level. An ecosystem recovers at a natural recovery rate, which depends on the magnitude and the toxicity of the release. This rate can be accelerated by human action (resource restoration). The optimum level of restoration depends on a dynamic consideration of these processes (Ward and Duffield, 1992).

Figure 14.4 illustrates a simple time profile that describes the loss of natural resource asset services after a damage event. The baseline service could be reached after a time span either at the natural recovery rate or through a combination of natural and restoration processes. Furthermore, depending on the allocation of property rights, society could demand compensation for damage events, which would require restoration beyond the baseline level of service. Because clean-up activities represent the largest component of social costs of restoration of damage caused by a serious event (e.g., release of hazardous materials), accurate estimation of these costs is an essential element of the economic analysis of damage restoration. By definition, in addition to the actual expenditures necessary for restoration, estimation must include an assessment of the social opportunity cost of restoration (i.e., what goods and services must be foregone as a result of the restoration). A further element is the timing of these costs over future periods. Finally, analysis must estimate the level and timing of the social value of restoration.

Given estimates of profiles of social costs and benefits, economic analysis of restoration can determine the appropriate level and timing of restoration to ensure that a maximum level of net social welfare is achieved. A critical element is the rate at which society prefers present to future consumption. When the social discount rate increases, the optimal timing of restoration activities must be accelerated. Timing of restoration will also be influenced by the form of compensation required to be paid for the damage. Monetary payments or in-kind compensation can be involved.

A rich history of public policy to affect restoration of contaminated sites is available for the United States where, in 1980, a law was passed to establish a fund (i.e., the Superfund) to pay for restoration of sites. This law, known as the Comprehensive Environmental Response, Compensation, and Liability Act (CERCLA) put in place an institutional mechanism for restoration of sites that offered inadequate private incentives for restoration. This law was later amended in 1986 by the Superfund Amendment and Reauthorization Act (SARA). The objectives of these laws were twofold. First, the legislation put in place a system of liability for releases of hazardous substances. Second, the legislation put in place a fund for financing the clean-up or restoration of polluted sites. The liability mechanism of this legislation establishes a means through which private and public suits for damages could be brought against individuals or organizations responsible for the pollution. Damages awarded are paid to the Superfund and used to finance clean-up and restoration of sites. Furthermore, this legislation challenges the courts to define what is meant by restoration through its requirement for a definition of what constitutes damages. Subsequent cases have resulted in definitions that view damages as including not only the costs of restoration of a site, but also the lost value of direct use, indirect use, and nonuse values.

14.4.2.2 Prevention Issues

While restoration involves correction of damage caused by soil contamination, the stocks of pollutants can also be altered by changing the flow of pollution. This possibility follows directly through what was labelled above as preventive public policy. Here, two approaches are feasible.

The first approach involves incentive-based strategies to alter the flow of pollution. These strategies involve changing incentives through the use of taxes or subsidies, or changing the level of use of polluting inputs, production practices, or the level of pollution through the use of regulatory standards. In general, the efficiency of the various strategies depends on the ability of policy makers to estimate both the marginal costs and benefits associated with the contamination process. When the effects of contamination are uncertain, the marginal benefits of contamination clean-up are difficult to estimate and the incentive levels are difficult to establish. While a variety of schemes have been developed to accommodate the difficulty in estimating marginal costs of clean-up (e.g., tradable permits), marginal benefits still remain difficult to evaluate. Standards suffer from the same problems as economic incentives on this point. In both cases, the level of the instrument must be set to optimize social welfare. Where information is inadequate, incomplete, or uncertain, substantial errors can arise, and the implications of such errors will mean reducing the welfare improvement achieved.

The second approach relies upon the establishment of liability for the effect of pollution, that is, the level of contamination. This approach involves the specification of property rights for society for an uncontaminated site. Establishment of private liability can have an effect on soil contamination through the operation of private markets for the site's ownership or for goods produced on the site. This is feasible when detection of contamination is relatively cheap compared to the value of the site. In that case, potential buyers of the site can inspect it to determine the level of contamination and estimate the restoration costs. Bids for the site would be directly affected by the estimated restoration costs. However, where private detection of contamination is costly, this approach is not expected to be feasible. In that case, public intervention is required. It is of interest to note that the liability approach can be linked to other strategies. For example, liability can be limited to cases where standards are exceeded or recommended production practices are not used. In that case, private liability can be limited and a transfer to the public sector is required. Again, in incomplete information, the use of such approaches may result in disappointing improvements in social welfare.

14.4.3 Farming and Wetland Protection

For many decades, wetlands were regarded largely as nuisances. They were drained, cleared, filled, flooded, contaminated with toxins and nutrients, and exploited for whatever resources could be extracted from them (Turner and Jones, 1991). By far, the most important economic sector absorbing wetlands has been agriculture. In comparison, urbanization and the development of infrastructures were only responsible of a limited percentage for wetland conversion to other uses. In the light of the agricultural commodity surpluses and of the financial burden of farm support, the conversion of wetlands into additional farmland has become increasingly inconsistent because they produce numerous benefits for society. Some benefits are either irreplaceable if lost or can only be replaced at a very high cost. In general terms, wetlands regulate water flows, store, filter, and purify water, buffer the effects of storms, and provide an essential habitat for flora and fauna.

14.4.3.1 The Cotentin Case (Bonnieux et al., 1995)

The Cotentin wetlands in Lower Normandy (France), which extend over 37,500 hectares, are partially a manmade creation. The effective management of water flows, through hydraulic schemes, began in the first half of the eighteenth century. It continued during the nineteenth century with drainage works and land conversion to agriculture resulting in a reclamation of 7000 hectares of polders and shallow tidal lands. But, after 1950, due to the postwar agricultural revolution, the economic interest in these wetlands compared to more favoured regions decreased. Simultaneously, intensification on the best fields and abandonment of the less productive lands occurred leading to an upsetting of the equilibrium between the environment and agriculture. Abandoned and intensified wetlands are suffering a decline of species associated with these particular habitats. Moreover, the rural landscape, which is a historical creation resulting from a delicate balance between wet and dry land on the one hand, and between the influence of nature and man on the other hand, is threatened by the growing of scattered scrub.

The predominant farming activity is the raising of dairy cows and beef cattle. The trend toward more intensive dairy farms in recent years has led to the increase of silage and maize crops at the expense of hay and permanent pasture. At the same time, increased fertilizer use changed the character of grasslands and led to a reduction in the area suitable for breeding birds. All these changes, occurring both in dry and wet land, radically altered landscape and wildlife value and threatened the preservation of habitats.

Table 14.3 shows the large array of benefits provided by the Cotentin wetlands. The social benefits of wetland conservation include three value categories. First, the wetland ecosystem structure (plants, animals, fish, soil, and water) yields benefits that are of direct use value to society, for example, wildlife support, recreational fishing and hunting, scenic landscape, and so on. Second, the interaction between wetland hydrology, soil, and vegetation determines a set of processes that provide an indirect use value. This includes flood control and groundwater recharge as well as life support and assimilative capacity. Wetlands also yield an existence value which is not derived from direct or indirect use.

The consumption of some goods may be linked with benefits in some ways. Consumers may favour quality products in their purchasing decision; beneficiaries pay for hunting and fishing: 25,600 hunting licenses (including 10,500 licenses for waterfowl) and 3000 fishing licenses are sold in the market; scenic beauty is partially moneterized through eco-tourism and similar activities. Nevertheless, market pricing is of a very limited extent, and wetland owners generally cannot capture all the bene-

TABLE 14.3. The provision of benefits in the Cotentin wetlands, France (Source: Adapted from Bonnieux et al., 1993)

Value categories	Benefit sources	Policy measures
Direct use value	Rural characteristics	Labelling quality products
	Wildlife: waterfowl, birds, fish, etc.	Country side protection: hedge replanting, drainage of channels and ditches, etc.
	Landscape quality: variety in shape and colour, cultural heritage, etc	Management prescriptions and good agricultural practices
Indirect use value	Flood control and groundwater recharge	Good agricultural practices
	Life support	Protection and enhancement of biodiversity
	Assimilative capacity	Waste recycling, compost using, surface impoundment, etc.
Existence value	Wildlife of specific interest	Habitat protection

fits of their resource for their own use or sale. Flood control and groundwater recharge benefits accrue to others who live in small towns and villages around the Cotentin wetlands; life support and assimilative capacity benefits as well as habitat protection benefits cannot be commercially exploited, and so on.

For most benefits provided by the Cotentin wetlands, market exchange mechanisms are inappropriate. The prevalence of the public good attributes of these benefits suggests a possible role for government intervention in order to correct the market factors that had encouraged wetlands conversion. This issue is related to the recognition of the joint-product relationship between benign farming practices and environmental goods. This has led to the implementation of specific schemes such as the Environmentally Sensitive Area (ESA) agreement. CAP reform accompanying measures provide the legal framework for paying farmers for the provision of environmental goods. These payments are in line with the ESA schemes, which make the interconnection between environmental and agricultural policy explicit.

To protect this area of significant botanical as well as ornithological interest, a Natural Regional Park (NRP) was established in May 1991. The NRP extends over 125,000 hectares, including 25,000 hectares of low-lying land formed by the flood plains of five rivers. To prevent environmental undesirable changes, part of the NRP was designated as an ESA. Up to now, the ESA has been restricted to an 8000 hectare wetland area. Within the NRP lie nine sites of high ecological interest. Their designation is concerned with the protection of peatland areas and wetland habitats. Due to their major role as a habitat for migratory birds, these wetlands have been designated as a Ramsar site. Moreover, several hunting reserves are managed under the supervision of both the Ministry of Agriculture and the Ministry of the Environment. Public purchase has also been used to protect 180 hectares of polder in the coastal zone of the NRP.

The general objectives for the ESA are to: (i) support dairy farming and cattle rearing in the existing grassland area; (ii) conserve wet grassland areas, ditches, drainage channels, and their associated landscape; and (iii) improve the ecological value of the wet grassland and associated habitats. To achieve these objectives two schemes have been implemented.

The first scheme is offered to local communities to manage commons. During the first campaign (1991–93), 1528 hectares out of 4000 hectares of eligible commons were entered into this scheme. The second scheme is offered to individual farmers. For the same period, 154 farmers have joined the scheme for a total of 1635 hectares. The uptake level is 41 percent in terms of acreage. No target level has been officially set. Nevertheless, these uptake levels can be considered as a success in comparison with similar cases in other regions and countries.

The second scheme is based on voluntary five-year agreements with farmers. Its functioning is straightforward since all farmers are offered the same price for accepting the same conditions to protect and enhance the environment. Typically, the conditions relate to the acceptance of constraints on farm operation. Four levels of management prescriptions were offered to farmers. Accordingly, eligible farmers are faced with two choices. First, a dichotomic choice as they are asked to accept or reject voluntary participation in the scheme. Second, as farmers enter the scheme, they have to choose the relevant tier.

The farmers' role as guardians of the countryside is firmly acknowledged. Direct payments through standard agreements are justified because farmers accept constraints on their activities resulting in an increased level of public benefits. These schemes represent a policy mechanism aimed at inducing changes in farming activities to increase social welfare. The payment is the price for purchasing or renting rights by the NRP which represents society (Crabtree, 1993).

14.5 Concluding Comments

It is now well recognized that agricultural production can adversely affect environment, for example, by depleting or contaminating water supplies and increasing soil degradation. Some adverse effects can be attributed to agricultural policies that distort output and input prices. For a long time, the basic CAP instrument has been market price support, which encourages production, but penalizes consumers and has resulted in EU surpluses in most temperate farm products. The use of this instrument has over time become increasingly costly in terms of budgetary expenditures, complicated administration, distortion of resource use, international trade tensions, and detrimental environmental effects.

For grains and oilseeds, the reform has changed the direction of the CAP in shifting the burden from the consumers to the taxpayers, via the mechanism of targeted compensatory payments. But it is clear that the reform was not primarily designed to solve the ill-defined environmental problems that modern farming may create. The final decisions adopted in May 1992 appear far less ambitious than the initial proposals of the Commission of the European Union as regards the reorientation of European agriculture toward meeting the long-run objectives of competitiveness, resource conservation, and sustainable development.

In the past, environmental policies were mainly designed and implemented at regional levels. The agricultural sector, in the EU and other industrialized countries, has been largely exempted from the application of the PPP, and market-based strategies have not, so far, played a large role. This is due to various factors, including mainly the nonpoint nature of discharges and the political power of the farm lobby. The problem is now considered as sufficiently serious and transnational in character to require a wider approach (Haley, 1994). Agricultural and environmental policies can be complementary or in conflict depending on the characteristics of the problem and the types of policy instruments (Antle and Just, 1991). The conflict in policy objectives should obviously be addressed through efforts to integrate agricultural and environmental programmes. As noted by Munk (1995), the issue of the integration of environmental objectives into sectoral policy has to be considered in the more general context of the reinstrumentation of agricultural policies (i.e., the shift from market price support toward more decoupled forms of assistance to farmers) and the level of policy responsibilities.

References

Anderson, M. NAFTA and environmental quality: Issues for Mexican agriculture. Sullivan, J., ed. *Environmental Policies: Implications for Agricultural Trade.* USDA-ERS, Foreign Agricultural Economic Report No. 252. Washington, DC: USDA-ERS; 1994:58–74.

Antle, J.M.; Just, R.E. Effects of commodity program structure on resource use and the environment. Just, R.E.; Bockstael, N., eds. *Commodity and Resource Policies in Agricultural Systems.* Berlin: Springer-Verlag; 1991:97–128.

Baumol, W.J.; Oates, W.E. *The Theory of Environmental Policy*. Cambridge: Cambridge University Press; 1988.

Bonnieux, F.; Gloaguen, Y.; Rainelli, P.; Vermersch, D. Paiements directs et préservation de l'environnement: Application au cas des zones humides de Basse Normandie. Lehman, B.; Popp, H.W.; Stucki, E., eds. *Direct Payments in Agricultural and Regional Policies: Proceedings of the 30th EAAE Seminar*; 1992 November 11–13; Chateau-d'Oex, Switzerland. Konstanz: Hartung-Gorre; 1993:199–216.

Bonnieux, F.; Rainelli, P.; Vermersch, D. A consistent policy for the reduction of nitrate pollution from livestock rearing. Albisu, L.M.; Romero, C., eds. *Environmental and Land Use Issues: An Economic Perspective*. Kiel, Germany: Wissenschaftsverlag Vauk Kiel KG; 1995:287–300.

Buchanan, J.M.; Tullock, G. Polluters' profits and political response: Direct controls versus taxes. *Amer. Econ. Rev.* 65:139–47; 1975.

Commission of the European Union, Directorate-General for Economic and Financial Affairs. EC Agricultural Policy for the 21st Century. *Eur. Econ.* 4; 1994.

Crabtree, R. Effectiveness of standard payments for environmental protection and enhancement. Lehman, B.; Popp, H.W.; Stucki, E., eds. *Direct payments in Agricultural and Regional Policies: Proceedings of the 30th EAAE Seminar*; 1992 November 11–13; Chateau-d'Oex, Switzerland. Konstanz: Hartung-Gorre; 1993:144–156.

Field, B.C. *Environmental Economics: An Introduction*. New York: McGraw-Hill Inc.; 1994.

Haley, S. Assessing environmental and agricultural policy linkages in the European Community. Sullivan, J., ed. *Environmental Policies: Implications for Agricultural Trade*. USDA-ERS, Foreign Agricultural Economic Report No. 252. Washington, DC: USDA-ERS; 1994:102–112.

Kageson, P. *Economic Instruments in European Environmental Policy.* Stockholm: European Environmental Bureau; 1993.

Kuch, P.; Reichelderfer, K. The environmental implications of agriculture support programs: A United States perspective. Becker, T.; Gray, R.; Schmitz, A., eds. *Improving Agricultural Trade Performance Under the GATT*. Kiel, Germany: Wissenschaftsverlag Vauk Kiel KG; 1992: 215–231.

Mahé, L. P.; Rainelli, P. Impact des pratiques et des politiques agricoles sur l'environnement. *Cah. Econ. Sociol. Rural.* 4:9–31; 1987.

Munk, K. J. The scope for integrating agricultural and environmental policies. Unpublished paper presented at the conference on agriculture and environment; 1995 May 4–5; Club de Bruxelles, Brussels.

OECD. *Water Resource Management: Integrated Policies.* Paris: OECD; 1989.

OECD. *Economic Instruments for Environmental Protection*. Paris: OECD; 1990.

Pigou, A. *The Economics of Welfare*. 1st ed. London: Macmillan; 1920.

Randall, A. *Resource Economics: An Economic Approach to Natural Resource and Environmental policy.* New York: John Wiley and Sons; 1987.

Reichelderfer, K. Environmental protection and agricultural support: Are trade-offs necessary? Allen, K., ed. *Agricultural Policies in a New Decade: Resources for the Future and National Planning Association.* Washington, DC: Resources for the Future and National Planning Association; 1990:201–230.

Rendleman, C. M.; Reinert, K. A.; Tobey, J. A. Market-based systems for reducing chemical use in agriculture in the United States. Environ. Resour. Econ. 5:51–70; 1995.

Sidgwick, H. *Principles of Political Economy*. New York: Macmillan; 1887.

Turner, K.; Jones, T., eds. *Wetlands, Market and Intervention Failures: Four Case Studies.* London: Earthscan Publications Ltd.; 1991.

Ward, K.M.; Duffield, J.W. *Natural Resource Damages: Law and Economics.* New York: John Wiley and Sons; 1992.

Zilberman, D.; Mara, M. Agricultural externalities. Carlson, G.A.; Zilberman, D.; Miranowski, J. A., eds. *Agricultural and Environmental Resource Economics.* New York: Oxford University Press; 1993:221–267.

Part VI
Systems and Land Planning Analysis

15

Land Use Allocation and Environmental Impact Assessment in Land Planning

Germán Glaría and M. Angeles Ceñal

15.1 Introduction

Physical planning with an ecological basis appears in response to the need of solving conflicts arising from land use. It is a systematic analysis of land suitability and alternatives of use, which are the basis for the selection of the best choices to put into practice. The best choices are to be understood as those fulfilling social needs in the present, always bearing in mind that the future of new generations is not to be jeopardized.

15.2 Methodological Outline

Formulation and assessment of alternatives constitute a basic element of land planning. Without alternatives there is no planning; there will be only the application of technology to a specific situation. For instance, technological studies leading to commercial reforestation of a given species in a given place cannot be regarded as planning. Planning involves knowing the possible alternatives of use of a site and choosing the best use according to physical and socioeconomic criteria.

The formulation and assessment of alternatives are based on methodological processes that differ from one another but have a common purpose. The general approach in the formulation of alternatives has four phases: (i) Preparation—definition of objectives; (ii) analysis of the environment; (iii) diagnosis; (iv) planning—analysis, assessment, and selection of alternatives, including environmental impact assessment (EIA).

Subphases can be distinguished in each of these phases, as shown in Figure 15.1.

This chapter focuses on the diagnosis and planning phases of the general approach. Of the four phases cited, the first two have a general nature and can be integrated in any planning process. By way of contrast, the studies necessary for the diagnosis (third phase) depend on the objectives of the planning activity. Finally, the formulation of alternatives (fourth phase) responds, essentially, to two approaches: (i) knowing the best location for a specific use (e.g., the location of a reservoir); and (ii) knowing the best alternative of use for an area.

15.3 Diagnosis

15.3.1 Preliminary Considerations

The purely descriptive information collected in the inventory in phases one and two constitute the basis for land diagnosis. This,

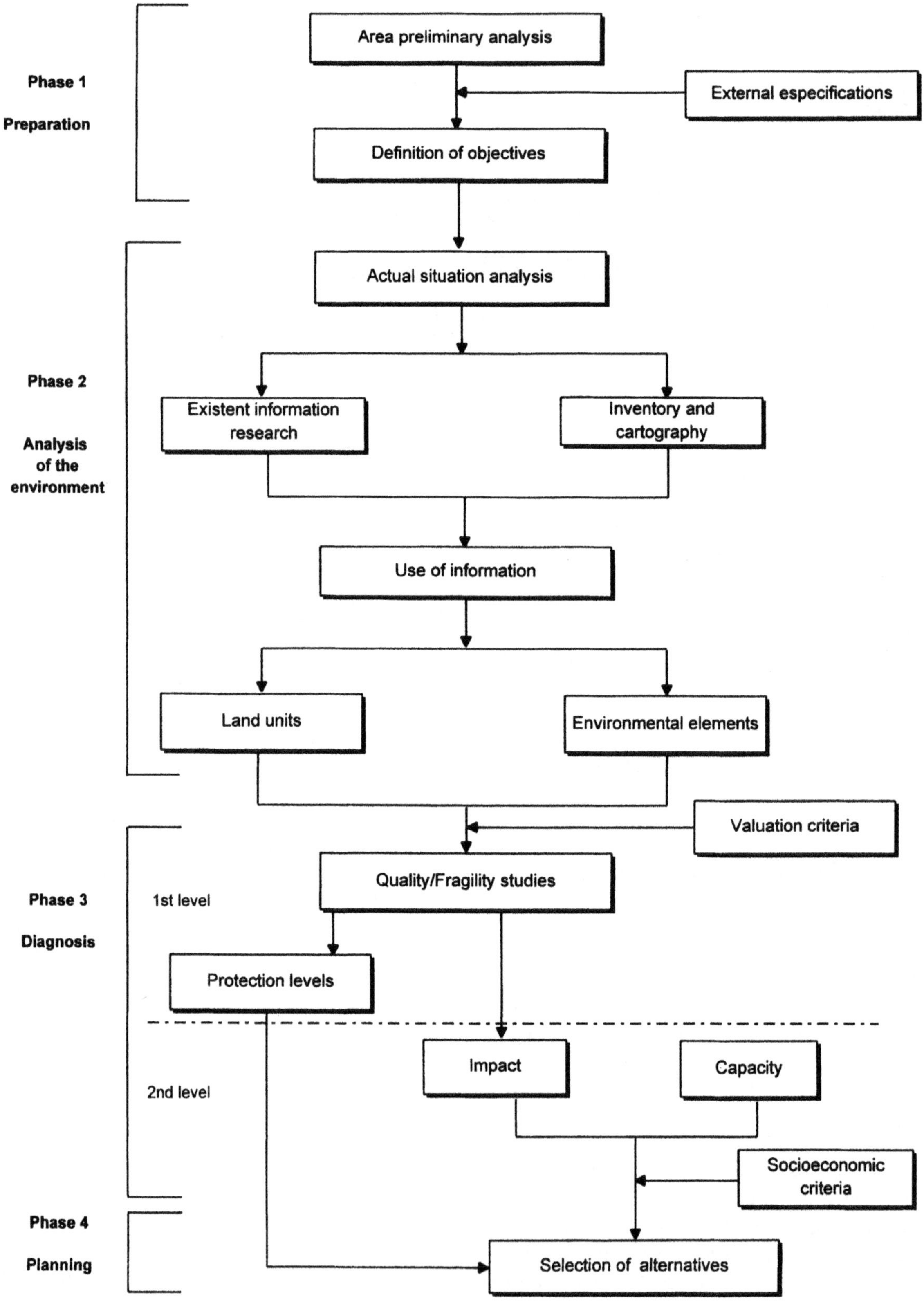

FIGURE 15.1. Simplified outline of the planning process.

according to the general planning process, requires a supply of information with two levels of generality:

(i) First, there is a diagnosis of the present environmental situation

(ii) Second, the diagnosis includes proposed activities according to the original objectives of the plan, in order to assess the relationships between these activities and the present environment. These activities can be taken as a starting point to pose definite planning alternatives.

15.3.1.1 Information Analysis by Elements

The diagnosis by elements (variables of the environment) is based on an inventory made of thematic maps of elements and risks. The information, which is of a descriptive nature (types of soil, vegetation, etc.), will be assigned a value according to the objectives considered in the diagnosis. Therefore, for each of the land points, we go from a vector representing the types of the different elements present in it to a scalar that corresponds to the overall value assigned previously:

$$\overline{P_i} = (P_{i1} \cdots P_{ij} \cdots P_{im}) \text{ becomes } P_i = V_i$$

where:

V_i = global assessment of the point with regard to the objective considered

P_{ij} = value of element j present at point i

15.3.1.2 Land Units

The primary information from the inventory by elements can be enhanced, without resorting to new sources, by looking for its internal structure and endeavouring to describe the land according to all or, at least, a group of its elements. In other words, this means going from a piece of information expressed in many thematic maps to a single map which is a combination of all. Thus, the environment would be organized in classes, each of which would correspond to a different combination of each element type. This classification would correspond to a description of a multivariant situation irrespective of any assessment; thus, the classes created are different by definition and they reflect exactly the land variety.

In some cases for large territories in which the basic information is scarce, the use of land units has proven to be very useful when these have been created out of a minimum number of key elements.

Land units have been used since the beginning of the century. The two basic lines are the genetic approach and the territorial or landscape approach (Glaría, 1980). The origin of the former dates back to the growth of physical geography in the nineteeth century. The work of botanists, geologists, and edaphologists, who arranged their classifications according to genetic criteria, influenced those studies. The areas obtained from that point of view were vast regions generally delimited by climactic characteristics and later subdivided by geomorphological features.

In the territorial or landscape approach, units are defined by characteristics that are recurrent in the area. Christian and Stewart started developing their studies following this line of research at the end of the World War II (Christian and Stewart, 1953). Their work had a remarkable influence upon planning schemes developed in various countries. The approach takes the concept of "land system" as a departure point. It presents a recurrent model of topography, soil, and vegetation and represents areas of a certain homogeneity into which the land is divided. These areas are intended to make up a synthesis of each element's most remarkable features. Due to their own homogeneity, land units are landscape units because they reflect visually all internal relationships.

15.3.2 First Level of Diagnosis: Quality and Fragility Analysis

15.3.2.1 Environmental Assessment

In a first level of diagnosis, the inventory refers to intrinsic features of the environment that are related to the conservation value of its different elements, either separately or globally. It usually meets two basic criteria: (i) quality; and (ii) fragility.

Quality is understood as the conservation value of a certain resource or a given land. It involves different aspects, such as ecological quality, productive quality and landscape quality. Each type of quality is measured according to specific characteristics; thus, the ecological quality of a plant community can be determined by: (i) its singularity, diversity, and integrity; (ii) the productive quality, by means of the edaphic quality; and (iii) the landscape quality, by means of the view-shed characteristics.

Fragility is understood as the possibility of undergoing a deterioration process. It usually refers to ecological and landscape aspects but also to behaviour under certain risks (e.g., possibility of erosion, pollution of aquifers, etc.). Therefore, fragility is an intrinsic quality of the land, although it is also connected with the development of activities.

15.3.2.2 Levels of Protection

The classification of land by levels of protection is widely used in urban planning and in the planning of natural spaces because one of its aims is knowing what uses or what intensity level of uses are allowed at each point of land.

The common way to define levels of protection is by carrying out studies on the quality and fragility of the land with regard to agroforestry productive values, ecological values, and landscape values, usually concerned with visual fragility.

The analysis of each of these aspects will result in a classification of the land that will be reflected on a map. As it is shown in Figure 15.2, the combination of the aspects results in a final map reflecting levels of protection and showing the uses that can be developed at each level.

15.3.3 Second Level of Diagnosis: Relationships Between Activities and the Environment

The next step in the diagnosis is to assess the relationships between activities and the environment. When the influence of the environment upon a specific activity is analysed, there arises the concept of capacity, whereas the effect of the activity on the environment is connected with the concept of impact (MOPT, 1993).

15.3.3.1 Capacity

Among the various definitions of the term capacity, the one that will be used here is that of the land's suitability to withstand the development of an activity according to its natural characteristics. The consideration of possible impacts is not included because those will be analysed separately.

There are many ways to deal with a study of capacity, and the approaches have been divided into: (i) those based on separate elements; and (ii) those based on homogeneous land units.

(i) *Development outline of a study of capacity that takes a database made up of inventory elements as a starting point.* Information is defined by a geographically referenced data matrix, such as the following:

$$\begin{pmatrix} \bar{x}_1 & \cdots \\ \bar{x}_i & \cdots \\ & \bar{x}_n \end{pmatrix} = \begin{pmatrix} x_{11} & x_{1j} & x_{1m} \\ x_{i1} & x_{ij} & x_{im} \\ x_{n1} & x_{nj} & x_{nm} \end{pmatrix}$$

where:

$\bar{x}_i$ $(i = 1 \cdots n)$ = "point" or elementary reference surface of the information
m = number of elements
n = number of points
x_{ij} = type present in point i of element j

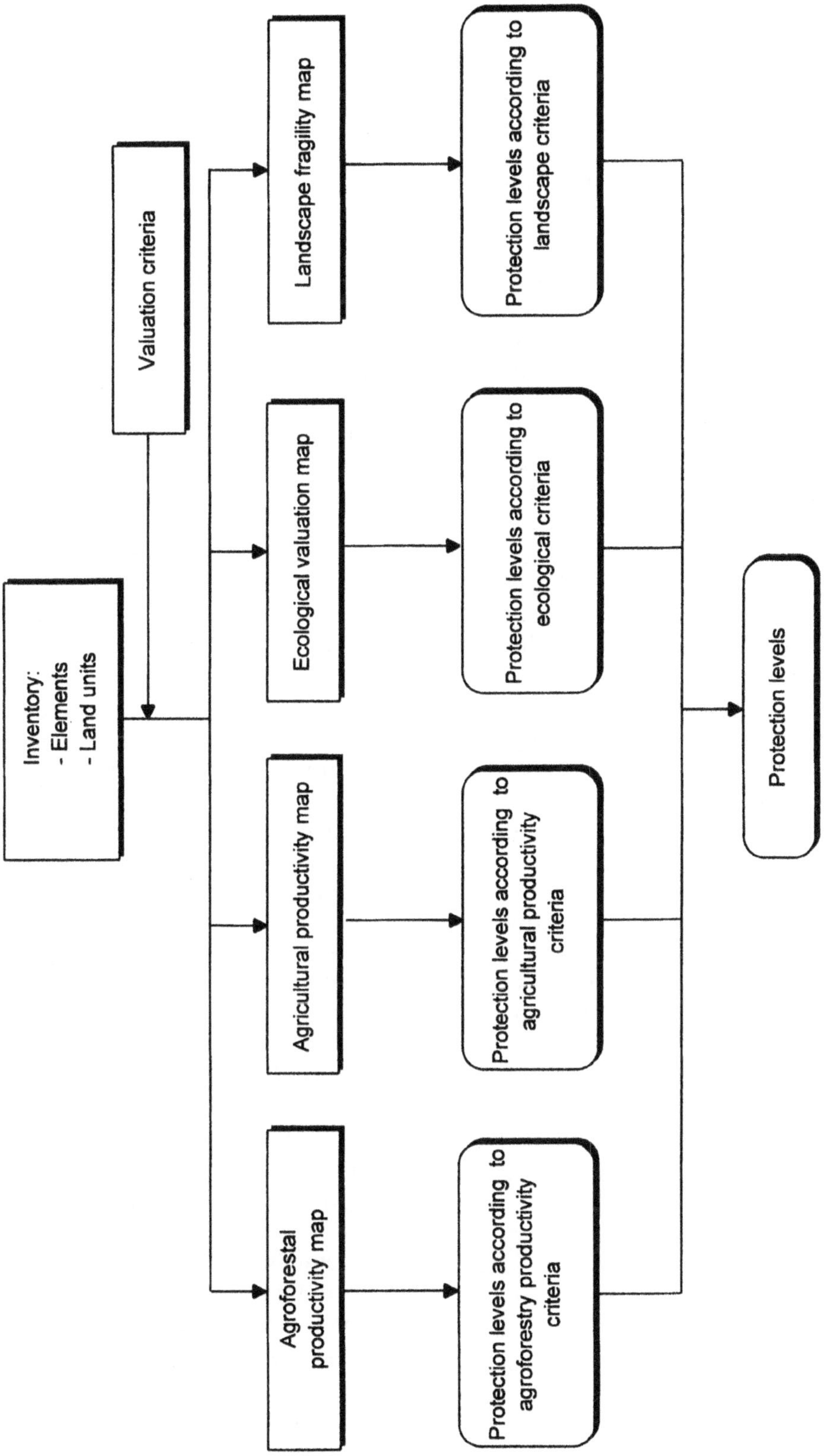

FIGURE 15.2. Outline defining levels of protection.

A generic point $\bar{x}_i$ $(x_{i1} \cdots x_{ij} \cdots x_{im})$ would be oak wood, quartz sands, or flat land, which are the types of vegetation or lithology, respectively. The model has two clearly differentiated stages: inventory assessment and integration of partial assessments. When the activity that is the object of the study has been defined accurately, the following actions must be performed:

- Selection of those e_i $(i = 1 \cdots m)$ elements of the environment providing the most relevant information on the activity development. Each element will contain a variable number of e_{ik} types.
- Assessment of the relative importance of the elements according to the information they provide about the capacity of the environment.
- Assessment of the relative importance of the types of each element according to the capacity they contribute to the development of an activity. Among them there will be types giving information about the impossibility of carrying out an action, either because it is not feasible from the technical point of view, or because the present use prevents a change of activity. These will exclude the area in which they are situated from the development of the action. Thus, if one were to analyse the land capacity to be irrigated in order to have its agricultural productivity improved, irrigation would not be feasible at certain points due to the gradient.
- When partial assessments of elements and types have been done, all the information has to be integrated in a unique score expressing the capacity at each point. The methods of integrating the information may differ. An additive or multiplicative algorithm of the type:

$$C_i = \Sigma e_j t_{ij}$$

is often used, where:

C_i = capacity of point i
e_j = weight of element j
t_{ij} = weight of type j in point i

The final result of the study will be a classification of land in terms of capacity scores:

$$\begin{pmatrix} C_1 \cdots \\ C_i \cdots \\ C_n \end{pmatrix} = \begin{pmatrix} e_1 t_{11} & \cdots & e_j t_{1j} & \cdots & e_m t_{1m} \\ & & e_j t_{ij} & & \\ e_1 t_{n1} & \cdots & e_j t_{nj} & \cdots & e_m t_{nm} \end{pmatrix}$$

The assessments are grouped in classes in order to interpret and use the results more easily.

(ii) *Capacity study outline based on homogeneous land units.* From the point of view of contents, the outline is similar to the one depicted when the information is split into elements. Because the land is described by means of homogeneous units, the assessment will be carried out by starting from the characteristics that can be deduced from each unit.

15.3.3.2 Formation of Capacity Classes

With regard to the formation of classes, there are two basic aspects that ought to be considered: (i) the criteria according to which the classes are arranged; and (ii) the number of classes.

There is no general recommendation on how to define classes, because this depends on the objective of the work. The use of variable or fixed intervals is very common. The classes of variable intervals may respond to a specific rule to which the scores obtained will have to adjust. The most common case is the use of fixed intervals in which the number of types is determined.

Regardless of the criterion followed, it is important to take into account either the real rank of the scores obtained or the rank of scores derived by weighting types and elements. Thus, for instance, if the objective of study is to learn about the capacity of a single use for different areas (this would be the case of using the study of capacity as a source of information for a future reforestation plan), then it would be enough to know the scores of capacity obtained. Nevertheless, we would not know if the areas with greater score were really areas of high potential capacity for that use. It is fundamental to take this aspect into account when comparing the capacity between various uses, which is the most general situation in a planning study.

15.3.3.3 Impact

An impact study varies depending on the stage of the project in which the study is carried out. In the planning stage, the basic objective is land classification in levels of impact. Nevertheless, in the following stages, and after the specific location for an alternative use has been selected, the impact study will identify, assess, and, where appropriate, suggest measures in order to correct the alterations produced by each action of the project.

In the planning phase that is being dealt with, the way to carry out an impact study is not determined by any fixed premise, although, like with studies of capacity, two approaches can be pointed out: (i) the assessment of impacts caused by activity upon each element of the environment; and (ii) the analysis of impacts upon land units.

15.4 Contents and General Methodology of Environmental Impact Studies

15.4.1 Introduction: Legal Background

Environmental awareness began to be a major concern in the 1960s. In the first stage, it was concerned about natural resources being polluted, destroyed, or exhausted. Later on, the ecological balance and landscape integrity were among environmental concerns. Awareness gradually pervaded the decision-making organs of society, which started coming to terms with the notion that the economic models of the 1960s had been overcome and that policies protecting natural resources, integrated with socioeconomic policy, were needed.

These policies required assessment instruments that analyse, not only monetary cost–benefit relationships, but also the environmental changes. Environmental impact assessments or studies are one of these instruments.

The NEPA (National Environmental Policy Act) dates from December 31, 1969 in the United States and is the first to make these studies obligatory in a systematic way. It has inspired the regulations of other countries on environmental assessment (Wood, 1995) and has also resulted in the development of various impact assessment methodologies and techniques. The NEPA does not define what projects have to be assessed from the environmental point of view. As a guideline, it points out that, whenever a federal agency intends to carry out an important action with a significant effect on environmental quality, a detailed estimate of environmental impact has to be undertaken.

The contents of an environmental impact report, known as EIS (Environmental Impacts Statement), according to the NEPA include, either implicitly or explicitly, the major aspects of current impact assessment regulations. Thus, an EIS should contain the following: (i) the environmental impact of the proposed action; (ii) any adverse environmental effects that cannot be avoided should the proposal be implemented; (iii) alternatives to the proposed action; (iv) the relationship between local short-term uses of the environment and the maintenance and enhancement of long-term productivity; and (v) any irreversible and irretrievable commitments of resources that would be involved in the proposed action should it be implemented.

From the methodological point of view, the enactment of the NEPA brought about a series of identification and impact assessment techniques. In spite of many of them being universal, it should not be forgotten that each project, on each site, constitutes a separate case with minor or major differences. There is not a single methodology of environmental impact studies.

The NEPA has strongly influenced the European Economic Community Directive 85/337/EEC on the environmental impact assessment of certain public and private works (Hildebrand and Cannon, 1993). This directive, which has been applied in the European Union countries since 1988, came into being, not only due to an awareness of the need to assess and control negative environmental impacts of certain projects, but also in order to prevent conditions of unfair competition among the EU member states, which derived from the different environmental legislation applicable when the directive was enacted (CEC, 1977, 1979).

With regard to what type of projects should be subject to the regulation, the directive lays down that it is compulsory for all member states to subject those projects that might have a significant impact upon the environment to an environmental impact analysis before they are authorized. The directive lists a number of projects which are always subject to an EIA, while the subjection of others depends on their nature, dimensions, or location (CEC, 1985).

As to what an EIA consists of, in Article 3 the directive points out: "The environmental impact assessment will identify, describe, and assess in an appropriate manner, in the light of each individual case, and in accordance with the Articles 4 to 11, the direct and indirect effects of a project on the following factors: (i) human beings, fauna, and flora; (ii) soil, water, air, climate, and landscape; (iii) the interaction between the factors mentioned in the first and second points; and (iv) material assets and the cultural heritage."

15.4.2 Levels of Detail and Objectives of Environmental Impact Studies

EIS are mechanisms to assess projects in any of their phases. A project is a process that starts with a definition of objectives, continues with the planning of activities, and undergoes a series of stages that are different for each type of project until a building project is reached from which the work can be implemented. The execution of the work does not mean the end of the project; on the contrary, the project continues in the operation and management phases until the end of its useful life. Each of these phases is subject to an assessment/decision process that allows progression from one stage to the next. An environmental assessment is required along with the technical and economic ones.

In the same way each phase of the project process has its own objectives, the environmental assessment will have specific contents related to the level of detail in that phase of the project. Therefore, we can talk about different types of environmental impact studies.

The impact study in the planning phase requires studies of fragility or vulnerability and studies of quality. The analysis of fragility enables us to foresee the possibility each environmental element has of suffering from deterioration or its capacity to absorb changes. The knowledge of quality enables us to assess the importance of change with regard to an element affected by an alteration.

Having chosen the best alternative in the planning phase, a final solution will be worked out in the draft and project phases. Thus, the next phases will have a successively greater level of detail in the information being used. Eventually, the necessary level of detail is attained that expresses all the information on a document (executive project), which will be taken as a starting point for the execution. The impact studies of each phase will deepen the analysis of impact assessment and identification, the development of mitigation measures, and also of the environmental monitoring plan, all of which will become part of the assessment and decision-making processes at the end of each phase.

15.4.3 Methodological Outline of Environmental Impact Studies

15.4.3.1 Basic Considerations

As has been discussed already, EIS have objectives that vary depending on the project phase in which the EIS has been carried out. However, a general outline can be established with regard to the contents of the study: (i) description of the project and its actions; (ii) environmental inventory and description of the key ecological interactions; (iii) impact identification and assessment; (iv) setting up protection and mitigation measures; and (v) setting up an environmental monitoring plan.

The development of this outline varies depending on whether it refers to an EIS belonging to the planning phase or to the project phase. Figures 15.3 and 15.4 show the general development of an EIS in each of these phases. Both are relatively similar despite the fact that they are applied in successive phases. Hence, the EIS corresponding to the project phase will carry out a thorough analysis of the selected alternative's environmental effects. These effects were already evident in the study related to the planning phase. Besides, mitigation measures with the same level of detail as the rest of the actions of the project will be proposed. An environmental monitoring programme will be devised in order to set up a mechanism that guarantees the execution and effectiveness of the mitigation measures already planned.

15.4.3.2 Contents of the EIS Phases at the Planning Level

If we focus on the planning level, impact studies can be carried out in two phases. In the first phase, a preliminary impact study will be carried out to remove from an action in those areas where it would not be feasible from an environmental point of view, according to the existing information, or to point out that an alternative is not acceptable from the environmental point of view.

It all amounts to an attempt at avoiding, from the beginning, significant or critical impacts, that is to say, any alteration that involves an irreversible effect on a highly valuable resource or element. An example of a critical or significant impact would be to alter the habitat of a species in danger of extinction with no possibility of mitigation measures and resulting, therefore, in the decrease of the population of that species.

Having excluded the areas or alternatives with a significant impact, the EIS of the selected alternatives will be carried out in the next stage.

The following subsections are devoted to the different phases and their contents.

15.4.3.3 Project Analysis

The EIS will start with an environmental analysis of the technical characteristics of the activity or activities in question. This analysis must include, at least, information about the following aspects: (i) location; (ii) need of materials that may cause adverse impacts when they are obtained; (iii) forecast and location of waste disposal; and (iv) secondary actions necessary for the execution of the work (e.g., approach roads).

15.4.3.4 Analysis of the Environment

The objective of analysing the environment is to learn about the current state of each alternative plan's spatial area and to make an estimate of that area's development without the plan. Thus, in the impact assessment phase there will be enough information to compare the quality of the environment if the plan is not executed with that if the plan is executed. The result of this comparison between two hypothetical situations is, precisely, the expression of the environmental impact.

It is not easy to foresee the evolution of the environment, especially when it is a long-term forecast and the area is subject to strong human pressure. This factor would be the cause for future interventions in the environment, which cannot always be foreseen because they keep changing depending on the socioeconomic situation at that moment and on land management policies. However, the approach to the future situation must be as accurate as possible.

The study of the environment requires only information about the elements that can be significantly affected by the different alternatives of a plan. Excessively large inventories referring to aspects that are not relevant to impact assessment should be avoided. For each element there will be an analysis of all

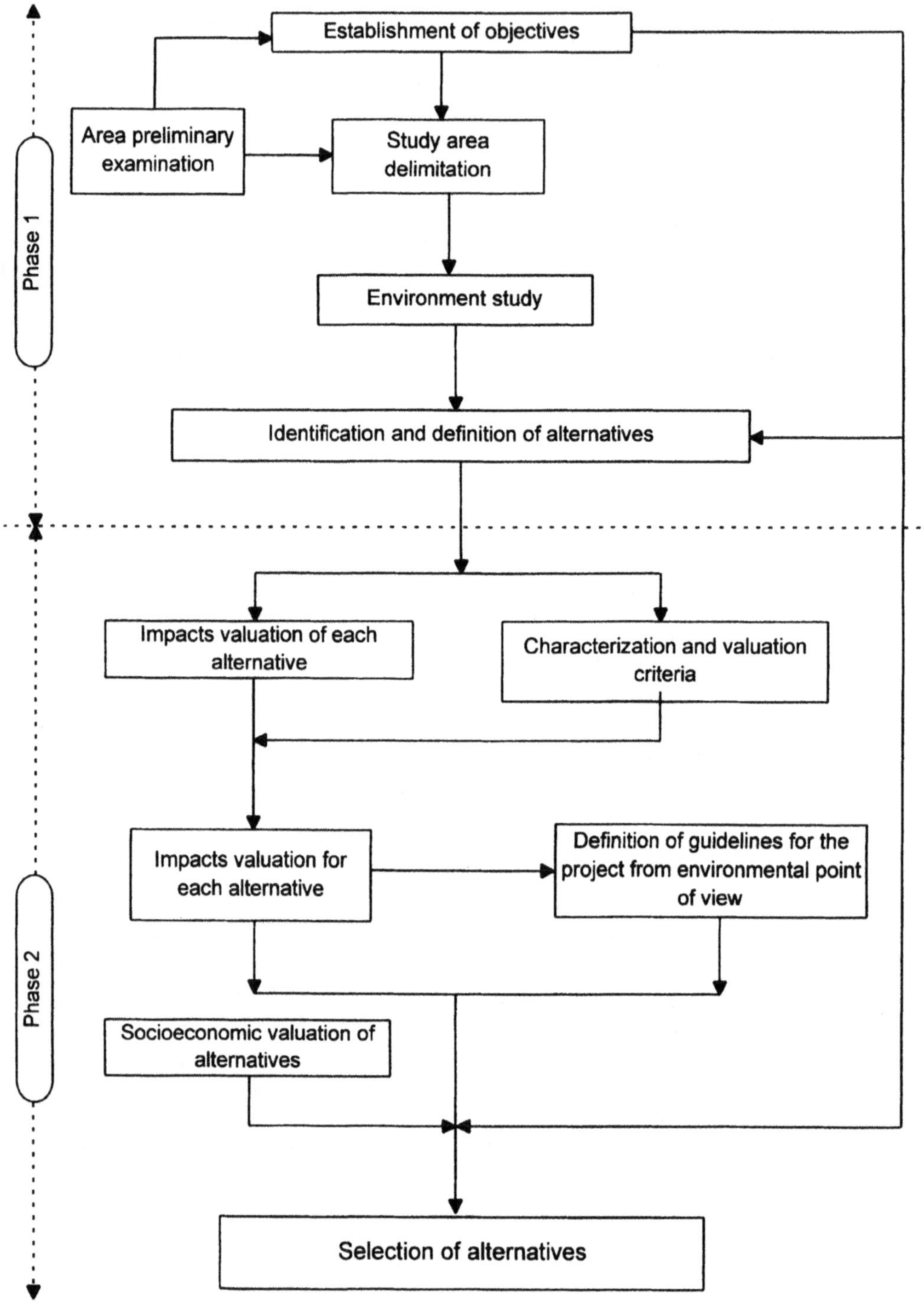

FIGURE 15.3. Outline of the contents of an EIS corresponding to the planning phase.

those characteristics, qualities, or indexes acting as impact indicators. For example, if the element affected by the execution of an irrigation plan is vegetation, the impact indicators could be the area affected, the scarcity of the plant communities or its position in the evolutionary scale.

In addition, other aspects acting as impact

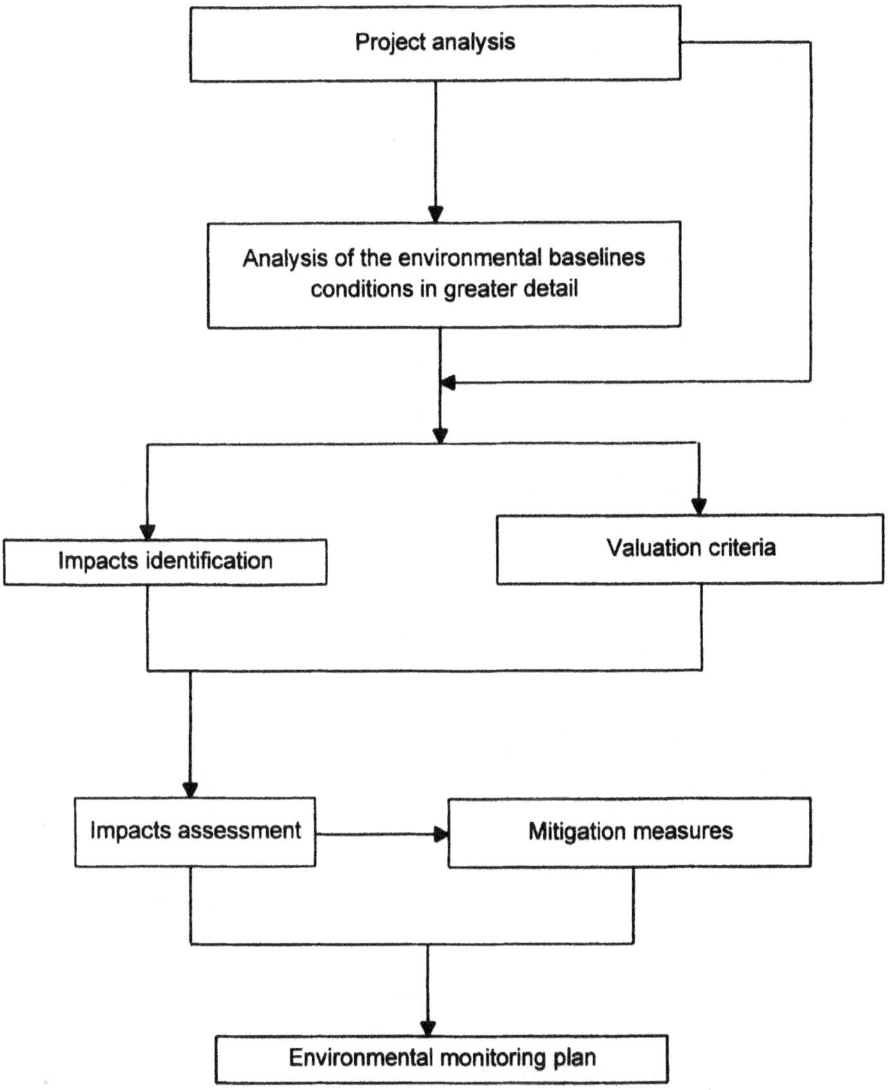

FIGURE 15.4. Outline of the content of an EIS corresponding to the project phase.

modifiers will be included in the inventory. Thus, with regard to traffic noise, the sound level in decibels A, Leq dB(A), measured from the front of houses, could be an indicator. The relief acts as a modifier of acoustic impact; for instance the Leq increases in proportion with the gradient of the road.

Finally, those elements that, without being indicators or modifiers, are determinant for the development of mitigation measures can also be the object of an inventory. As an example, we can mention certain climactic variables which supply the necessary information to select species for plantations and which have been taken into account in a project of measures mitigating the impact of an open-cut mining activity.

For each aspect considered, the study must comprise all the area or scope where the effects of the action are evident. Hence, the area to be studied will depend on the characteristics of the project and the elements considered. The area of study is usually different for each of the elements. Thus, while it may be large in the case of landscape, because it includes all the action view-shed, the area of study corresponding to soil may be more limited.

Action \ Element	Soil	Water quality	Flora	Fauna	Landscape
Building operations	●		●	●	●
Dam presence	●		●	●	●
Approach roads	●		●		●
Extraction of materials	●	●	●	●	●
Flooding of the basin	●		●	●	
Reservoir presence		●		●	●
Changes of the river flow		●	●	●	
Oscillation of the water level					●

FIGURE 15.5. Simplified matrix of the impact identification of a reservoir project.

15.4.3.5 *Identification of Impacts*

After the characteristics of the environment are known and the plan determining the characteristics of the actions that may have environmental impact has been analysed, the alterations or impacts can be identified with the following methods:

(i) *Checklists*. These are standard lists of possible impacts which derive from each specific action of a project. The user of a checklist can select those impacts that actually take place in the case being studied. The main advantage of these lists is that they help examine all the set of effects in a systematic way. Table 15.1 is an example of a checklist showing potential impacts of a reservoir project.

(ii) *Cause–effect matrixes*. The lists of possible impacts of an action become two-dimensional matrixes, in which the second axis lists potential actions required by implementation. Two-dimensional matrixes present the same problems as the checklists because they are, essentially, a set of lists for different actions which are presented jointly. Figure 15.5 shows an example.

(iii) *Components interaction matrixes*. In order to solve some of the problems of the checklists and the cause-effect matrixes, such as not showing the connections between impacts, many authors have devised other matrix models which underline the interaction among different effects (Ross, 1976; Bisset, 1980). In these models a list of environmental features that will undergo some alteration are confronted with each other, and the cases of interrelation will be set down in the matrix.

(iv) *Interconnection networks*. They identify impacts by underlining, not only direct but also indirect effects and the relationships between them. These relationships are usually presented graphically by interconnection lines between both types of effects, thus, creating a network from which the name above derives (Sorensen, 1971).

TABLE 15.1. Example of a checklist. Impacts caused by one of the actions (flooding of a reservoir basin) derived from the construction and exploitation of a reservoir.

Loss of wild habitat
Loss of soil as a productive resource
Loss of the uses of the soil
Microclimatic changes
Thermal stratification of the water
Eutrophication
Loss of historical-artistic resources
Modification of accessibility
Depopulation of human centres
Etc.

(v) *Systems diagrams*. Some authors have suggested applying energy flux diagrams to impact analysis in ecosystems. Thus, the effects on the environment correspond to changes in the energy fixation and flux processes (Odum, 1971; Odum and Odum, 1976; Gilliland and Risser, 1977). These methods are useful for the description of impacts centred on ecosystem productivity changes, but problems arise when identifying impacts that cannot be accounted for in terms of energy fixation or flux.

Any of these techniques can be based on (Westman, 1985): (i) the scientific theory available in each specific case; (ii) the analysis of similar cases (where there are any) providing information about the impacts on the environment whenever the characteristics of the environment and those of the action are similar to the case studied; (iii) use of abstract formulations or models representing reality (models that underline the evolution of phenomena or the development of processes in time and space are quite useful; e.g., models of pollutants atmospheric diffusion, of noise transmission, of potential erosion quantification); and (iv) laboratory experiments or experimental field studies when there is not enough knowledge of the environmental response to an action.

15.4.3.6 Characterisation of Impacts

The effects of the proposed action or activity on the environment have a number of other features that help identify and describe them. Among them, the following can be mentioned.

Sign refers to whether the effect is positive or negative. In general terms, when we talk about impacts, we think of negative effects, but there can also be positive effects. However, we must not mistake positive effects for objectives of the action or activity.

There are many other attributes of impacts that can be used. We can talk about *direct* effects and *indirect* or *secondary* effects to indicate whether the alteration of a certain element or environmental variable is a direct consequence of the action causing the impact, or whether it results from the alteration of other elements or environmental variables. Since the environment is a system of interrelations, an alteration on any of its parts will have an impact on the rest of them (Holling, 1980). We can also mention *synergistic* effects when the combination of two or more effects produces one that is greater than the sum of each of them considered separately.

With regard to the way impacts change with time, three types of effects can be pointed out: *short-term, medium-term, and long-term effects* according to the time passed since the action generating the impact took place until its consequences are felt. There are also *cumulative effects* in which the extension in time of the cause of the alteration involves a gradual increase of the impact.

The *variation of impact in space* should also be taken into account, showing whether the variation increases or diminishes when the distance from the impact-generating source is augmented and whether it occurs in the same site, in the surroundings, or far from the action causing the impact.

We can also consider the *probability of occurrence* or the reversibility or irreversibility of impacts. *Reversible impacts* are those in which it is possible to go back to the situation prior to the impact occurrence; irreversible impacts are those in which returning to the original situation is either impossible or extremely difficult. Sometimes, positive restoration actions can be taken into consideration; thus, we talk about *recoverable* and *irrecoverable* effects in the same way as reversible and irreversible effects.

Finally, we have to mention the *magnitude* of the impact, which is different from the *meaning* of the impact. The magnitude of the impact would be quantitative (i.e., the area that has been altered, volume of polluted water, etc.) and because of this it is an objective measure, whereas the meaning of an impact is subjective. Magnitude is not always directly connected to meaning (e.g., the destruction of 1 ha of scrub which is a habitat of interest may have a greater meaning than the destruction of 10 ha of a less interesting type of scrub).

TABLE 15.2. Characteristics of the impacts

Positive/negative nature	It indicates whether the effect is positive or negative.
Direct/indirect effects	It indicates whether the effect is directly the consequence of the alteration or whether it derives from the alteration of a different resource or variable.
Cumulative effect	The effect gradually increases as long as the impact action continues.
Distant or close effects	It indicates the spatial relationship between the effect and the impact action.
Short-term, medium-term and long-term effects	It indicates how the effect varies in time.
Synergy	The accumulation of effects is greater than the sum of the parts.
Risk	It indicates the probability of impact occurrence.
Reversibility and recoverability	It indicates whether it is possible or not to return to the situation prior to the impact occurrence, either by natural or artificial means.
Magnitude	It indicates whether it is a major or minor alteration in terms of area altered, polluted volume, etc.

Table 15.2 is a summary of the descriptive characteristics of impacts.

15.4.3.7 Impact Assessment

After the effects on the environment have been identified and described, they must be assessed. The result of the assessment expresses the meaning of the future alteration from the point of view of environmental quality.

For this reason, before the impact assessment, it is also necessary to know or to estimate the environmental quality value of the resources that have been studied in the inventory phase by considering the "0" situation, that is to say, the situation if the proposed action is not carried out. Then, an estimate can be made after the action has been carried out, taking into account the characteristics of the impact. In this way, the importance of the alteration (i.e., the change of quality) can be assessed.

Quality is measured by indicators which can be quantitative (e.g., the ppm scores of phosphorous, O_2, etc., used with regard to water) or qualitative (e.g., singularity, fragility, etc., with regard to vegetation). The impact value can be made to coincide with a qualitative or quantitative assessment scale.

Qualitative assessment establishes qualitative categories (e.g., high, medium, and low). Quantitative assessments usually turn qualitative categories into a numerical score by using various methods that can either derive from an initial arrangement of qualitative categories, or can involve simply assigning scores to qualitative categories lacking a previous arrangement.

Regarding cause–effect matrixes, one of the first and best known is the one published by Leopold et al. (1971). It is a cause-effect matrix in which lines and columns of the proposed actions are confronted with environmental elements or variables that might be affected. In the reference matrix 100 actions with a potential impact were listed against 88 environmental elements that might be affected. When a specific action results in an alteration of an environmental element, it is indicated in the corresponding line/column intersection, in this case, by means of a quantitative expression of two characteristics of the impact: (i) magnitude, defined as the degree, the area, or the quantity of the effect; and (ii) importance, which is related to the meaning of the alteration for humans. Figure 15.6 shows an example of this type of assessment.

In the case of interaction networks, the impact value of each of the actions taking place in the environment is made up of the combination of partial values of direct impacts and of impacts deriving from them. Each of these values has to be somehow weighted by the probability of occurrence with regard both to direct and indirect impacts (Rau and Wooten, 1980) or by other methods (Ross, 1976; Bisset, 1980).

Impact assessment can be expressed either with regard to every single environmental element (individually), or as a whole, by using a single score which summarizes the whole impact. An assessment by elements allows us to know about impacts on elements and also provides us with information about the actions of the projects that have to be modified in order to lessen or prevent the impact. If the impact is inevitable, this kind of assessment allows us to know where we should be directing monitoring measures. The disadvantage of this method is that it does not provide information about the impact of the action or project on the environment as a whole. Methods that express the impact as a whole are useful in order to select alternatives, because it makes much easier the choice of an alternative with a minimum impact.

Impact assessment by elements can be qualitative or quantitative. Although the use of techniques expressing the impact value in a quantitative way is usually recommended, this is not always possible due to the varied nature of the environmental elements affected and the techniques for the determination of environmental quality value.

The usual way to obtain a global impact assessment is by using a weighted sum of partial impact assessments. Hence, we must bear in mind that when adding up scores, the most significant impacts become blurred, regardless of the technique used. It does not seem advisable to use this method when there are critical impacts. A critical threshold is one that if crossed will cause irreparable loss or damage.

There are many procedures and criteria to assign weights and add up impacts (Canter, 1977; Canter and Hill, 1979; Golden et al., 1979; Bisset, 1980; Rau and Wooten, 1980; Clark et al., 1981; Hollick, 1981). Thus, in the case of Leopold's matrix, where a score is assigned to the magnitude of the impact and another one to its importance, the latter would represent the weighting of the impact according to the meaning of alteration for man. In the case of interconnection networks, apart from assessing magnitude and importance, the probability of the impact is also taken into account; in this case, it is the probability of occurrence and the importance that

Action \ Element	Soil	Water quality	Flora	Fauna	Landscape
Building operations	4 / 2		2 / 6	3 / 6	2 / 3
Dam presence	3 / 2		2 / 4	4 / 8	3 / 5
Approach roads	2 / 3		4 / 6		4 / 6
Extraction of materials	1 / 1	4 / 5	4 / 8	3 / 9	4 / 7
Flooding of the basin	7 / 6		7 / 8	4 / 9	
Reservoir presence		2 / 2		2 / 1	4 / +4
Changes of the river flow		1 / 2	4 / 6	5 / 9	
Oscillation of the water level					5 / 7

Magnitude / Importance

+ : Positive impact

FIGURE 15.6. Simplified matrix of the impact assessment of a reservoir project.

weight the magnitude score. In both cases, the impact global score is the sum of partial impact scores, qualified by the importance and/or probability assigned to each alteration.

One of the pioneering and most widely spread systems of global impact assessment is the Environmental Evaluation System (Dee et al., 1973) developed in Batelle Laboratory in order to assess the impact of hydraulic projects. Within this system the environment is divided into components to which a weight deriving from their contribution to the global quality of the land is assigned. The quality score of each component is measured with parameters that are specific for them (e.g., DBO, pH, etc., for water; or vegetation diversity for aesthetic components). These scores are transferred to a common scale of environmental quality, so that the same score in the scale indicates the same relative quality of each of the aspects considered. An example of these transformation functions is shown in Figure 15.7. The score of the impact is worked out by comparing the global quality score of the area for environmental baseline conditions and the global score deriving from the action. Both the weights assigned to each parameter and the function transforming parameter scores into environmental quality scores are determined by teams of experts.

15.4.3.8 Mitigation Measures

An EIA is not yet finished with the assessment of impacts. It is also necessary to study possible mitigation measures of expected impacts taking into account their effectiveness and feasibility from the technical and economic points of view. The application of these measures allows alterations to be minimized, so the final result of the action will only produce those impacts that can not be avoided; in other words, they are "residual impacts."

The definition level of mitigation measures will vary depending whether the action is being analysed in the planning phase or in the project phase. In the planning phase, these measures are simple guidelines that must be taken into account in later phases. We have to bear in mind that the adequate choice of a planning alternative is the best way to diminish impacts, because the best mitigation measure is to prevent them. Likewise, including the guidelines that have been set up for the mitigation measures in the project design implies the consideration of environmental aspects as another variable in the decision-making project, and is not just to be taken into account afterward, when it is already too late to correct the impacts of the action.

15.4.3.9 Impact Assessment Monitoring

The impact assessment monitoring programme will set up a system that guarantees the fulfilment and efficacy of protection and mitigation measures contained in the environmental impact study. This programme must also meet the following objectives: (i) to enable checking the accuracy of impact forecasts; (ii) setting up a mechanism to identify unexpected alterations in the impact study and devise appropriate mitigation measures; (iii) control of residual impacts (i.e., those that can not even be prevented by mitigation measures) by setting up the necessary warning signals; and (iv) to check that the execution of the work corresponds to the project which is the object of the impact study.

The results of the monitoring programme will serve as a basis for a periodical inspection of the planning (Lee and Wood, 1980). Hence, the programme must be capable of adapting to changes that might take place in the course of time. It is also important to know what kind of system will be used in order to communicate the results to the appropriate organisation that decides on the planning alternatives. This information has to be included in the monitoring plan.

The impact assessment monitoring programme cannot be fully developed in the planning phase due to the level of detail in the definition of the action; however, the basic guidelines can be laid down. It is possible to point out the type of information that will be necessary and also the mechanisms for the communication of results plus the decision-making mechanisms affecting the plan. All this will make up the guidelines that will be further analysed in the following phases until a complete definition of visiting programmes,

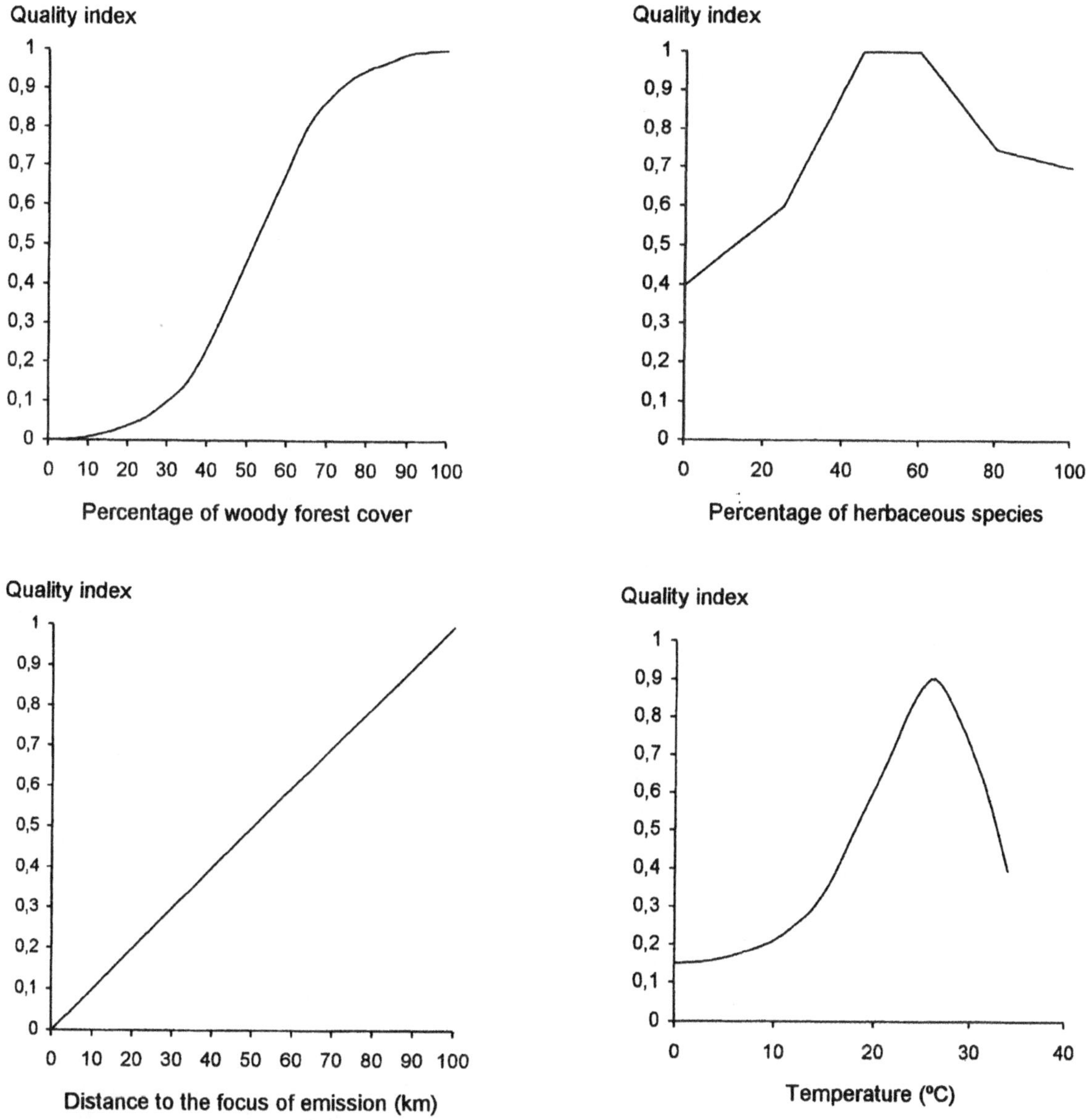

FIGURE 15.7. Examples of transformation functions.

data gathering, analyses, setting up of warning signals, and so on, will be established in the execution phase.

15.5 Assessment of Alternatives: Multicriteria Decision Process

15.5.1 Introduction

When the last phase of the planning process is reached, it is then the right moment to compare all the possible alternatives in order to make decisions or allocate uses. The information contained in the capacity and impact models exhausts the exploitation possibilities of the inventory; thus, it is not possible to make any further progress in the planning process if we only take into account environmental aspects. It will only be possible to provide, as additional information, the suitability of the location of a use by comparing the studies of capacity with those of impact. It is obvious that those areas with a minimum impact and maximum capacity will be the most suitable ones.

However, the information about the physical environment is not enough to classify the land in levels of suitability for the development of a single use. Additional information is necessary to achieve the final land allocation of uses. In general terms, the environmental information will have to be combined with socioeconomic information. The spatial compatibility between different alternatives will have to be analysed, and priorities will be set up according to regional policy criteria.

Until not long ago, the assessment of alternatives was in accordance with the classical model, that is to say, searching for the best solution according to a specific criterion. In short, it all amounted to discovering those solutions that were feasible (i.e., solutions in accordance with imposed restrictions) and to select the optimum one that answered to a specific function/objective.

However, in the field of physical planning (and also in the economic field and in any field that involves making decisions by an individual or by society) the assessment of alternatives is a multicriteria assessment. Generally, the criteria used in the assessment have different measurement units (financial profits, environmental impact, social impact, etc.), which are often incompatible with one another; therefore, we can not usually talk about an optimum solution in a multicriteria approach. In this sense, the concept of efficient or nondominated solution in the paretian sense is basic: a set of solutions is considered as efficient or nondominated solutions according to *n* criteria or objectives whenever any other solution aimed at improving the chances of fulfilling an objective brings about the impairment of the fulfilment, at least, of another objective or objectives.

The multicriteria-decision process involves a series of concepts and stages that are shown briefly in Figure 15.8.

Two discrete multicriteria techniques will be discussed. The first and oldest is the weighting method; it is the easiest to apply but has important limitations. The second is the ELECTRE method, which is probably the more useful of the two.

15.5.2 Weighting Method

The weighting method is an elementary multicriteria assessment technique that involves adding up each criterion partial assessments and weighting the importance of each of them. The best alternative would be that fulfilling the following:

$$\text{Max} \sum_{j=1}^{n} P_j f_j(a_i), \ \forall_i$$

where:

a_i = alternative i ($i = 1 \cdots m$)
P_j = weight of criterion j ($j = 1 \cdots n$)
$f_j(a_i)$ = assessment of alternative a_i with regard to criterion j

The Planning Assistance Through Technical Evaluation of Relevance Numbers (PATTERN) is an example of the application of this method and it operates in the following way:

(i) Setting of criteria (or partial objectives) in order to choose alternatives. Examples are reducing the financial costs of a project implementation or minimizing its environmental impact.

(ii) Determination of specific criteria for each partial objective; e.g., with regard to reducing the financial cost, a specific criterion would be the curtailment of expropriation, building, and exploitation costs.

(iii) Weighting. Weights are assigned to each criterion to reflect relative importance with regard to the project global objective.

(iv) Determining an index for each criterion, the score of which can be worked out in quantitative terms. Thus, "dollars" would be the index with regard to the financial cost of a project implementation. In the case of environmental impact, the index would result from the environmental impact assessment expressed in a numerical scale.

(v) Calculating the absolute value of each index in connection with the corresponding criterion for each alternative. The value is usually the result or consequence of previous studies, such as the study of financial costs or environmental impact.

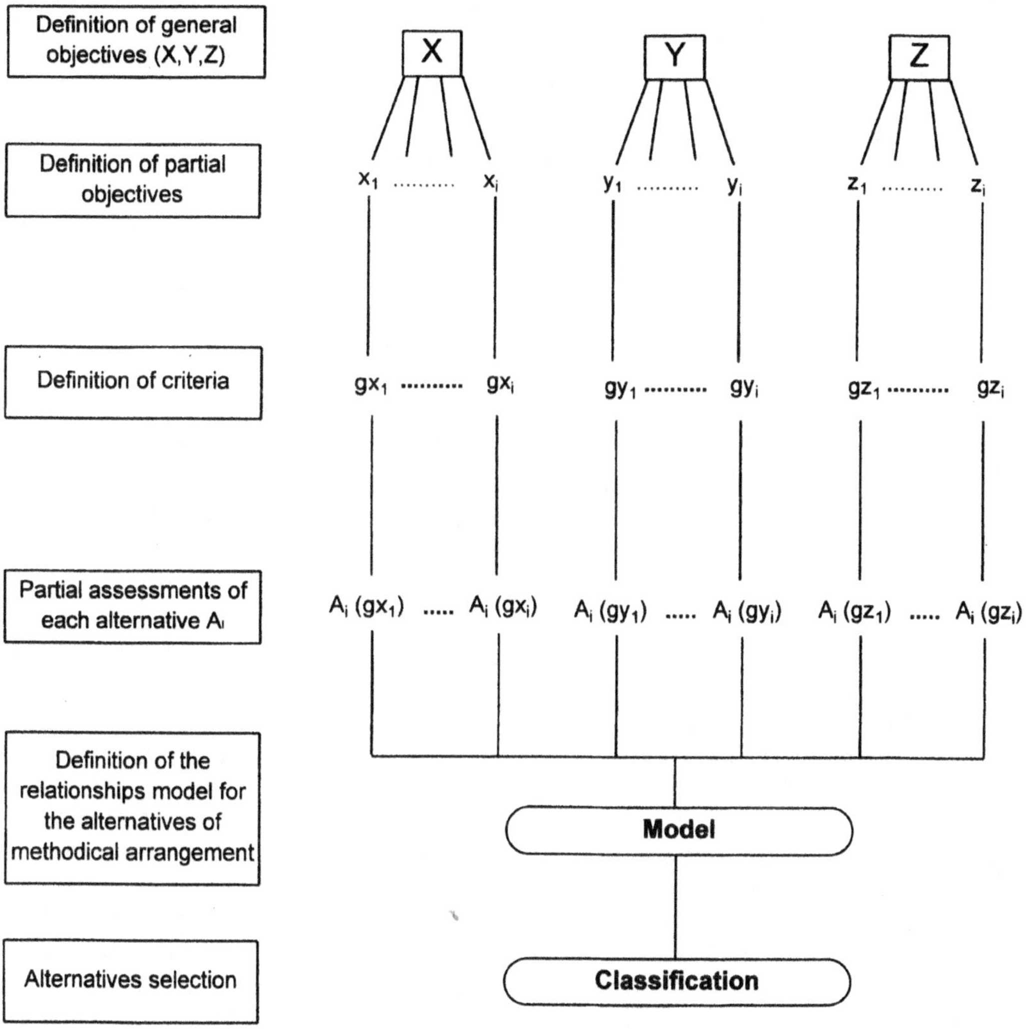

FIGURE 15.8. Outline of the multicriteria-decision process.

(vi) In order to establish a comparison, the former assessments have to be expressed in a homogeneous scale. Therefore, each of the former assessments (absolute ones) has to be expressed as a percentage with regard to the sum of the assessments of all alternatives in connection with the corresponding criterion. This way of expressing things is valid if the fulfilment of the criterion implies the maximization of the indicator. If, for instance, the environmental impact is to be minimized, the same procedure is applied but to the inverse valuation.

(vii) When the scores have been made homogeneous, the *pertinence index* of each alternative "*i*" is calculated by adding up the multiplication of the weights assigned to each criterion by the homogeneous assessments of alternative "*i*" with regard to each criterion:

$$I_i = \sum_{j=1}^{n} A_{ij} P_j$$

where:

I_i = pertinence index of alternative i ($i = 1 \cdots m$)
A_{ij} = homogeneous assessment of alternative i with regard to criterion j ($j = 1 \cdots n$)
P_j = weight assigned to criterion j

(viii) Selection of the best alternative. The pertinence indexes for each alternative

TABLE 15.3. Absolute assessment

	Criteria		
Alternatives	Building costs (000$)	Irrigated area (km^2)	Environmental impact (1–10)*
A	22,000	180	7
B	25,000	165	6
C	23,000	170	6

* 10: strongest impact..

TABLE 15.4. Homogenized assessment

	Criteria		
Alternatives	Building costs (000$)	Irrigated area (km^2)	Environmental impact (1–10)*
A	0.3525	0.3495	0.30
B	0.3102	0.3204	0.35
C	0.3372	0.3301	0.35

* 10: strongest impact.

express its contribution to the fulfilment of all criteria. Therefore, the best alternative will be the one with the highest index score, and the worst one will be that with the lowest score.

To illustrate this method let us assume that three alternatives (*A*, *B*, and *C*) are being considered for the layout of an irrigation canal (Tables 15.3, 15.4, and 15.5). The assessment criteria are the cost of building the canal, the extent of the area irrigated by the canal, and the environmental impact. The weight assigned to each of these criteria are 0.35, 0.35, and 0.30, respectively. The objectives are, on the one hand, to reduce the financial cost of building it and the environmental impact, and on the other hand, to maximize the irrigated area. As it is shown in Table 15.5, $I_c > I_A > I_B$, therefore, *C* is the best alternative.

The main disadvantage of this method, which adds up the scores without including any limitations, is that it does not take into account the meaning of low partial assessments should the rest of them be high. This means that if the environmental impact of an alternative is very high, this alternative can still be the best one if the assessments are high for the rest of the criteria. Thus, it is a procedure that only takes the "majorities" into account.

15.5.3 ELECTRE Method

ELECTRE (Elimination et Choix Traduisant la Realite) is a set of methods deriving from ELECTRE I, which was formulated by Benayoun et al. (1966). ELECTRE II, III, and IV were later developed (Roy, 1991). In order to introduce these methods, ELECTRE I and II will be mentioned.

15.5.3.1 ELECTRE I Method

This method divides the set of alternatives (*A*) into two subsets. One of them contains the favourable alternatives (A^+), whereas the other one comprises the nonfavourable (A^-). The smaller the former is, the easier the decision making. The most advantageous situation will be when (A^+) contains a single alternative.

This division is based on an outranking relationship among alternatives. This relationship determines when an alternative a_i is at least as good as another one a_j. There can be three types of relationships: (*i*) a_i outranking a_j (a_i R a_j); (ii) a_j outranking a_i (a_j R a_i); and (iii) a_i and a_j can not be compared (a_i R a_j and a_j R

TABLE 15.5. Calculation of pertinence indexes

Alternatives	Building costs (000$)	Irrigated surface (km^2)	Environmental impact (1–10)	Pertinence index (*I*)
A	0.1234	0.1223	0.09	0.3357 = I_A
B	0.1086	0.1121	0.1050	0.3257 = I_B
C	0.1180	0.1155	0.1050	0.3385 = I_C

* 10: strongest impact.

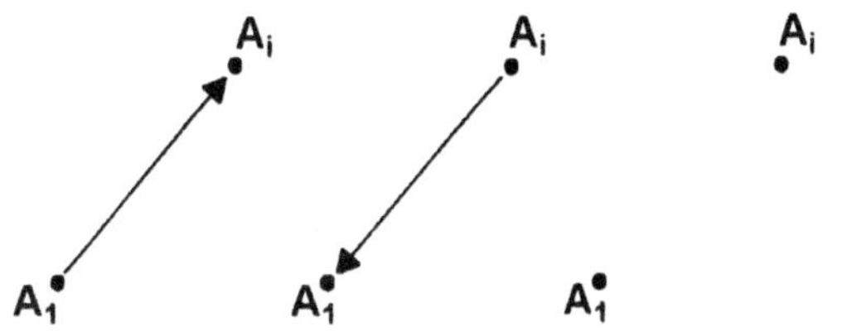

FIGURE 15.9. Outranking relationships.

a_j). A graphic chart of these relationships is shown in Figure 15.9 (Benayoun et al., 1966).

A graph $G = (A, R)$ can be drawn taking into account all the possible relationships between all the possible couples of alternatives, with A being the set of vertices (alternatives) and R being the set of arcs representing the relationships between a_i and a_j.

The outranking relationship is defined by means of concordance and discordance indexes. Given two alternatives a_i and a_j, a criterion g_k for which it is true that the alternative a_i is better than alternative a_j or equal to it, namely, $g_k(a_i) < g_k(a_j)$, is a criterion *concordant* with the hypothesis that a_i outranks a_j. Should the opposite occur, namely, $g_k(a_i) < g_k(a_j)$, the criterion will be *discordant* with the previous hypothesis. The concordance and discordance indexes derive from those concepts. The calculation of concordance and discordance indexes is as follows:

(i) *Concordance index*. In order to calculate the concordance score C_{ij} between two alternatives a_i and a_j, we do not only take into account the number of concordant criteria but also, and especially, the weight assigned to each of them. The expression of the index will be:

$$C_{ij} = (W^+ + W^=)/W$$

where:

$W+/a_i$ = sum of the criteria weights for which $g(a_i) > g(a_j)$
$W^=$ = sum of the criteria weights for which $g(a_i) = g(a_j)$
W^- = sum of the criteria weights for which $g(a_i) < g(a_j)$
$W = W^+ + W^= + W^-$

(ii) *Discordance index*. The discordance index D_{ij} between a_i and a_j is calculated by taking into account the width between the scores corresponding to criteria g_k for which $g_k(a_i) < g_k(a_j)$. We must not forget that not all the criteria have the same importance; therefore, when calculating the discordance, this fact should be taken into account, because a certain discordant width corresponding to a very important criterion does not have the same meaning as one corresponding to a less important criterion.

Hence, in its original formulation, which is being discussed here, the ELECTRE I method recommends choosing a constant interval between consecutive levels of the same scale and varying the interval from one scale to another in proportion to its weighting. Thus, the discordance index D_{ij} between two alternatives a_i and a_j would be calculated in the following way:

$$D_{ij} = h/hmax$$

where:

h = the biggest of the differences between the scores $g(a_i) - g(a_j)$ for discordant criteria

$hmax$ = total width of the scale with the largest difference between the maximum and the minimum scores

The discordance index D_{ij} can also be worked out in a less restrictive way by using the difference between the scores that come immediately before the largest. This would mean not taking into account the most discordant criterion, which introduces the concept of *level of severity* (s) indicating the number of criteria removed from the calculation of the discordant index (s = number of discordant criteria that have been removed +1). When the criterion with the largest difference of scores is removed, it is said that the level of severity equals 2 (it would equal 1 if no criteria was removed). The discordance index is generally expressed as $D_{ij}(s)$, according to the level of severity.

We say that an alternative (a_i) *outranks*

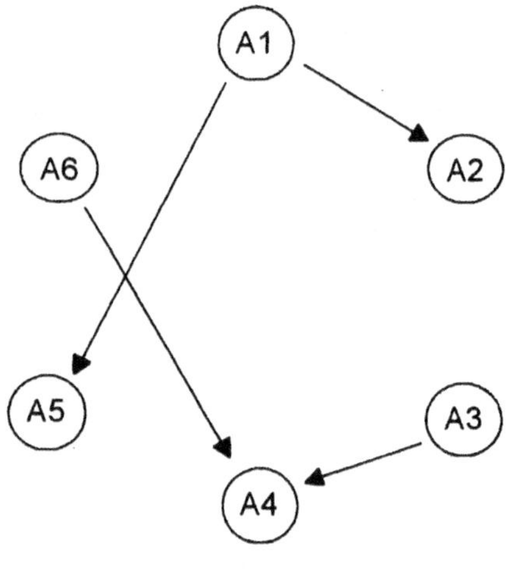

FIGURE 15.10. Example of delimitation of the kernel in a preference graph.

another one (a_j) when (i) the score of the concordance index, C_{ij}, is high, that is to say, when alternative a_i is preferred to a_j for quite a number of criteria and/or for those having more weight; and (ii) the score of the discordance index $D_{ij}(s)$ is low, in other words, when the difference $g(a_j) - g(a_i)$ does not express a considerable superiority of alternative a_j to alternative a_i for discordant criteria.

The formulation of these ideas requires defining two scores, p and q, the *concordance threshold* ($0.5 < p \leq 1$) and the *discordance threshold* ($0 \leq q < 0.5$). It is said that a_i outranks a_j ($a_i\ R\ a_j$) when the following conditions occur simultaneously:

$$C_{ij} > p \text{ and } D_{ij}(s) < q$$

We have seen that the outranking relationship can be represented by a graph $[A, R(p,q,s)]$ (Figure 15.10). The kernel of this graph will contain a subset of alternatives that are preferred according to the outranking relationship, although within this subset there is no organisation among them. Our attention ought to be drawn to the following points:

(i) There are no arcs among the kernel elements. Therefore, the alternatives of the kernel are not comparable with one another in this group. Although this may be seen as a practical limitation of the method, it reflects a real situation which, in some cases, results from not having enough information to compare alternatives.

(ii) Any alternative outside the kernel is dominated by at least another one from within the kernel.

(iii) The outranking relationship is not necessarily of a transitive nature. In other words, if $a_i\ R\ a_j$ and $a_j\ R\ a_k$, we cannot say that a_i outranks a_k. This statement makes sense if we bear in mind the great variety of criteria used in order to calculate the indexes.

(iv) If the concordance and discordance threshold scores are less restrictive (namely, if "p" gets closer to 0.5 and "q"to 0) and the level of severity (s) is lower, then there will be more arcs in the graph. At the same time the outranking relationship will lose realism and the graph's interpretation will be less clear.

For example, let us assume we are trying to select the best locations in order to build a reservoir for irrigation purposes. At the beginning we have six alternatives ($A1$, $A2$, $A3$, $A4$, A5, $A6$) and four assessment criteria with the following weighting of their importance:

$G1$, ecological impact (weight 3)
$G2$, social acceptance of the project (weight 3)
$G3$, landscape impact (weight 2)
$G4$, loss of soil as a productive resource (weight 2)

For each criterion there is a qualitative assessment scale of the following type: very high, high, medium, low, very low. If we turn these into quantitative scales and for the interval of each score scale we take into account each criterion's weight, we obtain the matrix shown in Table 15.6. The concordance C_{ij} and discordance $D_{ij}(1)$ indexes can be calculated with the scores of this matrix (Table 15.7).

When the C_{ij} and $D_{ij}(1)$ scores are known, in order to set up the outranking relationship all that needs to be done is defining the concordance (p) and discordance (q) thresholds. Assuming $p = 0.7$ and $q = 0.3$, we obtain the

TABLE 15.6. Matrix of weights and scores

	Criteria (weighting)			
Alternatives	*G*1(3)	*G*2(3)	*G*3(2)	*G*4(2)
*A*1	6	9	4	4
*A*2	3	6	4	4
*A*3	0	9	8	10
*A*4	12	9	6	6
*A*5	9	6	6	8
*A*6	12	6	4	2

TABLE 15.7. Concordance and discordance indexes

	Concordance matrix C_{ij}					
	*A*1	*A*2	*A*3	*A*4	*A*5	*A*6
*A*1	—	1	0.6	0.3	0.3	0.7
*A*2	0.4	—	0.3	0	0.3	0.7
*A*3	0.7	0.7	—	0.7	0.7	0.7
*A*4	1	1	0.6	—	0.8	1
*A*5	0.7	1	0.3	0.4	—	0.7
*A*6	0.5	0.8	0.3	0.3	0.6	—

	Discordance matrix D_{ij} (1)					
	*A*1	*A*2	*A*3	*A*4	*A*5	*A*6
*A*1	—	0	0.5	0.5	0.33	0.5
*A*2	0.25	—	0.5	0.75	0.5	0.75
*A*3	0.5	0.25	—	1	0.75	1
*A*4	0	0	0.33	—	0.17	0
*A*5	0.25	0	0.25	0.25	—	0.25
*A*6	0.25	0.17	0.67	0.33	0.5	—

ELECTRE graph shown in Figure 15.11. As we can see, the kernel of the graph has been reduced to two alternatives, *A*3 and *A*4, which outrank the rest.

15.5.3.2 ELECTRE II Method

The ELECTRE I method selects a set of alternatives that includes the most favourable ones and also the most difficult to compare. However, it is often necessary to arrange these alternatives, and this is the purpose of ELECTRE II (Roy and Bertier, 1971).

The basic outline of this method is similar to the one before. As a starting point we take a set of *n* alternatives $A(a_1 \cdots a_n)$ and of *m* criteria $G(g_1 \cdots g_m)$ the importance of which is expressed by a series of weights (P_i). Each alternative a_i is assessed according to each criterion g_k, the assessment being expressed in any numerical scale. This is different from ELECTRE I because it is not necessary to arrange constant intervals, which are proportional to the criteria's weight inside each scale.

The outranking relationship presents some differences. According to ELECTRE II, a_i outranks a_j when: (i) the sum of the criteria weights according to which a_i is at least as good as a_j is high enough (concordance condition); and (ii) for any criterion g_k according to which a_i is not better than a_j there is no big difference of assessment $g_k(a_i) - g_k(a_j)$ (discordance condition). In order to define both conditions, concordance and discordance thresh- olds, which are more or less severe, are set up, resulting in a strong or weak outranking relationship.

An alternative a_i *strongly outranks* a_j when the following three conditions are simultaneously met:

$W^+/W^- \geqslant 1$
$C_{ij} = (W^+ + W^=)/W \geqslant c_1$
$D_{ij}(k) = g_k(a_i) - g_k(a_j) \leqslant d_{2k}$ for every discordant criterion k

or, alternatively, when the following conditions are met:

$W^+/W^- \geqslant 1$

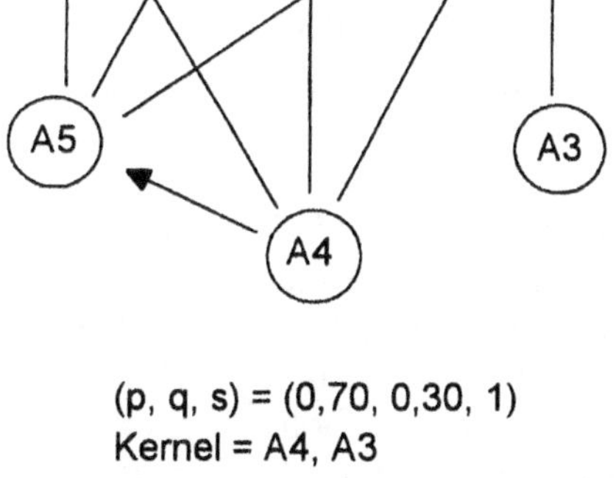

FIGURE 15.11. Preference graph (ELECTRE I Method).

$C_{ij} \geqslant c_2$
$D_{ij}(k) \leqslant d_{1k}$

There is a *weak outranking relationship* when the following three conditions are simultaneously fulfilled:

$W^+/W^- \geqslant 1$
$C_{ij} \geqslant c_3$
$D_{ij}(k) \leqslant d_{2k}$

where c_i values are *concordance thresholds* and verify that $1 > c_1 \geqslant c_2 \geqslant c_3 > 0$ and d_i values are *discordance thresholds* that have to be set up for each criterion by taking partial assessments $g_k(a_i)$ into account. It has to be verified that $d_2 > d_1 > 0$.

The outranking relationships described before allow us to draw a graph with two types of arcs, depending on whether it is a strong or weak outranking. From the analysis of the graph we can proceed to an arrangement of alternatives taking into account the length of paths (number of arcs in the graph) that converge on the vertices (alternatives). This is a direct arrangement. The best alternatives will be those that are not outranked by any other (length of paths = 0), whereas those on which the largest number of arcs converge will come last. Only strong outranking relationships will be used for this type of arrangement, and the weak ones will only be taken into account in order to break the deadlock. An inverse type of arrangement can be made by taking into account the length of paths that start on each vertex (or alternatives). The final arrangement can be defined as the average of direct and inverse arrangements.

If the outranking relationships of ELECTRE II are applied to the example of ELECTRE I, the following result is obtained:

(i) *Concordance and discordance thresholds.* The following concordance thresholds, which are often used in this method, have been set up: $c_1 = 3/4$, $c_2 = 2/3$ and $c_3 = 3/5$. With regard to discordance thresholds, d_{1k} and d_{2k}, they have to be set up for each criterion k by taking into account partial assessments of alternatives for these criteria. Because two types of scales have been used for partial assessments, the selected thresholds are:

Criteria G_1 and G_2:
d_{1G1} and $d_{1G2} = 5$, d_{2G1} and $d_{2G2} = 6$
Criteria G_3 and G_4:
d_{1G3} and $d_{1G4} = 4$, d_{2G3} and $d_{2G4} = 5$

(ii) *Concordance matrix.* The concordance matrix resulting from the comparison among alternatives for which $W^+/W^- \geqslant 1$ is shown in Table 15.8.

TABLE 15.8. Concordance matrix, c_{ij}

	A1	A2	A3	A4	A5	A6
A1	—	1	—	—	—	0.7
A2	—	—	—	—	—	—
A3	0.7	0.7	-	0.7	0.7	0.7
A4	1	1	—	—	0.8	1
A5	0.7	1	—	—	—	0.7
A6	—	0.8	—	—	—	—

TABLE 15.9. Discordance matrix, D_{ij}

	A1	A2	A3	A4	A5	A6
A1	—	$0 < d_1$	—	—	—	$6 = d_2$
A2	—	—	—	—	—	—
A3	$6 = d_2$	$3 < d_1$	—	12	9	12
A4	$0 < d_1$	$0 < d_1$	—	—	$2 < d_1$	$0 < d_1$
A5	$3 < d_1$	$0 < d_1$	—	—	—	$3 < d_1$
A6	—	$2 < d_1$	—	—	—	—

TABLE 15.10. Direct arrangement of alternatives

	Length of in-paths (strong preferences)	Length of in-paths (weak preferences)
1st: Alternatives 3 and 4	—	—
2nd: Alternative 5	1	—
3rd: Alternatives 1 and 6	2	1
4th: Alternative 2	5	—

TABLE 15.11. Inverse arrangement of alternatives

	Length of out-paths (strong preferences)	Length of out-paths (weak preferences)
1st: Alternatives 4	4	—
2nd: Alternative 5	3	—
3rd: Alternatives 3 and 1	1	1
4th: Alternative 6	1	—
5th: Alternative 2	—	—

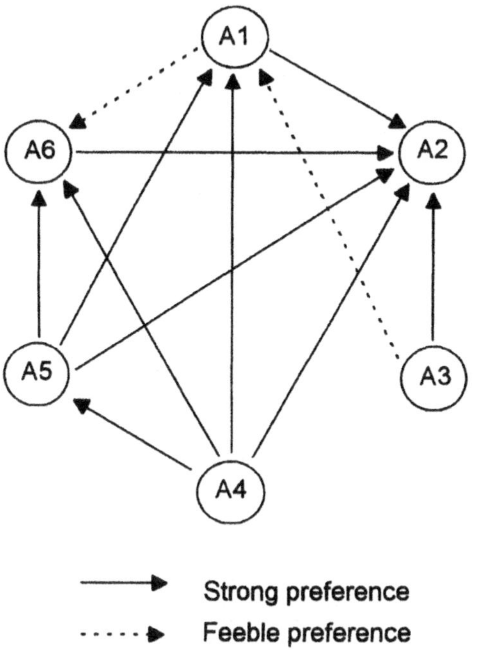

FIGURE 15.12. Preference graph (ELECTRE II Method).

(iii) *Discordance matrix*. As happened in the previous case, it is only calculated for alternatives for which: $W^+/W^- \geqslant 1$. The boxes of the matrix show the discordance score and whether it is less or the same as the corresponding d_1 or d_2 (Table 15.9). With these scores and the outranking relationships, the graph of Figure 15.12 can be drawn. From the analysis of the graph we can immediately infer the direct classification shown in Table 15.10. The opposite classification is shown in Table 15.11. If the average of both classifications is calculated, then the final classification obtained is: $A4 > A5, A3 > A1 > A6 > A2$.

As we can see, an almost complete classification is obtained with ELECTRE II, which is very useful in the resolution of planning problems, in which the interest is arranging the alternatives from the best to the worst. This marks a sharp contrast with respect to ELECTRE I. In the case of ELECTRE I, although it was known that the kernel was made up of alternatives $A3$ and $A4$, there was no way to know the arrangement between them or the arrangement among the other alternatives.

15.5.4 Other Discrete Multicriteria Methods

Apart from ELECTRE I and II, other methods based on the outranking relationship can be pointed out: ELECTRE III and ELECTRE IV (Roy, 1991) and PROMETHEE (Preference Ranking Organization Method for Enrichment Evaluations) (Brans et al., 1986).

Likewise, the AHP (analytic hierarchy process) can be pointed out (Saaty, 1980) plus other continuous multicriteria methods (Romero and Rehman, 1989; Romero, 1991) which can equally apply to discrete problems. This is the case, for instance, of goal programming and compromise programming.

References

Benayoun, R.; Roy, B.; Sussmann, B. ELECTRE: *Une Methode pour Guider le Choix en Presence de Vue Multiples*. SEMA, Direction Scientifique, Note de Travail No. 49. Paris: SEMA; 1966.

Bisset, R. Methods of environmental impact assessment: Recent trends and future prospects. *J. Environ. Manage.* 11:27–43; 1980.

Brans, J.P.; Vincke, Ph.; Mareschal, B. How to select and how to rank projects: The PROMETHEE methods. *Eur. J. Oper. Res.* 24:228–238; 1986.

Canter, L.W. *Environmental Impact Assessment*. New York: McGraw-Hill; 1977.

Canter, L.W.; Hill, L.G. *Handbook of Variables for Environmental Impact Assessment*. Ann Arbor, MI: Ann Arbor Science Publishers, Inc.; 1979.

Christian, C.S.; Stewart, G.A. *General Report on Survey of Katherine-Darwin Region*, 1946. Land Resources Series 1. Canberra: CSIRO; 1953.

Clark, B.D.; Chapman, K.; Bisset, R.; Wathern, P.; Barret, M. *A Manual for the Assessment of Major Development Proposals*. London: Department of the Environment, HMSO; 1981.

Commission of the European Communities. European Community policy and action programme on the environment for 1977–1981. *Official Journal of the European Communities* C139:1–46, June 13; 1977.

Commission of the European Communities. *State of the Environment: Second Report*. Brussels: CEC; 1979.

Commission of the European Communities. Council Directive of 27 June 1985 on the assessment of the effects of certain public and private projects on the environment. *Official Journal of the European Communities* C175:40–49, July 5; 1985.

Dee, N.; Baker, J.K.; Drobny, N.L.; Duke, K.M.; Whitman, I.; Fahringer, D.C. Environmental evaluation system for water resource planning. *Water Resour. Res.* 9:523–35; 1973.

Gilliland, M.W.; Risser, P.G. The use of systems diagrams for environmental impact assessment. *Ecol. Model.* 3:188–209; 1977.

Glaría, G. Las unidades ambientales. Estudio sobre la oportunidad de su definicion en una situacion real. Ph.D. Thesis, ETSI Montes, Madrid; 1980.

Golden, J.; Quellete, R.P.; Saari, S.; Cheremisinoff, P.N. *Techniques for Aiding in the Assessment Process. Environmental Impact Data Book.* Ann Arbor, MI: Ann Arbor Science Publishers, Inc.; 1979.

Hildebrand, S.G.; Cannon J.B., eds. *Environmental Analysis: The NEPA Experience.* Boca Raton, FL: Lewis; 1993.

Hollick, M. The role of quantitative decision-making methods in environmental impact assessment. J. Environ. Manage. 12:65–78; 1981.

Holling, C.S., ed. *Adaptive Environmental Assessment and Management.* Chichester: John Wiley; 1980.

Lee N.; Wood, C.M. *Methods of Environments Impact Assessment for Use in Project Appraisal and Physical Planning.* Department of Planning and Landscape, Occasional paper No. 7. Manchester: University of Manchester; 1980.

Leopold, L.B.; Clark, F.E.; Hanshaw, B.B.; Balsley, J.R. *A Procedure for Evaluating Environmental Impact.* U.S. Geological Survey Circular No. 645. Washington, DC: Department of the Interior; 1971.

Ministerio de Obras Publicas y Transportes (MOPT), ed. *Guiapara le Elaboracion de los Estudios del Medio Fisico.* Serie monografias. Madrid: MOPT; 1993.

Odum, H.T. *Environment, Power and Society.* New York: Wiley–Interscience; 1971.

Odum, H.T.; Odum, E C. *Energy Basis for Man and Nature.* New York: McGraw-Hill; 1976.

Rau, J.G.; Wooten, D.C., eds. *Environmental impact Analysis Handbook.* New York: McGraw-Hill; 1980.

Romero, C. *Handbook of Critical Issues in Goal Programming.* Oxford: Pergamon Press; 1991.

Romero, C.; Rehman, T. *Multiple Criteria Analysis for Agricultural Decision.* Amsterdam: Elsevier; 1989.

Ross, J.M. *The Numeric Weighting of Environmental Interactions.* Occasional paper No. 10 in Lands Directorate. Ottawa: Environment Canada; 1976.

Roy, B. The outranking approach and the foundations of ELECTRE Methods. *Theor. Decis.* 31: 49–73; 1991.

Roy, B.; Bertier, P. *Le Methode ELECTRE II.* SEMA, Direction Scientifique, Note de Travail No. 171. París: SEMA; 1971.

Saaty, T. *The Analysis Hierarchy Process.* New York: McGraw-Hill (version extended in 1988); 1980.

Sorensen, J.C. A framework for identification and control of resource degradation and conflict in the multiple use of the coastal zone. M.Sc. Thesis, Department of Landscape Architecture, University of California, Berkeley; 1971.

Westman, W.E. *Ecology, Impact Assessment and Environmental Planning.* New York: John Wiley; 1985.

Wood, C.M. *Environmental Impact Assessment: A Comparative Review.* Harlow: Longman; 1995.

16

Systems Analysis As a Tool for Rural Planning

Wolf Dieter Grossmann and Juan Bellot

16.1 Why Do You Need A Systems Perspective?

For decades, the analytic approach has been dominant for the study of many processes. This approach takes into account the parts of the process with many details and then assembles the parts into a plan or answers according to a theory or a blueprint. This is the engineer's method, and most scientists and planners rely on it. It is useful when we are building a house or a bridge. It is not useful when we are dealing with natural systems for which there is no theory, no blueprint, not even a complete list of parts. In this case, we need another approach that takes the problem as a whole. Hence, in recent decades many systems approaches have been developed (Forrester, 1969, 1971; Patten, 1972; Holling, 1978; Odum, 1983) to overcome this situation. A systems approach prefers a more holistic, "global," or top-down view of the phenomena, even at the cost of losing details.

These two approaches (analytic and systemic) are not opposites. Quite the contrary. If they are used in a complementary way the results are better than those from only one approach. Sometimes emphasis is put on the wholeness of a system. Other times emphasis is placed on the analysis of parts. The Rural Planning Course in the Mediterranean Agronomic Institute of Zaragoza (IAMZ) relies on both approaches. From the first chapters of this book, we have presented techniques and concepts to understand in detail, and with an analytic approach, the relevant processes taking place in the evolution of a region or county that are typical objects of our planning work. Each contributing discipline in the region contributes to understanding from its particular point of view, and with its specific tools. Nevertheless, the planner cannot reduce his work to a problems analysis from only one or a few points of view; rather, it is essential to tackle the problems from a global perspective as well, based on a systemic approach.

Systems analysis is a technique that allows us to analyse systems and simulate and evaluate their past and future behaviour. In the 1960s and 1970s, the pioneers named above applied the new tools to fields as different as ecology, economics, and engineering. Systems analysis is a tool capable of simulating interactions and changes that take place in a system considering basically the systems structure. However, the task is difficult. Many people have the capability to design and build systems because they have the ability to visualize their parts, functions, and external relationships, and thus they could predict the consequences of the systems structure. But without proper training and good tools, they do not have the ability to understand the behaviour or reactability of the complete, complex system.

The systemic approach is related to the general systems theory (GST) the major aim of which is to describe and present systems in a formal way. The founder of the systems theory was Ludwig von Bertalanffy, who in 1945 defined systems theory as a mathematical description of systems in nature (von Bertalanffy, 1971). Thus, GST comprises the von Bertalanffy theory as a subunit. Systems analysis is a part of the systemic approach, because in its early phases of application it analysed systems through descriptions of their parts, relations and interconnections.

16.2 The Systems Concept

In GST a system is defined as any interconnected set of elements that are coherently organized around some purpose. In other words, a system is a group of parts that are interacting in manifold processes. This kind of system is the object of rural planning.

The rural or land system or ecosystem is the study system in the context of GST. It is large in components, in their relationships, and in spatial area projection. Also it has many complex processes that work on different spatial scales and temporal rates, interconnected in some cases with external processes (Odum, 1983).

By applying the system concept of GST we try to explain a part of the universe selected for study. Our system is surrounded by an environment with which it interacts. The boundary of the system is somewhat arbitrary because a system is a construction of the human mind so that it is most amenable to analysis as a problem, a configuration or a dynamic.

Based on these purposes, usually, there are logic rules for deciding what to include and what to exclude because the components must interact in significant ways to produce the observed behaviour. The systems components may be tangible, such as land, water, population, and forest in an ecosystem, gears in a machine, or individual organisms in a population. They may also be abstract, like legal rules, laws in a society, words in a language, or ideas in a philosophical system. The task of choosing an appropriate system for investigating a particular problem or behaviour is inseparable from the consideration of underlying organization and complexity.

GST states that all systems are composed of *elements*, *interconnections*, or relationships, and a *purpose*, or sometimes several purposes. It is easy to see the elements in a system, because often they are visible or tangible items such as land, water, or vegetation. The relationships are not directly visible or tangible, and sometimes one fails to notice them. Relationships are the rules and information flows that hold the elements together. Relationships do not capture our attention as elements do, but they are observable. Some of them are physical flows (e.g., water, food, money, etc.); most are information flows (e.g., the signals that are sent to each decision point within the system to control the elements and physical flows).

If relationships are difficult to see, purposes are even more difficult. A system purpose is also information, but it is not necessarily spoken or written or expressed in any explicit way, except through the functioning of the entire system. Purpose can be deduced from the systems behaviour. The purpose of a system may not be the same as the purposes of elements within the system.

The system is defined by all three characteristics and, to understand the system, one must understand all three. A system has an integrity or wholeness, and an active set of mechanisms to maintain that integrity. For years, we have known that: "the whole system becomes more than the sum of its parts when there are interactions" (Kitching, 1985). Managing a system, whether it is a resource base or an economy, requires understanding the integrity, responsiveness, and purposes of the whole system.

This is different from just understanding the sum of its parts (Swartzman and Kaluzny, 1987). Rural planning needs these kind of approach, because its objective is to manage the rural system and to improve its performance, mitigate problems, and increase the economy, population and life quality, while maintaining the system sustainability.

Rural systems are generally open, and only for very few processes can they be considered closed. These open systems exchange matter, energy, and information with the environment and, consequently, their elements change with time. In this case the system is considered *dynamic*.

If all the elements in the system and their relationships are immutable, the system is called *static*. Rural and living systems are never static, due to the multiple infections and perturbations received from the outside environment. In general, all the rural systems used in rural planning are dynamic, and consequently their aim is to arrive at a dynamic equilibrium of their components with the resource stocks.

This *dynamic equilibrium* is reached when the inflows and outflows in the system are the same, and the levels of resource stocks are not changing in time. In a dynamic system, the changing elements are called *state variables*, and their evolution in time is described by the different values that the variables take in time. This evolution in time is called a trajectory, and can be represented as a function, a table of pairs of values, or a graph. The trajectory of a state variable in a system reflects the behaviour of this element, in response to internal or external events.

The planners, using the systems analysis approach, can show how the variables change as a function of other elements and time. Over time, successions of events trace out the pattern behaviour, reflecting the static or dynamic characteristics of the system which arise from the system structure responses to the events.

Knowledge at the level of behaviour gives some power to predict; knowledge at the level of structure is required for management. People in general, and planners in particular, try to manipulate stocks by manipulating inflows, but the system also works if we manipulate outflows. With these approaches the behaviour can be different.

16.3 Complexity, Hierarchy, and Sustainability

Rural systems can be considered as complex systems, because of their large number of components, interactions, and land extension. Complexity increases as the number of components increases or, correspondingly, as the number of variables and parameters in a mathematical model describing the system increases. However, complexity must be categorized by more than just the number of components (Weinberg, 1975). The key criterion is the type of complexity involved in many observation sets. As a consequence, the ecosystems considered in rural planning must be conceived as "medium-number systems" and, according to Weinberg, it is suggested that for their analysis they are regarded as "organized complexity." Any success in analysing this ecosystem can only be achieved through careful and explicit consideration of the organization underlying the complexity.

The world, or at least the part that forms our rural system, is organized through subsystems aggregated into larger and larger systems. Most elements in the subsystems are connected, but not with equal strength. A system composed of subsystems or elements is called a hierarchical system (O'Neil et al., 1986). Each level in the hierarchy can be overridden by the next higher level, and is thereby under the constraint or control of the next higher level. The higher-level control, in a sense, is performing a more general strategy to which the more local strategy of the lower-level controls is subordinated. Each level of the hierarchy corrects for errors or inefficiencies in the level below. The hierarchy permits both flexibility and stability. If a subsystem like a population of organisms can take care of many of their own needs and purposes maintaining themselves, and simultaneously serves the needs of the larger aggregate system (e.g., land use, agriculture, industry, etc.), a stable, resilient and efficient structure results. In some cases the subsystem goals dominate at the expense of the total system goals; the resulting behaviour is called *suboptimization*. This behaviour, even in the case that can produce good results in a short time, needs to be neutralized because in the long term its effect can be negative to the whole system. The solution to this problem should be the use of another system property, that is, to have a greater central control of the system itself.

The system property called *decomposability* has a special application in rural planning. This property means that the relationships within each subsystem are usually much stronger than relationships between subsystems, and permit the separate study of subsystems, as well as the interactions among them. The systems economist Simon (1969) uses the term systems *nearly decomposable to distinguish those in which the subsyste*ms have a weak, but not negligible, interaction. In these nearly decomposable systems, it make sense to study and understand the linked subsystems as if they were separate units.

The near-decomposability of resource systems has enabled us to organize this book using the rural planning method followed by the IAMZ in its training course. Both the book and the course are organized by single-resource sections, before dealing with all resources as a whole. Each single kind of resource: water, soil, population, economic capital, and so on, has its own subsystem properties, which are necessary and useful to learn in their own right.

One of the most important concepts related to the behaviour of the system, and with the planning of their future, is the notion of *sustainability*. The major objective of any planning is to maintain the sustainability of resources as has been proposed in Chapter 1. Sustainability is the goal of any rural system, however, often, the competition for resources, the effects of perturbations or the normal resource depletion produce a system decline, after or without an overshoot period.

Overshoot is a term that means the system goes outside its normal limits without notice and without purpose. The overshoot is a common process in current systems. The causes are a quick movement, action, and change that move the system far from steady-state behaviour. In systems analysis, overshoot is produced by distraction in the control, wrong data, or wrong information, feedback delays, slow responses to controls, and so on, as long as there are stocks higher than the normal steady-state storage. Consumption of stored resources at a rate over the input average produces a reduction of stocks over time, eventually their decline, and, finally, system collapse.

After a period of overshoot, a decline of the process and the system normally follows. Frequently, the system has controls to reduce the overshoot phase, but the inertia of the system change or the slow response could delay the control reaction and the solution arrives at the wrong time (Meadows et al., 1991).

Examples of this pattern can be easily observed in nature. This could be the case of a water stock, used by a group of farmers, and farm production. If the consumption rate is higher than the recharging rate, the trend along time is the emptying of the stock and consequently the decline in the productivity, after an overshoot characterized by fast consumption and growth. On the contrary, when systems controls work correctly in time, the feedback loops reduce the fast consumption and the results show a small overshoot, but not followed by a collapse.

Sustainability comes from three different system properties: (i) resilience, (ii) stability, and (iii) adaptability.

Resilience is the ability of a system to recover or restore itself to its original or equilibrium state, after perturbations (in Holling's 1978 description: "to be able to bounce back after disturbance or destruction and to keep this capability"). To describe correctly the resilience of a system, the best way is to indicate the full range of perturbation and how long the system can support the perturbation outside the normal values maintaining its capacity to recover. Resilience is a very important property of the systems or subsystems, because it indicates their capability of resisting perturbations (intensity and duration) and shows the most sensitive elements in the systems that do not have the capacity to recover their normal level after the perturbation has ended.

Stability is concerned with the degree to which a system damps perturbations. If the system amplifies outside perturbation, it is unstable. A system could be stable but not resilient. That means that the response in the system is less important than the received external perturbation, but without full recovery of the initial position. A system can also be resilient but not stable, if it recovers its former state, but only after some fairly large oscillations.

The third relevant property of the system in relation to planning is *adaptability.* An adaptable system can adjust itself to new conditions. It has the ability to evolve or change its internal structure and behaviour, as its environmental conditions change. Most systems that are adaptable are also fairly resilient and stable. It is possible to observe stable and resilient systems returning to equilibrium after an external perturbation, but not being adaptable in shifting that equilibrium in response to new conditions.

Fortunately, most physical and biological systems are resilient, stable and adaptable. Sometimes, human management makes them different. With rural planning we will be searching for ways to increase productivity and sufficiency while maintaining or even enhancing resilience, stability, and adaptability. These aims could be considered contradictory, because resilience and stability reflect resistance to changes. Nevertheless, to increase productivity we introduce changes through the planning, based on the capacity of systems to adapt to the new conditions, without destroying the system.

The reaction of any system to maintain its trajectory inside the normal limits is called *counterintuitive behaviour*. This action is opposite to the adaptability of the system, and more connected with resilience and stability characteristics. All rural systems (natural and manmade), from the simple to the more complex, have many feedback loops of different lengths (with different time delays, different number of elements in them, etc.). If a planner tries to change the behaviour of the system, he usually only succeeds in overcoming the resistance by the short-term feedback loops. He then meets increasing resistance by the middle-length feedback loops, and his good intentions will, in a somewhat longer run, be ruined by long-term feedback loops.

16.4 Systems, Models, and Simulation

The major reason to use models is the complexity of the system, derived from the complexity of relationships in their elements, the variable characteristics of each element, and the apparent unpredictable effect of humans on the environment and the system. The manager or rural planner, needs a logical and organized representation of these relationships, and the use of models is helpful because they involve multiple dynamic interactions among a wide range of elements. Another reason to use models is that many of the system components change or progress on very different time scales; some of them change in a day, others in weeks, and others in years or centuries. The models provide a way to integrate and contrast the compatibility of the information from the real system. In this sense, the model is a tool to synthesize all the information that we have in a simplified representation of a system.

A collection of data on a system might be considered a model, but it is one of the most unclear and unusable. The treatment of these data introduces clarity and permit us to summarize some of the interrelations. If the time dimension is also taken into account, this model is converted from a static to a dynamic form. However, these models remain *descriptive*, and only show the existence of relations between the elements of a system without any explanation of causal interactions.

Explanatory models are useful in rural planning because various levels of organization or knowledge are distinguished. The different levels of integration may be classified, according to the size of the system, as individuals, popu-

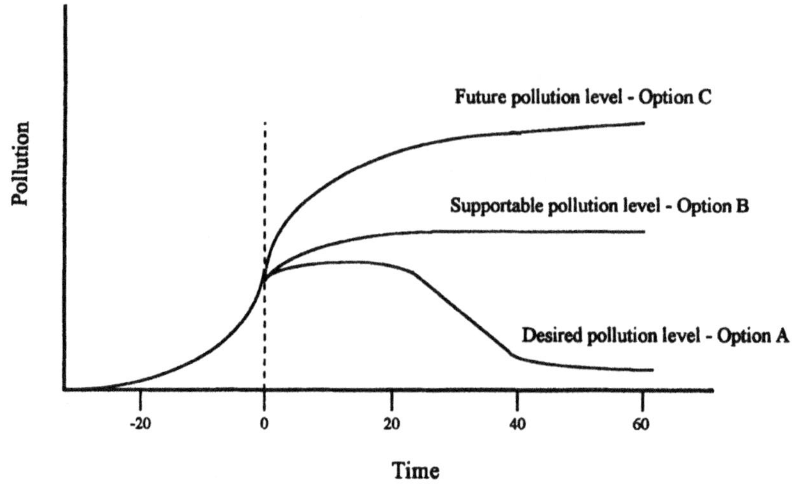

FIGURE 16.1. Water pollution reference mode.

lations, communities, ecosystems, regions, countries, and biosphere. Explanatory models tell us what research has to be carried out for at least two integration levels. The lower integrated level will be the explanatory level, and the upper level in the model is the one to be explained. If there are discrepancies between the results of the model and reality, the model may be adjusted to obtain better agreement. To resolve discrepancies between model and reality, it may be possible to use experiments focused on aspects of the model that are suspicious. These aspects are then studied at the explanatory level. On the basis of this new information, elements of the model may be replaced by improved versions, after which the results of the model and reality may be compared again.

Planning on a regional scale could employ whole system models constructed of large integrated socioeconomic and ecological models. Due to the complexity of the system and of the model, this attempt may fail to give valid predictions. It is convenient to look for an optimum size of a system and model to have a useful *description-oriented* model.

A model can be simple if it tries to answer a few specific questions and thus needs only involve a few processes. When we focus our attention on the problems of a region, we need to know the nature of the problems, how to resolve the actual problem, or how to decrease future problems. This type of approach tends to be more successful and more useful than the previous one, and is called a *problem-oriented* model.

All problems consist of two temporal parts: (i) the actual and past behaviour of the system, and (ii) the desired behaviour of the system. If there is a problem it is because the actual behaviour is different from the desired. Problem analysis will often use a graph over time of the most important variables of the system, called a *reference mode.* In practice, it is recommended to begin a system analysis with a reference mode. Figure 16.1 shows the reference mode of water pollution in an aquifer. Over the past forty years the trajectory reflects the increasing pattern of this variable, and, for the future, several possibilities are represented in the way of three options. From these we can plan a strategy to drive the system to the desired level.

16.5 Dynamic Simulation Models

There are several groups or kinds of models that are used frequently. Dynamic models, matrix models, stochastic models, multifactorial models, optimization models, game theory models, and chaos theory models are the best known. This list is not exhaustive and in many cases the models use a combination of more than one kind of mathematical application (Jeffers, 1982). The dynamic models are considered useful in the first steps of systems analysis of environmental problems. These models concentrate attention on the basic relationships of the system and define the variables and subsystems relevant to the planner. However, in the last steps of modelling, other more mathematically complex models may be useful.

In general terms, a model that can be evaluated over time is named a dynamic model. This type of model is quite common in fields like systems dynamics models, as developed by Forrester (1961, 1969), the energy-related models of Odum (1983), or the adaptive environmental assessment and management (AEAM) approach (Holling, 1978). Dynamic simulation models (DSM) are based on the assumption that the state of each system at any moment can be quantified, and that changes in state can be described by mathematical equations such as rate or differential equations.

To construct these types of simple models, only elements that are directly relevant and connected to the problem are included in a problem-oriented approach. To introduce these methods we need to understand the following concepts (Aracil, 1986):

(i) *State variables*. These are the major elements or system components. They are variables that describe stocks like the amount of biomass, the number of animals, the amount of nitrogen or water in soil, vegetation, people, money, food, and so on. Other examples of stocks are capital, resources, and goods in the transport system or machine equipment.

(ii) *Driving variables* (also called forcing functions). These are variables outside the system boundaries, which influence it. They represent the influence of external factors on the system, but the factors are not influenced by the processes within the system.

(iii) *Rate variables*. These are the major cause of the dynamics in the system. It is rate variables that change state variables. Their values are determined by the state and driving variables according to rules that are based on knowledge of the physical, chemical, biological, and socioeconomic processes that take place in the system.

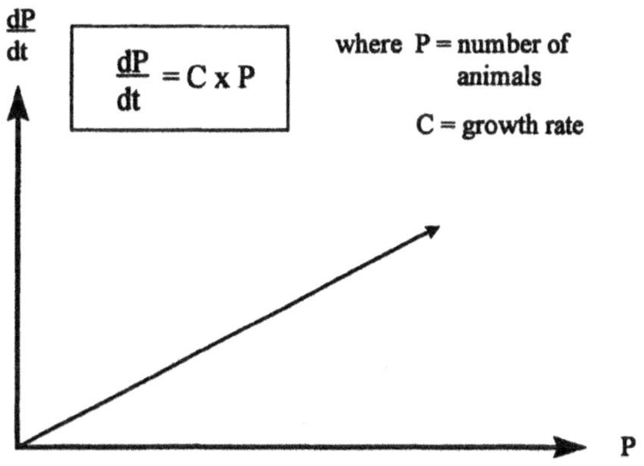

FIGURE 16.2. Example of a rate variable as a function only of the state variable.

The model run starts with initial values for the state variables. These initial conditions are used to calculate the rate variables. The new values of the rate variables, in turn, are used to calculate the changes of the state variables over time. During the modelling processes, the ultimate status of the system is not the only feature of interest; we are also concerned with its behaviour over time. Thus, the rate of change of state variables in time must be known, as well as the direction of change. If these rates have clear patterns, they may be formalized by means of rate equations or differential equations (Spain, 1982).

For in- stance, the rate equations for an elementary system are given in Figure 16.2. In this case, as in many others, the rate variables have the dimensions of the state variable per unit of time. The rate variable is a function of the state variable *P*. When the rate variable expressed in general terms as the derivative *dP/dt*, depending on the state variable *P*, a feedback loop exists, that is, the state of the variable determines its rate of change and hence its subsequent state. This feedback takes place continually, and is a way to reflect the interactions between system components.

16.5.1 Flow Diagrams: Stocks, Flows, and Feedback Loops

Flow diagrams are used to visualize feedback loops, rate and state variables, and, more generally, our knowledge about a system. They illustrate the most important elements and relationships of a system and are a qualitative model of a system. The flow diagrams may be useful at the start of a project to facilitate the identification of relevant state variables and rate formulation. They are also a way to make the system and characteristics of a model easily accessible.

Forrester (1961) and Odum (1972) provide two different sets of symbols to build such diagrams, with rules for drawing or interpreting them. Both languages are quasi equivalents, with similar symbols for the same objectives, but with differences due to their original field of application: ecology in the case of Odum and economics for Forrester. Probably the symbolic language of Forrester is more useful and appropriate for the socioeconomic system involved in rural planning. The Odum symbols are better adjusted to ecological systems, but are more complicated and they do not allow us to model more than Forrester's symbols. In any case, both sets of symbols can be used for the same objective, depending on the preference of the planner. Additionally, most modellers use flow diagrams without elaborated symbols, which are better for initial understanding, but are worse when actually building the system models. The two authors usually start with flow diagrams, translate these into systems diagrams and build the model from these.

A sequence of stocks and flows is the most simple representation of a system structure. The rate variables represented by a valve in a flow act as a faucet adjusting the flows into or out of stocks. As mentioned before, rate variables could be a function of the stocks or state variables, or be independent of them (Leffelaar, 1993). When the rate variables depend of the state variable, a feedback loop exists. A feedback loop effect can be positive or negative. Figure 16.3 represents the example used before, showing the feedback loop diagram.

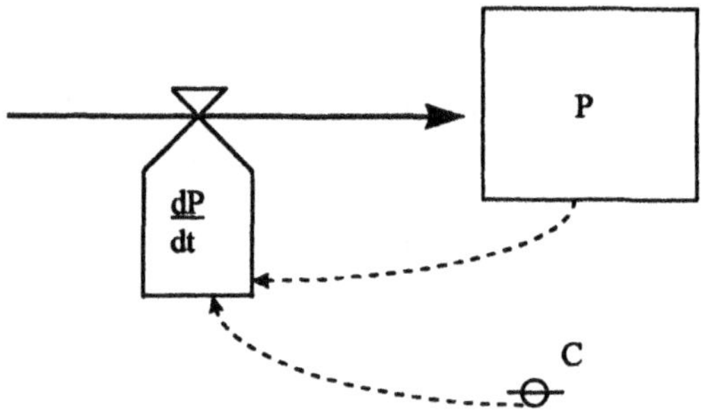

FIGURE 16.3. Animal population growth.

In summary, a feedback loop is a chain of causal connections from the level of a stock, through a set of decisions and actions dependent on the stock level, and back again to a change or correction in the level of the stock. Feedback loops take information from the stock, in order to affect the physical flows into or out of the stock. In practice, the simplest representation of a feedback loop is shown in Figure 16.4.

As an example of a flow diagram for a regional system we can see Figure 1.4 from Lugo et al. (1983) in Chapter 1. This figure shows a generalized diagram of all the subsystems of the Ebro River Valley (Spain), illustrating the role of diagrams as a planning tool for data gathering and discussion. As can be observed in this Odum diagram, different subsystems are involved on a regional scale: (i) the agricultural subsystem including cattle and other range animals; (ii) natural ecosystems; (iii) the urban subsystem; (iv) the industrial subsystem; and (v) the subsystem of water irri-

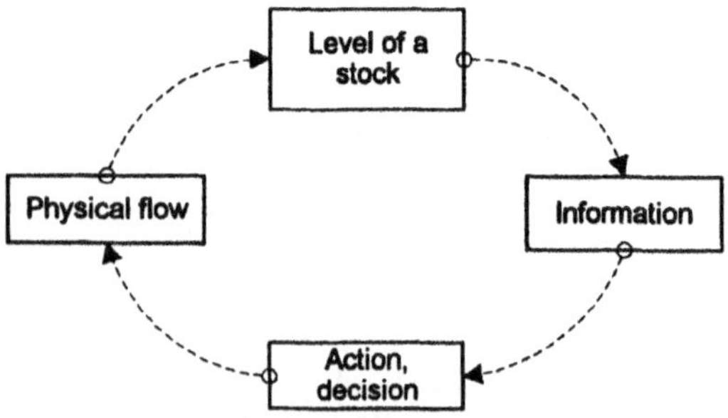

FIGURE 16.4. A simple feedback loop: information on a stock controls the action, which in turn affects a flow that changes the stock.

gation. In the processes of interconnecting these subsystems, we need, in addition to a water budget, a land use submodel, inputs and outputs of energy, materials and goods, and the various storages where products of the activities of the subsystems accumulate. These storages include agricultural and natural products, manufactured goods, wastes, water, and money.

16.5.2 Causal Loop Diagrams

A causal loop diagram is one more advanced step over the flow diagram. The causal loop diagram is created by incorporating the positive or negative sign at the end of the arrows. It is an extremely practical way for communicating ideas about the influences. By convention, arrows indicate a single-direction, from cause to effect, as can be seen in Figure 16.5. In the example given in this figure a higher population means more births "+," and more births mean an increase in population "+." A plus (+) sign in the middle of this feedback loop indicates that an increase in the first variable (population) "feeds" back on itself with more increase. Correspondingly, a minus (−) sign in the middle of the loop indicates that an increase in the first variable (population) feeds back on itself with a decrease through the intervening variables. The convention is such that the sign is always determined if the first variable is increased. As a rule, a global positive effect (+) is obtained if the number of negative arrows in the loop is even; otherwise the overall effect is negative.

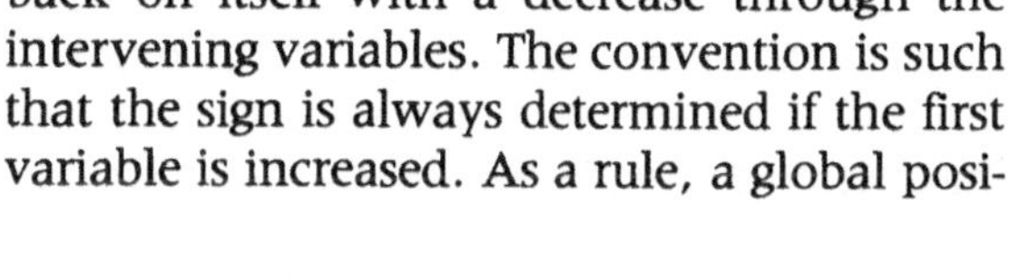

FIGURE 16.5. A simple feedback system with two feedback loops. A positive sign "+" next to the arrow head indicates an increase in the variable where the arrow begins, causing an increase in the variable where the arrow head is located. Correspondingly, a negative sign "−" indicates that an increase in the first variable causes a decrease in the second variable.

Rural systems as social and natural complex systems consist of hundreds of interlocking control mechanisms or feedback loops that maintain stocks within desirable levels, in order to assure the constancy and sustainability of flows. Negative feedback loops are self-correcting, and positive feedback loops are self-reinforcing or explosive. As a consequence, in the sustainable real systems, positive loops never operate very long without interruptions from negative loops.

16.6 Modelling Simple Relationships

Although systems analysis is much more embracing than just knowing how to produce models, in many cases it will be useful to have an operative model to simulate the systems behaviour. With a model it is possible to forecast future system scenarios. In this context, modelling is the process of building a model to make simulations of the system function, and simulation is the result of a model run.

During the simulation process, the model checks some hypotheses, and the results are answers in the form of state variable levels and behaviour. The modelling process can be summarized in five points: (i) identification of components, structure, and behaviour of the system involved with the problem; (ii) construction of the flow diagram of the system; (iii) functional analysis of the system components, as regards their role in the system context; (iv) anticipatory analysis of the system trend—the change concept; and (v) the trajectory of resources, people, and other components.

Natural and human systems can respond in many different ways over time. As we know, some system variables increase with time, others remain steady or fluctuate, and still others decline. The fate of a system depends on the supply and kind of energy, matter and information available for the system and on the way in which the system components are organized.

16.6.1 Renewable Resources and Population Relationship

The type of response of a state variable based on a renewable source that can supply as much as is needed is *exponential growth*. It is one of the simplest models between two variables and represents density-independent growth.

There is no limit to the growth of the state variable if the food supply is maintained. Odum (1983) refer to this situation as "growth on an energy source with constant pressure." In reality, perpetual exponential population growth is impossible. Normally, a resistance to growth comes from the system, as interference within the population, increasing mortality rates, or spatial difficulties for obtaining resources. However, during the early stages of growth, when demand of energy and other resources is small, resources will be available at constant pressure, and the growth may be exponential.

When the exponential responses are progressively neutralized by the resistance of the system or the environment, we have a logistic response of the state variable, which represent density-dependent growth. Logistic growth is common in nature and socioeconomic systems. One example of this could be the human population and its city services. In this case, when a population is small, the resources (e.g., food, energy, services, etc.) are abundant and the growth of population is exponential during the first periods. However, when services or resources are crowded and the population is larger, the individuals, their activity, and actions begin to interfere with life quality. The interference accelerates in proportion of the number of interacting people, until the population stops growing and the number remains constant in time. Figures 16.11 and 16.12 illustrate the flow diagram of this growth model and the trajectory of their state variables.

The logistic response is of special interest in rural planning, because it introduces a kind of trajectory which arrives at a maximum value of the state variable that can be supported by the system without problems. This maximum value is called *carrying capacity*. Carrying capacity is the maximum level of our variable in agreement with the sustainable development, which is the goal of planning. Sometimes, the self-control capacity of the system is delayed, and the growth to carrying capacity follows an oscillating trajectory around the maximum desired level before it arrives at the equilibrium. This behaviour pattern forms part of the typical logistic response of a state variable, when a feedback lag exists in the response of the equation.

Another kind of carrying capacity could be imposed by the resource, when inflow to the system is controlled by the capacity of the users. In this case, the trajectory of the state variable is similar to the logistic response, but it is controlled by the resource inflow rate. In the long run, the state variable arrives at a *steady state* showing a new type of equilibrium in the system. Planners can use this pattern to express the limiting capacity of resource use, even in the case that it could be renewable and the state variable density-independent.

Human systems frequently use all types of resources (renewable or nonrenewable), and try to do it at a maximum rate. For many years cities have used electrical energy produced by hydroelectric power from rivers. When cities grow, new power plants are built along the river until all the energy from the mountain water flow is being used. When this point is reached, new power plants cannot increase the energy obtained from the rivers. From this point of view the system can only maintain a steady state adjusted to the available energy. These sources may be called renewable because they flow continuously and are not totally used. In contrast, coal, oil, and petroleum are gone when they are totally used and act as nonrenewable resources. The earth may produce some more fossil fuels, but this needs millions of years.

16.6.2 One Depletable Resource and a Population

The simplest model of the interactions between a human population and a resource consists of just one resource and one popula-

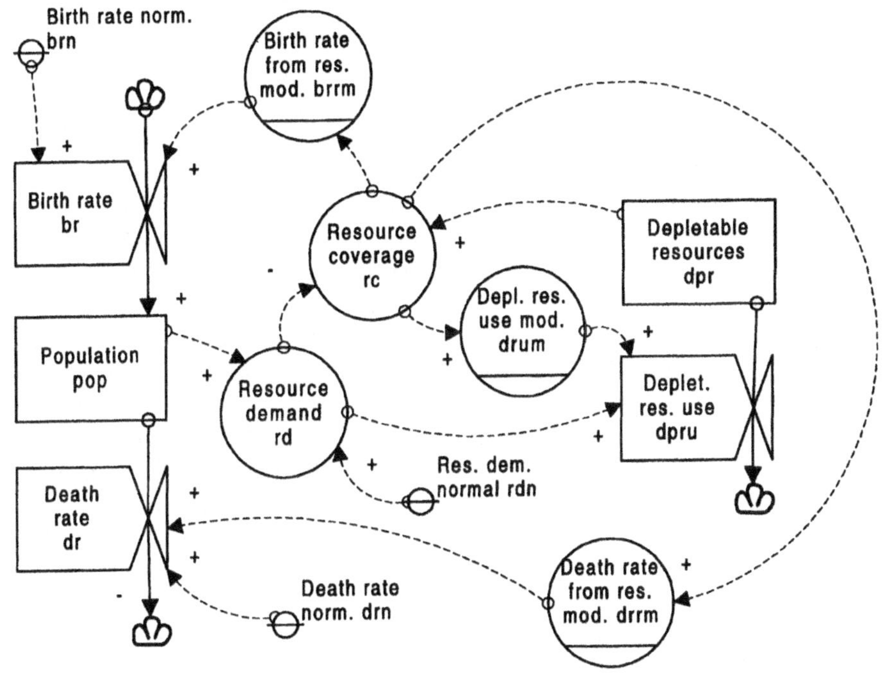

FIGURE 16.6. Systems diagram of a population that depends on one depletable resource.

tion. Here two assumptions are made: (i) the resource is crucial for the population, and (ii) the resource can be depleted or consumed at a rate higher than its rate of reproduction or regrowth or replenishment. Increasing scarcity is possible for most resources if the population continues to grow, for example, land area, stocks of fish, forests, crude oil, minerals, or water. The population could either directly depend on a resource, such as water or food, or indirectly depend on it, for example, through income from selling a resource such as crude oil or minerals to buy food.

In the beginning, when the population is small, the income from the resource suffices to buy food and services for all, and exponential growth of the population results. Population growth causes increasing demand, increasing scarcity, and, eventually, the reserves of the resource begin to decline, extraction becomes more difficult, and income drops. At this stage a feedback loop begins to depress birth rate and to increase death rate. Hence the growth of the population slows down. As the reserves continue to get smaller, the population number no longer goes up but instead decreases until not much is left of the resource, and the population cannot survive. The population must migrate. This pattern of growth is frequent in animal populations living on temporary food reserves (e.g., water, leaves, etc.).

Figure 16.6 illustrates the flow diagram of this growth model. As can be seen in this figure, a larger population number causes higher resource demand (variable *rd*). Resource demand also increases with the specific demand per person (resource demand normal, *rdn*). A higher resource demand increases the use of the (depletable) resource (variable *dpru*). The resource is not replenished; hence it decreases all the time, depending on the rate of use. As the resource is consumed, the resource coverage is going down.

Resource coverage is usually calculated in how much longer this resource will last at the present consumption rate; hence the resource coverage is the ratio of the resource and the resource demand. One example of resource coverage is the world food supply that usually lasts for one year. This is just enough to have sufficient reserves until a new harvest from the Northern Hemisphere becomes available, as this hemisphere has the largest agricultural area, and to have all the food in the processing industries, in the transport systems and in shops and households. In 1995, this supply dropped to 10 months due to a bad harvest in the United States. This caused an increase of famine in some parts of the world. In general, when the coverage is low, only a part of the demand can be fulfilled; that means high demand but little supply.

This model uses a table function (depletable resource use modifier, *drum*) which is a so-called modifier, a very effective technique with wide applicability. Here, the value of this function decreases if the resource declines. Figure 16.7 shows its graph. This is a nonlinear relationship between the availability of the resource, expressed by the variable resource coverage, *rc*, and the variable resource use modifier, *drum*. The demand (variable resource demand, *rd*) in its effect on consumption is modified with this table function (hence the name modifier) by simply multiplying the demand with the value of this function. Hence, if there is plenty of the resource compared to demand, all of the demand is met. Actually, the demand is here multiplied by a factor of 1.4, not by 1, because if supply is ample, the use of the resource tends to be careless. For example, today the production of 1 kilowatt per hour of electricity needs 10 percent of the primary energy which was necessary at the turn of the century. Thus, if supply is good, use is wasteful.

The numerical values in Figure 16.7 are very typical for consumption of food because food does not have a long lifetime and is usually consumed rapidly. The data would be quite different if the resource would be crude oil, for example. But whatever the type of resource, there is almost always a sinusoidal (S-shaped) relationship between supply and degree of demand fulfillment.

The relationship between a consumer and a resource is an example of a predator-prey relationship. The best-known predator and prey model, the one by Lotka–Volterra (Ebeling and Peschel, 1985), assumes a linear relationship. Although this form is mathematically simpler than the relationship used here, the Lotka–Volterra form is ecologically wrong and results in misleading dynamics. It means that with increasing supply the consumption increases proportionally even if the number of predators is constant—which cannot occur, and that with decreasing supply the consumption goes down only proportionally. In reality, if only very few individuals of a species are left they can hide in the best hideaways. If only a small fraction of the initial supply of a resource is left, then this resource becomes diluted and it is difficult to locate and to extract.

Hence, in ecosystems, if the resource becomes scarce, consumption goes down faster than supply. If supply is increasing but number of predators or customers is constant, consumption will reach a ceiling and will not increase above that. Therefore, a sinusoidal rela-

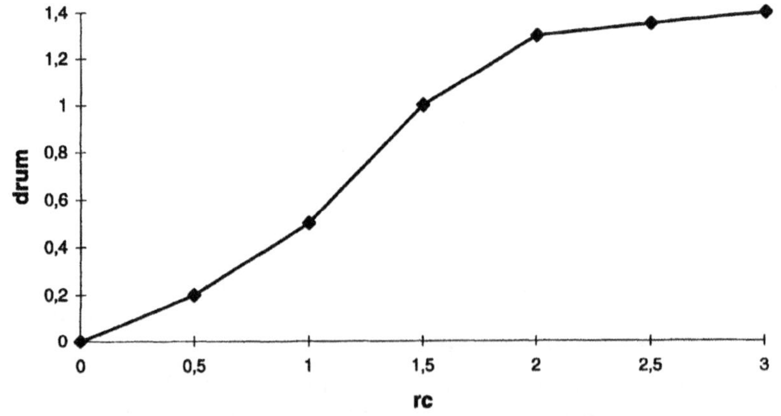

FIGURE 16.7. Graph of the table function "depletable resource use modifier" (*drum*).

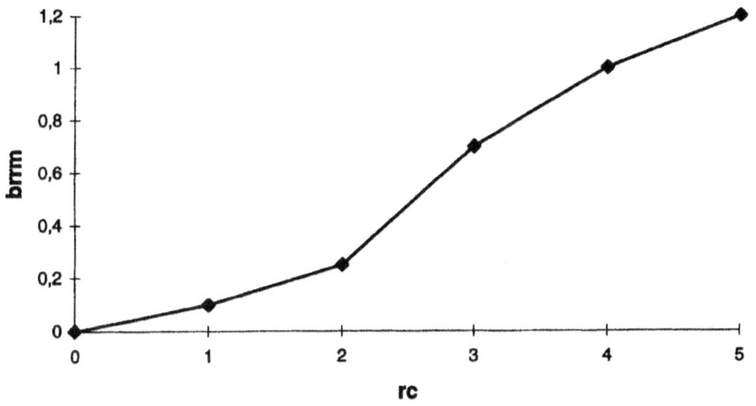

FIGURE 16.8. "Birth rate from resource availability modifier" (*brrm*): the modifier that adapts the birth rate to the supply of the crucial resource.

tionship like that in Figure 16.7 is correct. This relationship, however, is not always valid for human-nature interactions, a fact emphasized by Odum (1984). If a resource is scarce, prices go up. But whatever the price and the costs are to extract or hunt, there are always individuals who could pay any price. This indeed happens for some very rare resources. For example, incredible sums are paid for the horn of the rhinoceros or for large falcons for hunting. In comparison, animals and plants are limited in their capabilities such as speed in hunting or ability to extract water from dry soil.

Here it is assumed that the resource is crucial for the existence of the population; hence, with declining supply the birth rate declines (variable birth rate from resource modifier, *brrm*) and the death rate is increasing (variable, death rate from resource modifier, *drrm*). These effects are also represented by modifiers; the graphs of the table functions for birth and death rates depending on resource coverage are depicted in Figures 16.8 and 16.9. Figure 16.8 shows that at low supply (low values of *rc*) the modifier takes very low values and eventually becomes zero. In Figure 16.9, *drrm* value is somewhat below 1 for high values of resource supply *rc*. That means that the death rate is slightly below average (the normal death normal is multiplied by this modifier). At low rates of the resource supply, the death rate increases dramatically as it is multiplied by high values of the modifier (up to the 8-fold of the normal death rate).

With these three modifiers the model is complete to run. An "inner mental model" of

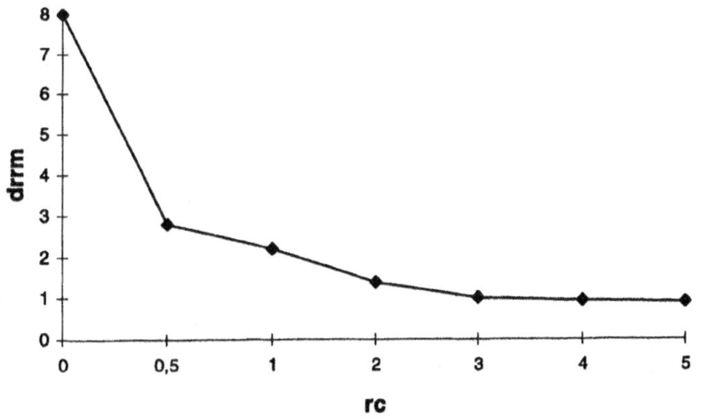

FIGURE 16.9. Graph of the table function "death rate from resource modifier" (*drrm*).

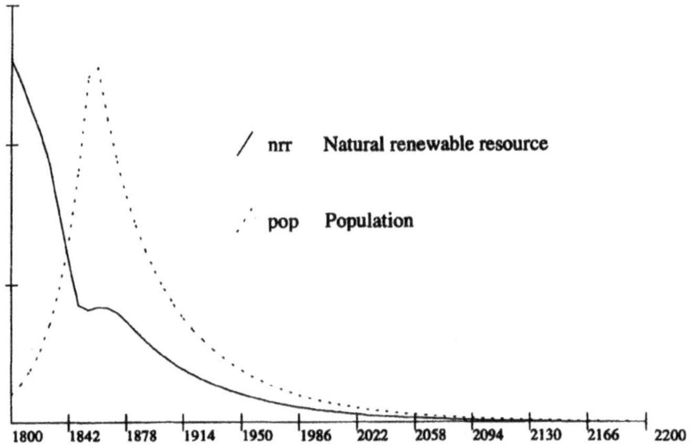

FIGURE 16.10. Trajectory over time of the two main state variables, population and natural resources remaining.

the expected model behaviour, the so-called *reference mode*, is necessary prior to a model run. The reference mode should be graphed so that it can be compared with the outcome of the run. Any deviations reveal either an error in thinking, in the model or an unexpected result. All three are important insights. Here the reference mode is of an exponential increase of the population in the beginning, which triggers an increasing decline of the resource. Shortage of the resource will at first slow the growth of the population causing it to peak and then to collapse. Accordingly, the decline of the resource will also slow but the resource will be depleted. Figure 16.10 depicts exactly this pattern. Please note the initial exponential increase followed by a leveling off. This "logistic" pattern is very typical for most biological growth processes. If data on increase of wood biomass, for example, do not exhibit this pattern, they are most likely wrong.

16.6.3 More Than One Resource

Populations can usually rely on more than just one resource. If several of the resources are renewable, this may increase the stability and sustainability of the population. The simplest case is a system with one renewable and one nonrenewable resource (Figure 16.11). This model links three subsystems: (i) a human population; (ii) a subsystem of nonrenewable resources; and (iii) one on renewable resources.

The model of renewable resources, like the one on nonrenewable resources, has a consumption rate. Additionally, it has structures for the reproduction of the renewable resource. Basically, the interaction of resources with the population is identical to that of the previous system, but not its dynamics. Again, after an initial exponential growth period of a small population, the nonrenewable resource is depleted. The remaining renewable resource may be consumed at a rate higher than the replacement rate so that it becomes depleted too. In this case the dynamics are identical to that of the previous system. If use of the renewable resource is more careful, the population may still overshoot (due to the nonrenewable resource) but afterwards decline to a sustainable level and a dynamic equilibrium may result. But here, a third type of dynamics is possible. If the consumption rate of the renewable resource is too high, it may come close to depletion, resulting in a decline of the population. A low population number and with it, low consumption, may allow a recovery of the renewable resource which in turn supports new growth of the population causing a too heavy use of the renewable resource, and so on.

Some human societies may be examples of this type of system. Our economic system has

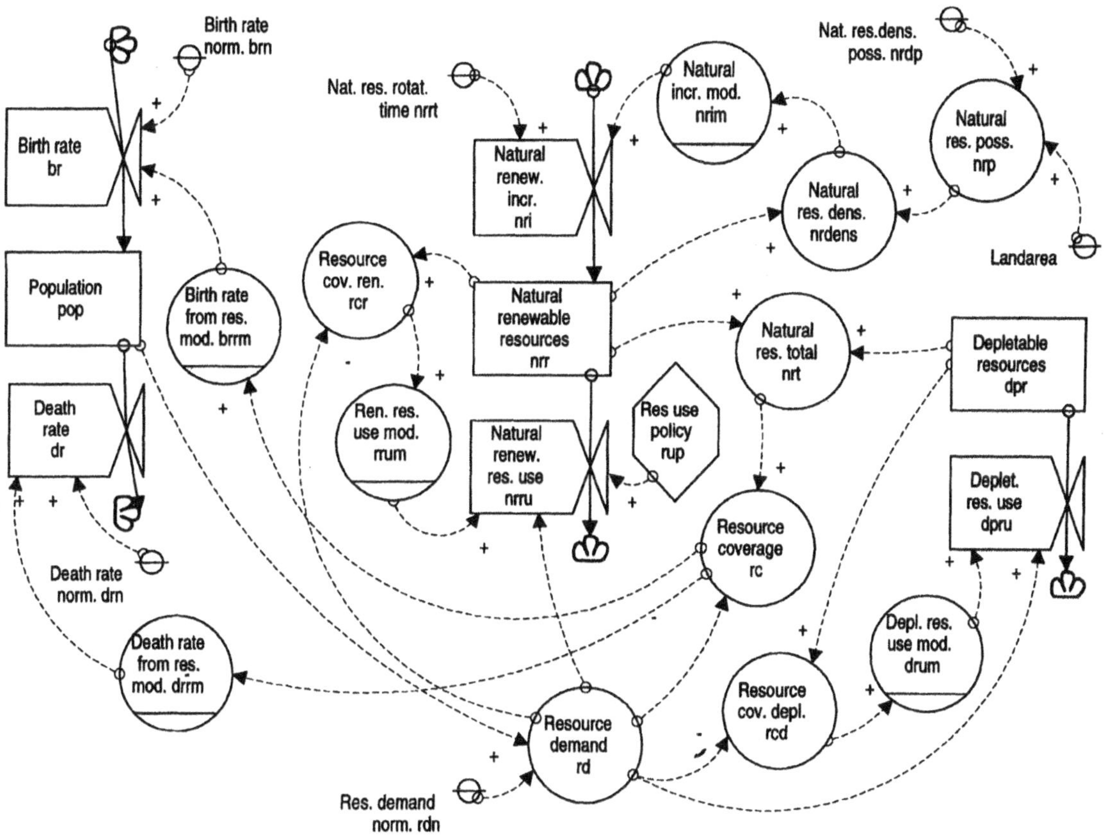

FIGURE 16.11. Systems diagram of the model with one renewable and one depletable resource.

been growing on many nonrenewable resources such as fossil fuels and minerals, as well as on renewable resources such as wood, top soil, surface water, or agricultural harvests. The nonrenewable resources are consumed one after the other. They may also become unavailable because their use may be banned for ecological reasons; for example, use of fossil fuels may interfere so severely with the global climate that a convention on greenhouse gases similar to the convention on CFCs may occur. Donnella Meadows (1984) predicts that this will happen soon. In any case, human socioeconomic systems may decline to a level that is in a dynamic steady state with available renewable resources.

Here, the numbers in the submodel on the renewable resource are such that they depict a forest. Hence, the reproduction process is named "increment." The model structure is valid also for other renewable resources, but not the data. This submodel reveals the typical logistic growth pattern described in Figure 16.12, if growth starts from a very low density. Here, the rotation time is fairly short, about 50 years, which is typical for fast-growing tropical trees.

This pattern demonstrates that with very little harvest there is no increment of wood biomass (right part of figure, biomass is fairly constant after year 60). The highest increment is at about 50 percent of maximum forest biomass (year 30). Thus, sustained harvests are highest if they push biomass down to about 50 percent of the possible maximum. In real forest systems the biomass can increase for very long periods of time because the stems of trees can increase for centuries. Thus, the model is simplified also with the assumption that there is a maximal biomass.

In the next model run the forest will be fully stocked in the beginning. The reference mode for the population and the renewable resource could be a careful harvesting will decrease the

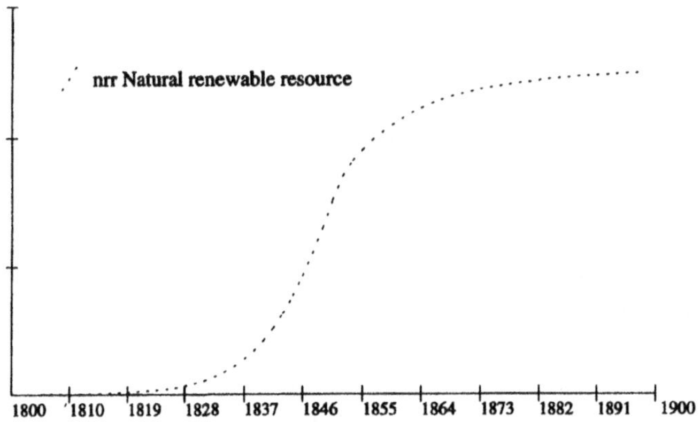

FIGURE 16.12. Growth pattern of the renewable resource used in the model.

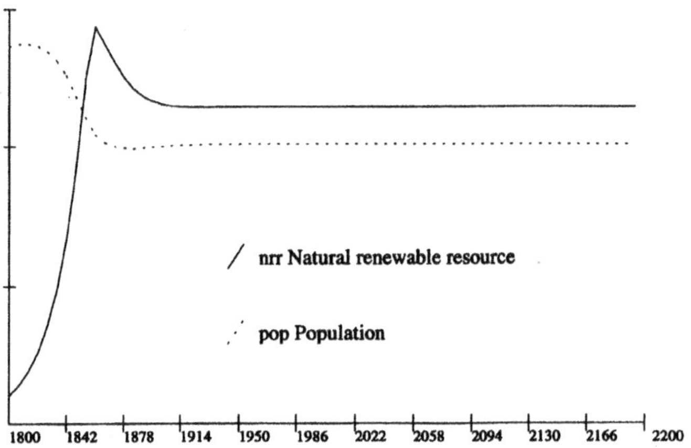

FIGURE 16.13. Basic run with careful use of the renewable resource.

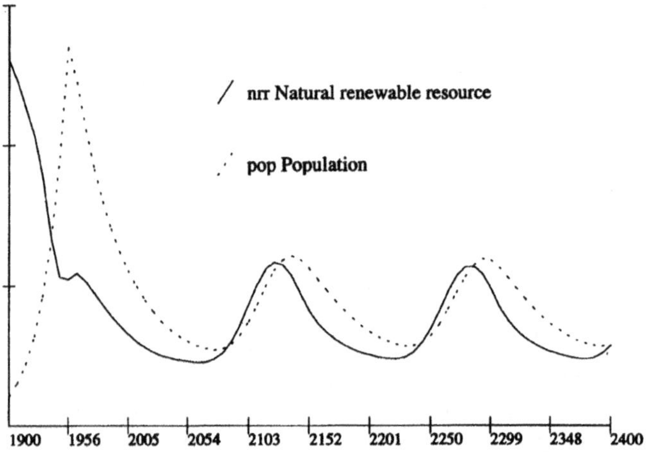

FIGURE 16.14. Dynamics of the model if harvesting is a little bit too high.

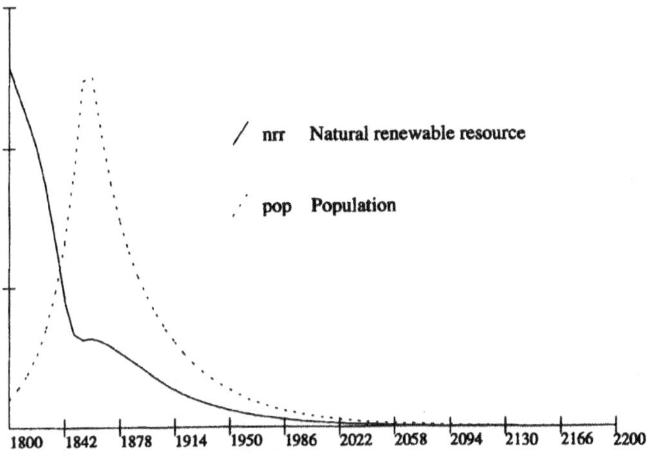

FIGURE 16.15. Run with a careless policy leading to extinction of the resource and of the population.

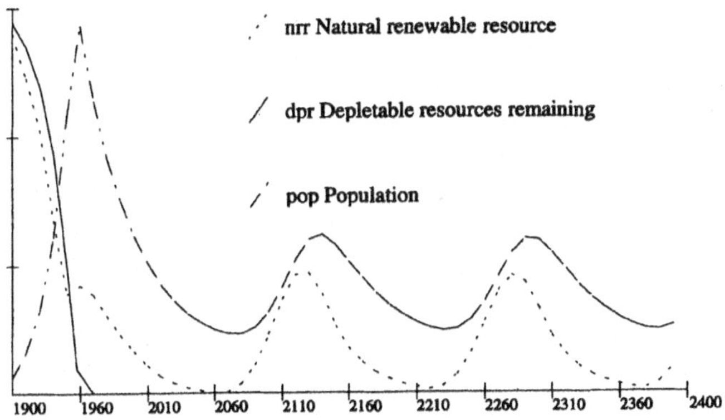

FIGURE 16.16. An example of how depletable resources trigger the overshoot of the population. Hence, they are a mixed blessing.

initial biomass (if the forest was fully stocked), but increase the increment. The population will increase until no more can be sustained from this resource. Compare this conclusion with the actual model run in Figure 16.13. In this run, it turns out that the population initially overshoots the maximum number. This is due to the fact that a resource shortage affects the population number with a delay.

What happens, if the harvesting policy is less careful? Figure 16.14 is a model evaluation with harvesting that is a little bit too high, controlled by the policy variable resource use policy, *rup*, which is here increased by a factor of 2 compared to the previous evaluation. The population overshoots and begins to deplete the renewable resource because increment is lower than harvest. When the population is sufficiently reduced, the natural resource recovers very rapidly, which in turn allows another recovery of the population, which in turn uses too much of the resource, and so on.

A nonsustainable harvesting policy could allow harvests that are higher than regrowth. In this case the renewable resource is threatened. For the next model run the parameter resource use policy, *rup* (depicted in Figure 16.11) is increased by another factor of 3. Figure 16.15 demonstrates the model run.

The final Figure 16.16 illustrates how the availability of large resource amounts in the beginning triggers a huge population growth which, after depletion of the depletable resource, nearly causes a total collapse of the population. As so often is true in ecology, bone and bane (here a good resource supply and its disastrous consequences) are close neighbours.

16.7 A General Structure of an Age Group Model with Different Applications

The population model used above is too simple for most applications. A reasonable population model needs different age groups: children above 4 or 5 years of age (depending on the country) attend school; children below this age do not. In many countries, children between 16 and 20 do not have a job, are not married, and may attend advanced schools or training for a job. Adults from 21 to 45 usually have children and a job; adults above 45 no longer get children but have a job. People above 65 (or 68, depending on the country) are retired. The structure of such a model is depicted in Figure 16.17.

The variable "birth rate normal" should depend on other variables, because historically it changes very much over time. In the simplest case, the one depicted in Figure 16.17, it is made dependent on time, but it could depend on other variables. In the past ten years, fertility rates have gone down in almost all countries of the world in a parallel pattern, although starting at different values. In the model evaluation of Figure 16.17 a human population is depicted with such a declining birth rate. If this development continues—and there are no reasons why it should not—country after country will hit its maximum population number, and afterward its population will slowly begin to decline. According to projections from the U.N., the global population could reach its maximum around the years 2040–2050. Afterward, country after country may experience lower ecological pressures due to a decreasing population, but may simultaneously face the problem of an aging population with few children, few young people to care for the old, too few people to care for tourists, and recreationists, and so on.

The next diagram (Figure 16.18) shows an age group model for a population. Only one category of mortality is included: mortality of the oldest age group. Birth, of course, can directly increase only the population between 0 and 5 years of age. One-fifth of the population between 0 and 5 years moves to the next age group per year, as there are five years in the first state variable *pop*0.5. Correspondingly, one-ninth of the children in the second state variable (*pop*6.15) move to the third age group, those between 16 and 20 years of age,

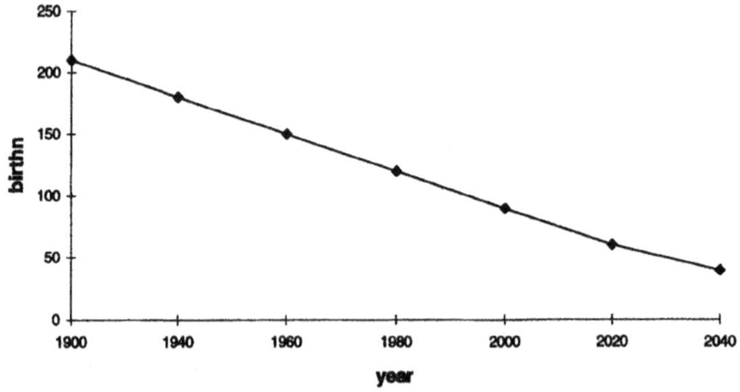

FIGURE 16.17. "Birth rate normal of age class 21–45" (*birthn*): assumption on development of births per year per 1000 women between 21 and 45 years of age over time (years 1900–2040). Divide the data specified in this graph by (70 × 2)/24 which is about 6, to get the numbers matching the normal convention, *brn*, births normal per 1000 inhabitants per year.

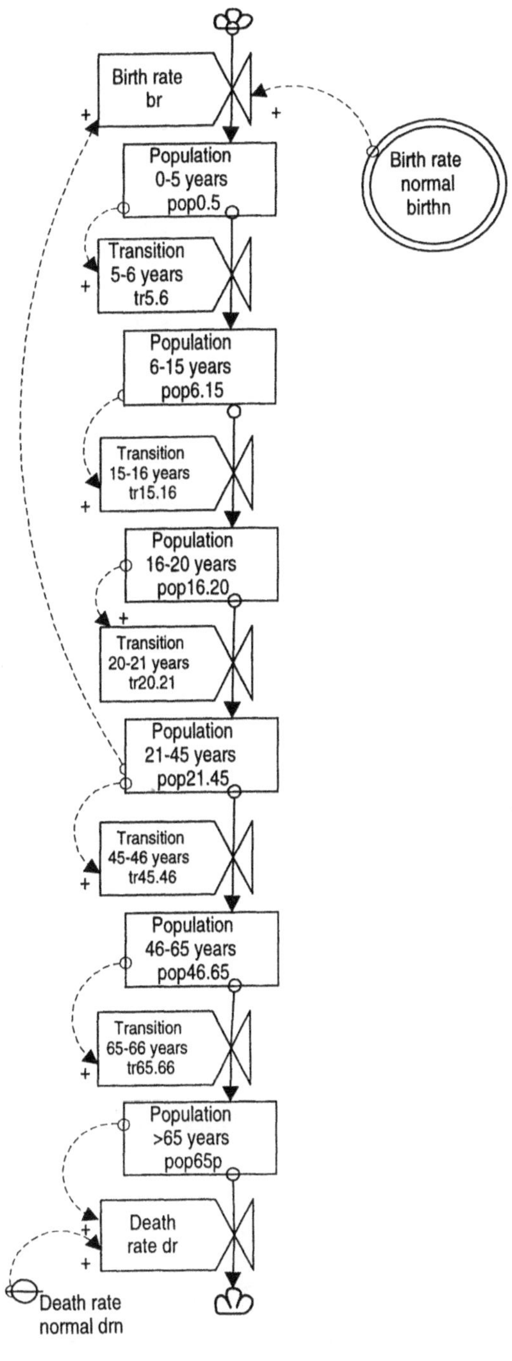

FIGURE 16.18. Population model.

and so on. The birth rate is the product of the table function "birth rate normal" times the population number in the state variable *pop*21.45. Hence, no teenage pregnancies occur in the region for which this model was built (Indonesia). This is true for Indonesia as it is a Muslim state, but it would be different for the United States where teenage pregnancy is high. The model for the United States would need another connection from population 16–20 years of age with its own specific fertility rate.

The next two diagrams (Figures 16.19 and 16.20) depict evaluations of this model. Figure 16.19 shows only three of the age groups for clarity purposes. All age groups exhibit the parallel increase in numbers in the beginning. Afterward, the number of children decreases first; with an appropriate delay decreases the number of people who work and only in the end the number of retired people becomes smaller. Figure 16.20 shows how the number of children per family goes down from 3 in the beginning to about 1.3 in the end, a number below reproduction level.

This pattern of approaching an equilibrium of population size, with an increasing fraction of old people and afterward a decline of the human population, is valid in a similar form for almost all countries of the world in a parallel pattern, although distributed over time due to different levels of fertility rates in the present time.

The basic model structure of different age groups is valid for quite different systems. The similarity to forest management is obvious, if the management is based on age groups of trees. A common scheme could be: trees of 0–5 years ("thicket"), 6–20 years, 20–60 years, and above 60 years. However, here mortality is extremely high in the first age group. In Germany, for example, 20,000–30,000 young trees may be planted per hectare. In the age group 20–60, less than a thousand are left and are further decreased by preharvests. In the end, only sixty trees may remain on one hectare.

The same systems diagram is also applicable to technical infrastructure of a region. This system is highly relevant for investment decisions, as infrastructure of different age groups needs very different levels of maintenance. Demand for maintenance may be very low in the first few years for many types of technical infrastructures, but afterward may increase

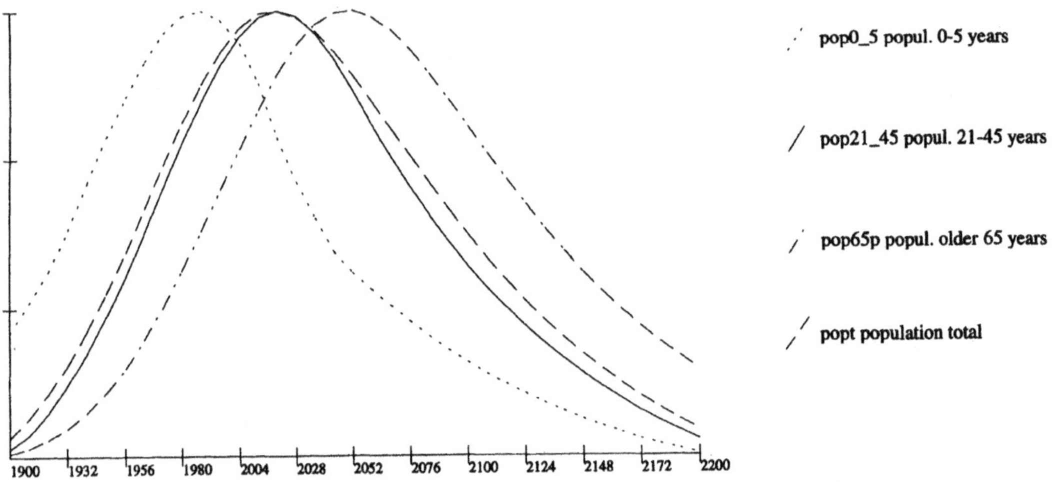

FIGURE 16.19. Evaluation of the outcome of present trends in declining birth rates from 1900 to 2200 with a population maximum at around the year 2020. The ratio of old people to children is beginning to increase after about year 1980, when number of children has peaked.

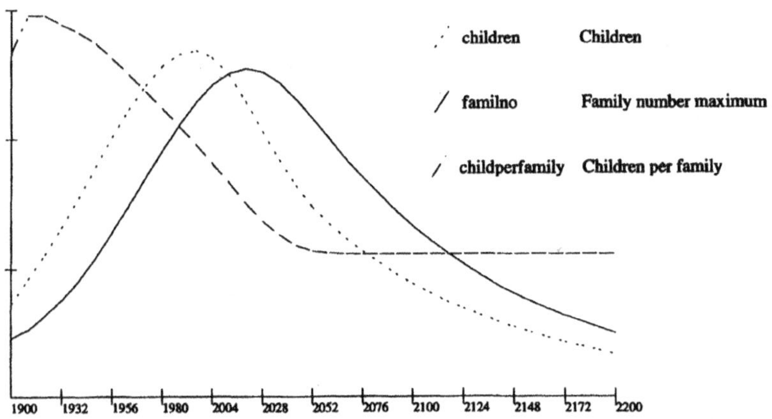

FIGURE 16.20. Effects of a declining population for number of children, number of families and number of children per family.

rapidly. Additionally, costs of maintenance can also increase dramatically with age of the infrastructures. Thus, if all infrastructure is built within a very short period of time, all of the infrastructure becomes older in parallel and maintenance costs can explode and become prohibitive. Thus, a model of this type can be used to support decisions on investments in infrastructure.

16.8 Main Components of a Socioeconomic Model on Land Use Change

Some main drivers of land use change are the necessity to produce food either for living or as a cash crop which, combined with a growing population, leads to a conversion of forest into

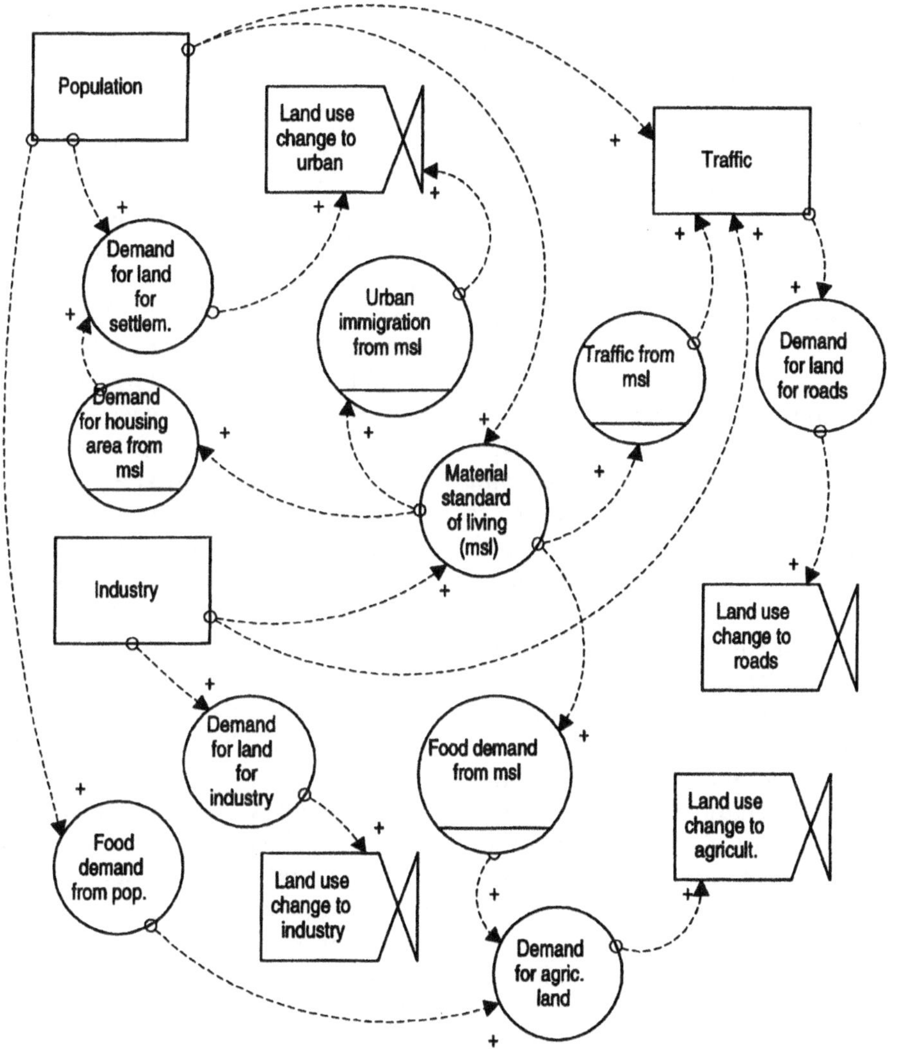

FIGURE 16.21. Some main driving factors of land use change.

agriculture. Many forests are destroyed by overharvesting for rapid income, and due to the extension of settlements, economic areas, and infrastructure. Other drivers of land use change are connected to increasing wealth in its aspect of the "material standard of living""as opposed to nonmaterial wealth due to intangibles such as the entities listed by Boyden (1976) in his case study on Hong Kong (e.g., "existence of social backup groups").

Factors of the material standard of living (i.e., average income per person as well as regional income) are depicted in Figure 16.21.

The explanation of this figure is as follows. A larger *population* needs more food and a larger land area for their settlements and infrastructure. The *material standard of living* (*msl*) increases demand for several types of land:

(i) A richer population wants larger houses, more apartments, and larger gardens and parks around buildings; thus, with increasing msl, cities expand.

(ii)A richer population tends to eat more, which increases demand for food and thus for agricultural land area, and it eats more meat—

the production of one energy unit of meat needs the fivefold to fifteenfold amount of primary food, depending on the animal species.

(iii) A richer population has more cars per inhabitant and, also, the mileage per car may go up—both factors increase demand for traffic infrastructure.

(iv) A richer population attracts more immigration from the outside which causes changes of land use in all categories listed above.

As regards *industry*, the larger the industrial investments, the more land area is needed. Additionally, more industry increases traffic that needs additional land. Furthermore, a larger industry needs more resources. Both production of biological resources and mining of depletable resources cause land use changes that are enlarged if the economy grows. On the other hand, *traffic* needs roads. Their construction causes land use change. Roads, in turn, improve accessibility of remote areas and thus allow new settlements that cause additional land use changes. Traffic is both a prime driver of land use change and also a second-order driver because it is itself driven by changes in population, material standard of living, and industry.

There are several factors that modify the effect of these main drivers. These are:

(i) *Effectiveness of resource use*. For example, the amount of food produced per hectare of agricultural land increased several fold due to use of high-yield grains, fertilizers, and biocides. However, production of these agricultural goods needs additional inputs by these agricultural substances, which in turn cannot be produced without land use changes. Most of the present high-yield policies are not sustainable.

(ii) *A main modifier of present land use drivers are lifestyles and attitudes*. Increasing environmental concern was a main reason for protection of manifold areas. Simultaneously, appetite for new adventures such as hang gliding, ballooning, mountain biking, parachuting, or whitewater canoeing extend the disturbance of man on nature into ever new, ever more remote areas.

16.8.1 Systems Analysis of Migration

Migration of population and of economy is a main driving factor of land use change. Immigration into a region by both companies and people depends on the attractiveness of the particular region in comparison with competing regions. If the regional attractiveness is lower than that of competing regions, people or companies will leave (negative value of the variable immigration in the systems diagram). The attractiveness of a region for population is defined by the regional quality of life which is a multifaceted, soft entity. About 1000 components were listed in a publication on "indicators of the quality of life" by UNESCO (1978). But only some of these are really important to influence migration. The idea of modifiers is also used here to control immigration. In an analysis of reasons for immigration it usually turns out that only very few factors are crucial.

Although these factors are soft, our own experience (Grossmann et al., 1996, model "Regio") shows that they can be calibrated to reproduce historic dynamics of population numbers and of tourist stays in a reasonable way. In our workshops with regional experts they could always very rapidly put into an ordering the comparative strengths and weaknesses of their region compared to the most important competing regions. They could also easily order these components according to their relative importance. This allows construction of models that include soft factors, if they are important.

In Figure 16.22 those components that are named in most empirical studies on the relative advantage or attractiveness of regions are put together. Immigration is modified by attractiveness. In the normal situation, an equal number of people leave as arrive, so that migration does not change the number of inhabitants. If the region is more attractive than its environment, more people arrive than leave (this idea of relative, opposed to absolute, attractiveness was already developed by Forrester for his model "Urban Dynamic"") (Forrester, 1969). These factors are now grouped according to their short-term relevance. It will be explained later why the relative long-term relevance of these factors is very different. This makes policy decisions difficult and models valuable. The groups are as follows:

(i) *Group 1: housing and jobs.* Availability of housing, expressed by the variable "housing ratio," which means number of houses divided by number of inhabitants and availability of jobs (variable "job ratio"). Most politicians and planners attribute the highest importance to the two factors, housing and jobs. But some far-sighted citizens do not so much evaluate the present situation but also the future economic outlook, expressed by the variable "state of the economy"—is it forward or backward oriented, that is, is it more based on software industry or on shipyards? Less wealthy prospective immigrants look at the regional services (e.g., public housing).

(ii) *Group 2: health care, schools, regional services.* Whereas politicians underline the importance of housing and jobs, managers want to know whether the health services are good or advanced, for example, whether hospitals and medical specialists are located in the region and whether good schools and universities are available in the region (variables "regional health services" and "regional schools"). They may also look at other regional incentives, such as libraries or low-income taxes. If a region is not able to attract good far-sighted managers, its economy will decline in the long run and jobs will be lost. Thus, ulti-

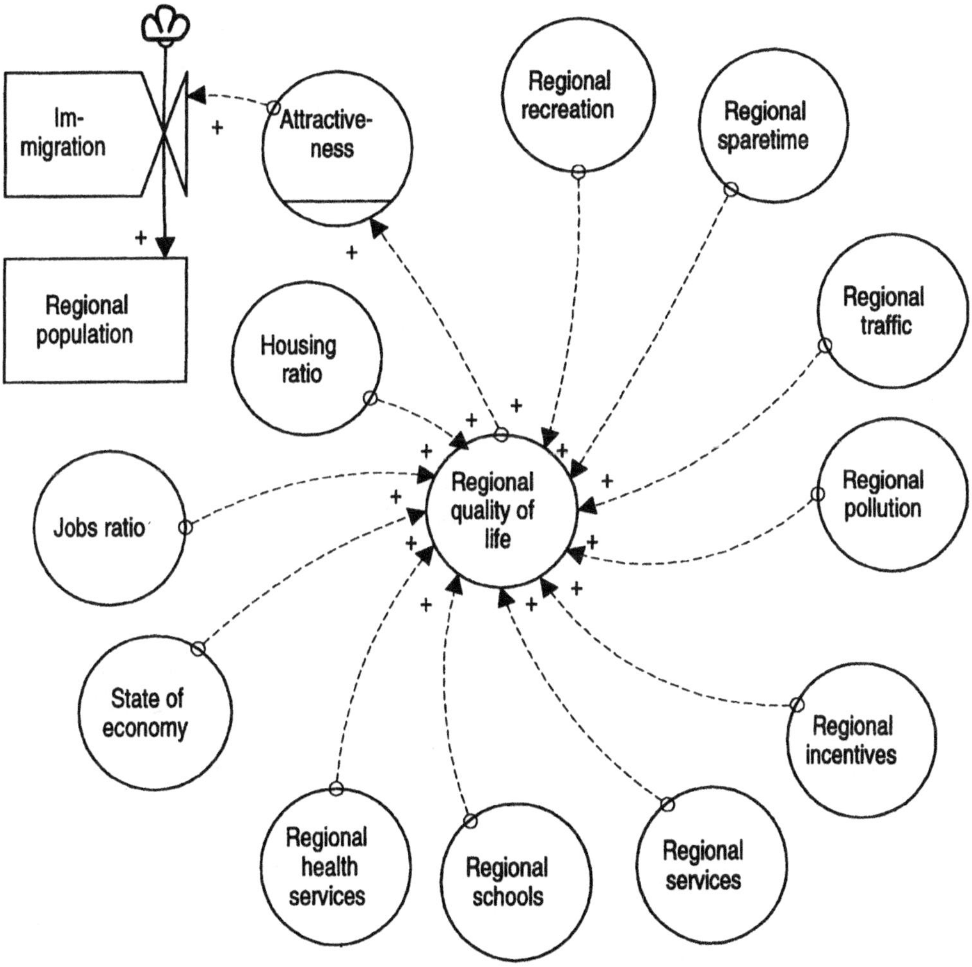

FIGURE 16.22. Main components of regional attractiveness for people to migrate into this region.

mately and in the longer perspective, good schools and attractive health services may be more important so that a region will attract people.

(iii) *Group 3: environmental quality.* Pollution and traffic are the two factors that can most adversely affect the environmental quality of a region. Excessive traffic pollutes, causes noise, endangers people, and decreases the attractiveness for outdoor activities. Environmental quality is a factor that can depopulate a region, meaning that most people who have alternatives leave (e.g., people who are respected experts or have a university education). This happened to the Ruhr district in Western Germany after its coal- and steel-based economy lost its appeal in the late 1960s. Without these crucial people the Ruhr district became unable to build up a more future oriented, appealing economy and, even today, although pollution is greatly decreased it is still in economic difficulties and suffers from its former bad reputation. Another example: When the chemical industries in the city of Linz (Upper Austria) declined in the 1980s and had to make many people redundant, the population was no longer willing to accept the pollution caused by the remaining industry. Thus, the declining industry had to meet new, rigid environmental legislation in addition to its financial difficulties. Now old industries are, on a worldwide scale, beginning to lose their appeal and competitiveness to new, information-based economies. The story of outmigration from such regions, so well-known from old industrial sites in all developed countries, will recur.

(iv) *Group 4: outdoor recreation and spare-time possibilities.* Almost all evaluations of crucial factors confirm that new economies are set up by people who are mobile and prefer those regions where they meet like-minded people, have reasonable support (see Figure16.23 on main components to foster growth of an information based economy), and in particular, where they have great possibilities for outdoor

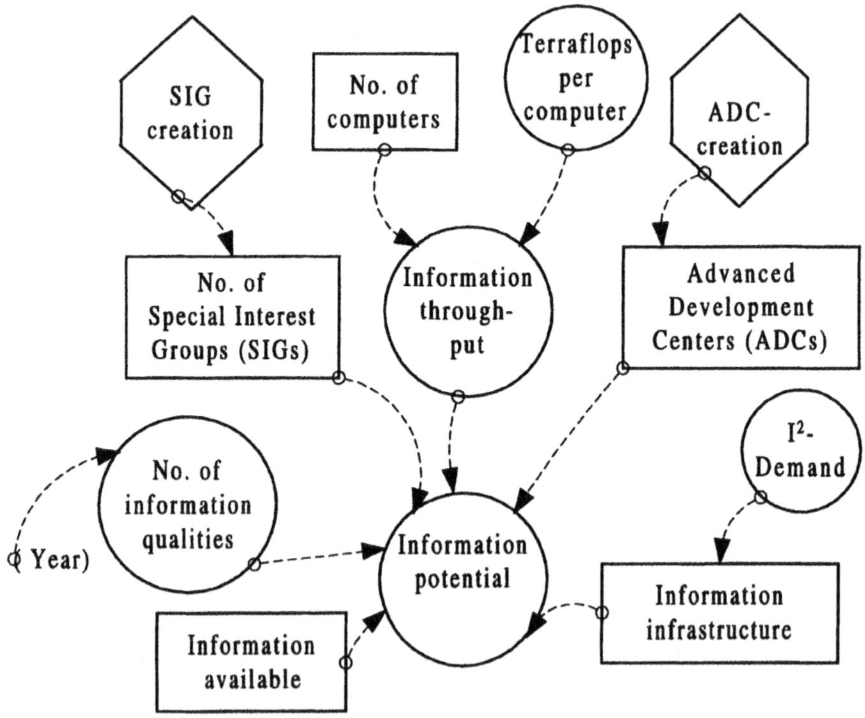

FIGURE 16.23. Main components of the "information potential" which indicates how much can be done with information through computers and networks.

recreation and, also, a little bit less important, a rich spare-time offering (e.g., Hall and Preston, 1988).

Hence, regions like Lausanne in Switzerland or the Mediterranean region in France around Nice advertise their great outdoor environment to attract new business. One evaluation of an innovative, information-based economy demonstrated that these companies and their staffs are extremely demanding with respect to environmental and urban quality (Multhaup and Grossmann, 1997). The evaluation of regional success stories demonstrates that jobs can be created and houses can be built if the right people are available and if some conditions are met. Hence, the most important factors are not housing and jobs, but people and their well-being. This fits in quite well with what Inamori, head of Kyocera, states: the most important task of a company is to support the human development of its employees (Senge, 1990). In the short run, however, housing and jobs are crucial. Long- and short-term benefits may adversely affect each other. Systems analysis can help to get such issues right.

16.8.2 Modelling of the Relationships for Regional Attractiveness

All components of the regional quality of life are expressed as table functions that take numerical values between zero and a positive maximum depending on the relative quality of the respective component. The table functions express the relative weight of this component as attributed by the judgements of the experts. All the table functions are multiplied to get the regional quality of life. Thus, if one table function takes a very high value, it increases the value for the regional quality considerably. On the other hand, if one table function takes the value zero, it suppresses all good contributions by all other components, because the product is zero too, if one factor is zero. If one component takes the value of 1, it does not change the present value of regional attractiveness. Hence, the value of 1is a normal value; a value for a component that is just normal in comparison to competing regions, not good and not bad.

For long-term evaluations the highest value for the table function for outdoor recreation has to be much higher than the highest value for housing. Thus, for example, the table function for outdoor recreation may take values between 0.3 and 10, the table function for housing between 0.9 and 2. Hence, none of these table functions can decrease the regional quality to zero, but lack of outdoor recreation decreases the overall quality far more than lack of housing. Ample supply of housing, on the other hand, cannot contribute as much to the regional quality as can outdoor recreation (maximum values of 2 and of 10, respectively). Obviously, it does not make sense to attribute negative values to several of the table functions, because the product of two negative values becomes positive. However, for modelling outmigration if attractiveness is low, the table function for attractiveness can take negative values. By definition, it should take negative values if its argument—the product of all components of regional quality—is less than 1.

16.9 Land Use in the Future and Its Most Important Drivers

The most sweeping changes in land use and land cover in the future will be due to changes in four driving forces: (i) population; (ii) "Agenda 21"; (iii) effect of information and communication technology—transition into Information Society; and (iv) economy.

16.9.1 Population

Major changes in population are due to further increase, a global peak at around the year 2050, and ageing occurring in country after country within the next fifty years, as discussed above with the model with six age groups. Additionally, large parts of the world are becoming more wealthy with effects discussed above in relation to the material standard of living, whereas, simultaneously, marginalized groups are exerting more pressure on land to be able to make a living, which increases soil erosion and desertification.

16.9.2 Agenda 21, or "Global and Regional Sustainability"

Manifold changes are implemented to allow a global sustained living of the human global population. Many more changes of economy and lifestyles are necessary. Sometimes these changes are slow and sometimes they are rapid. Some of the most important changes with an impact on land use include exclusive reliance on renewable, no longer on depletable, resources. The model evaluations with one depletable and one renewable resource described above show in a nutshell the dangers due to an unsustainable build-up of population numbers fuelled by depletable resources. Renewable resources as a base include renewable energy (e.g., solar, water, wind). The problem of global climate change due to "greenhouse gases" is closely interrelated to the consumption of too many resources. Even a renewable agriculture could still contribute too many such gases, in particular emissions of methane from both cattle and rice paddy fields. Thus, some land use changes could be necessary here. There is an urgent need to do with far less traffic both for worldwide trade and travel as well as within regions. Again, this will directly affect land use for traffic infrastructure as well as the structure of agriculture and of settlements.

16.9.3 Transition to the "Information Society"

The emergence of the Information Society is a worldwide phenomenon. Its impacts on land use are manifold. It changes the economy, supports new lifestyles, and thus affects two of the main factors shaping landscapes. Moreover, it has other effects on land use and landscape design:

(i) Rural areas are no longer remote. The meaning of geographical distance becomes transformed; its importance shifts. Thus, rural areas can offer jobs that until now could not exist in rural areas.

(ii) Former problems of rural areas (e.g., critical size of human population to preserve schools, to have an adequate medical treatment, etc.) can be overcome.

(iii) The still growing problems of large settlements can cause a decline of attractiveness of larger cities. Connect this with the increasing attractiveness of rural areas to get a new picture of urban-rural interactions.

(iv) Material transport, travel, and cars become less important. Is the present urban sprawl the precursor of new sustainable distributed patterns of human settlements?

The transition is driven by several factors (Figure 16.23), all of which grow very rapidly, interconnected with one another. The "information potential" is defined as the capability to use more and better information more effectively to do things better, differently, or to become able to do new things. Accordingly, an Information Society is technically based on the information potential. The information potential is composed of and grows with: (i) the number and speed (or throughput) of computers; (ii) the size and capacity of networks ("information infrastructure"); (iii) the amount of available digital information; and (iv) the number of people capable of and equipped to use these new possibilities.

In Figure 16.23 this is indicated by the boxes "Special Interest Groups" (SIGs) and "Advanced Development Centres" (ADC). People meet in SIGs to help one another to understand and use the information potential. SIGs are very helpful to support a population in the transition to a new mode of doing things. ADCs are groups equipped with very good software that help to solve very tough problems, that filter and refine informations, and that help other people to upgrade their capabilities in using the information potential. The term "number of information types" refers to the number of different types of information that can be digitally processed, such as numbers, texts, databases, audio, images, video, sounds, virtual reality, and imaging (visualization, simulation). All of these factors increase very rapidly.

16.9.4 Economy

The economy is in the middle of a shift away from being information poor toward being information rich. Economic effects have been mostly positive; the number of jobs (OECD, 1994) has grown considerably during the last thirty years, e.g., by 80 percent in the United States and in Oceania, or by 50 percent in Japan. Western Europe is lagging behind with just 8 percent growth in jobs, but it may now remove its structural barriers toward innovations.

Some of the most important factors behind this transition towards an Information Society are put together in Figure 16.24. In this figure the importance of an attractive outdoor environment was outlined above ("amplification through resonance with environment"). Major factors are the social and the psychological base. The readiness in the United States to accept innovations and to support entrepreneurs are major growth factors for the new economy (Spectrum Strategy Consultants, 1996).

The main focus in the infrastructural base

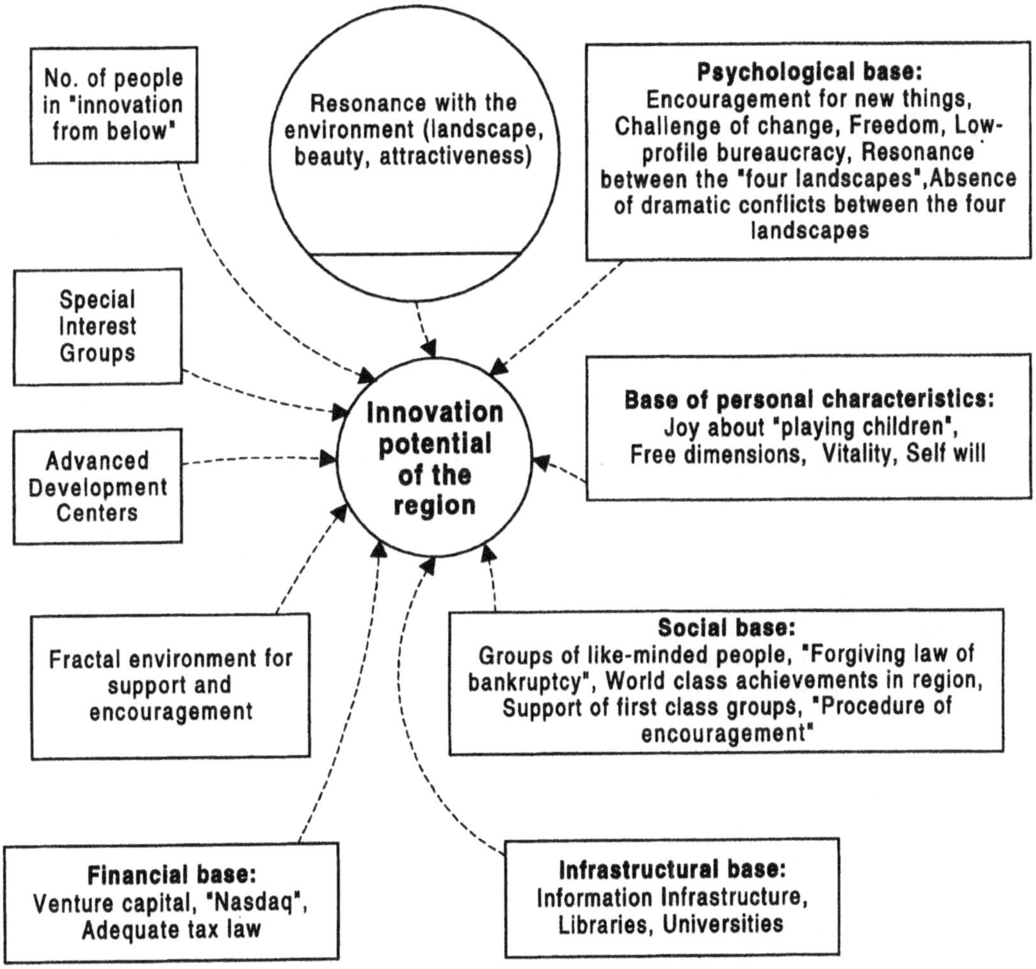

FIGURE 16.24. Main components to foster growth of an information based economy.

shifts away from traffic networks toward networks for effective exchange of information. Support infrastructure becomes even more important than in the past such as (electronic) libraries or (digital) universities. As regards the financial base, lack of a stock market for small high-tech companies (like the Nasdaq in the United States) is cited as the only factor that still prevents Bangalore in India (the Indian centre of software development) from "exploding" like the famous Silicon Valley in the United States. Europe has just founded its own "Easdaq."

16.10 Situation of Developing Countries with Respect to the Information Society

What is the situation in the developing countries? Almost all are way behind as regards their communication infrastructure and the number of phones per inhabitant. The number of Internet servers is low even in all of the Asian Tiger countries compared to Europe. The Internet is a worldwide network of about 90 million people who are connected through about 10 million servers (computers of all sizes from very big mainframes down to big PCs) so that they can exchange all types of information effortlessly and almost instantaneously on a worldwide scale. The Internet is the fastest growing technology in human history with growth rates between 500 and 1200 percent within three years in many countries

But the analysis throughout this chapter has confirmed the importance of people. This importance was confirmed by Nicholas Negroponte (1995), head of the Multimedia Lab at the Massachussetts Institute of Technology, who asked: "Which country is in a better position for the Information Age: Germany, where more than 50 percent of the population are above 45 years of age, or Mexico, where more than 50 percent are below 20 years?" Young people are much more open to learn and understand the new possibilities than older people. The fraction of younger people is much higher in all developing countries than in developed ones. Hence, awareness for the present changes may be the most important factor. Technology, even lack of infrastructure, is an obstacle that can be overcome quite rapidly. For example, about five global satellite networks are now under construction, which will mean ample availability of telecommunication virtually everywhere. Indeed, information processing and programming is rapidly moving into developing countries, for example, to India (Bangalore was cited above) or to Vietnam (which produces part of the programming that was contracted by India from worldwide clients).

Thus, to an extent, developing countries can leapfrog developed ones. Increasingly, the "First World" and the "Third World" coexist with each other in almost all countries. Thus, the transition into the Information Society occurs simultaneously in all countries. Only very few countries still do not have an Internet access (see the map by the "Internet Society" on the global availability of the Internet).

Therefore, it is becoming ever more important that all countries, developing as well as developed, preserve and cultivate their most important assets: people and the environment. Information-based economies are much more environmentally benign than smokestack economies. Outdoor environment is a crucial factor for global environmental survival, but also for the growth of new economies. Hence, the still ongoing devastation of cultural landscapes is sabotaging the build-up of new economies and the preservation of factors crucial for global competitiveness. New coalitions between people, economy, and the environment are emerging that need to be understood, nourished, and built. The process is new, complex, and perplexing. System models are an indispensable tool to elaborate favourable options and policies.

References

Aracil, J. *Introducción a la Dinámica de Sistemas.* Madrid: Alianza Universidad Textos; 1986.

Bertalanffy, L. von. *General System Theory.* Harmondsworth: Penguin; 1971.

Boyden, S. *Hong Kong: A Case Study on a City and Its Inhabitants.* Paris: UNESCO; 1976.

Ebeling, W.; Peschel, M., eds. *Lotka–Volterra Approach to Cooperation and Competition in Dynamic Systems.* Berlin: Akademie Verlag; 1985.

Forrester, J. *Urban Dynamics.* Cambridge, MA: MIT Press; 1969.

Forrester, J. *World Dynamics.* Cambridge, MA: MIT Press; 1971.

Forrester, J. *Industrial Dynamics.* Portland, OR.: Productivity Press; 1961.

Grossmann, W.D.; Maierhofer, E.; Kasperidus, H. *Dynamische Modelle zur Okonomisch–Okologischen Zukunftsgestaltung (Dynamic Models for Shaping the Socioeconomic–Ecological Future).* Berlin: Deutsches Umweltbundesamt; 1996.

Hall, P.G., and Preston. *The Carrier Wave.* London: Unwin; 1988.

Holling, C.S. *Adaptive Environmental Assessment.* New York: John Wiley and Sons; 1978.

Jeffers, J.N.R. *Modelling.* London: Chapman and Hall; 1982.

Kitching, R.L. *Systems Ecology: An Introduction to Ecological Modelling.* New York: University of Queensland Press; 1985.

Leffelaar, P.A., ed. *On Systems Analysis and Simulation of Ecological Processes: With Examples in CSMP and Fortran.* Dordrecht: Kluwer Academic Publishers; 1993.

Lugo, A.; Patiño, M.P.; Mayoral, O.; Pizarro, D.; et al. *Models of Energy Flow for Rural Planning in the Ebro River Watershed.* Options Méditerran-éennes, Serie Etudes 1983-II. Zaragoza: CIHEAM; 1983.

Meadows, D. *Lecture at the Potsdam Institute for Climate Research (PIK).* Potsdam; 1984.

Meadows, D.; Meadows, D.; Randers, J. *Más Allá de Los Límites del Crecimiento.* Madrid: Ediciones El Pais; 1992.

Multhaup, T.; Grossmann, W.D. *Fünf Fallbeispiele Informationsbasierter Wirtschaft: Ihre Ansprüche an Mitarbeiter und Folgerungen für Umweltqualität und Regionale Attraktivität.* Bericht, Leipzig: UFZ-Bericht, UFZ-Leipzig; 1997.

Negroponte, N. Balance of trade in ideas. *Wired* 4:188; 1995.

Odum, H.T. An energy circuit language for ecological and social systems: Its physical basis. Patten, B., ed. *Systems Analysis and Simulation in Ecology;* Vol. 2. New York: Academic Press; 1972: 140–212.

Odum, H.T. *Systems Ecology: An Introduction.* New York: John Wiley and Sons; 1983.

Odum, H.T. Lecture at the International Institute for Applied Systems Analysis. Luxembourg; 1984.

OECD. *OECD Jobs Study.* Paris: OECD; 1994.

O'Neil, R.; De Angelis, D.; Waide, J.; Allen, T.A *Hierarchical Concept of Ecosystems.* Princeton, NJ: Princeton University Press; 1986.

Patten, B., ed. *System Analysis and Simulation.* New York: Academic Press; 1972.

Senge, P. *The Fifth Discipline.* London: Doubleday; 1990.

Simon, H. *The Science of the Artificial.* Cambridge, MA: MIT Press; 1969.

Spain, J.D. *BASIC: Microcomputer Models in Biology.* Reading, MA: Addison-Wesley; 1982.

Spectrum Strategy Consultants. *The Development of the Information Society: An International Analysis* (to order contact HMSO). (Can also be reached on the Internet, at the EU server); 1996.

Swartzman, G.L.; Kaluzny, S.P. *Ecological Simulation Primer.* New York: Macmillan; 1987.

UNESCO. *Indicators on the Quality of Life.* Paris: UNESCO; 1978.

Weinberg, G. *An Introduction to General Systems Thinking.* New York: John Wiley and Sons; 1975.

17

Rural Planning: A Case Study

Juan Bellot, Frank B. Golley, and M. Teresa Aguinaco

17.1 Introduction

Rural planning is an instrument through which human activities are linked to the rural area. It is a projection in space and time of the social, cultural, environmental, and economic policies of a society. Social and economic development of the population in the rural areas is continually taking place, which leads to changes in use of the soil and development of new activities. These changes create a conflict for the managers of the territory who have to choose between the "natural uses" and the perceived objectives or needs of landowners, farmers, town planners, environmentalists, and so on (Haber, 1990). For it to be useful, all planning must envisage conflicts of use and act as a regulating mechanism of the system, capable of anticipating changes (Margalef, 1974). Therefore, for planning we have to analyse the problems with an integrated approach, coordinating the activities involved in all the sectors: agriculture, forestry, hydrology, conservation, industry, services, and so on (Bertrand, 1978; Grossmann, 1991).

This concluding chapter presents a case study, which applies the methods and principles developed earlier in the book to a concrete example. This approach presents basic system planning as a rational, continuing process of decision making, so that we can achieve a harmonious distribution of human activities, regulate the uses of the soil, develop sustainability within the capacity of the region, and thus make conservation compatible with development.

Given the limited space in this chapter, it is not possible to develop all aspects of the methods with sufficient amplitude to cover all of the possibilities that might be present. Our objective is to apply each methodological step to at least one resource or use and to evaluate some of the possible alternatives. Nevertheless, we have tried to explain the method of data analysis and to obtain conclusions so that the reader has a guide and example of the application of the method and principles that can be used for other activities and resources. In this way we integrate the foregoing chapters in the same way that a field project integrates theoretical lectures.

17.2 Description of the Case Study Region

Though many practical studies have been carried out since the first Rural Planning courses of the IAMZ (Junquera et al., 1977; Aït-Bachir et al., 1997) not all have been equally successfully; nor have the results illustrated all the possibilities offered by the work method.

It may have been possible to approach this

practical case in a very different way, showing the application of the work methods independently in stages, applying them at each phase to different areas as an example. But, although this approach would have been useful, we considered it better to choose a target area and make an integrated application of all the planning phases, including the limitations derived from its characteristics, resources, and problems.

The area chosen for this case study belongs to the comarca of Moncayo in Aragon (Spain). It is situated in the extreme northwest of the province of Zaragoza, between 41°44' and 42° north latitude and 1°14' and 2°02' west longitude. It is composed of twelve municipalities and covers 423.6 km^2 (García Manrique, 1960; Pellicer, 1986). It has good communications with the principle urban centres in Zaragoza, Navarra, and Rioja. Although 15 percent of its surface area is a protected natural park (the Moncayo), the area is characterized by a predominantly rural economy, small agricultural holdings, and an ageing population, showing a clear trend toward depopulation (Higueras, 1984). This situation is common in the majority of the rural areas of Spain.

The general objective of the case study is to proportion the distribution of the uses and activities in the area so that the natural resources could persist and thus maintain the human population, at a higher quality of life, which would imply an increase in income and an improvement in the people's well-being.

17.3. General Methodology

The method presented here is based on a systems approach to the study area (Grossmann, 1983), that is to say, on a hierarchical concept of the zone, integrating the subsystems of the system of interest (physical, socioeconomic, legal, and institutional), forming a part of a more amplified system (Golley and Bellot, 1991). It is necessary to characterize and understand the dynamics, the flows, and their internal and external relationships. The methodological scheme used in this case study is illustrated in Figure 17.1, which is based on the framework of concepts of Bertrand (1968), González Bernáldez et al. (1973), Ramos (1979), Gómez Orea et al. (1982), Sachs (1982), Gómez Orea (1985), and Haber (1989). It shows a series of steps that develop sequentially for clarity of presentation, recognizing that in a planning activity they would not necessarily be sequential but iterative, because it is a cyclic process of continual feedback that leads us to a decision. Each one of the phases has concrete objectives, but all together, the aim is to gather the information, plot it spatially, and submit it to statistical processing to be used in the overall soil use allocation in order to resolve problems of the zone.

17.3.1 Inventory

The first step is the inventory. The inventory consists of the collection and study of the information necessary to identify and describe the systems and subsystems, allowing a selection of the elements and variables to understand the system structure and function within the overall objectives of the study. Although a systems approach assumes an integration of data and method, in the inventory phases we will examine separately the subsystems of the system of interest and lastly analyse the interactions between the distinct subsystems.

17.3.1.1 Physical Subsystem

The physical subsystem is characterized by the following, most significant, elements:

(i) climate, focusing on those climatic conditions that limit and control the human activity;

(ii) lithology, in order to know the nature and properties of the geology as well as the permeability and resistance of their materials;

(iii) soils, especially to understand their aptitude for various uses in agriculture, animal husbandry, forestry, and industry;

(iv) vegetation, to identify the current and

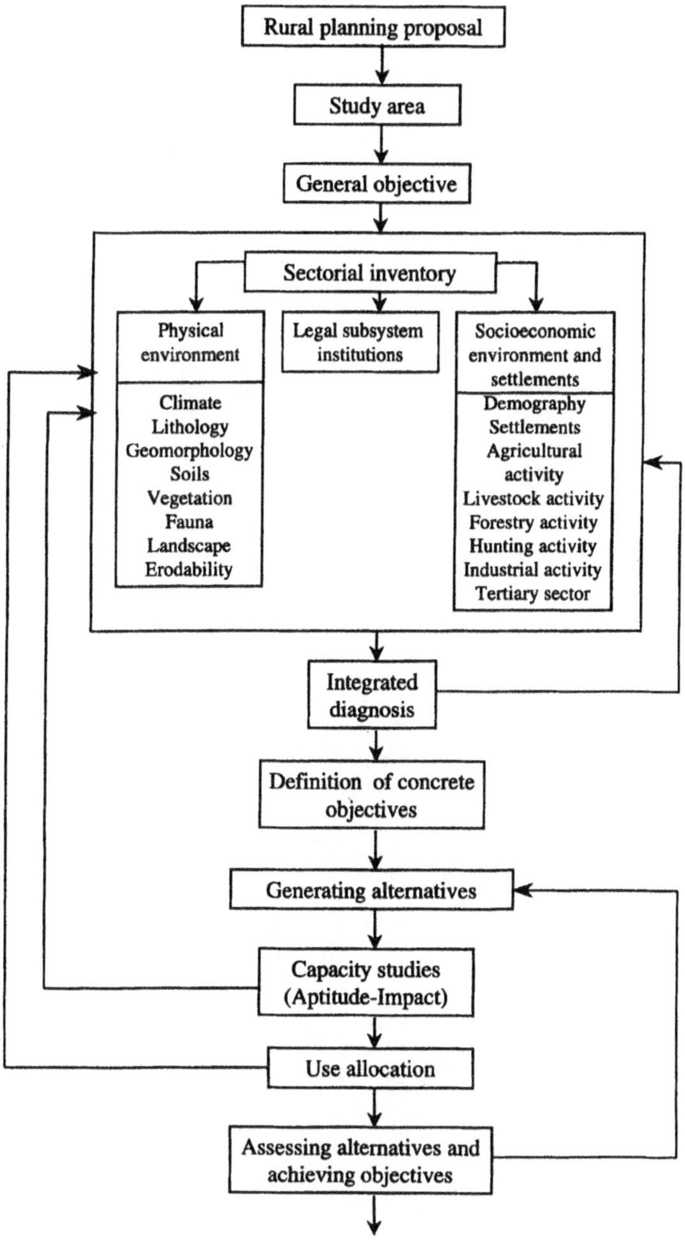

FIGURE 17.1. Methodological diagram proposed for the development of the land planning project, containing the main stages of the work and the recommended sequence.

potential plant formations that might support rational development;

(v) hydrology, to permit evaluating the hydric resources available in the zone;

(vi) topography, so that we can understand the relief in the zone and how it will influence activities and developments; and

(vii) the fauna in order to locate areas requiring protection or management for fishing and hunting.

These elements are known as direct variables and they may be obtained by observations in the area or from appropriate maps. In

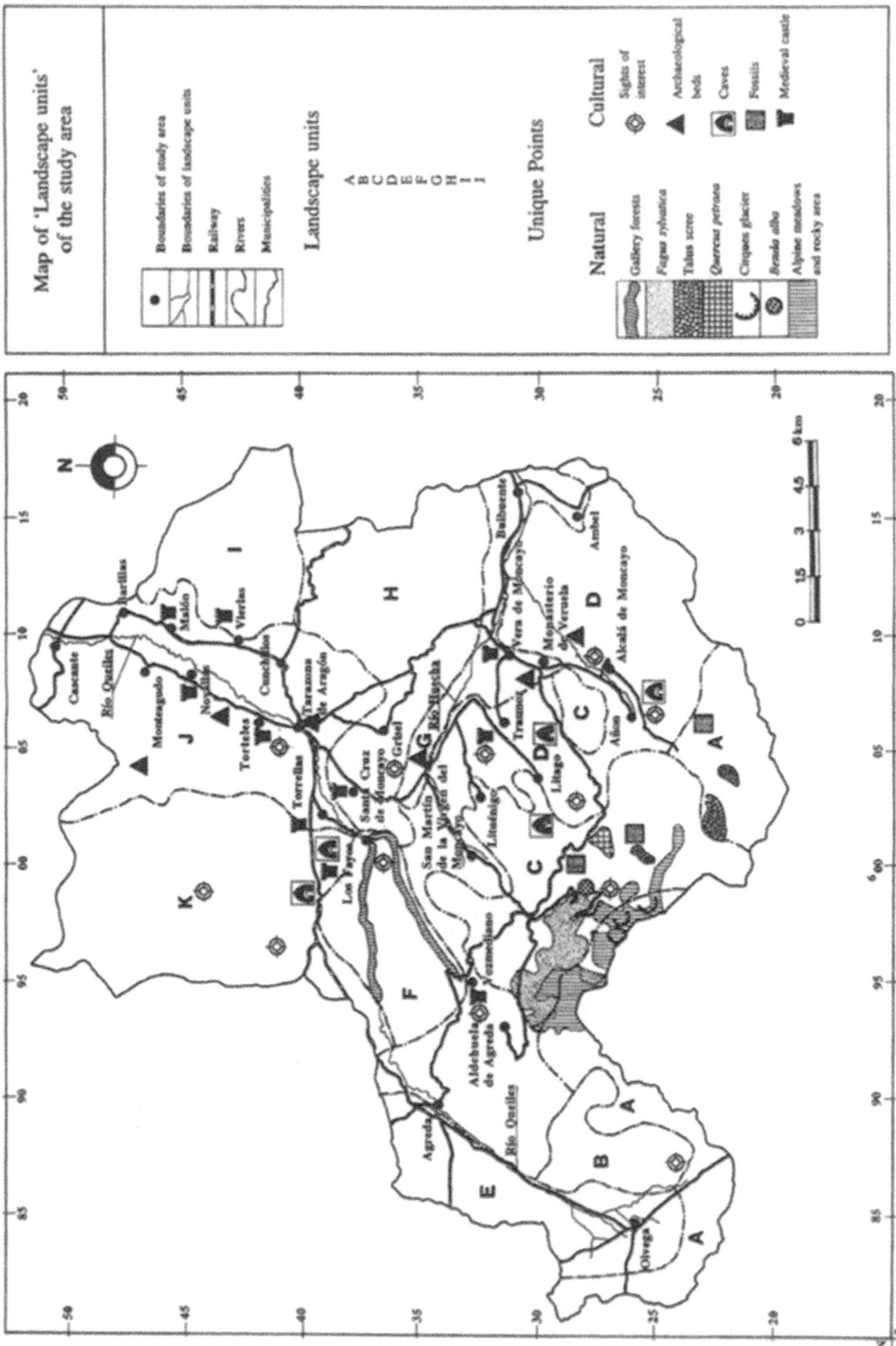

FIGURE 17.2. Example of the thematic cartography drawn up during the inventory phase. Map of landscape units, unique points, water courses, roadways, and settlements in the study area.

unmapped zones it is possible to use methods discussed in Chapters 3–9 of this book, or to consult with experts and use methodological handbooks (Ramos, 1979; MOPT, 1993). In fact, the ideas expressed in Chapters 3–9 are necessary for interpretation and evaluation when allocating soil uses. Other elements of the physical environment can be deduced from those listed above. These are called "deduced variables." Their practical value is higher due to their integrative character and therefore are more appropriate for making decisions. In this group we might include geomorphological factors, land capability classes, erodibility of soils, risk of inundation, landscape, and so on (Strahler, 1974; Soil Survey Staff, 1975; Forman and Godron, 1986; Bryan, 1990).

Depending on the final objective, the information must be obtained at the most adequate scale. Logically, the larger scale is better, but it is not always possible to use a large scale for all variables (1:5000 or 1:10,000). The combination of the variables requires that the same scale be used for all of the cartography. In general, the most useful scale is 1:50,000, used for showing the topography in most countries, and to a lesser extent, 1:100,000 or 1:200,000 (see Chapter 3). When the information is not complete or is not updated it may be possible to combine the cartography with aerial photography. These photographs may be obtained easily in most countries as they are the basis for commercial and cadastral cartography. On the other hand, it may be possible to request flights if the data are old or if more focused information is required. Using the most modern techniques of remote sensing (Belward and Valenzuela, 1991) and synthetic aperture radar (SAR) imagery (Madsen et al., 1993) it is possible to ground our database of the physical subsystem at an adequate scale to realize the objectives of planning. These satellite images are expensive and processing is difficult, but clearly this is the methodology of the future, when we will have updated images on different scales from all parts of the earth and are able to exchange them via the Internet (Chapter 2).

In routine work, in this inventory phase, we elaborate all of the necessary thematic cartography relevant to the zone that is required for our decision making. Figure 17.2 is an example of the thematic maps that describe the zone of the case study, utilizing the available published information, the visits to the terrain, and the techniques described in Chapters 3–9. This information is later digitized and converted into raster or vectorial format for the Geographic Information System (GIS) program in the following way:

(i) A squared network is superimposed on each thematic map of the study area. Each cell represents a surface of variable size as a function of the capacity of the GIS program, preferably small in size because it allows greater precision at the time of decision making. ARC-INFO, SURFER, IDRISI, and so on, are available programs in the market that can be applied in this use. In our case, we use SIG-MOD program (Alloza, 1995), a raster type of information system, which was developed for the IAMZ rural planning course and offers many applications and teaching possibilities. Its limitation is the maximum number of cells that may be represented (6400). If the study area is large, each cell necessarily represents many hectares.

(ii) All of the information contained on a thematic map is coded, assigning values for each type of information (i.e., levels of variation of an element) and representing the corresponding thematic legend. The assignment of value is based on different criteria, especially because each cell represents considerable spatial heterogeneity. The most common approach is to assign a value to the cells representing the most abundant type within them, or the type existing in a particular and predetermined vertex (corner cell), or a particular type when it is present in the cell and is a unique variable or has a linear distribution.

Figure 17.3 shows the vegetation map of the study area obtained following the steps described.

17.3.1.2 The Socioeconomic Subsystem: Society, Economy, and Settlements

The inventory of the socioeconomic subsystem involves the characterization and description of the demography, social organization,

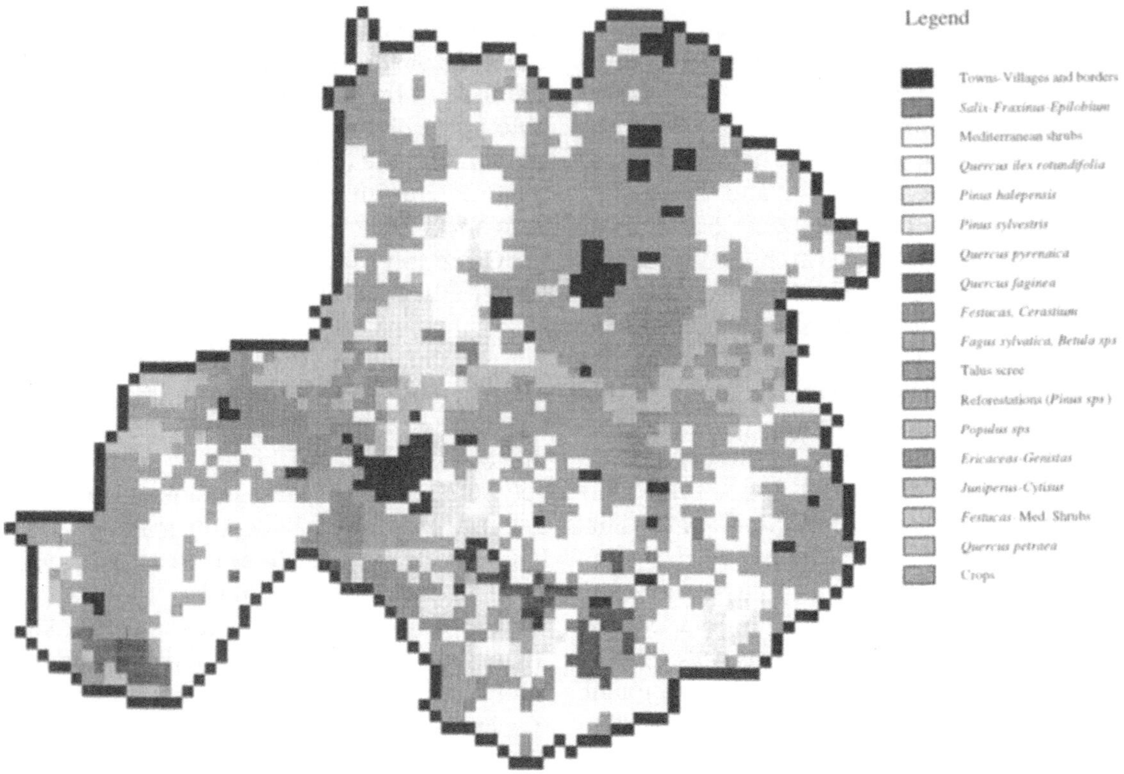

FIGURE 17.3. Map of land occupation by types of vegetation, generated using SIGMOD program following the coding of the vegetation map. Basic map in raster form to be used by the GIS program.

and economic activities of the region. It may also include a description of the established settlement systems, their structure and spatial organization, hierarchies, and so on, including studies of the comarcal and regional connections and the external areas of influence. We utilize information from a variety of sources for the socioeconomic subsystem: with geographical references (e.g., analysis of aerial photographs, field records, maps of land use and developments of infrastructures and official data) and those without a spatial reference (e.g., census and agricultural statistics, bank, industry and commercial data, economic indicators, etc.). The most relevant data in this phase could be elaborated and interpreted following Chapters 10–14 and general books (Chorley and Hagget, 1971; Pressat, 1972; Clusa et al., 1979).

The most relevant demographic aspects are those that refer to the quantification and spatial distribution of the population: (i) internal structure (e.g., sex ratio, rates of males/females, population pyramid, dependency ratio, etc.); (ii) its dynamic analysis (e.g., birth rate, mortality, ageing); (iii) its spatial mobility (e.g., emigration, immigration); and (iv) the demography trends.

The dynamic population analysis should also, wherever possible, include the daily commuter rate, including origin and destination. These daily journeys normally go beyond the comarcal borders and motivate some of the most frequent part-time activities. Knowledge of these rates of activity is important in order to evaluate the variable "population." Ergograms and an analysis of labour needs during the different months of the year will enable supply to be compared with demand and furthermore to consider introducing complementary activities to their annual rhythm. All of these parameters are necessary to understand the situation of the population, the reasons for its state, and the perspectives for its future. It is also important in the subsystem to understand the social organization of the population and the dominant institutions or groups. Close contact with these agents and local leaders facilitates the transfer of results and may add to the social implications of the planning project. Chapters 10 and 12 are dedicated, to a great extent, to the analysis of such systems and the techniques of determining the problems and aspiration of the population through its leaders and representatives.

It is important to consider the expectations of the local people and their capacity of being involved in planning projects and develop the alternatives that could be proposed. Public participation would therefore be a convenient tool to use, during both the inventory phase as well as further phases, including diagnosis of the area and proposal of alternatives (Goodey, 1976).

The population is the basis of its economic activity, therefore it is necessary to know the economically active population, the employment and unemployment rates, and other parameters in order to understand the potential of the region for development. The detailed analysis of economic activities of a zone is usually based on three classical sectors: (i) the primary, corresponding to the work of agriculture, pasture, and forestry; (ii) the secondary, dedicated to industrial work; and (iii) the tertiary, which deals with services.

In the primary sector we are interested in knowing the structure of the properties, types of crops, livestock, types of vegetation, actual production levels, circles of marketing and forms of management. In the secondary sector we are interested in knowing the existing industries, the structure of the enterprises, and the evolution of the market at the local, regional, national, and international levels. The services of the this sector are of special importance and are the most beneficial in terms of improving the potential of the territory. When more development occurs in the zone the largest percentage of the gross interior product goes to the service sector. Figures 17.4–17.6 and Tables 17.1–17.4 show some of these data that were obtained directly from the sources or were elaborated.

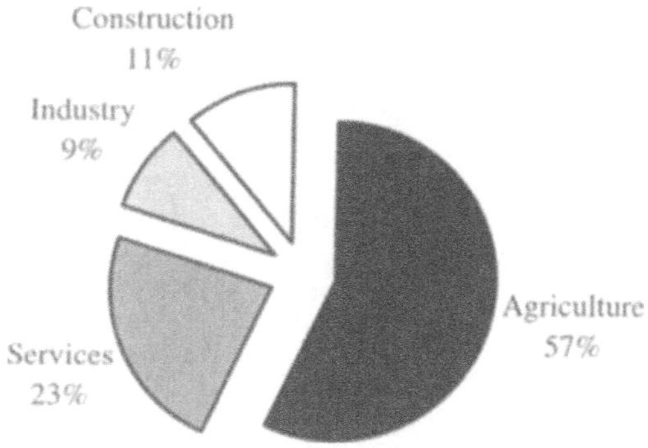

FIGURE 17.4. Percentage distribution of the population employed in the area between the main sectors of production (data from IAE, 1994).

17.3.1.3 The Legal and Institutional Subsystem

The third subsystem involves legal and institutional factors which make all plans a rational instrument of public action and orientation of private initiative. For this reason it is necessary to understand the legal and institutional context that organizes and administers the society. We must know the general laws and dispositions applied to a concrete area, as well as the competent public institutions (Valcárcel-Resalt and Troitiño, 1992).

For the alternatives of the plan to be convincing and actually be implemented, it is necessary to demonstrate that they fit into the legal and administrative structure, are viable technically and financially, fit into the social and economic context, as well as that they have support from private initiatives. The lack of consideration of many of the characteristics of this subsystem is the cause of difficulties of implementation of many territorial plans, because they raise legal and administrative difficulties and interfere with political realities. The inventory of this subsystem should not be exhaustive, nor should it be made before knowing the alternatives because that would imply that all the revisions of legal and administrative rules must be studied, which at this point is neither relevant nor useful. We are

FIGURE 17.5. Percentage distribution of the income in the study area, according to (left) the original sector (primary, secondary, and tertiary) and (right) within the primary sector (agriculture, livestock farming, and forestry) (data from IAE, 1995).

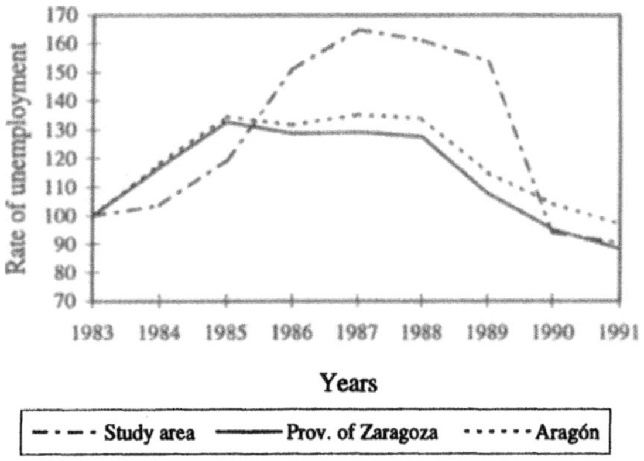

FIGURE 17.6. Evolution through time of the unemployment rate in the study zone, in the province of Zaragoza and in the autonomous community of Aragon, between 1981 and 1991 (unpublished data from the municipal town halls).

TABLE 17.1. Data of surface population, density, vegetative growth, fecundity index and migration rate for each municipality of the study area, Zaragoza province, Aragón region and Spain (data from IAE, 1994)

Muncipality	Surface (km²)	*De facto* inhabitants	*De jure* inhabitants	Density (inhabitants /km²)	Annual vegetative growth	Fecundity Index	Migration rate (past 10 years)
Alcalá	13.7	166	166	12.12	−10	–	0.00
Añón	64.1	345	345	5.38	−17	0.39	11.43
Los Fayos	3.9	164	174	44.61	−8	0.18	9.52
Grisel	14.5	46	59	4.07	−4	0.33	86.21
Litago	15.3	174	189	12.35	−9	0.42	13.27
Lituéñigo	11.2	124	125	11.16	−7	0.19	29.94
San Martín	5.1	285	314	61.57	−9	0.43	6.08
Santa Cruz	3.9	96	102	26.15	−8	0.37	17.86
Tarazona	243.8	10,638	10,557	43.30	81	0.44	82.34
Torrellas	2.6	307	347	133.46	−12	0.42	28.44
Trasmoz	18.1	63	80	4.42	−4	0.00	9.52
Vera	27.4	479	498	18.17	−4	0.38	49.26
Zaragoza province	17,243.5	861,329	837,327	48.56	–	–	–
Aragón region	47,645.8	1,221,546	1,188,817	24.95	–	–	–
Spain	504,782.0	38,872,268	38,872,268	77.01	–	–	–
Study area	423.6	12,887	12,956	30.59	–	–	–

TABLE 17.2. Degree of fragmentation and distribution of plots of agricultural and forest lands in the municipalities of the region (data from INE, 1991)

Municipality	Number of plots	Irrigated lands (ha)	Rainfed lands (ha)	Rainfed + irrigated lands (ha)	Average plot size (ha)	Abandoned lands (ha)	Total forests per municipality (ha)	Timber forests (ha)	Ligneous forests (ha)	Open forests (ha)
Alcalá	1,233	1,085	115	1,200	0.97	154.4	50	50	0	0
Añón	1,227	487	267	754	0.61	234.5	2,600	1,500	900	200
Los Fayos	407	14	187	201	0.49	83.8	168	50	118	0
Grisel	953	780	363	1,143	1.20	—	—	—	—	—
Litago	1,300	327	57	384	0.29	298.0	594	200	384	10
Lituénigo	1,273	335	85	420	0.33	84.4	474	70	400	4
San Martín	400	118	100	218	0.55	132.5	105	102	3	0
Santa Cruz	390	40	175	215	0.55	—	—	—	—	
Tarazona	9,633	9,203	3,879	13,082	1.36	521.9	2,445	1,852	344	249
Torrellas	157	39	147	186	1.18	—	—	—	—	—
Trasmoz	1,165	640	277	917	0.79	303.7	464	0	460	4
Vera	2,745	1,311	470	1,781	0.65	18	390	0	390	0
Study area	20,883	14,379	6,122	20,501	0.98	1,831.2	7,290	3,824	2,999	467

TABLE 17.3. Data of population and income disposable per municipality and per capita

Muncipality	Inhabitants	Income disposable (ptas)	Income capita (ptas)	Relative order
Alcalá	166	59,082,443	355,918	5
Añón	345	103,876,119	301,090	3
Los Fayos	174	72,728,932	417,982	7
Grisel	59	16,556,080	280,612	2
Litago	189	64,368,019	340,572	4
Lituéñigo	125	53,758,527	430,068	10
San Martín	314	82,692,549	263,352	1
Santa Cruz	102	42,741,612	419,035	8
Tarazona	10,557	4,209,476,549	398,738	6
Torrellas	347	146,682,962	422,717	9
Trasmoz	80	35,115,213	438,940	11
Vera	498	240,436,759	482,805	12
Study area	12,956	5,127,515,764	395,764	

TABLE 17.4. Calculation of the Gini Index $[GI = (P_i - Q_i)/P_i]$ for the area. The result obtained $(GI = 0.024)$ indicates a high degree of equal distribution of income between municipalities. Q_i = cumulative percentage of income ranked in ascending order. P_i = cumulative percentage of population for each Q_i defined

Municipality	Population (%)	Income (%)	P_i	Q_i	$P_i - Q_i$
Alcalá	1.28	1.15	1.28	1.15	0.13
Añón	2.66	2.03	3.94	3.18	0.76
Los Fayos	1.34	1.42	5.28	4.60	0.68
Grisel	0.46	0.32	5.74	4.92	0.82
Litago	1.46	1.26	7.20	6.18	1.02
Lituénigo	0.96	1.05	8.16	7.23	0.93
San Martín	2.42	1.61	10.58	8.84	1.74
Santa Cruz	0.79	0.83	11.37	9.67	1.70
Tarazona	81.49	82.10	92.86	91.77	1.09
Torrellas	2.68	2.86	95.54	94.63	0.91
Trasmoz	0.62	0.68	96.16	95.31	0.85
Vera	3.84	4.69	100.00	100.00	0.00
Study area	100.00	100.00	438.11	427.48	10.63

looking for a general approach at this stage that can be more precise when we get to the point of evaluating alternatives.

17.4 Diagnostics

The diagnostic analysis is an interpretation of the current situation in the area, in view of its history and its potential for future change. It has to define clearly the problems of the zone in order to deduce the objectives, to which rural planning will give solutions (Gómez Orea, 1994). The diagnosis may be integrated or sectorial. In our case we will take a sectorial approach because it allows a better description of the problems and possible causes. The inventory of the physical and socioeconomic systems of the *comarca* of Moncayo, selected as a case study, permit us to interpret the current situation in the zone and identify the major problems (Atucha et al., 1993), which can be synthesized in the points mentioned below.

17.4.1 Diagnostics of the Physical Environment

(i) A limitation of the hydric resources in a large part of the area, in spite of the total yearly precipitation being sufficient to cover the demand. The problem is due principally to deficiencies in irrigation network systems and to the high permeability of the limestone substrate.

(ii) Low aptitude of the soils for agricultural production and a lesser versatility for cultivation, characteristic of an agriculture with relatively low productivity. The causes are the semi-arid climate and the relief.

(iii) Increase in the surface of abandoned land with scarce vegetation and steep slopes. The lack of tillage may affect erosion positively or negatively depending on the zone.

(iv) The natural vegetation is clearly regressive, due to an increase in reforestation pines and degraded shrubland and a reduction in beech, oak, and evergreen oak forests (i.e., autochthonous vegetation).

(v) Poor incomes from forests, despite 64 percent of the area being mountains. The major cause may be the low profit of its resources, because the matorral and natural forests contribute only with the 17 percent of the gains from hunting and firewood. Other uses, such as wood production and grazing, are only practised in small areas.

(vi) Low profit of the hunting resources of the zone in spite of the large areas of suitable vegetation.

(vii) A necessity to increase the present area of the existing natural park, and to create a buffer area to guarantee conservation and increase its protection.

(viii) Elevated risk of erosion in the zone of greatest relief and low vegetation cover, or in areas that have been subject to inadequate use.

17.4.2 Diagnosis of the Social and Economic Subsystem

(i) The ageing of the population, due to negative demographic growth and high emigration of the active population.

(ii) Scarcity of jobs and ageing of the active population, meaning high rates of dependency.

(iii) Very low average levels of income per capita.

(iv) Low density in the majority of the municipalities, with a concentration of population and services in only one city (Tarazona).

(v) Low-yielding agricultural production, which mainly serves as a means for subsistence rather than economic activity. This is due, not only to harsh environmental conditions, but to a series of economic and political factors that cause an increase in the abandonment and excessive division of land as well as dispersion of properties.

(vi) Insufficient use made of range resources since livestock farming could be extended to other zones.

(vii) Weak importance of the industrial and service sectors in much of the territory, because they are concentrated primarily in Tarazona (47 percent of the population and 76 percent of the municipal value).

(viii) Deficient technology in the food

industry, with only small enterprises of wine and olive oil.

(ix) Weak tourist services, which could be extended and increased to other points of interest.

(x) Deficiencies in the homogeneous distribution of the sanitary infrastructure, schools, houses, and so on, and other deficiencies related to the lifestyle (e.g., water, light, telephone) throughout the zone.

17.4.3 Strengths, Weaknesses, Opportunities, and Threats (SWOT) Matrix

Another way to carry out the diagnostic analysis is the SWOT matrix, an integrated methodology (Gómez Orea, 1994). In this case a cross-tab of double input is used to obtain the strategies to be applied in the area. The inputs in rows are occupied by strengths and weaknesses, and the inputs in columns correspond to the opportunities and threats. The intersection points permit to define in a practical way the strategies to solve the problems.

17.5 Analysis of the Systems

The inventory provides a descriptive and static vision of the zone. The next step is to utilize during the diagnostic analysis the systems concept described in Chapters 1 and 16. Systems analysis is a technique used to obtain a holistic and dynamic view, showing the interactions between the subsystems, emphasizing the interdependent relationships between the elements of the physical environment, the social structure, and human activities (Senge, 1990). When the relationships between such activities are known, an accurate and integrated diagnosis can be made of the problems in the zone, for an adequate assignation of land use (Haber et al., 1984; Grossmann and Eberhardt, 1991).

The techniques of analysis and interpretation of this system developed in Chapter 16 are especially useful at this stage in the process for elaborating a diagram of the system flows (Figure 17.7). This simplified diagram of the *comarca* of Moncayo reflects the most relevant components in the system (e.g., the physical environment, agricultural, forest and pasture soil, population, livestock, fauna, industry and services, tourism, etc.) and the flows of direct influence between them, with auxiliary variables (e.g., level of income, rates of population growth and ageing, division of fields, land tenure, etc.). Even though a flow diagram of this type seems to be complex in its design, it is possible to identify many of the key variables in the system, the dominant influences, and the main direct and indirect flows between the variables.

When proposing actions to change the dynamics of the study area, these diagrams permit us to see the indirect consequences that some proposals may have. Although they do not directly affect a particular element, they may have an influence on it. On occasions, this helps to avoid unwanted effects because a positive effect on one component could have a negative effect on others.

17.6 Outlining of the Specific Objectives

In light of the problems derived from the diagnosis, and considering the flows and interactions detected in the systems analysis (Figure 17.7), we define the objectives and begin the planning project. Our objective is to better the quality of life of the inhabitants of the comarca, orienting the development in an integrated way with the goals of conservation and rehabilitation of the natural values.

From the diagnosis we have deduced two fundamental problems of the zone. The first set relates to the physical environment and locates unfavourable physical conditions and poor management of the resources. The second involves the social and economic subsystem and identifies the presence of a reduced, ageing, and declining population, with low returns from productive activities. The consequence is

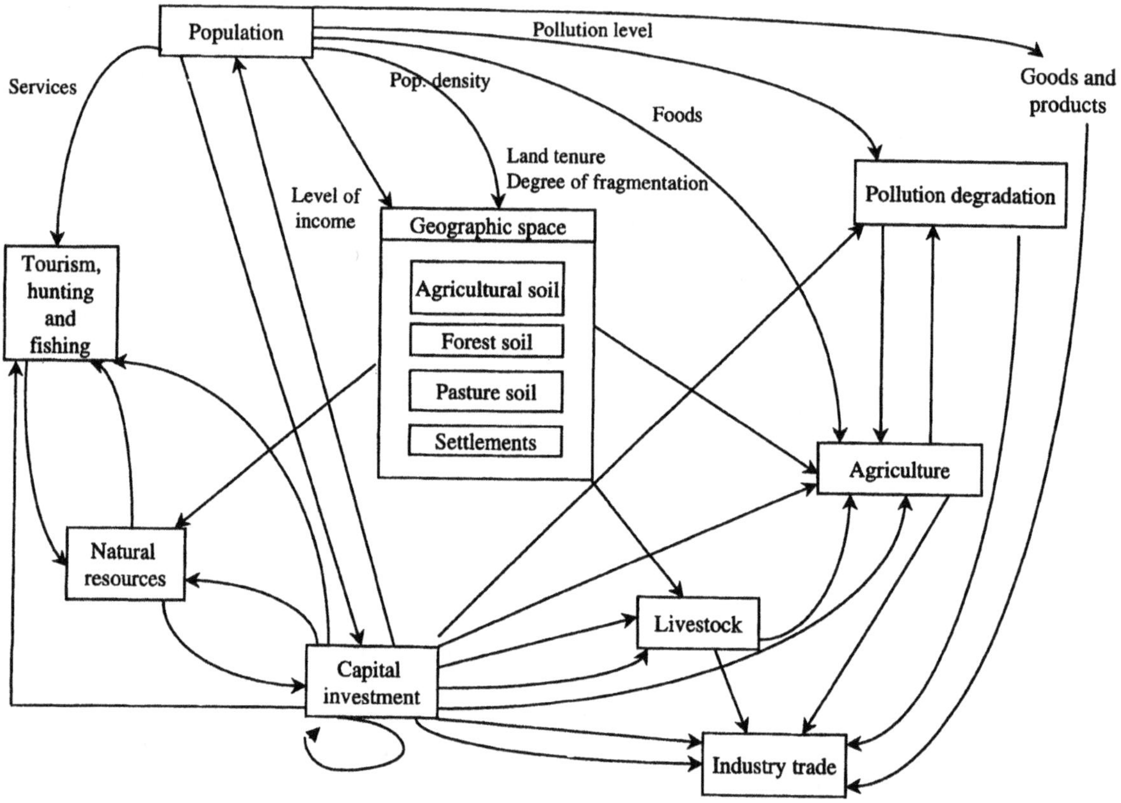

FIGURE 17.7. Simplified flow diagram of the study area system, showing the main elements and links.

a disorganization in the social and the natural environment. As a solution, there are two general subobjectives: (i) guarantee adequate management that permits the natural resources to persist; and (ii) maintain the population providing a higher level and quality of life, which implies an increase in income and an improvement in the quality of the environment.

These general subobjectives can be broken down into four more specific objectives:

(i) *Increase in production*. Establish lines of action that reactivate the agricultural sector and strengthen the socioeconomic situation of the population, in the general context of multiple use, integrated with the potentialities of the region.

(ii) *Protection and conservation of natural resources*. Improve the natural vegetation, soils, fauna, and so on, considering the ecological, economic, and social criteria;

(iii) *Recreation* (increase the tourism and attractiveness of the zone for recreation). Contributing to the expansion of Moncayo Natural Park, leading to an increase of its scientific and cultural values and its potential for recreational activity; and

(iv) *Infrastructure development*. Increase the availability of the ecotourism facilities and hydrological resources.

17.7 Generation of Alternatives

The generation of alternatives is the most creative phase of planning, when actions to solve the problems of the zone can be proposed. However, the risks are recommending: (i) nonviable alternatives; (ii) alternatives causing undesired effects; (iii) spatial overlapping

TABLE 17.5. Proposal of general alternatives to be developed in the study area, related to the objectives of the work to which each alternative contributes

Alternatives	Objectives			
	OBJ-1	OBJ-2	OBJ-3	OBJ-4
A-1 Introduction of new crops (rainfed and irrigated)	×			
A-1.1 Expansion of aromatic and medicinal plants				
A-2 Introduction and improvement of agro-food industries	×	×		
A-3 Increase the multiple use of natural areas (truficulture, beekeeping and mushroom picking)	×	×		
A-4 Increase the reforestation areas for exploitation	×	×	×	
A-5 Seedling and planting areas to prevent or mitigate erosion		×	×	
A-6 Land restoration of abandoned areas	×	×		×
A-7 Use or preservation of outstanding areas		×	×	
A-8 Improve accommodation and communication facilities for ecotourism			×	×
A-9 Increase the present surface of the natural park		×	×	×
A-10 Improvement of irrigation: networks and distribution systems	×			×

alternatives; and (iv) proposing alternatives that may counteract the effects of others. The first two types of risk are solved by the so-called justification of alternatives phase, and the last risks are resolved by the assessment of alternatives.

Considering the information from the inventory, the diagnosis, and the systems analysis, we next propose alternatives that would actually allow us to meet our objectives and that would represent changes in the global dynamics of the system, and in the current distribution of land uses. Within the context of the objectives defined above, we show in Table 17.5 the proposed alternatives and which objectives are satisfied.

17.8 Justification of Alternatives

With the justification of alternatives we turn to arguments of a technical, social, or economic nature that endorse the appropriateness of an alternative for solving specific problems of the area. In our study case, due to the long list of alternatives proposed, we have opted to justify only one of them by way of example: the cultivation of aromatic and medicinal plants (A-1.1).

Through this alternative we propose establishment of new crops that would partly substitute traditional crops that are in clear regression or have less economic rentability. In this way, agricultural activity can be kept as a source of income and work. Besides, this plan permits advantage to be taken from the optimal agroclimatic conditions for these crop species.

Justification is based, among other reasons, on the fact that a number of these species are well adapted to the territory. They grow without difficulty in highly stony soils with low rainfall. Nevertheless, it is advisable to select carefully the species to be planted, taking into consideration their qualities and physical requirements, in order to take maximum advantage of their potential production and increase in profitability. The potentiality of these crops in the study zone is endorsed by the abundant spontaneous aromatic and medicinal flora. However, it is also possible to introduce species that are genetically improved and capable of being cultivated with modern techniques, making them a real economically viable alternative to improve income from agriculture and environmental conservation.

The utilization and consumption, both national or international, of products derived from these plants has increased greatly in recent years, which predicts an interesting future for these activities (Burillo and Padró, 1989). For this and other reasons, the scientific knowledge of these species and the cultivation techniques has advanced over the past years. A long series of tested plants is available, with reliable data on their requirements, costs, yields, and so on, permitting an economic assessment to be made of this alternative.

The economic evaluation of this alternative requires knowing the production cycle of the crop over a period of eight to ten years. Its cultivation is mechanical, with careful mineral fertilization and weeding every two years. Harvesting may be mechanical or manual. Burillo and Padró (1990) show in a study of a similar area an estimated yield of essential oils ranging from 9 to 80 l/ha, depending on the cultivated species, the mean values being 30 l/ha. It is important to notice that the production in the first two years is low, then stabilizes for about ten years. The data on the gross margin return as well as the net margins, considering production costs and the expenses incurred in drying and/or distillation of the fresh matter, may be found in Burillo and Padró (1990). The global costs of crop establishment were estimated at 130,000 ptas/ha, due to soil preparation, plantations, and seeds. The management cost during the first ten years is about 340,000 ptas/ha. So the average gross benefit during this period can be estimated at 150,000 ptas/ha per year, which permits an annual net margin of 100,000 ptas/ha.

Based on these experiences, we know that the net margin may vary as a function of the crop development and the level of mechanization. In order to maximize the income of the plantation, the optimal size of a farm must permit an acceptable level of mechanization, a volume of green matter big enough to incur minimal costs and enough processed product (oil or dry matter) to be marketed. In addition, this alternative can produce a synergistic effect on other activities, such as bee-keeping, tourism, and so on, besides improving the soil protection in areas with high a risk of erosion.

17.9 Assessment of Land Capacity for the Proposed Alternatives

To determine the capacity of a territory for the proposed alternatives we have opted to use the approach called "integration at the end of processes" (Gómez Orea, 1994), applying the method of aptitude impact (Villarino, 1978), which is similar to the methodology described in Chapter 15. While other methods (IMGRID, aptitude, capacity, etc.) exist in the bibliography (Gómez Orea, 1992), we have opted to use the aptitude-impact method due to its pedagogic clarity.

After the inventory, selection is made of the elements of the environment that condition the development of each activity. For some alternatives that require concrete conditions to start, the study of capacity consists of identifying the areas or patches that provide these conditions, thus obtaining a single capable class and considering the rest of the area as inapt.

For other alternatives we classify the territory in capacity classes that reflect the different degrees of acceptability the territories have for supporting a given alternative. This requires weighting the types of elements to be considered.

The capacity of a territory to support an activity or alternative not only depends on its aptitude, but also on the possible impacts of the activity. For this reason it is also necessary to carry out environmental impact studies, following a reasoning similar to that for aptitude, and described in Chapter 15. When the impact of an activity is intolerable for certain cells we have an excluded zone for the activity. On the contrary, when the activity does not have an impact, the whole area can support it, and therefore its capacity will depend only on the levels of aptitude (Rau and Wooten, 1980).

The most normal situation is when different grades of impact occur in different cells. In this case we obtain a classification of the territory in classes of impact that reflect grades at which the activity affects the characteristics of the territory. As with aptitude, for impact studies we may also propose weighting the impact classes of each element. Finally, by comparing aptitude and impact for one alternative, we can determine the capacity of the cells for the development of that particular alternative (Villarino, 1978). The ideal localization is that which has the maximum aptitude and minimum impact. However, in reality this combination is seldom found. On the contrary it is more common to find a combination of the different classes of aptitude and impacts, giving the planner the responsibility of choosing those zones that are most appropriate to the alternative.

It is also the responsibility of planners to decide the localization of the possible alternatives among one of the many areas exhibiting the capacity for it. When the area presents various zones with a high or very high capacity, it is necessary to use the other type of information collected in the inventory phase for decision making. The distances from watercourses, settlements, or infrastructures may be favourable or negative factors when deciding on one site or another, according to the alternative. The participation of the population or of local leaders may be used at this phase for decision making and site selection, using the criteria of minimum distances to spot points or routes of interest, protected surroundings, and so on.

As an example from our study case we will obtain maps of capacity for the alternative sites for aromatic plant crops.

17.9.1 Aptitude for Cultivating Aromatic Plants (A-1.1)

To identify the cells apt for growing aromatic plants, five elements have been used: altitude, land capability, soil, slope, and vegetation. Each of these elements has different types which facilitate with a different degree this alternative. Each type receives a value, which, weighted by the specific weight assigned to the element (from 1 to 5), will give the final value for the alternative in each type. The combination of these elements gives us aptitude values, summarized in Table 17.6.

Considering altitude, we have given highest importance to the zones of intermediate altitude, excluding the highest and lowest zones, and a relative weight of 3. The land

TABLE 17.6. Calculation of the final aptitude value of the area in order to locate the alternative "aromatic crops." The elements involved are shown together with their relative weight, the value of each class or type within the elements and their final values for the calculation of aptitude.

	Type value	Relative weight	Final value
Altitude			
< 600	5	3	15
601–700	5	3	15
701–800	5	3	15
801–900	4	3	12
901–1000	2	3	6
1001–1200	1	3	3
> 1200	100	3	300
Land capabilities classes			
Classes I and II	5	4	20
Class III	4	4	16
Class IV	3	4	12
Class V	2	4	8
Class VI	1	4	4
Class VII	1	4	4
Class VIII	0	4	0
Slopes			
0–5%	5	3	15
6–10%	3	3	9
11–20%	2	3	6
21–30%	0	3	0
> 30%	100	3	300
Vegetation			
Low-profit crops	5	4	20
High-profit crops	5	4	20
Pastures	5	4	20
Open shrub	4	4	16
Middle dense shrub	4	4	16
Dense shrub	4	4	16
Afforestation			
(*Pinus halepensis*)	3	4	12
(*P. sylvestris*)	3	4	12
(*P. uncinata*)	3	4	12
Holm oak (*Quercus ilex*)	4	4	16
Q. robur and *Q. petraea*	2	4	8
Q. faginea and *Q. pyrenaica*	2	4	8
Beech (*Fagus sylvatica*)	2	4	8
Soils			
Calcareous fluvisols	5	5	25
Calcareous regosols	3	5	15
Calcareous cambisols	4	5	20
Acrisols	3	5	15
Lithic leptosols	1	5	5
Distric luvisols	2	5	10
Distric cambisols	2	5	10

TABLE 17.7. Final levels or categories of aptitude for the alternative "aromatic crops," derived from the reclassification (at equal intervals) of the summation of final values in each cell of the elements involved

Value intervals	Aptitude classes	Aptitude
1–19	Very low	1
20–38	Low	2
39–57	Medium	3
58–76	High	4
77–95	Very high	5
> 95	Excluded	0

capability classes act by limiting the alternative in the areas with higher classes, and therefore the zones with classes II and III will be considered more apt, being assigned a relative weight of 4. Slope and vegetation also influence the cultivation of aromatic plants. The slope can be considered as a limiting parameter only in the steepest zones and the most valuable areas are the lowest slopes that facilitate cultivation. The value of this weighting is 3. The vegetation carries a weight of 4, and the types of vegetation established have been valued according to the facilities offered by the different plant formations for the introduction of aromatic crops. The soils map has been considered in relation to the quality and fertility of soils to be cultivated with a weight of 5, indicating its importance to the alternative (see Table 17.6).

The operations to obtain the maps of aptitude are made on base maps in the form of a matrix, which may be manipulated with a GIS program (e.g., the SIGMOD). These programs will allow reclassification of the types in each element using the assigned values of aptitude for the type. Later, this map is multiplied by a scalar number that represents the relative weight of the element in the weighting process. The result is a new thematic map with the final value of aptitude for each element.

The aptitude maps for all elements will then be summed in order to obtain a map of total aptitude for each alternative. In our case, this map gives us values between 1 and 95, besides the superior values (> 95), which have been derived from excluded classes. Finally, these values (from 1 to 95) have been reclassified in five aptitude classes, which conform to the final map of aptitude for this alternative. This latter reclassification may be at regular or irregular intervals depending on the criteria followed. In our case, we recognize the classes of intervals shown in Table 17.7.

17.9.2 Impact of Aromatic Plants

Although some specific characteristics of the environment may be changed, this alternative is considered to cause a low impact. However, the introduction of aromatic crops will have a variable impact in three elements: soils, vege-

TABLE 17.8. Calculation of the final value of the impact of alternative "aromatic crops" in the area. The elements involved as well as their relative weight, the value of each class or type within the elements and their final values for the calculation of impacts are shown.

	Type value	Relative weight	Final value
Soils			
Calcareous fluvisols	1	2	2
Calcareous regosols	2	2	4
Calcareous cambisols	1	2	2
Acrisols	2	2	4
Lithic leptosols	5	2	10
Distric luvisols	3	2	6
Distric cambisols	3	2	6
Risk of erosion			
Type I: very low erosion	1	2	2
Type II: low erosion	2	2	4
Type III: medium erosion	3	2	6
Type IV: high erosion	100	2	200
Type V: very high erosion	100	2	200
Vegetation			
Low-profit crops	1	5	5
High-profit crops	5	5	25
Pastures	2	5	10
Open shrub	1	5	5
Middle dense shrub	2	5	10
Dense shrub	3	5	15
Afforestation			
P. halepensis	2	5	10
P. sylvestris	3	5	15
P. uncinata	4	5	20
Holm oak (*Quercus Ilex*)	4	5	20
Q. robur and *Q. petraea*	4	5	20
Q. faginea and *Q. pyrenaica*	5	5	25
Beech (*Fagus sylvatica*)	5	5	25

TABLE 17.9. Levels of impact, derived from the reclassification of the final values obtained in each quadrat when summing the elements involved

Value intervals	Types of impact	Impacts
1–8	Very low	1
9–16	Low	2
17–24	Medium	3
25–32	High	4
33–41	Very high	5
>41	Excluded	0

tation, and erosion risk. Regarding soils, we have a new use that may reduce soil fertility. As for vegetation, the impact consists of the loss of the original crop and forest covers. As regards the erosion risk, this means an alteration of the present state, which may or may not improve the conditions for protection. With these criteria we have reclassified the maps of soils, vegetation, and erosion risk with values for each type shown in Table 17.8, and weighted with 2, 5, and 2, respectively. The sum of these three weighted maps produces the values that are reclassified into five classes of impact (Table 17.9).

For all the proposed alternatives that have a spatial position in the study area, we obtain maps of aptitude and impact following the procedures described above.

17.9.3 Elaboration of Capacity Maps for Alternative A-1.1

After we obtain the maps of aptitude and impact for the alternative, we elaborate the maps of capacity by combining them. This process must be undergone for each alternative. A matrix of criteria is constructed for assigning classes of capacity, using the class of impact and aptitude in a table with double input as shown in Figure 17.8. The capacity values obtained in each cell from the cross-tab of both maps has been reclassified into five classes of capacity with variable intervals depending on the criteria of the planners. As we can see, the classes of higher impact 40 and 50 have been included in the classes IV and V of the matrix because they are nearly incompatible with the objectives of sustainability and rational uses of the environment. Also, we consider with capacity III the classes of aptitude 1 and 2 because it reflects a minimal vocation for the proposed alternatives. Figure 17.8 shows a possible classification using all the values. In other cases, more restricted criteria can be applied, such as that used in Ayache et al. (1993) or that shown in Gómez Orea (1994).

To locate the alternative, we recommend the use of maximum capacity cells (class I),

Impact \ Aptitude		Very high 5	High 4	Medium 3	Low 2	Very low 1
Very low Positive	10	15 / I	14 / I	13 / II	12 / III	11 / III
Low Indifferent	20	25 / I	24 / II	23 / II	22 / III	21 / III
Medium Very unfavourable	30	35 / II	34 / II	33 / II	32 / III	31 / III
High Unfavourable	40	45 / IV	44 / IV	43 / IV	42 / V	41 / V
Very high Unacceptable	50	55 / IV	54 / IV	53 / IV	52 / V	51 / V

Magnitude / Capacity level

FIGURE 17.8. Matrix of capacity levels for the alternative "aromatic crops," derived by crossing aptitude and impact classes. Level I is the maximum capacity and V the minimum capacity. As can be seen, the five impact categories are multiplied by 10, to distinguish them from the impact and aptitude combinations. If this transformation is not used, a value of 6 could come from the sum or combination of aptitude 4 + impact 2, but also a combination of aptitude 2 + impact 4. By multiplying the impacts by 10, the combinations are aptitude 4 + impact 20 = 24; however, aptitude 2 + impact 40 will equal 42.

and after the following classes (II, and with a lesser guarantee class III). The latter two classes (IV and V) are usually rejected.

17.9.4 Global Model for Land Capacity: Assessing Uses and Resolution of Conflicts

When we have realized the territory's capacity to admit the different proposed alternatives, we have to find the optimum combination to support the activities in order to obtain a synthesis map. Each cell of the synthesis map has been assigned to one of the uses proposed, which gives the best capacity. To obtain this map we first carry out a study of the coexistence between alternatives, followed by the resolution of conflicts.

17.9.4.1 Study of the Coexistence of Alternatives

This phase is developed through the matrix of relationships and compatibility between the proposed alternatives (Bergamín et al., 1995). With this matrix we try to define the possibilities of coexistence in time and space of the alternatives, expressed in the following terms: (i) compatible alternatives (C), that is to say, those that may coexist in time and space without prejudicing the other alternative (e.g., production of aromatic plants and bee-keeping); (ii) incompatible alternatives in time and space (N) (e.g., dry and irrigated agriculture); and (iii) alternatives that may be compatible in the future (F), that is to say, which present a close incompatibility but which are based on the fact that one of them is undergoing development toward a determined level in the presence of the other alternative and does not affect it (e.g., afforestation with sheep husbandry).

The matrix of compatibility between alternatives is shown in Table 17.10. There are some uses that may coexist now or in the future, but we need to examine each cell to resolve conflicts of overlapping and propose only one use per cell.

17.9.4.2 Resolution of Conflicts

From the capacity maps for each alternative we can analyse those alternatives that coincide in space. In our case study, considering only classes I and II, we observe that from the 2785 cells that conform the *comarca*, there are 660 cells appropriate for only one of the alternatives, 547 cells for two alternatives, 305 for three alternatives, and 29 have the capacity to support four alternatives. The rest of territory cannot be guaranteed to be appropriate for any alternative. That is to say, there are only 660 cells where there is no conflict in uses and 881 could support two, three, and four potential uses.

By superimposing the capacity maps for all the alternatives, it can be shown that in the spaces with conflicts two situations arise: (i)

TABLE 17.10 Spatial compatibility matrix among alternatives. As explained in the text C = compatible uses, N = noncompatible uses, and F = noncompatible uses at present but compatible in the future

	A-1	A-1.1	A-2	A-3	A-4	A-5	A-6	A-7	A-8	A-9
A-1	=									
A-1-1	F	=								
A-2	N	N	=							
A-3	F	F	N	=						
A-4	F	F	N	C	=					
A-5	N	C	N	F	C	=				
A-6	C	C	N	C	C	C	=			
A-7	N	N	N	C	F	C	C	=		
A-8	N	N	N	N	F	N	F	N	=	
A-9	N	C	N	C	C	C	C	C	N	=
A-10	C	C	N	N	C	C	C	N	N	N

coexistence of alternatives with a distinct level of suitability (365 cells); and (ii) coexistence of alternatives with the same suitability grade (36 cells). In the first case, the conflict is solved by assigning the use that has the highest suitability. The second case is resolved by a matrix that reflects the strategy of the planner team.

As indicated in Chapter 15, evaluation of alternatives is done using techniques of the unique or multicriteria decision-making method, the latter being the most commonly used. Chapter 15 describes some multicriteria methods for evaluating alternatives: the weighting method and the ELECTRE (elimination et choix traduisant la réalité) method.

Here, as an example, we develop global valuation and compatibility of alternatives using the Planning Assistance Through Technical Evaluation of Relevant Numbers (PATTERN) method proposed by Ramos and Weddle (1975) based on the weighting method. In this case the actions of conservation and protection take precedence over production and, in turn, the actions producing the major incomes and employment are given priority.

The experts panel evaluates each alternative with respect to two criteria: environmental impact (EI) and economic rentability (ER), each having a value from 1 to 10. The first criteria (EI) expresses the grade of impact that the development of each alternative supposes. ER is an estimator of the importance and significance of each alternative in the generation of economic resources. As our intention is to maximize the incomes and minimize the impacts, we use the values of ER directly and the inverse of EI values.

The valuation by the experts, calculation steps, and the pertinence index (PI) is shown in Table 17.11, in which: (i) EI and ER are average values of the experts panel; (ii) IEI% is the index of the inverse of environmental impact in percent with respect to the global impact from all the alternatives IEI% = $(100/\text{EI})/\Sigma\text{IEI}$; (iii) ER% is the economic rentability index in percentage with respect to the global income ER% = $(100 \times \text{ER})/\Sigma\text{ER}$; and (iv) the pertinence index (PI) is the sum of the products between the IEI% and ER% with their relative weight PI = (IEI% $\times$ P_1) + (ER% $\times$ P_2). In this case, the relative importance of EI and ER with regard to the global objective of the project was 0.6 and 0.4, respectively.

TABLE 17.11. Calculation of the Pertinence Index (PI) for every alternative, in order to maximize economic profitability and minimize the environmental impact. EI = environmental impact; ER = economic rentability; IEI = inverse of environmental impact; PI = pertinence index

Alternatives	EI	ER	IEI	IEI%	ER%	PI
A-1	5.0	7.5	0.20	4.33	13.16	7.9
A-1.1	2.0	6.0	0.50	10.82	10.53	10.7
A-2	8.5	7.0	0.12	2.60	12.28	6.4
A-3	5.0	5.5	0.20	4.33	9.65	6.5
A-4	3.0	3.0	0.33	7.14	5.26	6.4
A-5	1.0	1.5	1.00	21.65	2.63	14.0
A-6	2.0	2.5	0.50	10.82	4.39	8.2
A-7	2.5	4.5	0.40	8.66	7.89	8.3
A-8	6.5	7.0	0.15	3.24	12.28	6.9
A-9	1.0	4.5	1.00	21.65	7.89	16.1
A-10	4.5	8.0	0.22	4.76	14.04	8.5

The highest PI is the most desirable alternative, and with these values of PI a second matrix of double entries will be developed, called the matrix of conflict resolution (Table 17.12). The results indicate that the expansion of the natural park, the actions to mitigate the erosion, and aromatic crops are the most convenient for the area.

Whenever two alternatives coexist in a cell and have the same degree of capacity, the final selection must be in accordance with the evaluation reflected in the matrix of resolution of conflicts. However, we must not forget that in specific cases, even though only a single use is chosen, various alternatives may coexist. The final result of the process has been the elaboration of a map of future uses in order to introduce the prospect of change (Figure 17.9).

Multicriteria linear programming is another method that could be used for locating alternatives (Romero and Rehman, 1989). The iterative conditional model algorithm (ICM) applied by Alier et al. (1996) in Spain enables the best combination of soil uses and production technology to be located considering the different objectives and priorities given to them. In the same line of methodology, the

TABLE 17.12. Matrix of conflict resolution between pairs of alternatives, after applying the values of the pertinence index calculated for the area

	A-1	A-1.1	A-2	A-3	A-4	A-5	A-6	A-7	A-8	A-9
A-1										
A-1.1	A-1.1									
A-2	A-1	A-1.1								
A-3	A-1	A-1.1	A-3							
A-4	A-1	A-1.1	A-4	A-3						
A-5	A-5	A-5	A-5	A-5	A-5					
A-6	A-6	A-1.1	A-6	A-6	A-6	A-5				
A-7	A-7	A-1.1	A-7	A-7	A-7	A-5	A-7			
A-8	A-1	A-1.1	A-8	A-8	A-8	A-5	A-6	A-7		
A-9	A-9	A-9	A-9	A-9	A-9	A-9	A-9	A-9	A-9	
A-10	A-10	A-1.1	A-10	A-10	A-10	A-5	A-10	A-10	A-10	A-9

FIGURE 17.9. Final map situating land uses for the spatial distribution of all the alternatives proposed in the areas of greater capacity and after all conflicts of use have been resolved.

interactive multiple goal linear programming (IMGLP) model is another alternative used in the allocation of soil uses (van Ittersum and Rabbinge, 1996).

17.10 Extent to Which Objectives Are Reached

Although it is difficult to evaluate the effects of the alternatives, it is necessary to verify the possible and positive consequences of the plan. The chapters dedicated to the economic study of planning have a great utility in this work, although we do not want to consider only the economic values of the alternatives. Nevertheless, an economic analysis gives a sense of reality to the proposals because of the feasibility of the cost–benefit analysis, the availability of labour and the accessibility of possible financing. Not only do we have to account for the market values of the products, but also for the natural resource values, even though their valuation is much more difficult (Chapters 13 and 14).

In our case study, we have estimated that with our planning proposal the increase in gross margin of the productive activities arise 48.2 percent over the actual value (506 million ptas), corresponding to an increase in agriculture of 51.1 percent and 37.3 percent in livestock. On the other hand, the effects on jobs are also considerable, creating 75 new positions, namely 26 in agriculture, 39 in livestock husbandry and 10 in forestry.

Regarding the valuation of the changes of uses in natural resources, we have analysed the social and economic effects of new uses of natural space (increasing the size of Moncayo Natural Park, for example) in the development of the zone. Bergamín (1994), Loureiro (1994), and Rebolledo López (1994) develop a complete work with these aims, using evaluation methods such as structural analysis, cost–benefit analysis, input output analysis, multicriteria analysis, and environmental intangible benefits (e.g., travel cost and contingent valuation).

The results indicated that increasing the park size has major consequences for the *comarca,* moreso than its present use. The cost–benefit analysis of the recreational function has been made through the travel cost (implicated market) and contingent valuation (simulated market) methods. The costs have been calculated on the basis of public investment, daily payments, and the compensation payments for the decline of forest exploitation. The benefits are calculated from the actual net values (ANV) under different prospective uses (without variation in the number of visitors, with an increase of 3 percent during the 12 first years and increasing 5 percent during the 25 following years) and two rates of discounting in the amortization (3 percent and 5 percent). In all cases the ANV offers positive values. The best estimate of ANV is higher than 25,000 million ptas for the implicit market, and 7200 million for the simulated market (scenario 3 and a rate of 3 percent). Finally, concerning the effect on the comarca, it is estimated that 48 percent of the money received will be invested in the comarca, which is reflected by an increase of employment mainly in the service sector.

17.11 Simulation of Future Scenarios

As indicated in Chapter 16, systems analysis may be seen as a useful tool for the construction of simulation models which permit us to project the future evolution of our systems under new alternatives (Hijmans and van Ittersum, 1996; Rabbinge et al., 1996; van Ittersum and Rabbinge, 1996). A simple model of the management of the hydric resource permits the consequences of use and demand systems to be simulated, relating to the sustainability of the resource as well as to the relationship with the economic consequences on agriculture and industry. Other models may simulate the wildlife management of forest zones, the evolution of the tourist sector associated to the zone, etc. (Grossmann, 1991).

This experience gives us confidence that we can analyse complex systems and use these analyses to know in advance the behaviour of the future state of the rural landscape. In summary, the objective will be to simulate future effects of the changes we have introduced in the system with the proposed alternatives and test how long their effects will be displayed or evidenced.

References

Aït Bachir, S.; Giovanardi, F.; Lichaa El-Khoury, D.; Mema, M.; Rodríguez Pérez, J.R. Propuestas para el desarrollo sostenible integrado: Daroca. Unpublished study, CIHEAM-IAMZ; 1997.

Alier, J.L.; Cazorla, A.; Martínez Falero, J.E. *Optimización en La Asignación Espacial de Usos del Suelo: Metodología, Casos de Aplicación y Programa Informático.* Madrid: MAPA, Secretaría General Técnica; 1996.

Alloza, J.A. SIGMOD. Sistemas de información con integración de bases de datos y modelos de planificación territorial. M.Sc. Thesis, CIHEAM-IAMZ; 1995.

Atucha, J.L.; Ben Hadj Ali, H.; Echeverría, J.L.; Kristensen, M.J.; Ríos, J.; Rózpide, M.; Sahin, S. Nuevas orientaciones para el uso integrado de los recursos naturales en la comarca del Moncayo. Vols I and II. Unpublished study, CIHEAM-IAMZ; 1993.

Ayache, F.; Bergamín, G.; Loureiro, M.P.F. de C.; Icarán, C.; Núñez, E.; Ortaçesme, V.; Ouamane, K. Propuestas para la ordenación integral de diez municipios del Campo de Borja. Unpublished study, CIHEAM-IAMZ; 1993.

Belward, A.; Valenzuela, C., eds. *Remote Sensing and Geographical Information Systems for Resource Management in Developing Countries.* Dordrecht: Kluwer Academic; 1991.

Bergamín, G. Influencias socio económicas de las actividades del parque natural de la Dehesa del Moncayo sobre el desarrollo local. M.Sc. Thesis, CIHEAM-IAMZ; 1994.

Bergamín, G.; Icaran, C.; Loureiro, M.P.F. de C.; Ortaçesme, V.; Ouamane; K.; Ayache, F.; Núñez, E. Propuestas para la ordenación integral de diez municipios del "Campo de Borja." *Información Técnica Económica Agraria* 91V(2): 71–97; 1995.

Bertrand, G. Paysage et géographie physique globale: Esquisse méthodologique. *Revue Géographique des Pyrénées et du Sud-Ouest* 39: 249–272; 1968.

Bertrand, G. Le paysage entre la nature et la société. *Revue Géographique des Pyrénées et du Sud-Ouest* 2:239–58; 1978.

Bryan, R.B., ed. Soil erosion: Experiments and models. *Catena Supplement* 17; 1990.

Burillo, J.; Padró, A. Desarrollo experimental de plantas aromáticas y medicinales en la comunidad autónoma de Aragón. *I Jornadas Ibéricas de Plantas Medicinales Aromáticas y de Aceites Esenciales;* 1989 July 12–14; INIA, Madrid. Madrid: MAPA-INIA; 1989:39–48.

Burillo, J.; Padró, A. El cultivo de plantas aromáticas y medicinales. *Surcos Revista Técnica* 22: 32–37; 1990.

Chorley, R.J.; Hagget, P., eds. *La Geografía y Los Modelos Socioeconómicos.* Madrid: Instituto de Estudios de la Administración Local; 1971.

Clusa, J.; Duque, J. L.; Oria, J.; Russinés, J. Estudios Económico-financieros de Los Planes Generales CEOTMA. Madrid: CEOTMA; 1979.

Forman, R.T.T.; Godron, M. *Landscape Ecology.* New York: John Wiley and Sons; 1986.

García Manrique, E. *Las Comarcas de Borja y Tarazona y El Somontano del Moncayo: Estudio Geográfico.* Madrid: CSIC, Instituto J. Sebastián Elcano; 1960.

Golley, F.B.; Bellot, J. Interactions of landscape ecology, planning and design. *Landscape Urban Plan.* 21:3–11; 1991.

Gómez Orea, D. *El Espacio Rural en la Ordenación del Territorio.* Instituto de Estudios Agrarios, Pesqueros y Alimentarios, Serie Técnica No. 6. Madrid: Instituto de Estudios Agrarios, Pesqueros y Alimentarios; 1985.

Gómez Orea, D. *Planificación Rural.* Madrid: MAPA/Editorial Agrícola Española S.A.; 1992.

Gómez Orea, D. *Ordenación del Territorio: Una Aproximación Desde el Medio Físisco.* Serie Ingeniería Geoambiental. Madrid: Instituto Tecnológico Geominero de España/Editorial Agrícola Española S.A.; 1994.

Gómez Orea, D.; Jiménez Beltrán, D.; López de Sebastián, J. Esquema conceptual para la gestión del medio ambiente. *Ciudad y Territorio* 2:35–46; 1982.

González Bernáldez, F.; Sancho Royo, F.; García Novo, F. Analyse des réactions face au paysage naturel. CIHEAM, ed. *La Maîtrise des Ressources Naturelles.* Options Méditerranéennes No. 17. Paris: CIHEAM; 1973:66–81.

Goodey, B. *Participation du Public à la Planification au Royaume-Uni.* Strasbourg: Conseil d'Europe, Conseil de la Coopération Culturelle; 1976.

Grossmann, W.D. Systems approaches towards complex systems. Messarli, P.; Stucki, E., eds. *Colloque International MAB-& Les Alpes, Modele et Synthese*. Fachbeiträge der Schweizerischen MAB-Information. Vol. 19. Bern: Bundesant für Umweltschutz; 1983:25–57.

Grossmann, W.D. Model and strategy driven geographical maps for ecological research and management. Risser, P.G., ed. *Long-Term Ecological Research: An International Perspective* [Scientific Committee of Problems on the Environment (SCOPE)]. Chichester: John Wiley and Sons; 1991:241–256.

Grossmann, W.D.; Eberhardt, S. Geographical information systems and dynamic modelling: Potentials of a new approach. *Ann. Reg. Sci.* 26:53–66; 1991.

Haber, W. Using landscape ecology in planning and management. Zonneveld, I.S.; Forman, R.T.T., eds. *Changing Landscapes: An Ecological Perspective.* New York: Springer-Verlag; 1989:217–232.

Haber, W. Basic concepts of landscape ecology and their application in land management. Kawanabe, H.; Ohgushi, T.; Higashi, M., eds. Ecology for Tomorrow: Proceedings of an International Symposium; 1989 November 2–4; Shiga, Japan. *Physiology-and-Ecology-Japan* (Special number) 27:131–146; 1990.

Haber, W.; Grossmann, W.D.; Schaller, J. Integrated evaluation and synthesis of data by connection dynamic feedback models with a geographic information system. Brandt, J.; Agger, P., eds. *Methodology in Landscape Ecological Research and Planning.* Roskilde, Denmark: Roskilde Univ. Center; 1984.

Higueras, A., coordinator. *Geografía de Aragón.* Vol. V. Zaragoza: Editorial Guara S.A.; 1984.

Hijmans, R.J.; van Ittersum, M.K. Aggregation of spatial units in linear programming models to explore land use options. *Neth. J. Agr. Sci.* 44:145–162; 1996.

Instituto Aragonés de Estadística (IAE), ed. *Censo de Población de Aragón, 1991: Provincia de Zaragoza.* Serie Información Estadística de Aragón. Zaragoza: IAE, Diputación General de Aragón; 1994.

Instituto Aragonés de Estadística (IAE), ed. *Estructura Territorial de la Economía Aragonesa, Año 1992: Renta Comarcal (Espacio Programa).* Serie Información Estadística de Aragón. Zaragoza: IAE, Diputación General de Aragón; 1995.

Instituto Nacional de Estadística (INE). *Censo Agrario 1989.* Madrid: INE; 1991.

Junquera, C.; Martínez, P.; Crespo, L.; Rodilla, C.; Pérez, E.; Aznar, T. Acondicionamiento de la zona norte del somontano del Moncayo. Unpublished study, CIHEAM-IAMZ; 1977.

Loureiro, M.P.F. de C. Valoración económica de bienes ambientales: Aplicación del método de coste de viaje al parque natural de la Dehesa del Moncayo. M.Sc. Thesis, CIHEAM-IAMZ; 1994.

Madsen, S.N.; Zebker, H.A.; Martin, J. Topographic mapping using radar interferometry processing techniques. *IEEE Trans. Geosci. Remot. Sen.* 31(1):246–256; 1993.

Margalef, R. Human impact on transportation and diversity in ecosystems: How far is extrapolation valid? *Proceedings 1st International Congress Ecology;* The Hague. Wageningen: Pudoc; 1974: 237–241.

MOPT. *Guía Para la Elaboración de Estudios del Medio Físico. Contenido y Metodología.* Madrid: MOPT; 1993.

Pellicer, F. *Moncayo, Borja, Tarazona.* Cuadernos de Aragón. Zaragoza: Editorial Trazo; 1986.

Pressat, R. *Demographie Statistique.* Paris: Presses Universitaires de France; 1972.

Rabbinge, R.; van Lattesteign, H.C.; Smeets, P.J.A.M. Planning consequences of long-term land use scenarios in the European Union. Jongman, R.H.G., ed. *Ecological and Landscape Consequences of Land Use Change in Europe.* ECNC Publication Series on Man and Nature. Vol. 2. European Centre for Nature Conservation; 1996:17–35.

Ramos, A., coordinator. *Planificación Física y Ecología. Modelos y Métodos.* Madrid: EMESA; 1979.

Ramos, A.; Weddle, A.E. *I Curso de Planificación Integrada del Paisaje Forestal.* Madrid: ICONA; 1975.

Rau, J.G.; Wooten, D.C., eds. *Environmental Impact Analysis Handbook.* New York: McGraw-Hill; 1980.

Rebolledo López, D.C. Valoración económica de bienes ambientales. Aplicación al parque natural de la Dehesa del Moncayo. M.Sc. Thesis, CIHEAM-IAMZ; 1994.

Romero, C.; Rehman, T. *Multiple Criteria Analysis for Agricultural Decisions.* Amsterdam: Elsevier; 1989.

Sachs, I. *Stratégies d'Ecodéveloppement.* Paris: Editions Ouvrières; 1982.

Senge, P.M. *The Fifth Discipline: The Art and Practice of the Learning Organization.* New York: Doubleday/Currency; 1990.

Soil Survey Staff. *Soil Taxonomy*. Agriculture Handbook No. 436. Washington, DC: Soil Survey Staff; 1975.

Strahler, N.A. *Geografía Física*. Barcelona: Omega; 1974.

TRAGSATEC. *Plan de Ordenación de Recursos Naturales de la Comarca del Moncayo*. Zaragoza: Diputación General de Aragón; 1990.

Valcárcel-Resalt, G.; Troitiño, M.A., eds. *Desarrollo Local y Medio Ambiente en Zonas Desfavorecidas*:.Monografía de la Secretaría de Estado Para las Políticas del Agua y del Medio Ambiente. Madrid: MOPT; 1992.

Van Ittersum, M.K.; Rabbinge, R. Explorative land use studies at the regional level. Castro, E.; Kruseman, G., eds. *Policies for Sustainable Land Use in Costa Rica*. San José, Costa Rica: Ediciones Guayacán Centroamericana S.A.; 1996.

Villarino, T. Modelo para la ordenación integral de espacios naturales y su aplicación a la Sierra de Calderona. Ph.D. Thesis, Politechnical University of Madrid, ETSIM; 1978.

Index

GPSR Compliance

The European Union's (EU) General Product Safety Regulation (GPSR) is a set of rules that requires consumer products to be safe and our obligations to ensure this.

If you have any concerns about our products, you can contact us on ProductSafety@springernature.com

In case Publisher is established outside the EU, the EU authorized representative is:

Springer Nature Customer Service Center GmbH
Europaplatz 3
69115 Heidelberg, Germany

Batch number: 10755805

Printed by Printforce, the Netherlands